CRITICAL COMPONENT WEAR IN HEAVY DUTY ENGINES

CRITICAL COMPONENT WEAR IN HEAVY DUTY ENGINES

P.A. Lakshminarayanan
Ashok Leyland, India

Nagaraj S. Nayak
Sahyadri College of Engineering and Management, India

John Wiley & Sons (Asia) Pte Ltd

This edition first published 2011

Registered office
John Wiley & Sons (Asia) Pte Ltd, 1 Fusionopolis Walk, #07-01 Solaris South Tower, Singapore 138628

For details of our global editorial offices, for customer services and for information about how to apply for permission to reuse the copyright material in this book please see our website at www.wiley.com.

Library of Congress Cataloging-in-Publication Data

Lakshminarayanan, P. A.
Critical component wear in heavy duty engines / P.A. Lakshminarayanan, Nagaraj S. Nayak.
p. cm.
Includes bibliographical references and index.
ISBN 978-0-470-82882-3 (hardback)
1. Internal combustion engines. 2. Machine parts–Failures. 3. Mechanical wear. I. Nayak, Nagaraj S. II. Title.
TJ788.L35 2011
621.43028′8–dc23

2011020808

Print ISBN: 978-0-470-82882-3
ePDF ISBN: 978-0-470-82883-0
oBook ISBN: 978-0-470-82884-7
ePub ISBN: 978-0-470-82885-4
Mobi ISBN: 978-1-118-08296-6

Set in 9/11 pt Times by Thomson Digital, Noida, India

Contents

List of Contributors

Y.V. Aghav
Kirloskar Oil Engines Limited, Pune, India

X. Fernandez
Tekniker-IK4, Avenida Otaola 20, 20.600 Eibar, Spain

Elena Fuentes
Tekniker-IK4, Eibar, Spain

A. Igartua
Tekniker-IK4, Avenida Otaola 20, 20.600 Eibar, Spain

I. Illarramendi
CIE Tarabusi, Urkizu Auzoa 58, 48140 Igorre, Spain

Kedar Kanase
Kirloskar Oil Engines Limited, Pune, India

M.N. Kumar
Kirloskar Oil Engines Limited, Pune, India

P.A. Lakshminarayanan
Ashok Leyland Limited, Engine R&D, Hosur, Tamilnadu 535126, India

Lawrence G. Ludwig
Schaeffer Manufacturing, Saint Louis, MO, USA

R. Luther
Fuchs, Petrolub AG, Friesenheimer Straße 17, 68169 Mannheim, Germany

M.V. Ganesh Prasad
Ashok Leyland Limited, Hosur, India

M.A. Ravichandran
Kirloskar Oil Engines Limited, Pune, India

M. Woydt
Federal Institute for Materials Research and Testing (BAM), D-12200 Berlin, Germany

Preface

Material wears away at the surface of an engine part when rubbed by another material under pressure. The total relative distance travelled, friction and hardness of the materials and the force determine the mass of metal removed by wear. A powerful law was formulated by Archard in 1950 describing the simple relationship between these parameters. This law is repeatedly applied in this book in different forms to estimate the wear quantitatively. Wear increases the tight clearances carefully built into a diesel engine and its critical auxiliaries. The increase in clearances beyond the design limits multiplied by a factor (usually $\sim$2) affects the efficiency of the engine parts and simultaneously increases the noise from it. When in excess, wear debris (produced at the surfaces of the critical parts) forms a very tiny fraction of the weight of the whole engine, but makes the entire machine useless. Thus, when a part reaches its critical wear limit, we can say the life of the part is finished. Hence, the study of wear phenomena in engines is important to answer philosophical questions on the life of an engine. To help reading and make the chapters more or less self-contained, some of the basic ideas on wear and friction are repeated keeping the context in mind.

Up to Chapter 10, the wear of critical parts is explained, based on observations, to understand the nature of wear with an aim to estimate wear quantitatively using models.

It appears, as shown in Chapter 2, that the bottom overhaul life is directly proportional to the square of the size of the engine and inversely to the load factor and the mean piston velocity. The larger the engine, the longer is the life of the engine. The critical parts that determine the life of an engine are the valve train, piston, cylinder liner rings and bearings. The majority of the book relates to a heavy duty normal four stroke diesel engine working off and on road.

Valve train wear affects sealing of the gases in the cylinder. They wear out due to corrosion and fretting at the seat. Archard's law is applied to estimate the guidelines for the tolerable micron-sized fretting movement and consequent wasting of the surface. Due to valve seat wear, the valve lash reduces. When the lash is consumed completely, the gases in the engine leak out past the valves that remain a crack open. Gas flow at high temperature cuts the seat and reduces the efficiency of the engine. In Chapter 3, a method to calculate the critical wear mass, beyond which the wear rate would become catastrophic, is given.

In Chapter 4, the cam wear due to high contact stress and relative velocity is explained. Calculation of Hertzian stress and the high wear zones on the cam are graphically explained. Severe boundary lubrication exists, for example, at the valve tip, rocker toe and the cam follower. A method to calculate the wear rate is provided.

The piston rings are starved of oil at the dead centres. The rings seek succour from the oil stored in the valleys of the rough liner surface. The oil not only helps in lubricating and separating the ring from the liner, thus preventing physical contact, but also evaporates, so absorbing the latent heat from the rings and hence cooling them. The design of roughness of the liner can be plateau honed or normally honed. The latter produces more debris in the initial stages and the surface slowly turns plateau. Plateau honing can be of different types. Kragelskii devised a method to calculate the wear of normally rough surfaces. In order to apply this powerful method to plateau honed liners in modern engines, a concept of 'plateauness' is invented in Chapter 5 and an equivalent normal roughness is derived and plugged into the Kragelskii model. Several types of honing are studied successfully for normal wear.

All the moving parts in an engine, with the exception of the pistons, grow very little relative to their mating parts because their expansion coefficients are nearly equal and the rise in temperature is rather small. However, a piston does not retain its cold shape while running, and hence the running clearances are affected. In addition, carbon, silica and oil deposits reduce clearances and affect the mechanics at the surface. Therefore, it is important to know the shape of the piston accurately when cold for manufacturing the piston and when hot for proper functioning. Both axisymmetric and fully three dimensional models with a powerful set of boundary conditions applicable to all types of diesel pistons in general are described in Chapter 6.

Chapter 7 describes wear problems when the top land of the piston touches the liner. It leads to polishing of the bore when the roughness created intentionally by honing the liner surface is completely removed and lubrication fails. The bore wear distorts the liner shape and the rings are unable to seal the oil travelling upwards into the combustion chamber. Though a separate chapter is dedicated for heavy fuel operation, the example chosen is for a heavy fuel engine. A calculation method is provided to enable the correct description of the top land and skirt profiles. For large engines running on heavy fuel, a scraper ring is the optimum solution to solve bore polishing problem, without losing performance.

Not only are the basic design of the shape, roughness and hardness of various surfaces important for the long life of an engine, but so is control of abrasives entering the engine. These are either produced by combustion and wear, or dust inhaled from the environment through filters that are less than 100% efficient. The higher the dustiness of the environment the more is the mass of dust (quartz) allowed by the filters. Chapter 8 describes the three-body wear in different piston ring grooves by inputting the measured sizes of dust collected in the grooves into the model. Every engine seems to have a critical dosage of dust that ends the life of the piston rings and the liner earlier than that determined by normal aging. It is expressed in terms of grammes of dust per litre of swept volume of the engine and seems applicable universally to all types of diesel engines.

Combustion of diesel fuel forms sulfur oxides which themselves are not corrosive. However, the water of combustion condenses on the cooler walls where the dew point is reached and forms sulfuric acid by dissolving the oxides. The acid is spread by the oil and gases flowing to different parts of the engine from the combustion chamber, causing corrosion of bearings, valve seats and steel parts like piston rings, liner and turbochargers. A narrow window on the coolant temperature, when maintained, could control the corrosive wear, as described in Chapter 10.

The sealing of gases by the piston ring is so important that coatings and materials are developed to work at high temperatures and pressures. During development, they undergo tribological tests outside the engines in engine-like conditions. The oscillation–friction–wear (SRV) device is fully exploited to characterize novel piston ring materials and coatings such as physically vapour deposited, high velocity oxi-fuel deposited coatings or plasma coatings. This enables choice of the best of a number of designs, processes and combinations of materials (Chapter 11).

Bearings hold a thin film of oil which enables transfer of high loads through the power train in a diesel engine. Chapter 12 describes the duties of the bearings, the lubrication regimes, the orbits of shafts in the journal, their tribology, characterization tests and types of bearings. Corrosive, abrasive and adhesion wear are described in detail. A concept of tribo-ecology is brought in to help resolve the dilemma faced by the designer while selecting a bearing for a given application.

In Chapter 15, what drives the changes in diesel engine oil specifications, and the roles of the governments, emissions regulations, OEM and the consumer, are discussed. The engine oil requirements, like easy starting and pumping, preventing wear, reducing friction, protection against rust and corrosion, minimizing deposits, better fuel economy, sealing high pressures, non-foaming and keeping the engine parts clean, are explained. An insight into the process of developing oil performance standards in keeping with the requirements is given. The role of base oils, how their properties are affected by the manufacturing method and additives in the oil, is explained. The emission treatment after the engine and post injection have profound influence on the design of oil, with less ash, sulfur and phosphorus indicating a paradigm shift in engine oil technology. Then telescoping into the future, the development of new generation of oil is debated.

Fuel injection equipment does not function in an engine at such a high temperature as the piston but is under extraordinary fuel pressure, as explained in Chapter 14. To save the pressure from leakage, the clearances between the metal parts are designed tight. With the advent of higher emissions norms, designers seek higher injection pressures and the clearance is reduced correspondingly. This increases the demand on the fuel filter to remove finer particles with higher efficiency. Water ingress in fuel in humid areas of operation leads to corrosion, and hence must be separated efficiently. The quality of fuel itself is changing with the reduction in sulfur (to contain emission), which reduces the lubricity of fuel hand in hand. Additivating fuel to satisfy the requirements is demanding attention of the fuel manufacturers.

Almost all of the entire commerce between nations is transported by ships that are powered by large diesel engines. Very large diesel engines work on a two-stroke cycle. The large engines invariably work on heavy fuel to save cost. In Chapter 15, the wear phenomena particular to large heavy fuel engines is considered. High sulfur, melting eutectic of catalyst fines (vanadium) and sea salt (sodium), carbon formation and acidification are some of the key characteristics that cause corrosive and other types of wear. If the thermodynamic parameters of the engine, as well as the quality and condition of oil and fuel, are not tightly controlled, premature wear can cause deterioration and bring the life of an engine to an early end. Parts of valves, turbochargers, piston and rings are described regarding corrosive and abrasive wear.

Chapter 16 is a peep into the complex world of filters. Air, fuel and oil have to be highly filtered to reduce the sand, carbon, water, metals, bacteria and other harmful material content that can cause severe wear. In most of the engines, these particles reduce the life of an engine that would otherwise age normally.

Acknowledgements

Our work at the Indian Institutes of Technology, in Madras and Delhi, helped us research into the wear of critical parts of heavy duty diesel engines. Opportunities were tremendous at Ashok Leyland Ltd and Kirloskar Oil Engines Ltd to come face to face with wear problems and solve them. We are thankful to Professor P.A. Janakiraman, Professor M.K. Gajendra Babu and Mr A.D. Dani for their support while tackling the problems. We are thankful to Mr K.L.S. Setty for encouraging us and continuous interaction and suggestions while writing this book.

Professors A. Igartua and X. Fernandez (TEKNIKER-IK4, Eibar, Spain), M. Woydt (Federal Institute for Materials Research and Testing (BAM), Berlin, Germany), R. Luther (FUCHS, Petrolub AG, Mannheim, Germany) and I. Illarramendi (CIE TARABUSI, Igorre, Spain) charmed us by accepting our request to contribute on the subject of piston rings. We are grateful to them.

We are indebted to Professor Elena Fuentes (Tekniker-IK4 Foundation, Spain) and Mr Kedar Kanase (Kirloskar Oil Engines Ltd., Pune, India) for capturing the difficult subject of engine bearings.

Oil is at the foundation of tribology of all the relatively moving parts in an engine. It transfers force and torque at pressures of high magnitude. The chemistry and function of oil in the hot and chemically active environment is complex. We are at the cusp of oil technology that is different from the past. New developments of ash-free, highly dispersant and long-life oils are created to withstand different types of oil stresses. We thankfully acknowledge the contribution of Professor Larry Ludwig (Schaeffer Manufacturing) on this important subject.

Filters of oil, air and fuel play a more and more important role in containing the wear of critical engine parts. We are thankful to our colleague Mr Ganesh Prasad for writing a capsule chapter on filters.

We thank our erstwhile colleagues at Kirloskar Oil Engines Ltd, Mr Yogesh Aghav, Dr M.N. Kumar and Mr M.A. Ravichandran, for sharing their experience and for the in-depth study of the special subject of wear in heavy fuel engines. We remember with gratitude the kind support received from S.E.M.T Pielstick during the study.

We gratefully acknowledge the encouragement by Mr Seshasayee, MD of Ashok Leyland to write this book. We thank the publisher John Wiley & Sons, and the editors James Murphy and Shelley Chow for giving us the opportunity to work on the subject of our passion and giving us all types of support to bring this book to reality. We would like to extend our special thanks to Mr. Kevin Dunn for going through all the chapters painstakingly and offering valuable comments.

P.A. Lakshminarayanan
Nagaraj S. Nayak
7 February 2011

Part I
Overture

1

Wear in the Heavy Duty Engine

1.1 Introduction

Stringent standards in different countries for both nitrogen oxides (NO_x) and particulate emissions from diesel engines are becoming unified with national boundaries becoming blurred (Figure 13.1). Emissions increase with engine wear. The easily observable consequences of wear in a diesel engine are the increases in smoke and consumption of fuel and oil. An assured wear life of 8000 hours is expected for a typical off-road engine in the field. Similarly, the vehicles on the road are expected to have an assured wear life of a million kilometres or more. Running many engines in the laboratory or multiple vehicles in the field may help to establish the reliability of the engines and their life. However, it is too late to recognize a problem during these trials. To satisfy reliability, after every failure in the field the engineer is asked to run more vehicles for a geometrically increasing number of testing hours. To prove the engine reliability could take up to three years in the normal course; this is longer than the life of newer emission regimes, and hence it is costly to repeat reliability experiments. In addition, the estimation of life, even by modern reliability methods, is straddled with an error of 50%.

Modern design procedures are detailed and well established against gross failures like fatigue, breaking and loosening. Therefore, only the micron sized failures at the surfaces that result in wear affect the life of an engine. Apart from the basic design aspects that are grounds for 'normal' wear or aging, extreme loads, environments like a dusty atmosphere plus quality of maintenance, fuels and oils bring an earlier end to the life of an engine. If the engine that has lasted its wear life is studied, only less than 0.01% of the total mass of the engine has been wasted away. In other words, more than 99.9% of the mass has survived when the parts are replaced at the end of their wear life. If the normal and abnormal wear is understood early based on the application or design of an engine, it is possible to incorporate features and protections in the design so that the engine is able to complete its expected life with sufficient margin.

1.2 Engine Life

In a diesel engine, the critical parts that wear out are the piston and ring assembly, liners, valve train, valves and bearings. The minimum lives of these parts determine the life of an engine. In other words, the wear life of an engine can be described as the bottom overhaul interval when all the worn parts are changed. The overhaul lives of various engines of different sizes (Chapter 2 references) are plotted in Figure 1.1. The points in the figure are normalized for a common load factor of 0.8 (Chapter 2). Apparently, the life of an engine seems to be a direct function of the power produced by the engine

Critical Component Wear in Heavy Duty Engines, First Edition. P.A. Lakshminarayanan and Nagaraj S. Nayak.

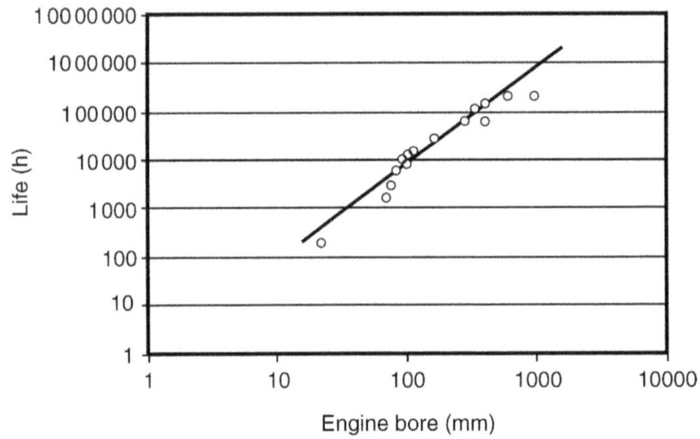

Figure 1.1 Dependency of wear life on engine size

per cylinder, which in turn depends on the size of the engine. The graph shows dependency of life of an engine on its bore.

1.3 Wear in Engines

1.3.1 Natural Aging

At a given load and speed, the pressure and the relative velocity of the pairs of wear parts in an engine vary cyclically undergoing varying degrees of lubrication. All the wear parts in the engine are separated by a thin film of oil. During most of the cycle, the hydrodynamic pressure generated at the wedge shaped interfaces (Figure 1.2) of the pairs of wear parts is sufficient to lift the surfaces beyond their micron sized roughness. However, within a cycle, the thickness of the oil film could drop so low that the pair may come in close contact in some zones. In addition, during starting and stopping in the normal life of an engine,

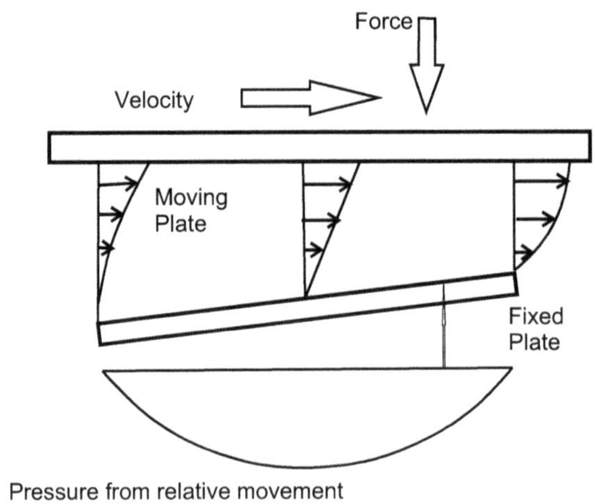

Figure 1.2 Lifting pressure generated by flow of oil in the wedge between relatively moving plates

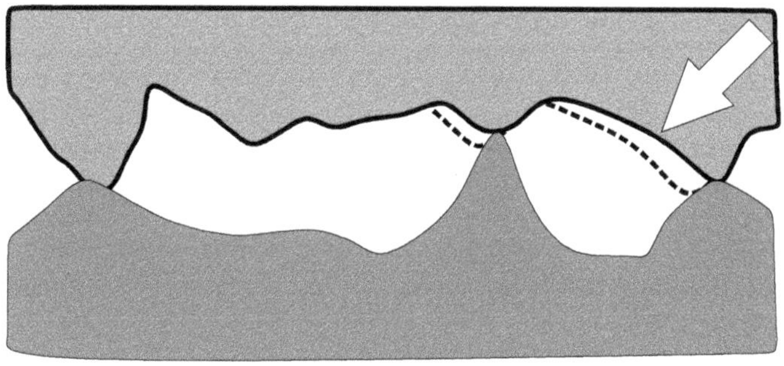

Figure 1.3 Microscopic view of the asperities of bearing surfaces in contact, where the local load is not normal to the nominal surface

the parts run at a relative velocity so small that the asperities of the two wear surfaces come in contact and very high contact pressures are generated at the roughness peaks. Since the peaks are random in occurrence and shape, an element of probability is introduced in their loading and in the direction of loading on the peaks, which is not always normal to the nominal surface (Figure 1.3). The pressure bends the roughness peaks in the local direction of application of load. Even if the local stress is less than the plastic limit of the material, the asperities break away in pieces after many cycles of load, due to fatigue. If there is a special affinity between the metals of the surfaces, it is possible that metal from one surface is transferred to the other surface causing adhesive wear. Such an eventuality is not unusual in the main and connecting rod bearings, which are highly loaded every engine cycle. Therefore, wear takes place when the surfaces separated by a thin oil film come in contact during operating conditions.

1.4 General Wear Model

The wear of a wide range of material combinations has been studied in unlubricated conditions in detail for different loads and relative speeds of wear surfaces (Archard and Hirst,1956). As a broad classification, two contrasting mechanisms of wear have been observed. In nearly all experiments, and for all types of wear mechanism, once equilibrium conditions are established at the surfaces, the wear rate ($kg\,s^{-1}$) is independent of the apparent area of contact. The wear rate is proportional to the load with the same surface conditions. In practice, this simple relation is modified because the surface conditions depend on the load. These rules of wear may be derived, on basic grounds, from the experimental results, or from more detailed theoretical calculations on the different wear parts of an engine.

1.5 Wear of Engine Bearings

The bearing overlay of about 20 microns thickness is an alloy of tin, antimony, copper and lead. This alloy has some important properties, namely, the ability to absorb lubricating oil on the surface and hence provide lubrication when the bearing is starved, the ability to absorb small particles of dirt (embeddability) without increasing the wear of the bearing or crankshaft and seizure resistance, so that it does not weld itself to the crankshaft material even under extreme loads or high speeds. Lastly, it must be able to be run slightly out of alignment without wearing out. Also, the material enables some kind of healing in the event of mild seizure and the material is transferred back to the parent bearing. The parts are heated by friction and also by heat transfer from the hot environment (e.g. hot combustion gases). The copious amount of oil flow between the surfaces enables washing away of both the debris and the heat. These wasted metal particles are found in the oil, sump and the oil filter.

1.6 Wear of Piston Rings and Liners

Usually, today there are three rings in engines that have a swept volume less than three litres per cylinder. Larger engines have four or five rings and they are always above the piston pin. Rings below the piston pin are out-dated, as the arrangement seriously interferes with the supply of oil to the rings above the piston pin and causes excessive wear of the rings and the piston skirt. The three piston rings are nearly always made of three different materials. The oil control ring is the bottom ring and this has significant amounts of lubricant but its job is to limit the lubricant reaching the compression rings above it and this makes life a little difficult for the compression rings. Therefore, the compression rings have to work with very little oil. The lower compression ring is usually an alloy of chrome, while the top ring is usually an alloy of cast iron. In some engines, steel rings are used at the top. Wear is prevented by a thin layer of oil on the surface and oil adsorbed in the surface of the bore and piston rings. The oil film temperature at the top ring is maintained by not only the coolant but, more importantly, also by evaporation of oil trapped in the rough valleys of the liner surface. The reciprocating nature of the piston transfers some lubricant from the cylinder wall to the piston ring and this stops the pistons and rings from seizing in the bore.

1.7 Wear of Valves and Valve Guides

Valves and valve guides, especially the exhaust valve and guide, survive in the harshest environment in an engine. They operate at temperatures ranging from below freezing to as high as 1000 °C. At the valve guide, the gas is at a pressure in excess of 0.3 MPa. The gas flow coats the inside of the guide with deposits of carbon. The guide and the valve seat conduct heat away from the valve into the water jacket to prevent it from melting. The guides are sometimes made of porous sintered metals which absorb oil to lubricate and cool the valve stems.

1.8 Reduction in Wear Life of Critical Parts Due to Contaminants in Oil

In a normal engine, the natural wear is far less than the wear due to contaminants in the oil, such as the metal debris, soot contributed by combustion and the dust from environment in which the engine is working. There are almost no roller or ball bearings inside an engine of a swept volume less than three litres per cylinder because they are more susceptible to damage by contamination as well as quite noisy and expensive to produce.

The combustion gases contain acid, soot and heat. At the walls of the liner, the oil is kept relatively cool by the coolant but it faces the intense heat of combustion protected only by flame quench distances and the flow boundary layer along the wall. At very abnormal conditions like knocking, the boundary layer can be torn and penetrated by the high temperature gases instantaneously (Heywood, 1988), and the oil can become over heated. Also, the soot of combustion in diesel engines disperses in the oil. In engines using exhaust gas recirculation (EGR) and late post injection to aid regeneration of diesel particulate filters, the problem is accentuated. Contamination of lubricating oil by diesel soot is one of the major causes of increased engine wear (Totten, Westbrook and Shah, 2003), especially with EGR technology used to curb nitrogen oxides (NO_x) emissions. The dispersion of the soot in oil is increased by additive packages containing zinc dithiophosphate (ZDP) for these engines (Sam, Balla and Gautam, 2007). Otherwise, in normal conditions, the oil degrades slowly by oxidation and sludge forms. The strong acids formed during combustion in the presence of water vapour dissolve in the oil and the acidity of the oil increases. The oil contains alkaline compounds to neutralize these acids. However, all the weak acids formed under heat in the oil are not neutralized, and hence oil in use is both acidic and basic at the same time. The acids cause corrosive wear. In addition, the air taken in by the engine contains dust particles of different sizes and only about 99.9% of it is filtered at the air filters. As the filter accumulates dust, its efficiency improves at the cost of increased loss in pressure across the filter element. Thus, in a new filter, 0.1% of the dust enters the engine and a small part of it is retained on the wet walls of the liner to be washed by the oil on the liner to the oil sump. The higher the concentration of the dust in the air entering the filter, the higher is the quantity of dust entering the engine as the percentage filtration remains the same for a given air filter. Excessive soot, dust and acid cause abrasive and corrosive wear of critical parts of an engine.

1.8.1 Oil Analysis

The periodic analysis of wear metals in the oil, therefore, diagnoses various parts that are wearing out in the engine and at what rate. It enables the user to take the right action in time. A comprehensive list of sources of wear metals in oil is given in Table 1.1 (Bentley Tribology Services, 2010; Totten, Westbrook and Shah, 2003).

Table 1.1 Sources of wear metals in oil

Metals	Wear Metal Source
Iron	cylinder liners, piston rings, valve train, crankshaft, rocker arms, spring gears, lock washers, nuts, pins, connecting rods, engine blocks, oil pump, gears, pinions, case hardened teeth, pump, motor, vanes, pump housing, cylinder bores and rods, pistons, bearings, cylinders, housing, shafts, piping
Copper	valve train bushing, wrist pin bushing, cam bushings, oil cooler core, thrust washers, governor, connecting rods bearings, valve gear train thrust buttons, oil cooler baffles, plates, thrust washers, bearings, wear plates, thrust washers, oil pump, thermostats, separator filters
Tin	valve train bushing, wrist pin bushing, cam bushings, oil cooler core, thrust washers, governor, connecting rods bearings, valve gear train thrust buttons, journal bearing pads (babbit), bearings, separator filters
Aluminium	engine blocks, pistons, blowers, oil pump bushings, bearings (some) cam bushings, oil coolers, pistons, aluminium complex grease contaminant, coolers, baffles, plates, housing, bearings, cylinder guides, wear plates, thrust washers, residue from synthetic media (alumina) filters
Chrome	rings, liners, exhaust valves, zinc chromate from cooling system inhibitor, shaft coatings, some special gears are chrome plated, cylinder liners, cooler tubes, baffles, plates, bearings, cylinder guides, wear plates, thrust washers, bearings, oil pump, oil coolers
Lead	main bearings, connecting rod bearings, red lead paint flakes
Silicon	engine blocks (alloying element with aluminium parts), ingested dirt from breathers, external sources, defoaming additive in lubricant, elastomeric seals, oil coolers, cooler tubes, baffles, plates, ingested dirt, silicone sealant, bearings, cooler (alloyed with aluminium)
Silver	valves, valve guides, cylinder liners, bearings, heavy alloying element for tool steel gears
Metallic additives	
Sodium	corrosion inhibitor additive, also indicates coolant leak into oil, can also be road salt, sea water, ingested dirt
Boron	corrosion inhibitor additive, anti wear/antioxidant additive, coolant leak, grease contamination
Magnesium	detergent/dispersive additive, can also be alloying element in steels
Calcium	detergent/dispersant additive, alkaline reserve additive for high sulfur fuelled engines, can be grease contamination
Molybdenum	solid/liquid anti-wear additive, alloy in bearing and piston rings
Barium	corrosion inhibitors, detergents, rust inhibitors
Zinc	anti-wear corrosion inhibitors, anti oxidants, alloying element for bearings, thrust washers, galvanized cases
Phosphorus	anti-wear corrosion inhibitors, antioxidant additives
Titanium	wear metal for aircraft engines, bearings, paint (titanium dioxide)
Vanadium	fuel contaminant, can also be alloying element for steel

1.9 Oils for New Generation Engines with Longer Drain Intervals

Oil is a part of the engine build like the connecting rod or the piston, unlike fuel and air which are consumables. The life of the oil is defined as the oil drain interval. The main driving force since 1990 for the development of API grades of oil is the concern over the environmental impact of diesel engine emissions. To further control emissions, lower limits on diesel fuel sulfur are set (reduction from 500 ppm sulfur to 15 ppm) (EPA, 2000). Demand for longer lasting oils as well as the concern over increased temperatures of the engine and oil sump due to current and future engine designs to meet these emissions standards have further driven the development of new engine oil service categories.

1.9.1 Engine Oil Developments and Trends

Selective catalytic reduction using ammonia from urea solution in conjunction with high pressure fuel injection enables achievement of very low nitric oxide and particulates levels along with the improved fuel efficiency by up to 10%. This technique is popular among most of the (truck, marine and electric power generation) engine designers in the world. Nevertheless, wary of handling urea solution, some engines are designed using a combination of cooled exhaust gas recirculation (EGR) at higher rates (30–35%) and exhaust after-treatment devices, such as catalytic diesel particulate filters and oxidation catalysts. There are some engines (Griffith, 2007) that use EGR gases drawn after the particulate filter and the clean gas does not induce engine wear. However, EGR engines commonly allow only unclean gases that cause wear. Therefore, the new generation of engine oils and diesel fuels are developed to provide durability of emission control systems, prevent catalyst poisoning and particulate filter blocking, while still offering optimum protection for control of piston deposits, oxidative thickening, oil consumption, high temperature stability, soot handling properties, foaming and viscosity loss due to shearing (Ludwig, 2007).

The diesel particulate filters are expected to operate for at least 240 000 km before they need cleaning. Engine emissions must comply for 696 000 km. The particulate filters are kept from clogging by active regeneration using electric or diesel heating or by passive regeneration using a catalyst (DECSE, 2000). The remaining residue and ash is blown against the exhaust flow and into a trap for disposal.

The new generation EGR engines generate more soot and experience higher peak cylinder temperatures due to the higher levels of EGR. This requires engine oil with improved oxidation resistance. A small amount of engine oil enters the combustion chamber and burns; its ash-like residue can lead to rubbing wear on the cylinder liner, causing the piston rings to not operate freely and, hence, higher oil consumption. In addition, metal oxide particles (ash) can be carried downstream with the exhaust to clog the diesel particulate filter.

In the exhaust stream, sulfur inhibits the effectiveness of the particulate filters by poisoning, and desensitizes the oxidation catalyst and the filter. It increases the conversion of sulfur oxides to sulfates, hence increases particulate emission and clogging of the filter. This can lead to reduced engine performance, due to increased backpressure and ultimately failure of the particulate trap.

Phosphorus in heavy duty diesel engine oils comes from the anti-wear agent zinc dithiophosphate (ZDP), corrosion inhibitors, friction modifiers and antioxidants. It can deactivate and damage the noble metal catalysts by coating on the active catalyst sites. As a result, harmful emissions, such as nitrogen oxides, carbon monoxide and hydrocarbons, increase.

To protect the after-treatment devices, the engine oil will have to contain lower sulfated ash, sulfur and phosphorus (SAPS), while still offering protection to the wear surfaces. SAPS are found in or derived from additives and base oils which help to extend oil drain intervals, base number (BN) retention, and protect against wear, oxidation, corrosion and piston deposits. However, lower sulfur in the new fuels is a blessing in disguise. Because of the use of ultra low sulfur diesel fuel in on-road applications, the metallic additives needed to maintain the required basicity (BN) of the oil will be correspondingly less, and thus the exhaust will have slightly lower ash content.

Considering all the points described above, API CJ-4 is developed to ensure protection of the after-treatment devices with chemical limits targeting the SAPS set for the first time ever for heavy duty diesel engine oils.

1.9.2 Shift in Engine Oil Technology

The low ash levels and the reduction in sulfur levels of the base oil and additives require replacing conventional metal-containing additives with alternatives low in metal content and sulfur, and in some cases ash-less. The use of these alternative additive chemistries has reduced the required basicity number from 10. For on-road diesel engines, this reduction in BN does not affect current oil life (drain interval) because of the use of ultra low sulfur diesel fuel (15 ppm maximum). Nonetheless, it can reduce oil drain intervals in diesel engines that will still be allowed to use 500 ppm sulfur fuels, for example, off road (Ludwig, 2007, Chapter 13).

The reduction in sulfur and volatility limits and the need for increased oxidation stability due to the increased thermal stress placed on the engine oil by the use of heavy EGR rates and after treatment resulted in an increased use of Group II, Group III and Group IV base stocks. The likely additional after-treatment devices, for example lean NO_x catalysts, lean NO_x traps, will further limit the chemistry of the oils to ensure compatibility of the catalyst and hence a number of severe tests are carried out to study the wear behaviour of critical parts (Whitcare, 2000) like cylinder liner, bearings, pistons and valve train.

1.10 Filters

Abrasive wear, pertaining to the internal combustion engine, has been the subject of numerous technical investigations, papers and articles. The majority of these presentations are in the area of filtration and the prevention of excessive or abnormal wear. Maintenance men must learn to determine if abrasive wear is present and then take corrective measures (Kolbe, 1969).

The contaminants from atmospheric air, combustion and degradation of oil are the most important in reducing the wear life of critical parts of a diesel engine. It is absolutely necessary to have improved filtration for air and oil filters to reduce engine wear (Fodor, 1979, 1982). The oil is operated by monitoring the contaminants within safe limits. Oil is one of the engine critical parts whose drain life must be very large. Therefore, filters play the most important role in maintaining the concentration of contaminants in the oil at a tolerable or economic level, as too large and fine a filter will be costly and unwieldy. The oil life is estimated when the acidity reaches the basicity or when the wear debris in the oil reaches set limits and the oil is ready for draining.

1.10.1 Air Filter

Airborne dust (quartz) is very abrasive and the most common cause of high wear of critical parts in engines. Studies have shown that engine life is a function of cleanliness of the air taken in (Sherburn, 1969). Concentrations and sizes of dust taken in by the engine determine wear life of an engine. Engine wear is produced by particles in the size range 1–40 μm and the most harmful particles are in the range 1–20 μm (Needleman and Madhavan, 1988; Treuhaft, 1993). Even in small amounts it can significantly increase the wear on piston rings and cylinder walls. By using high efficiency air and oil filters, engine wear can be significantly reduced (Jones and Eleftherakis, 1995). A part of the dust ingested by the engine eventually ends up in the oil. Silica in the oil should less than 25 ppm to limit its harmful effect. A new filter in modern engines has an efficiency of 99.8% and after about 20 000 km on the road this reaches 99.95%. In other words, the wear rate decreases fourfold, sometime after the new air filter element is placed.

1.10.2 Oil Filter

Similar to the air filter, the oil filter is a percentage filter, and hence a fraction of the dirt, soot and sludge inevitably passes through the filter to the bearing surfaces along with the oil. These particles initiate three-body wear and may result in seizure when caught in the small bearing gap but for the embeddability property of the overlay.

To enhance the wear life of the engine and the oil drain life, high capacity fine filters and bypass filters are helpful. A filter is considered nominally efficient at a certain micron level if it can remove 50% of particles of that size. In other words, a filter that will consistently remove 50% of particles 20 microns or larger is nominally efficient at 20 microns. Three times the reduction in the contaminants could be observed by using a ten micron bypass filter in conjunction with a main filter rated at 40 micron (Land, Shadday and Philips, 1979). Abrasive engine wear can be substantially reduced with an increase in filter efficiency. Compared to a 40 micron filter, engine wear was reduced by 50% with 30 micron filtration. Likewise, wear was reduced by 70% with 15 micron filtration (Staley, 1988). A filter is considered to achieve absolute filtration efficiency at a certain micron level if it can remove 98.7% of particles that size. Today, it is usual to have a 15–20 micron full flow filter with a tighter bypass filter. It was also reported by Staley (1988) that for the same oil quality, the wear of the main bearing was proportional to the contaminants and that improving the filtering improved the wear rate of main and connecting rod bearings by better than 95% and of the piston ring wear rate by 90%. In other words, if the contaminants are completely eliminated from the oil, the wear life of engine parts would enhance substantially.

1.10.3 Water Filter

Scale is formed in the coolant circuit and also there is precipitation of hard salts in the water. These are abrasive and wear out the seal in the water pump. Also, where the velocity of water flow is less and the passages are narrow, build up of debris could obstruct the coolant flow and cause heating up of the engine and oil film locally. Boundary lubrication related wear could be the consequence. A bypass filter in the water circuit solves these problems satisfactorily.

1.10.4 Fuel Filter

Advanced emission norms require higher injection pressures. However, higher injection pressures are obtained with very tight clearances between the metallic parts in the fuel injection pump and the injector so that the leakage through the clearances is kept low, even at higher fuel pressures. Correspondingly, the demand on the fuel filters for higher efficiency to limit small particles is high.

1.11 Types of Wear of Critical Parts in a Highly Loaded Diesel Engine

The book focuses on the life of the critical parts namely liner, piston, piston rings, bearings and components in the valve train affected by the adhesive, abrasive, corrosive and fretting wear.

1.11.1 Adhesive Wear

This form of wear occurs when two smooth bodies slide over each other or one surface adheres to the other. The strong adhesive forces arise out of intimate interaction at molecular level. In this process, the wear particles are pulled off from the softer surface and become welded to the harder surface. For example, cylinder bores will wear at the top ring reversal zone due to the adhesive action of piston rings under gaseous load.

1.11.2 Abrasive Wear

Abrasive wear occurs when a rough surface slides against hard particles, causing a series of scratches on the smoother surface, and the material from the softer surface is displaced in the form of wear particles. For example, a diesel engine running on heavy furnace oil undergoes wear of the piston grooves due to abrasive action of hard carbon particles produced by combustion.

1.11.3 Fretting Wear

Fretting wear arises when contacting surfaces undergo oscillatory tangential displacement of small amplitude. It is a type of wear because of the cyclic motion that produces a displacement (under high contact pressure) that is so small that it may be difficult to anticipate a large volume of wear debris. The wear of the inlet valve seat due to micron-scale displacement under highly loaded conditions illustrates fretting mode of wear in diesel engines.

1.11.4 Corrosive Wear

Virtually all materials except noble metals like gold or platinum corrode in the normal environment. The most common form of corrosion is oxidation. Most metals react with oxygen in air or water to form oxides. Abusive conditions, for example low or high temperatures, increase the rate of chemical reactions and the corrosive wear increases abruptly, leading to mechanical destruction of the surface layer due to sliding or rolling contact of two mating bodies. Typically, the greyish lapped appearance of liner surfaces under cold running condition is an example of corrosive wear.

References

Archard, J.F. and Hirst, W. (1956) The Wear of Metals under Unlubricated Conditions. *Proceedings of the Royal Society London, A*, **236**, 397–410.

Bentley Tribology Services (2010) Sources of Wear Metals in Oil Analysis. http://www.bentlytribology.com/publications/appnotes/app31.php (accessed 20 May 2011).

DECSE (2000) Phase 1 Interim Report No. 4: Diesel Particulate Filters – Final Report. Diesel Emission Control – Sulfur Effects (DECSE) Program, US Department of Energy/Engine Manufacturers Association/Manufacturers of Emission Controls Association, USA.

EPA (2000) Regulatory Impact Analysis: Heavy-duty Engine and Vehicle Standards and Highway Diesel Fuel Sulfur Control Requirements. EPA 420-R-00-26, Assessment and Standards Division, Office of Transportation and Air Quality, United States Environmental Protection Agency, Washington, DC, pp. 96–98.

Fodor, J. (1979) Improving Utilization of Potential I.C. Engine Life by Filtration. *Tribology International*, **12**, 127–129.

Fodor, J. (1982) Improving the Economy of I.C. Engines by Controlling the Contaminants Through Filtration. World Filtration Congress III, Downingtown, PA, pp. 707–711.

Griffith, R.C. (2007) Series Turbocharging for the Caterpillar® Heavy-Duty On-Highway Truck Engines with ACERT™ Technology. Technical Paper 2007-01-1561, SAE, Troy, MI.

Heywood, J. B. (1988) Internal combustion engine fundamentals. McGraw-Hill, p. 669.

Jones, G.W. and Eleftherakis, J.G. (1995) Correlating Engine Wear with Filter Multipass Testing. Fuels & Lubricants Meeting and Exposition, 16–19 October 1995, Toronto, Canada, Technical Paper 952555, SAE, Troy, MI.

Kolbe, J. F. (1969) Abrasive Wear – Identifications and Prevention. Technical Paper 690544, SAE, Troy, MI.

Land, O.H., Shadday, P.D. and Philips M.C. (1979) Diesel Engine Wear with Spin on – bypass lube oil filters. Technical Paper 790089, SAE, Troy, MI.

Ludwig, L.G. (2007) Heavy-duty Diesel Engine Oil Developments and Trends. *Machinery Lubrication*, http://www.machinerylubrication.com/Read/1036/diesel-engine-oil (accessed 20 May 2011).

Needleman, W.M. and Madhavan, P.V. (1998) Review of Lubricant Contamination and Diesel Engine Wear. Truck and Bus Meeting and Exposition, 7–10 November 1998, Indianapolis, IN, Technical Paper 881827, SAE, Troy, MI.

Sam, G., Balla, S. and Gautam, M. (2007) *Effect of diesel soot contaminated oil on engine wear*. Elsevier Science, Vol. 262 (9–10), pp. 1113–1122.

Sherburn, P.E. (1969) Air Cleaner Design – Present and Future. International Automotive Engineering Congress, 13–17 January 1969, Detroit, MI, SAE Technical Paper 690007.

Staley, D.R. (1988) Correlating Lube Oil Filtration Efficiencies with Engine Wear. Technical Paper 881825, SAE, Troy, MI.

Totten, G.E., Westbrook, S.R. and Shah, R. J. (2003) *Fuels and lubricants handbook: technology, properties, performance*, Vol. 1. ASTM International, West Conshohocken, PA.

Treuhaft, M.B. (1993) The Use of Radioactive Tracer Technology to Measure Engine Ring Wear in Response to Dust Ingestion. International Congress and Exposition, 1–5 March 1993, Detroit, MI, Technical Paper 930019, SAE, Troy, MI.

Whitcare, S. (2000) Catalyst Compatible Diesel Engine Oils, DECSE Phase II. Presentation at DOE/NREL Workshop Exploring Low Emission Diesel Engine Oils, 31 January–2 February 2000, Phoenix, AZ.

2

Engine Size and Life

Parameters like brake mean effective pressure, mean velocity of the piston, surface hardness, oil film thickness and surface areas of critical wear parts are similar for all the diesel engines. The mean piston velocity at the rated speed is nearly the same for all diesel engines. The mechanical efficiency normalized to an arbitrary brake mean effective pressure (bmep) is dependent on the size of the engine. The engine life seems to be proportional directly to the square of a characteristic dimension of the engine and inversely to speed and load factor for engines varying widely in sizes and ratings.

2.1 Introduction

The mechanics of the diesel engine is highly developed by constant research to give enhanced life. Quality of fuels, improvement in the surfaces of wear parts and introduction of computers to control the engine mainly for lower emissions have further improved the life. In a diesel engine, the critical parts that wear out are the piston and rings assembly, liners, cam followers, valves and bearings. The minimum of the lives of these parts determines the life of an engine.

Very large engines are checked for wear of critical components like bearings once every eight years (Royal Belgian Institute of Marine Engineers, 2005) and they could last at least three checks or about 25 years. It might be thought that opening-up inspections are required to monitor the bearing wear, but the wear of both white metal and AlSn40-lined bearings used in large engines is virtually nil (MAN B&W, 1996) in these large engines. It was also observed in very large engines that opening the assemblies of critical parts for inspection introduced dirt and distortions that actually increase the wear! Commercial vessels are typically heavily used, and their large engines are designed to operate for as many as 4000 to 6000 hours a year at the high engine loads (80% load factor) needed to push the vessel and its cargo through the water (EPA, 1999). The average life of large engines extends beyond 100 000 hours. For smaller engines working off road, the expected life is 8000 to 12 000 hours (EPA, 2002).

2.2 Engine Life

The wear life of an engine can be described as the bottom overhaul interval when all the worn out parts are changed. Maintenance manuals of a number of contemporary engines that vary widely in power produced per cylinder have been studied (mainly via the internet) In Appendix 2.A data on lives, engine parameters,

Critical Component Wear in Heavy Duty Engines, First Edition. P.A. Lakshminarayanan and Nagaraj S. Nayak.

engine sizes, wear rates and mechanical efficiencies of different engines are listed along with the sources of data.

2.3 Factors on Which Life is Dependent

The life of an engine is inversely proportional to the load factor, which is defined as a ratio of the average load at which the engine is running for a year to the full load. The load may be taken as the torque when the speed is varying. Smaller engines and engines used for fast boats, for example, have a lower load factor. Larger engines used in ships and off-road vehicles have higher load factor. For instance, a large main propulsion engine used in a ship has a load factor of 0.8 and a truck engine on road has a load factor of 0.4 in most of the applications where maximum road speed is 120 km/h. In developing countries, the engines are low powered to cope with the maximum speed limit of 60 km/h and with an eye on better fuel economy. In these circumstances, the load factor of the truck engines could be as high as 0.7. Similarly, the diesel engines running in emergency services are loaded only for about 300 hours a year (about an hour every day) for checking its working condition, and hence the load factor is nearly zero. On the other hand, in areas where the main source of electricity is only from the diesel generating set, the load factor is as high as 0.92. Therefore, the life of an engine could change very much, depending on the load factor. Since the engines (Appendix 2.A) are multi-cylindered from a single up to 16 cylinders, the study is made by normalizing the power per cylinder. Some manufacturers give engine life as the number of litres of fuel consumed before major overhaul is carried out. This is converted to number of hours for a load factor 0.8. The points in Figure 1.1 are normalized for a common load factor of 0.8 by using the $\textit{correction factor} = \frac{\textit{actual load factor}}{0.8}$. Power per cylinder is usually a function of engine size. The graph shows dependency of engine life on the physical dimension. Load factor, size and speed are hence the factors on which the engine life seems to depend.

2.4 Friction Force and Power

2.4.1 Mechanical Efficiency

The mechanical efficiency of an engine is defined at the rated power. It is the ratio of brake power to the indicated power. Figure 2.1 shows a plot of known mechanical efficiency of different engines against the

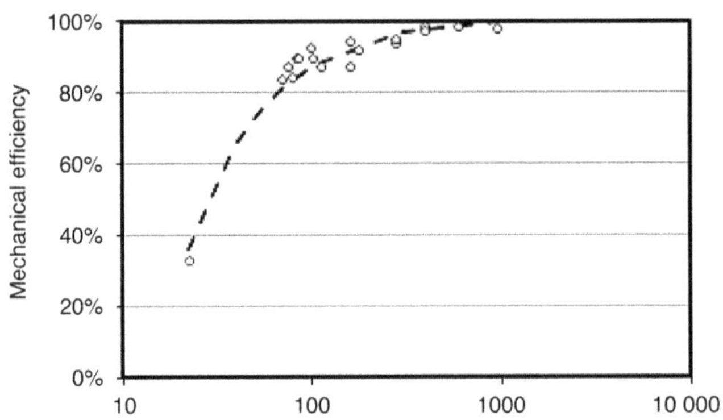

Figure 2.1 Increase in mechanical efficiency with bore and higher power per cylinder

engine bore, indicating the influence of the size of the engine on mechanical efficiency. Nonetheless, the relationship is highly nonlinear and can be unravelled as follows.

$$\begin{aligned}\text{Mechanical efficiency} &= \frac{\text{Brake power}}{\text{Indicated power}}\\ &= 1 - \frac{\text{Friction mean effective pressure}}{\text{Indicated mean effective pressure}}\end{aligned}$$

In a heavy duty diesel engine, the friction mean effective pressure is fairly independent of the load and is dependent only on speed. However, the mechanical efficiency is a strong function of the brake mean effective pressure (bmep). For example, the mechanical efficiency of a state-of-the-art highly loaded engine (bmep = 2.3 MPa) could be as high as 93% at full load and, for the same engine, 0% when idling. The engines in Appendix 2.A (Figures 1.1 and 2.1) are rated at bmeps varying widely from 0.6 to 3.1 MPa. Therefore, for the study of mechanical efficiency it is normalized to an arbitrarily chosen bmep of 1.7 MPa, so that the true effect of friction affecting the mechanical efficiency is clearly brought out.

When delved into in detail, it can be seen that the mechanical efficiency of larger engines is higher and, hence, the friction mean effective pressure is also less, indicating relatively less friction in larger engines.

2.4.2 Friction

Friction in an engine, either at the bearings, or between the piston rings and the liner, or in the valve train, is controlled mainly by hydrodynamic lubrication and partly by mixed lubrication for a small duration in a cycle. Thus, viscous friction is consuming most of the friction power. The friction power is proportional to the area of contact, viscosity and the clearances. The viscosity of the oil at the operating temperatures (narrow temperature range: 95–110 °C) of different parts is independent of the size of the engine. The Newtonian friction force and friction power can be written as follows.

$$\text{Friction force} = \mu A\frac{dV}{dy} \approx \frac{\text{viscosity} \times \text{Area} \times \text{velocity}}{\text{oil film thickness}} \tag{2.1}$$

where μ – the dynamic viscosity of oil, A = area of the wear surface and dV/dy is the velocity gradient of oil flow perpendicular to the surface.

$$\text{Friction power} = \text{Friction force} \times \text{velocity} \tag{2.2}$$

2.5 Similarity Studies

For similarity studies, two of the critical components (namely, bearings and the piston–ring–liner assembly) are discussed below.

2.5.1 Characteristic Size of an Engine

Many of the wear parts are dimensioned relative to the stroke (vertical dimension) or the bore (horizontal dimension). For small engines, stroke is nearly equal to or less than the bore. Table 2.1 shows the relationship of typical dimensions of critical parts scaling with the bore of the engine. For example, the bearing width is 0.46 times the shaft diameter (Daido Metal Company, 2010) and the shaft itself 0.67 times the bore.

Table 2.1 Sizes and clearances affecting friction scale, in general, with respect to engine bore

	With respect to shaft diameter	With respect to bore
Total height of all piston rings (Nural, 1983)		0.06–0.08
Diameter of piston pin		0.25
Height of piston skirt		0.48
Bearing width	0.46	
Oil clearance in crankpin and main bearings	0.82×10^{-3}	
Crank pin		0.67
Crank shaft journal		0.75

Hence, a good choice for the characteristic dimension (L) of an engine to pick for affinity studies of engine life could be engine cylinder bore (B).

2.5.2 Velocity

When the entire range of engines from the smallest of bores to the highest (~1000 mm) is considered, the stroke of very large engines is more than three times the bore:

$$\text{stroke} = \text{k} \times \text{Bore} = \text{k L}$$

where $k \approx 1 - 1.5$ for engines of bore < 400 mm and $k \approx$ up to 2.5 for very large bores.

The stroke and the engine speed determine the mean piston velocity:

$$\text{mean piston velocity} = \sim 2 \times \text{stroke} \times \text{engine speed}$$

The velocity of various wear parts in an engine scales with the mean piston velocity. However, on average, the mean piston velocity of most diesel engines at their rated speed is fairly invariant at 9.4 m/s[1] (Figure 2.2).

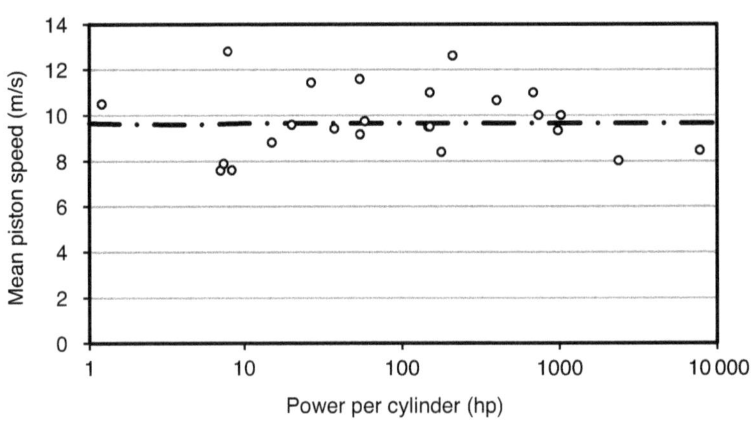

Figure 2.2 Mean piston speed is independent of horsepower per cylinder; bmep is in the range 8–25 kg/cm^2; bores in the range 26–1000 mm

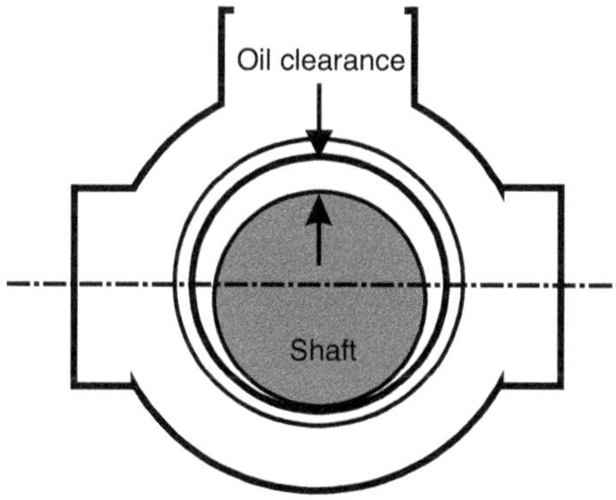

Figure 2.3 Minimum oil clearance

2.5.3 Oil Film Thickness

2.5.3.1 Oil Clearance in Bearings

The bearing clearances representing the oil film thickness bear a constant ratio with the shaft clearance. For example, the minimum oil clearances of the main and crank pin bearings (Figure 2.3) are nearly the same for all engines of bores ranging from 65 to 1000 mm. Similarly, the bearing width is 0.46 times the shaft diameter (Daido Metal Company, 2010):

$$\frac{\text{clearance of the main bearing}}{\text{journal diameter}} = 0.822 \times 10^{-3}$$

$$\frac{\text{clearance of the big end bearing}}{\text{diameter of the crank pin}} = 0.825 \times 10^{-3}$$

The shaft size, in turn, is proportional to the bore or stroke (characteristic length, L) in most of the engines. For instance, the main journal is 0.65–0.85 times the bore for many heavy duty diesel engines and the crank pin is 0.57–0.77 times the bore. In other words, the oil film thickness (bearing gaps) also scales with the characteristic size of the engine:

$$\text{oil film thickness} = \text{dy} \propto \text{characteristic dimension} = \text{L}$$

2.5.3.2 Piston Ring to Liner: Oil Film Thickness at Top Dead Centre (TDC)

The heights of piston rings that wear against the liner are in proportion to the bore (Nural, 1983). Similarly, the oil film between the liner and the piston ring is nearly dependent on the engine bore, as seen in Table 2.2 (Sutaria, Bhatt and Mistry, 2009; Takata, 2006; Saburi, Saitoh and Yamada, 2005; Taylor, 2004; Wolff and Wolff, 2009, Rakopoulos, Giakoumis and Dimaratos, 2007; Wong *et al.*, 2005):

$$\text{oil film thickness} \propto \text{characteristic dimension} = \text{L}$$

Table 2.2 Oil film thickness between the piston ring and the liner as a function of the bore

Bore (mm)	Film thickness (micron)
56	6
115	2
152	0.7
840	10
89.9	2.5
320	7
840	12

2.5.4 *Velocity Gradient*

The velocity gradient in the oil film contributing to Newtonian friction in the bearings and between the liner and the piston rings is inversely proportional to the film thickness (Equation 2.1). Since the velocities of the moving surfaces are scaled with the mean piston velocity (nearly a constant = 9.4 m/s), and since the oil film thicknesses are proportional to the bore of the engine, the velocity gradients in the oil films are proportional to the reciprocal of the characteristic size of the engine, that is, $\frac{dV}{dy} \propto \frac{1}{L}$.

2.5.5 *Friction Force or Power*

Therefore, all the wear parts of the engine, such as piston rings, bearings and liner, have a friction area that scales with the square of L. Thus, Equation 2.1 can be rewritten:

$$\text{Friction force} = \mu A \frac{dV}{dy} \propto \text{area} \times \frac{1}{L} \propto \frac{L^2}{L} = L \qquad (2.3)$$

Since the surface velocity of various running surfaces is a direct function of the mean piston velocity, friction power in Equation 2.2 reduces as follows:

$$\text{Friction power} \propto \text{Friction force} \times \text{mean piston velocity} \propto \text{Friction force} \propto L$$

2.5.6 *Indicated Power and Efficiency*

The indicated power per cylinder is a function of the swept volume, mean effective pressure and speed:

$$\text{Indicated power per cylinder} = \text{mean effective pressure} \times \text{swept volume per cylinder} \times \text{speed} \qquad (2.4)$$

where swept volume $\propto \text{Bore}^2 \times \text{stroke} \propto L^2 \times kL = kL^3$

Because the characteristic velocity is fairly a constant (9.4 m/s) at rated speed, it can be taken as independent of the size of the engine:

$$\text{engine rotational speed} \propto 9.4/(2L) \propto \frac{1}{L}$$

giving:

$$\text{Indicated power} \propto p \times kL^3 \times \frac{1}{L} \propto kpL^2 \qquad (2.5)$$

Using Equations 2.3 and 2.5:

$$\text{Mechanical efficiency} = 1 - \frac{\text{friction power}}{\text{Indicated power}} = 1 - \frac{\mathrm{m}}{\mathrm{k\,p\,L}} \approx 1 - \frac{\mathrm{m}'}{\mathrm{L}} \quad (2.6)$$

where m and m' are constants and p is the indicated mean effective pressure (mean effective pressures of all modern engines are mostly in the narrow range 17–26 kg/cm^2). The fit in Equation 2.6 seems to work when the bore is greater than 15 mm. The model for mechanical efficiency normalized to a bmep of 1.7 MPa appears satisfactory for a value of m' is 15 mm (Figure 2.4a). A cross plot of the actual mechanical efficiency (corrected to 1.7 MPa bmep) against the efficiency estimated using Equation 2.6 shows the fit is adequate (Figure 2.4b).

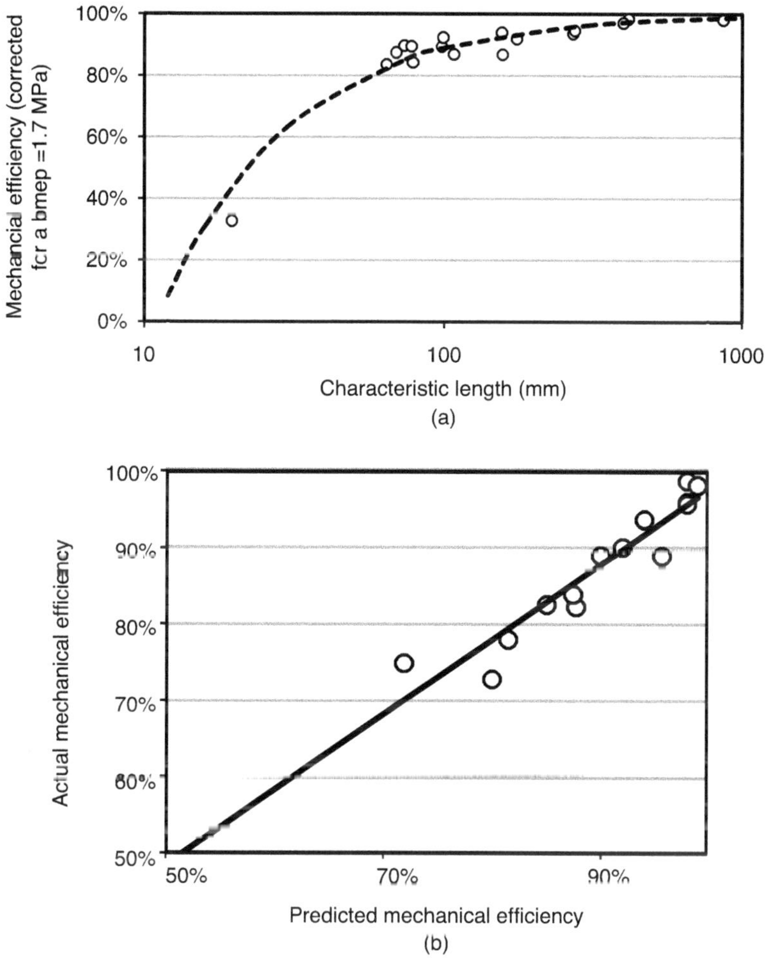

Figure 2.4 Mechanical efficiency corrected for a normalized bmep of 1.7 MPa for different engines at rated speed. (a) Dependency of mechanical efficiency on the characteristic size of engines; (b) comparison of calculated and actual mechanical efficiencies of engines

2.6 Archard's Law of Wear

Wear of relatively moving surfaces was neatly formulated by Archard and Hirst (1956) using a simple relationship for the adhesive, abrasive or fretting wear which takes place at unlubricated surfaces; it can be extended to nearly-lubricated (boundary lubrication) surfaces. The mass of wear debris emitted by the wear surface is a function of the force (F) acting on the surface and the distance (S) travelled relatively by the pair of wear surfaces, and is inversely proportional to its hardness (Equation 2.1). The type of wear and the environment determines the proportionality constant, *k*, in the Archard equation:

$$W = \frac{k}{H} F\ S \qquad (2.7)$$

Here, W = mass of wear debris emitted by the surface, H = hardness of the wear surface, F = force acting on the wear surface, S = relative distance travelled by the wear surface with respect to its associated surface and k = Archard's proportionality constant.

2.7 Wear Life of Engines

The mass of wear debris, W is given by:

$$W = \rho\, A\, \delta = \sim L^2\delta \qquad (2.8)$$

where δ = wear depth, A = nominal area of nearly lubricated surface which is undergoing wear (apparent area of contact) $\propto L^2$, and ρ = density of the material.

If t is the life time, then:

$$\text{the relative distance travelled by the wear surface in its lifetime} = v\ t \qquad (2.9)$$

v = is the relative velocity of wear surfaces which scales with the mean piston velocity.

From Equation 2.3:

$$\text{Force on the surface, } F = \text{Friction force} = \sim L$$

Substituting Equations 2.8, 2.9 and 2.3 in Archard's law (Equation 2.7) gives:

$$L^2\delta \propto L\ v\ t/H \qquad (2.10)$$

2.7.1 Wear Life

The tribology of the liner or piston ring and piston in engines of different bores, in general, is similar. For example, the hardness of the liners of modern engines of different bores is in the range 230–320 N/mm^2. The skirt contacting the liner is mostly of aluminium to contain the reciprocating weight. Similarly, the top ring groove is machined in either the forged steel piston crown or the cast iron insert in the aluminium piston. Thus, the finish and materials of all these parts are similar.

Aluminium–tin bearings rarely use an overlay. When an overlay is used on bearings (e.g. copper–lead–tin bearings) its fatigue strength is as important as its hardness. The fatigue strength increases with a decrease in overlay thickness. As a first approximation, the hardness of similar wear surfaces in an engine is the same (Appendices 2.B and 2.C). However, a sputtered overlay is an exception; its wear resistance is very high. These are used in engines which are rated at a higher bmep or in on-road engines using high rates of exhaust gas recirculation (EGR) to solve emission problems, that is those with unusually high unit loads (N/mm^2) with less bearing areas. However, the shaft and the bearings are expected to be finely aligned to avoid edge loading problems when (hard) sputtered bearings are used.

As explained previously, the typical piston velocity of all the engines is about 9.4 m/s at the rated speed. Therefore, the relationship in Equation 2.10 can be written as shown in Equation 2.11, assuming the hardness, materials, and hence the wear resistance of surfaces of all types of engines are in a narrow range:

$$t \propto L\,\delta \frac{H}{V} \propto L\,\delta \tag{2.11}$$

2.7.2 Nondimensional Wear Depth Achieved During Lifetime

The life for wear parts is determined when the nondimensional wear, Δ, which is defined as the ratio of wear depth to the characteristic size of the engine, reaches a critical limit ($\Delta_{critical}$). For example, the life of the liner can be determined by specifying the ratio of wear depth, δ, to the bore, B, that is:

$$\text{limiting ratio for life, } \Delta_{critical} = \frac{\delta_{critical}}{B}$$

2.7.2.1 Life of Liner

Cylinder liners worn beyond the maximum allowable limit should be replaced. The maximum allowable wear limits for engines can be found in the appropriate manufacturer's technical manual. In the absence of such specific information, the wear limits from the 'Diesel Engine Wear Limit Chart', available from the Naval Sea Systems Command of the USA (NAVSEA, 1992), can be applied:

(i) two-stroke cycle engines with aluminium pistons: 0.0025 mm/mm diameter
(ii) slow-speed engines over 455 mm (18 inch) bore: 0.005 mm/mm diameter, and
(iii) all other engines: 0.003 mm/mm diameter.

For example, Table 2.3 gives the rejection limits for the wear of the liner and rings for different bores of diesel engines. This is shown graphically in Figure 2.5. A value of 0.004 mm/mm appears to be reasonably allowable for bore wear for all types of engines.

Table 2.3 Wear nondimensionalized with respect to engine size when the parts serviced is the same for engines varying widely in size; for example, the wear of the bore of the liner is given in this table

Bore (mm)	$\delta_{critical}$ (mm)	Manufacturer's recommendation (mm/mm)	NAVSEA recommendation (mm/mm)
64.5	0.2	0.003	0.003
103	0.3	0.003	0.003
280	0.05	0.00018	0.003
400	1.5	0.00375	0.003
580	2.5	0.005	0.005
600	2.8	0.0047	0.005

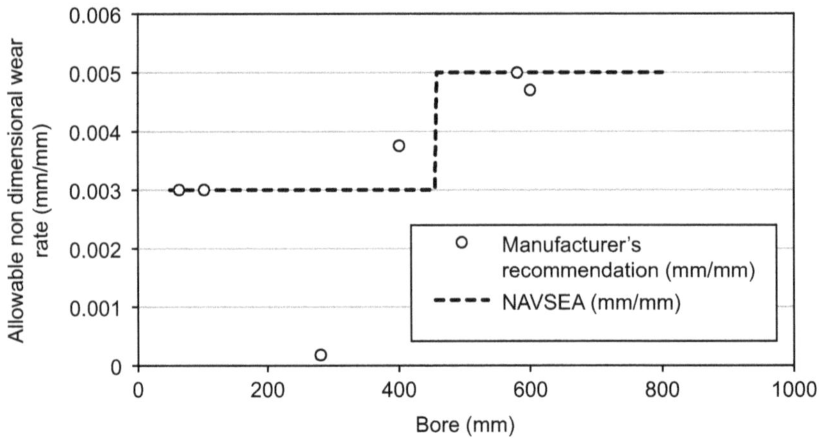

Figure 2.5 Nondimensional allowable wear limit specified by the manufacturer is similar to the NAVSEA specifications

2.7.2.2 Life of Piston Ring

The condemning limit of piston rings is usually when the 'closed gap' doubles the manufactured gap. Most piston ring manufacturers recommend a minimum end gap of 0.004 mm/mm times the bore diameter for the new top piston compression ring. The critical wear ratio then works out as 0.004 for the piston rings. Similarly, the critical wear ratio ($\Delta_{critical}$) for other important wear parts like the bearings are known to be independent of the size of the part itself.

2.7.2.3 Wear Life of Engine

Since all the dimensions of these critical parts scale with the engine bore (or stroke), it can be said that the critical wear ratio is independent of the size of the engine itself. So, as a first order approximation, the critical limit, $\Delta_{critical}$, can be modelled as invariant with the size of the engine for a critical component of interest. Expressing the life as the time (t) to reach the critical wear depth, $\Delta_{critical}(\delta_{critical} = \Delta_{critical} \times L)$, in Equation 2.11 gives:

$$\text{Life} = t \propto L\,\delta \sim L \times (\Delta_{crit}\,L) \sim L^2 \tag{2.12}$$

Figure 2.6 shows the wear life plotted as points against the characteristic dimension of the diesel engines of bores in the range 20–1000 mm. The smallest is an aero modelling engine running at 10 000 rpm having a bore of 26 mm and the largest is the main propulsion engine of characteristic size 965 mm rated at 105 rpm. In the backdrop, the rule of life (Equation 2.12) is plotted. The very large engine appears to fall short of the general rule. However, as the engine is younger than 25 years, any higher lifetime could not be plotted and hence appears to have a life lesser than the norm. Also plotted in the background is the bandwidth of +/− 35% deviation from the mean; almost all the engines fall within this band. Here, the distinction of the fuel type is not deliberated. One large natural gas engine is shown to have a longer life, as gas engines are relatively cleaner and wear at a lesser rate than diesel engines, especially

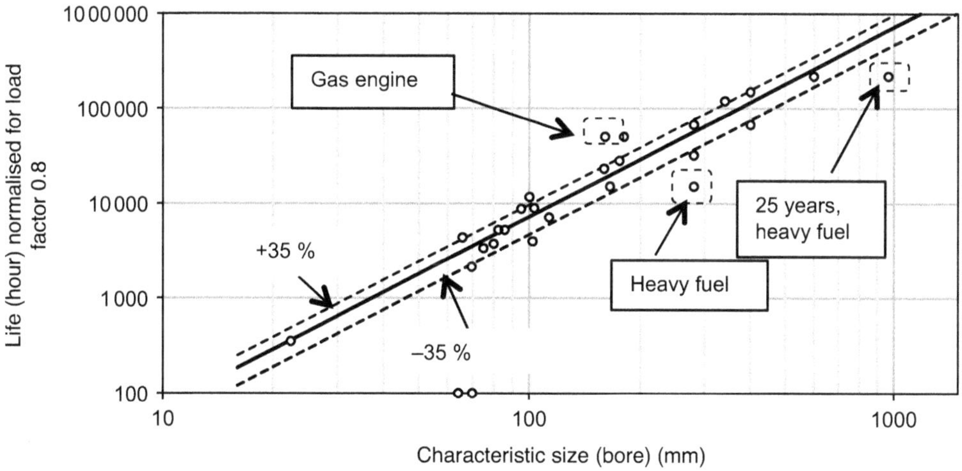

Figure 2.6 Increasing wear life with size of the engine without correcting for different piston speeds, with the rule of life in the backdrop

those running on heavy fuel oil (Chapter 15). If the mean piston velocity (v) is drastically different from the average 9.4 m/s, Equation 2.10 can be rewritten as:

$$t \propto \frac{L\,\delta}{v} \propto \frac{L^2}{9.4} \times \frac{9.4}{\text{mean piston velocity}} \propto L^2 \frac{9.4}{\text{mean piston velocity}}.$$

Applying the correction factor discussed in Section 2.3 for taking into account load factor, the life could be written as follows:

$$\text{Life} = 0.85 \times (\text{typical size})^2 \times \frac{0.8}{\text{Load factor}} \times \frac{9.4}{\text{mean piston velocity}} \tag{2.13}$$

In Figure 2.7 the life is replotted after correcting for the actual piston speed (corrected life = life given by the manufacturer × 9.4/actual piston speed). The points come somewhat closer to the bold line describing Equation 2.13. Figure 2.8 shows the correlation of actual top overhaul life and the life calculated using Equation 2.13 in a log-log graph.

2.8 Summary

The characteristic size of diesel engines varying widely in rated power can be taken as the engine bore. The mean piston speed of these engines at rated speed is about 9.4 m/s on an average. The mechanical efficiency (at rated speed) of an engine normalized to an arbitrarily chosen bmep of 1.7 MPa shows its strong dependence on the typical size of the engine. Larger engines have relatively less friction and hence last longer. A normally aging engine appears to have longevity (major overhaul interval, in hours) as a function of the square of the characteristic length (bore, in millimetres). In real life, load factor and mean piston velocity in operation could be different from the mean values and the life is corrected for the two factors. The accuracy of the rule of life is +/− 35%.

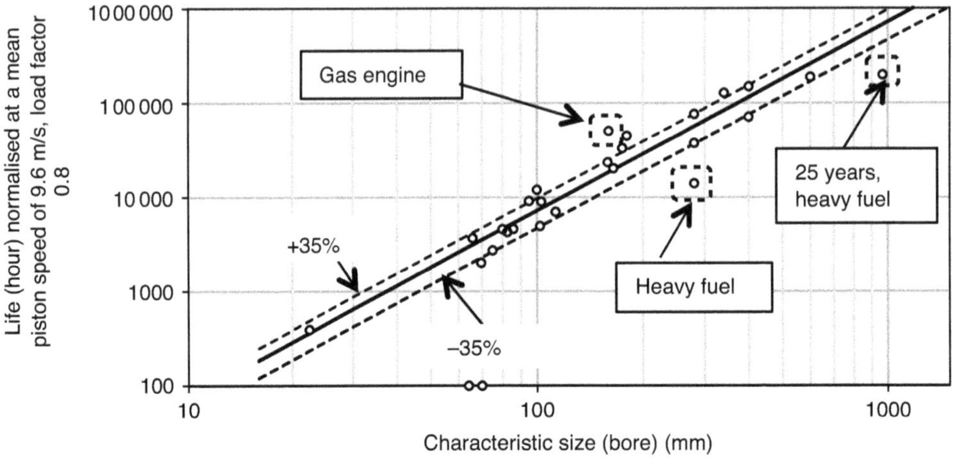

Figure 2.7 Predicted life of engines plotted against the characteristic size after correcting for the actual mean piston velocity at the rated speed, with the rule of life in the backdrop

The rule takes the philosophical view of the lifetime for which a properly designed engine would probably last under known operating conditions, if clean enough air and fuel are fed to the engine and oil as well as the coolant is maintained to avoid abnormal corrosive or abrasive wear of critical parts of the engine.

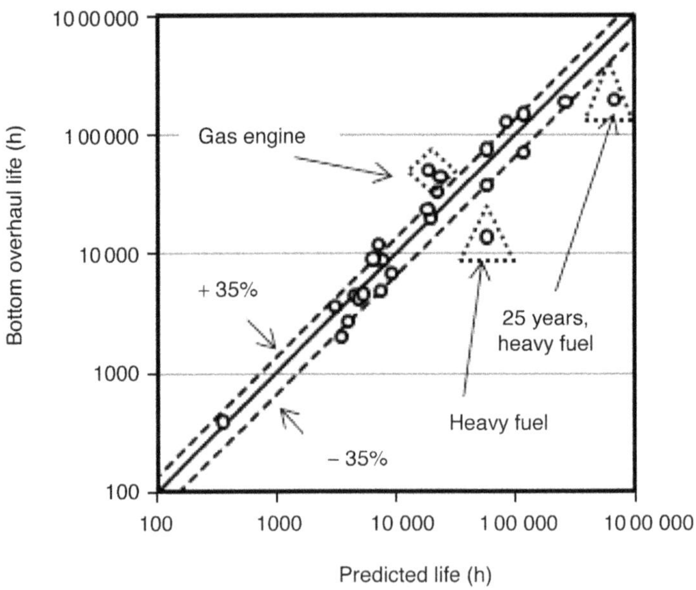

Figure 2.8 Correlation of actual bottom overhaul life and predicted life of the different engines shown in Figures 2.6 and 2.7

Appendix 2.A Engine Parameters, Mechanical Efficiency and Life

Engine type	Stroke/ cycle	Bore (mm)	Stroke (mm)	Volume (litre)	Speed (rpm)	sfc (g/kW h)	hp/cyl	Mechanical efficiency (%)	bmep (MPa)	Life (h)	Application
1	2	26	17	0.0075	10000	—	0.6	50	0.87	500	aero modelling
2	4	64	64	0.21	6000	—	7.8	—	0.57	—	off road
3	4	—	66	0.28	3600	—	7.3	—	0.64	5000	off road
4	4	70	70	0.27	3800	—	15.0	80	1.34	1700	car
5	4	75	76	0.34	3000	250	7.0	72	0.63	3000	pump set
6	4	80	98	0.49	3500	225	26.7	81	1.39	6000	light vehicle
7	4	83	76	0.41	3000	—	8.3	76	0.61	6000	off road
8	4	86	83	0.48	3000	—	9.9	76	0.62	6000	off road
9	4	95	127	0.90	2300		58	—	2.54	10 000	off road
10	4	100	125	0.98	2300	216	20	85	0.80	12 000	agricultural tractor
11	4	102	120	0.98	2900	—	54	87	—	8000	off road, on road
13	4	103	113	0.94	2500	203	37	88	1.43	14 300	on road
14	4	113	125	1.25	2200	203	54	88	1.78	14 300	on road
15	4	159	190	3.8	1500	184	149	96	2.38	26 500	off road, on road
16	4	160	190	3.8	1500		146	90	2.30	80 000	generating set
17	4	165	210	4.5	1800	200	210	—	2.34	30 000	marine, military
18	4	175	220	5.3	1500	204	150	92	1.70	25 000	generating set
19	4	180	280	7.1	900	—	178	—	2.50	50 000	generating set, marine, construction
20	4	280	320	20	1000	190	400	94	1.83	67 000	generating set, marine propulsion
21	4	280	330	20	1000	188	680	94	1.51	32 000	generating set, marine propulsion
22		340	530	45	600	—	734	—	2.43	120 000	generating set, marine propulsion
23	2	400	550	70	500	180	979	98	1.25	150 000	marine propulsion
24	4	400	530	63	600	179	1020	98	2.44	67 000	marine propulsion
26	2	600	2292	648	105	170	2360	99	3.12	219 000	marine propulsion
27	2	965	2439	1821	102	161	7780	98	1.88	219 000	marine propulsion

References: Aberli and McMillan, 2002; Brown, and Eckert;* Cummins, 2001, 2008; ECPlaza (Pielstick);* FG Wilson, 2009;* IMarE, 2007; Mabie Brothers;* MAN B&W, undated;* McGeary, Chew and Fogh, 2004; Salter, Smith and Roberts, 1954; Strome *et al.*, 2003; Van Wingen;* Walker, 2002 (* accessed in January 2011 on the internet).

Appendix 2.B Hardness and Fatigue Limits of Different Copper–Lead–Tin (Cu–Pb–Sn) Bearings

Bearing type		Composition of lining	Composition of overlay	Average Hardness: Bearing lining (HV)	Average Hardness: overlay (HV10) at: 20 °C	Average Hardness: overlay (HV10) at: 100 °C	Bearing unit load (N/mm^2): main	Bearing unit load (N/mm^2): big end	wear resistance[1] (relative)	Fatigue strength[2] (relative)	Tensile strength (strip form)[6] (N/mm^2)	Embeddability (relative)	SAE type
Cu–Pb–Sn	with electroplated overlay	Sn 3.5%, Pb 22%, Cu rem.	Sn 10%, Cu 2%, Pb rem.	70	12–15	9	42	52	X[2]	X	140	X	794
		Sn 2.25%, Pb 23%, Cu rem.	Sn 10%, Cu 2%, Pb rem.	85	12–15	9	52	62	X+ (5–10)%[3]	X+	120	X–	49 variant 1
		Sn 1.5%, Pb 26%, Cu rem.	Sn 10%, Cu 2%, Pb rem.	70	12–15	9	40	49	X– (1–2) %[4]	X–	180–200	X+	49 variant 2
	with sputtered overlay	Sn 10%, Pb 10%, Cu rem.	Sn 20%, Cu 2%, Al rem.	115	85–110	—	100	100	30 X[5]	X++[7]	180	X– – –	792

[1] Refer to wear graph (Figure 2.A.1) showing wear comparison between conventional overlay and sputter. However, there is marginal improvement in the wear resistance of conventional as well as sputter overlay based on the backing lining material and its properties. The harder the overlay the better is the wear resistance. The thinner the overlay the better is the fatigue strength of the overlay, probably better the wear resistance as well.

[2] X is standard performance, X+ is improved performance and X++ is excellent performance (e.g. sputtered bearing). Normally, the fatigue strength of the sputtered bearing overlay is much higher. Nevertheless, it is limited by the lining material on which it is deposited. So, many designers use sputtered overlays directly on steel itself to increase fatigue strengths to a very high level, however at the cost of the edge loading capability.

[3] Better wear resistance.

[4] Lower resistance.

[5] Better at same test loads.

[6] At 20 °C.

[7] 130–140 N/mm^2.

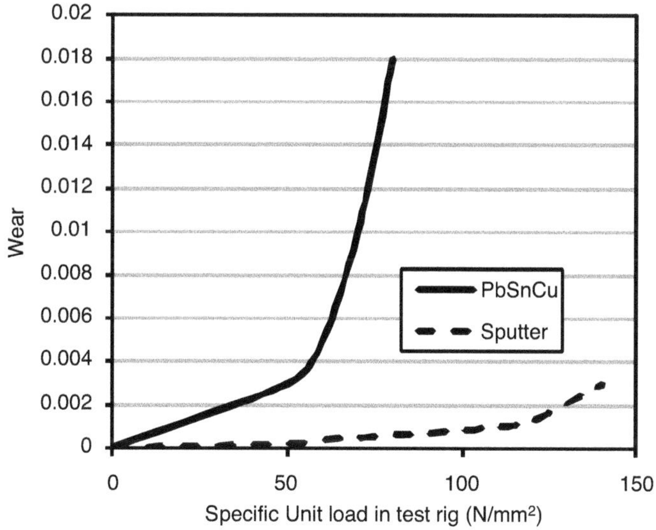

Figure 2.A.1 Wear rate against specific load at a rig

Appendix 2.C Hardness and Fatigue Limits of Different Aluminium–Tin (Al–Sn) Bearings[6]

Bearing type[3]	Composition of lining	Composition of overlay	Average Hardness: Bearing lining (HV)	Average Hardness: overlay (HV10) at: 20 °C	Average Hardness: overlay (HV10) at: 100 °C	Bearing unit load (N/mm²): main	Bearing unit load (N/mm²): big end	wear resistance (relative)	Fatigue strength at test rig (relative)	Tensile strength (strip form)[1] (N/mm²)	Embeddability (relative)	SAE type
Aluminium–tin	Sn 20%, Cu 1%, Al rem.	No overlay	40	—	—	32	38	Y–	Y[2]	117	Y	783[5]
Aluminium–tin–silicon–lead	Sn 12%, Si 2.5%, Pb 1.5%, Cu 0.7%, Sb 0.3%, Al rem.	No overlay	48	—	—	39	52	Y	Y+	137	Y–	788
	Sn 6.3%, Cu 1%, Ni 0.8%, Al rem.	No overlay	45	—	—	34	43	Y	Y	117	Y– –	770
Aluminium–tin–copper with overlay[4]	Sn 6.3%, Cu 1%, Ni 0.8%, Al rem.	Sn 15%, Cu 3%, Pb rem.	45	15–17	12	35	45	Y– –	Y	117	Y+	770[5]

[1] At 20 °C.

[2] Y is standard performance, Y+ is improved performance.

[3] Aluminium bearings will have three-layered structure steel backing, bonding foil and bearing lining. The bonding foil is either pure aluminium or aluminium alloy; the lining is of the compositions mentioned. Depending on the bonding foil design, the unit load, fatigue strength and wear properties can be influenced.

[4] Not much experience of aluminium bearings with overlays is available. However, the benchmark available for SAE 770 is included for reference. Also, sputtering is not common in Al-Sn bearings.

[5] It is known that SAE 770 has higher hardness than SAE 783 due to the presence of nickel and its lesser tin content. However, this can result in the higher wear of the journal or pin and hence for this material overlay plating is normally in use.

[6] Embeddability of electroplated overlay is better than aluminium bearings.

References

Aberli, K. and McMillan, J. (2002) Common rail at sea: The Sulzer RT-flex Engine. The Motor Ship Marine Propulsion Conference 2002, Copenhagen, Demark.

Archard, J.F. and Hirst, W. (1956) The Wear of Metals under Unlubricated Conditions, *Proceedings of the Royal Society London A*, **236**, 397–410.

Brown, D.T. and Eckert, B.O.The rotating piston more than 50 years on. *Ship en Werf 57ste Jaargang,* Nr 6, 329–333. http://www.knvts.nl/S&W%20archief/The%20rotatiing%20piston%20more%20than%2050%20years%20on.pdf (accessed 21 May 2011).

Cummins (2001) Data sheet QSK60-G3, curve FR6283. Cummins Inc., Columbus, IN.

Cummins (2001) Fuel-Efficient QSK Engines Optimized: Modular Common Rail Fuel System Enables Cleaner, More Efficient Combustion At High Engine Loads. Cummins Inc., Columbus, IN. http://cumminsengines.com/every/news/20080922US01.page (accessed 7 June 2011).

Daido Metal Company (2010) Half plain bearing design. Daido Metal Company Ltd, Japan. http://www.daidometal.com/english/products_eng/chapter01.pdf (accessed 21 May 2011).

EPA (1999) Final Regulatory Impact Analysis: Control of Emissions from Marine Diesel Engines. EPA420-R-99-026, United States Environmental Protection Agency, Office of Air and Radiation, Washington, DC.

EPA (2002) Median Life, Annual Activity, and Load Factor Values for Non-road Engine Emissions Modelling. EPA420-P-02-014, United States Environmental Protection Agency, Office of Air and Radiation, Washington, DC.

FG Wilson (2009) What is an authorized service centre – update. http://generators.blogs.com/fgwservice/ (accessed 22 May 2011).

Geist, M. (2001) The Sulzer RTA96C engine, Sulzer, Wärtsilä Switzerland Ltd.

MAN, B&W, Hydrodynamics of Ship Propellers. http://www.manbw.com/files/news/filesof1815/Hydrod propellers. pdf (accessed 22 May 2011).

MAN, B&W (1996) Emission Control: Two-stroke Diesel Engines. MAN B&W Diesel A/S, Copenhagen, Denmark.

McGeary, T., Chew, F. and Fogh, J.W. (2004) Investigations into Abrasive and Corrosive wear mechanisms of Pistons and Liners in large bore 2-stroke diesel engine cylinders. CIMAC Congress 2004, Kyoto, Japan. http://flamemarine.com/files/CIMAC2004.pdf (accessed 22 May 2011).

IMarE, (Ikatan Marine Engineer) (2007) Mengatasi Pencemaran. *Bulletin Marine Engineer,* **EDISI KE - XXXV,** 11-13. http://www.scribd.com/doc/43270339/imare (accessed 22 May 2011).

China Power Contractor Consortium, MTU Cummins, Perkins, Doosan, Volvo Diesel Generator sets and Their Spare Parts with shortest delivery Time and lowest Price. http://www.china-power-contractor.cn/MTU-Cummins-Perkins-Volvo-Diesel-Generator-from-6kw-to-2400kw.html (accessed 22 May 2011).

NAVSEA (1992) Engine man 2, Non-resident Training course. NAVSUP Logistics Tracking Number 0504-LP-026-7420, NAVEDTRA 14076, Naval Education and Training Professional Development and Technology Centre, Pensacola, FL, pp. 3–6.

Nural (1983) Kolben Handbuch. Alcan Aluminiumwerk Nuernberg Gmbh, Nuernberg, Germany. 1983.

Mabie Brothers, Perkins 900 series 37-48.5 kW. http://mabiebrothers.com/perkins/9_engine_family_descriptions/900/900.htm (accessed 23 May 2011).

ECPlaza, Pielstick Manufacturers, Buyers & Suppliers. http://www.ecplaza.net/search/0s1nf20sell/pielstick.html (accessed 23 May 2011).

Rakopoulos, C.D., Giakoumis, E.G.,and Dimaratos, A.M., (2007) Evaluation of Various Dynamic Issues during Transient Operation of Turbocharged Diesel Engine with Special Reference to Friction Development. Technical Paper 2007-01-0136, SAE, Troy, MI.

Royal Belgian Institute of Marine Engineers (2005) *Bearing up under the pressure, Marine propulsion.* Royal Belgian Institute of Marine Engineers, Niel, Belgium.

Saburi, S., Saitoh, Y. and Yamada, T. (2005) Tribology between Piston Rings and Cylinder Liners of Marine Diesel Engines. *IHI Engineering Review*, **38**(1), 11–18.

Salter, R.G., Smith, M.L. and Roberts, C.P. (1954) Study of miniature engine-generator sets, Part II. Investigation of Engines, Fuels, and Lubricants. WADC Technical Report 53-180, The Ohio State University Research Foundation, Columbus, OH, Wright Air Development Center, Air Research and Development Command, US Air Force, Wright Patterson Air Force Base, OH, Contract AF 18(600)-192, Project number 6058.

Strome, K., Moon, W., Sajdak, J. *et al.* (2003) Virginia Tech Shuttle Tanker, Ocean Engineering Senior Design Project, Aerospace and Ocean Engineering, Virginia Tech, Blacksburg, VA. http://www.aoc.vt.edu/~brown/VTShipDesign/2001Team2HiberniaFinalReport.pdf (accessed 23 May 2011).

Sutaria, B.M., Bhatt, D.V. and Mistry, K.N. (2009) Simulation and Modeling of Friction Force and Oil Film Thickness in Piston Ring – Cylinder Liner Assembly of an I.C. Engine. Proceedings of the World Congress on Engineering 2009, Vol II, 1–3 July 2009, London, UK.

Takata, R. (2006) Effects of Lubricant Viscosity and Surface Texturing on Ring-pack Performance in Internal Combustion Engines. MSc paper, Massachusetts Institute of Technology, Cambridge, MA.

Taylor, R.I. (2004) Transient Effects in Engines Operating at Steady Speeds and Loads, Transient Processes, in *Tribology* (eds G. Dalmaz et al.), Elsevier.

Walker, J. (2002) MTU and DDC extend 4000 series: new diesel features more cylinders of higher displacement for higher horsepower-Diesel Engine. *Diesel Progress North American Edition,* May 2002.

Wolff, A. and Wolff, N. (2009) Experimental verification of the model of piston ring pack operation of an internal combustion engine. *The Archive of Mechanical Engineering*, Vol. **LVI**, Number 1.

Wong, V., Tian, T., Moughon, L. *et al.* (2005) Low Engine Friction Technology for advanced natural gas reciprocating engines. Annual Technical Progress Report, Massachusetts Institute of Technology, Cambridge, MA.

Van Wingen, Partnership OEM with Perkins. E. Van Wingen NV, Evergem, Belgium. http://www.vanwingen.be/html/en/nieuws/artikel.php?id=1 (accessed 23 May 2011).

Part II

Valve Train Components

3

Inlet Valve Seat Wear in High bmep Diesel Engines[1]

The valve head deflects under large firing pressures in engines and there is a micron-sized displacement of the valve with respect to the seat in the head. In high brake mean effective pressure (bmep) engines, the wear of valve seats on the valve and in the cylinder head is dependent on seat load and the total relative fretting displacement after the valve has seated. In the new generation engines, the oil metered past the valve stem is just enough to lubricate the valve guide and not enough to lubricate the inlet valve seats. The exhaust valve seat is smeared by the oil mixed with the gases and, hence, fretting wear of the exhaust valve seat is tempered. However, the inlet valve suffers if fretting displacement is excessive. Wear in each cycle is modelled proportional to the product of load, the relative cyclic displacement and the friction coefficient. The total number of cycles to reach the critical wear limit is empirically obtained. Beyond the limit, the wear rate increases uncontrollably. The model is validated on an engine rated at 1.8 MPa bmep. Finite element analysis of the valve and the seat showed that the displacement of the valve seat could be correlated linearly with the deflection of the valve at the centre. In the study, by increasing the thickness of the valve head, and hence the stiffness of the valve, the relative displacement at the seat (after the valve has seated) is substantially reduced. A design norm for designing the valve head thickness for low fretting wear is given at the end of the chapter.

3.1 Introduction

Valve wear can be caused by design factors, namely, high seating velocity (Kelley and Musil, 1962), low hardness material and, most importantly, by fretting due to valve deflection under high cylinder pressure (Abduschelischwili and Kotjarow, 1965; Mettig, 1973; Hannum, 1960; Newton, Palmer and Reddy, 1953; Newton and Hoertz, 1952). In practice, additional seat distortion takes place due to poor functioning of cooling system, misaligned seating or improper tightening of the head. High engine speed, incorrect spring installation, seat grind-in, excessive tappet clearance, valve sticking and large stem-to-valve guide clearance will cause high seating velocities leading to wear of the valve seat. Durability tests on valve materials by Hori *et al.* (1989) showed that there is no remarkable difference in wear pattern found between valves made of sophisticated high wear resistant materials. Similarly, study of the wear mechanism of ceramic and metal valves in diesel engines by Updike (1989)

---

[1]All figures in this chapter are reproduced with permission from P.A. Lakshminarayanan, N.S. Nayak, Y.V. Aghav and A.D. Dani, "Solving inlet valve seat wear problem in high BMEP engines," SAE Paper 2001-01-0024, Symposium on International Automotive Technology (SIAT 2001) SAE Conference, 2001.

Critical Component Wear in Heavy Duty Engines, First Edition. P.A. Lakshminarayanan and Nagaraj S. Nayak.

concluded that metal valves offer excellent performance for heavy duty applications. However, ceramics seem to offer extended overhaul life for the valve train in alternative fuel applications where severe corrosion is a problem.

The work of Dooley, Trudeau and Bancroft (2000) showed that larger seat angles and higher friction at the seat result in high normal forces and hence shear stresses at the interface, which in turn increase the fretting movement between the valve and seat insert, particularly under nonlubricated conditions. Compared to the permanent distortion that arises from mechanical stresses set up during assembly of the engine, a periodic distortion occurs as a result of gas loads and thermal stresses encountered during engine operation (Giles, 1966). On the other hand, in a gas engine (Wiles, 1965) the exhaust valve and seat insert wear is more severe than in a diesel engine. When the valve overheats at high exhaust temperatures, it can cause pre-ignition, which accelerates the valve burning and results in loss of power and other destructive effects, and hence wear of the valve and the seat insert. The wear properties of the exhaust valve could be improved with the help of copper infiltration to decrease the friction coefficient (Oh *et al.*, 1991). The mechanism of shear strain causing major wear in the valve seat and the case of a self-lubricated material for the valve to reduce the wear are explained by Wang, Schaefer and Bennett (1995). However, the wear of the inlet valve seat seems to be most often due to distortion by design and incorrect seating in practice, as the valve seat is not lubricated by oil or diesel. On the contrary, the exhaust valve seat wears out through oxidation (Schaefer *et al.*, 1999), due to inadequate seat material or abusive engine operation.

3.2 Valve Seat Wear

3.2.1 Design Aspects to Reduce Valve Seat Wear Life

A typical configuration of valve and seat insert is shown schematically in Figure 3.1. The selection of valve seat angle is necessarily a compromise between flow considerations and seating pressures. The optimum seat angle of 45° is recommended for better flow capacity. However, an angle of 30° is beneficial in view of the reduction in wear of the inlet valve seat (Mettig, 1973).

There is another school of thought that recommends a flexible valve head for valve durability. A thinner section valve head and smaller fillet radii allow the valve head to conform more readily to a distorted elliptic seat to avoid wear due to fretting. Furthermore, for the smooth motion of the valve tappet to avoid low shock loading, the ramp between the base circle and flank of the cam must be carefully designed. This provides a safe margin that can accommodate the changes in length in the valve train that result from wear and temperature rise. Thus, providing an adequate coolant flow all the

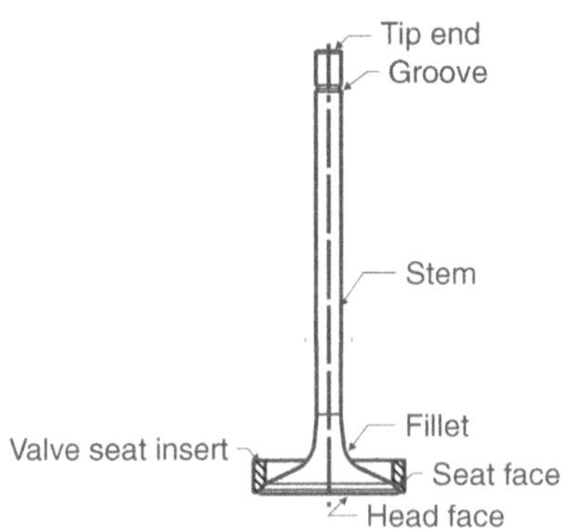

Figure 3.1 Schematic layout of valve and seat insert

way around the seat can reduce the valve seat distortion problem. Also, the coolant flow should be sufficiently high to avoid local boiling at valve seat insert zones.

3.3 Shear Strain and Wear due to Relative Displacement

The frictional shear stress on the seat surface is the cause of wear. The friction is dependent on the normal force (Rabinowicz, 1965; Kragelskii and Alisin, 1981). When there is a relative movement, it is in a nearly constant ratio to the normal force corresponding to the dynamic friction coefficient, f (Figure 3.2). The work done by the shear stress (beyond the plastic limit) goes to supply the surface energy that results in fracture into fine particles or debris.

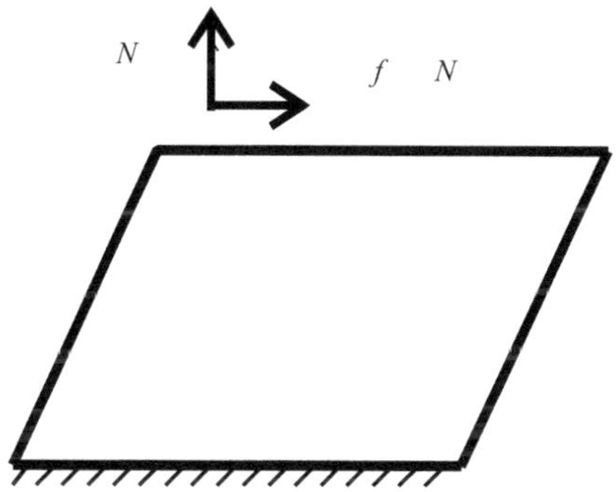

Figure 3.2 Shear strain schematic

3.4 Wear Model

For the sake of argument, the valve head can be considered simply as a beam on two supports, as shown in Figure 3.3 (Lakshminarayanan *et al.*, 2001). Usually, the deflection of the beam is considered too small to account for the relative movement between the support and the beam. However, the relative displacement, Z, for a given geometry of the beam at the valve seat becomes important in determining wear when the pulsating cylinder pressures are higher than 15 MPa. For a given geometry of the beam, the displacement is linearly proportional to the gas pressure, p, on the valve head:

$$Z \sim p \tag{3.1}$$

The wear of two bearing surfaces is proportional to the product of normal load and the total relative displacement (Archard and Hirst, 1956). The gas load on the valve head is transmitted to the small area of contact at the seat, which is amplified into seat pressure or normal stress. During the minute total relative displacement, Z, between the two wear parts, namely the beam and the support, a friction shear stress, τ, is created, which wears the parts. The shear is given by the friction coefficient times the normal stress. The number of load cycles during the wear life of the valve, n, gives the total displacement, nZ, of the shear stress. The deflection of the entire valve is an elastic regime, except the contact surfaces which undergo plastic shear if the normal load is excessive.

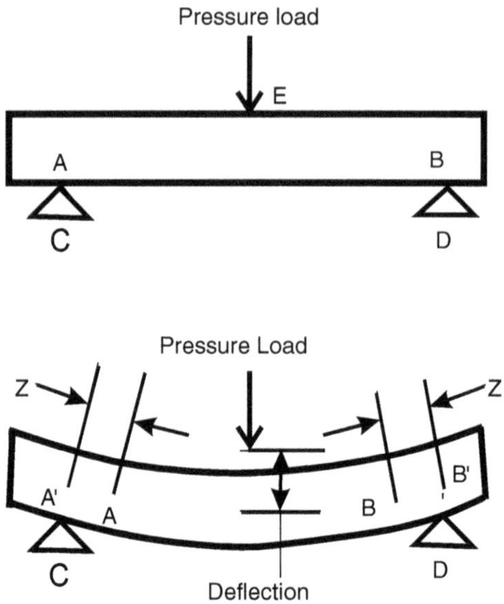

Figure 3.3 Valve head schematic (load simply supported on a beam)

3.4.1 Wear Rate

In the case of valves (Figure 3.4):

$$\text{The normal load on the valve head, } P = p \times \pi d^2/4 \quad (3.2)$$

$$\text{Total relative displacement during the life time, } x = nZ \quad (3.3)$$

where p = gas pressure on the valve head (N); d = diameter of valve head (m); and n = number of shear cycles.

$$\text{Normal pressure on the valve seat, } p' = P/\pi Dy \cos \alpha \quad (3.4)$$

$$\text{Shear stress, } \tau = fp' \quad (3.5)$$

$$\text{Number of shear cycles, } n = Nt/c \quad (3.6)$$

$$\text{Total wear per unit surface area, } W = k\tau nZ \quad (3.7)$$

$$= d^2nZ$$

where d = diameter of valve head (mm; c = 2 for a four-stroke diesel cycle and 1 for two-stroke diesel cycle; D = mean diameter of the seat (mm); f = dynamic friction coefficient; k = proportionality constant; N = engine speed (rpm); p = gas pressure on the valve head (N); t = time for which engine has run (s); W = wear (microns); y = width of actual seat area (mm); Z = relative displacement of valve (mm); and α = valve seat angle (degrees).

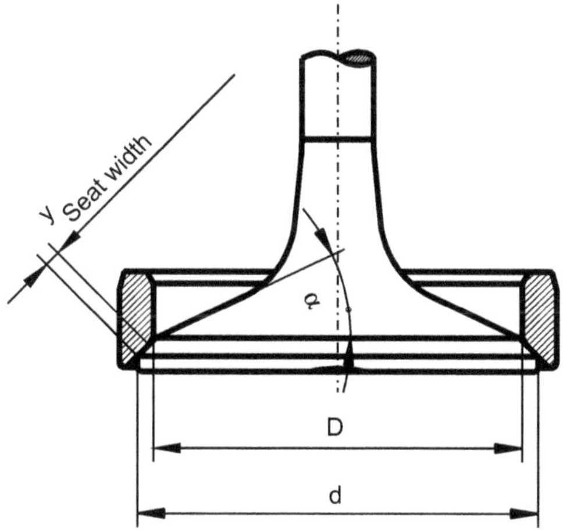

Figure 3.4 Typical dimensions of valve and valve seat

Equation 3.7 can be interpreted as an energy balance equation, where the work done by the shear is transformed to the surface energy of the metal debris. The term τnZ is the total work done per unit area by the shear force and W is the material removed per unit area due to wear. The wear rate, W/t, represents the increase in mean surface area due to the formation of debris after fracture into wear particles due to delamination. The constant, 1/k, represents the specific surface energy for fracture.

In diesel engines, the exhaust valve is usually of smaller diameter, d. Therefore, the fretting displacement at the seat is less than for the inlet valve. The oil escaping past the top ring of the piston and unburned diesel lubricate the exhaust valve; hence the friction coefficient, f, is low. These two factors make the exhaust valve seat more resistant to wear under identical cylinder pressure fluctuations, as seen in Equation 3.7. On the other hand, the inlet valve runs dry with modern valve stem seals that substantially reduce the leakage of oil past the valve stem. In the case of turbocharged engines, the leakage is further diminished by the blow-by from the port past the valve guide.

For a given geometry of the inlet valve, by combining Equations 3.1 and 3.7 the wear rate can be surmised:

$$W/t \sim Z^2 \tag{3.8}$$

The quadratic relationship between the wear rate and the micron-sized displacement at the seat emphasizes the need to contain the latter within limits to achieve satisfactory wear. The lack of any factor to represent roughness in the model given by Equation 3.7 suggests that only the initial wear rate can be predicted by the expression on the right hand side. The increase in roughness due to adhesion and delamination will accelerate subsequent wear at a higher rate.

3.5 Finite Element Analysis

The valve and the valve seat in the head are modelled axisymmetric in the finite element calculation, neglecting the circularly nonsymmetric form of the real case due to thermal distortion, and different errors in machining and assembly (Figure 3.5). Instead of considering the contact between the stationary seat and

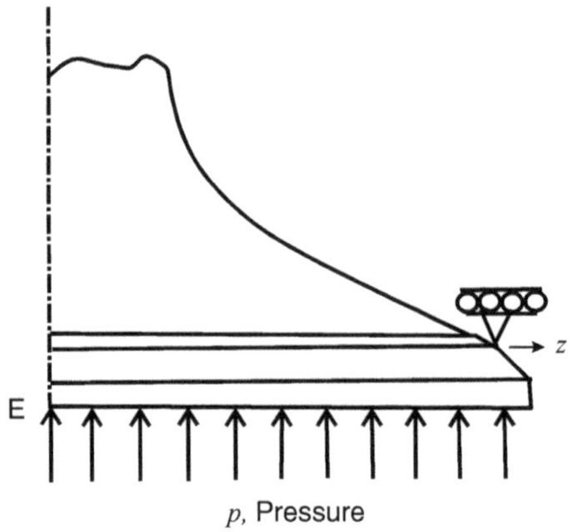

Figure 3.5 Axisymmetric finite element model (FEM) model of valve and valve seat in the head

the valve as a nonlinear problem, the node corresponding to the inner seat is constrained to move along the radial direction due to pressure load. Figure 3.6 shows the axisymmetric model of the half valve deflected under pressure load; the constraints are also shown by the triangular symbols pointing the constrained direction of relevant mode. For simplicity, the seat in the head was modelled infinitely rigid. The calculated displacement, Z, of the seat along z is correlated with wear.

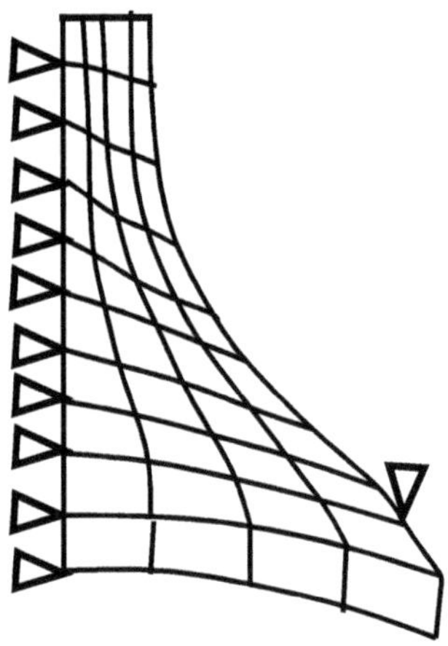

Figure 3.6 Axisymmetric model of the half valve deflected under pressure load (triangular symbols show restriction of movement in the pointed direction)

3.6 Experiments, Results and Discussions

A six-cylinder, 8.8 litre capacity engine was up-rated by 25% where the peak cylinder pressure increased from 11.5 to 15 MPa. An abusive endurance test at a 10% overload condition for more than 500 hours was conducted to predict the typical wear of various parts such as valves, seats and bearings that would take place in 3500 hours at an average continuous duty of 70% of full load.

Table 3.1 Study of combined wear of valve seats in the head and the valve by measuring valve recession in six cylinders, existing and improved designs

Design		Before endurance		End of endurance		
		Inlet	Exhaust	Inlet	Exhaust	Hours
Existing	Mean (mm)	0.70	0.85	1.03	0.92	110
	Std. Dev.	0.05	0.05	0.07	0.06	
Improved	Mean (mm)	0.70	0.83	0.78	0.88	550
	Std. Dev.	0.05	0.05	0.06	0.05	

Cam design: The inlet cam of the original engine had a ramp rate of 0.030 mm/cam degree at the maximum, which is sufficient for avoiding shock during valve seating up to a maximum engine speed of 2400 rpm. Any further reduction in ramp velocity will lead either to a ramp too gradual to properly set the valve lash or very large ramping angle.

3.6.1 Valve and Seat Insert of Existing Design

In only 110 hours, the combined wear of valve seats in the head and the valve of existing design was as high as 0.35 mm on the average in the up-rated engine (Table 3.1). The immediate result is the loss of valve lash, and hence the loss of gas tightness. The engine is observed to lose power or efficiency as the effective compression is lost (Table 3.2).

The wear is measured in two ways. The first method is to measure the sinking of the valve due to the increase in valve depth due to wear (Figure 3.7). The experimentally observed valve recesses, using the existing design at the beginning and end of the endurance test of 110 hours, are given in Table 3.1. The large increase in wear is plotted in Figure 3.8. Alternatively, the contour of the valve seat itself is plotted, as shown in Figure 3.9, to study the apparent digging (Table 3.3). Ridges on the seat surface can be seen in the scanning electron micrograph (Figure 3.10a). The ridges are highlighted in Figure 3.10b for clarity. The roughness as a result of high wear feeds itself and the wear rate accelerates.

3.6.2 Improved Valve and Seat Insert

When the hardness of the valve seat is increased by selecting an alternative material with the physical properties shown in Table 3.4, there is no noticeable improvement in the wear performance (Figure 3.11). Increasing the thickness of the valve head by two millimetres increased the basic stiffness against elastic

Table 3.2 Combined wear in endurance test with existing and improved designs at 1500 rpm and at overload

	Existing design		Improved design	
Endurance (hours)	Zero	110	Zero	550
Power (hp)	255	255	255	255
Specific fuel consumption (g/hp-h)	152	166	150	153
Smoke (Bosch Unit)	2.1	3.0	1.9	2.1

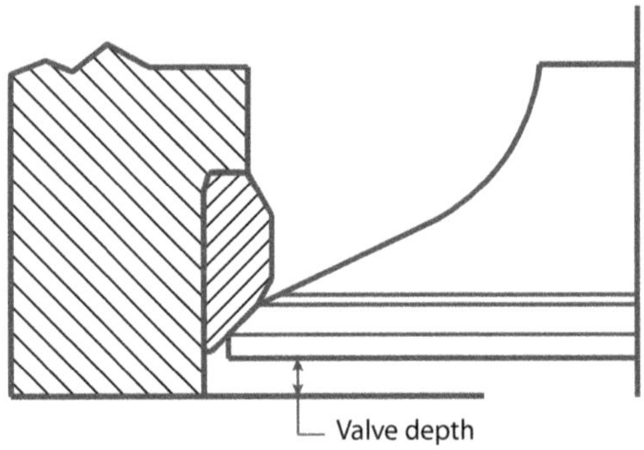

Figure 3.7 Measurement of seat wear by monitoring valve depth

cupping by about four times. The finite element model showed that the deflection at the centre was reduced to 6.9 microns and that the relative movement at the outer seat dropped to 1.3 microns.

The stiffness of the valve head was measured by pulling the valve stem with a force corresponding to the pressure on the valve head during running (Figure 3.12). A force of 35 000 N corresponds to a cylinder pressure of 15 MPa. Even though the actual deflection will be qualitatively less, under a uniformly distributed load due to gas pressure, this test verified the finite element calculation before conducting expensive engine experiments. The improved design is at least two times stiffer for a concentrated central load applied at the centre, simulating the pressure load (Figure 3.13). Further, during the redesign exercise the contact area was increased 1.8 times to reduce the contact pressure by as much.

When the valve head thickness for the same valve seat design is increased as explained, the running clearance between the piston and the inlet valve is less. To recover the clearance, an improved valve seat that was more recessed in the cylinder head was designed. To streamline the flow shooting through the gap between the valve and the seat, the geometry of the cylinder head below the seat was modified, as the combustion performance must be as good as that with the original valve, which was not as stiff. The important parameters, namely the swirl-number and the coefficients of flow, were measured in a steady flow rig (Thien, 1965). The flow characteristics

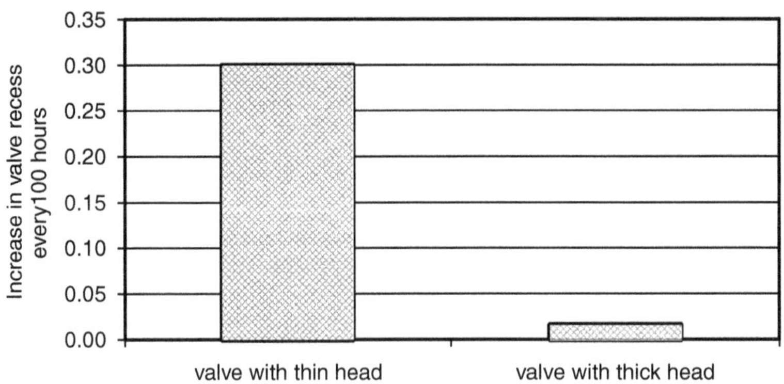

Figure 3.8 Improvement in wear because of improved design

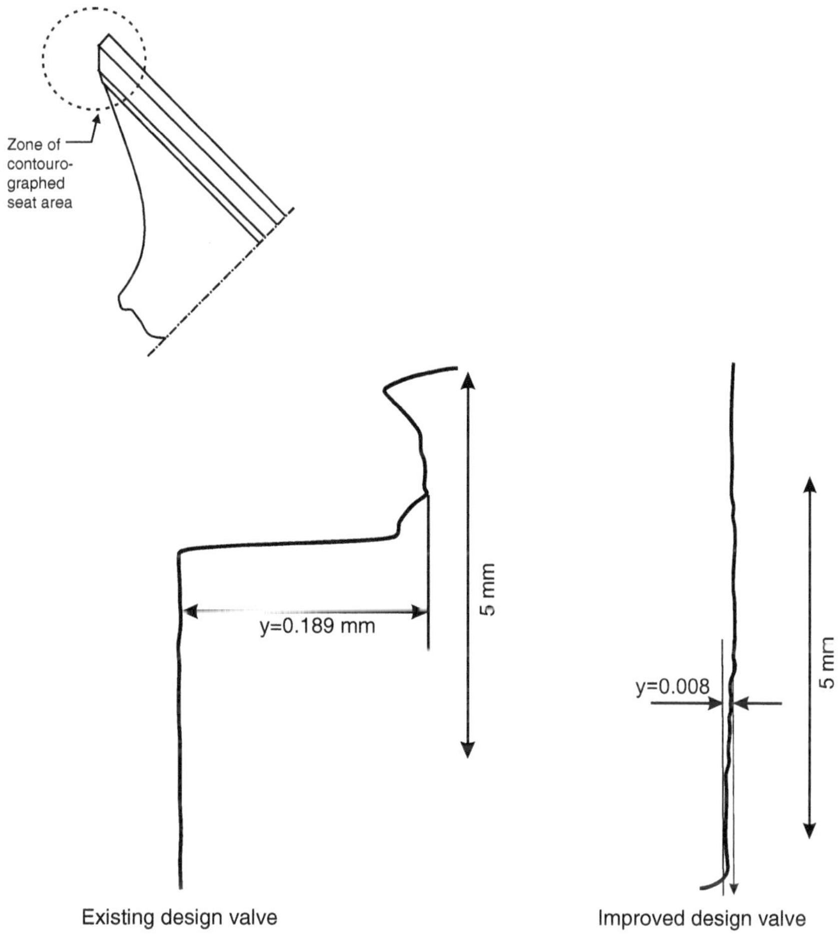

Figure 3.9 Contourograph of inlet valve seat indicating the wear of existing and improved designs. A finite element model of the existing design valve showed that the valve head deflection at the centre was 19 microns under a gas pressure of 15 MPa and that the corresponding relative displacement at the outer seat was 4.6 microns

at the steady flow rig are maintained when the improved valve system is introduced (Figure 3.14). The engine performance was identical with less stiff valve system (Table 3.2), where the performance is similar in both valve designs at the beginning of the endurance test. The performance remained same for the new design without any deterioration during the entire endurance experiment.

With the reduction of relative movement at the valve seat and the increased contact area, compared to the existing design the drop in actual wear rate (measured by contourograph) is about five times less in an engine for a period of operation of 110 hours under abusive overload. The wear results for 550 hours at overload are also given in in Table 3.3. The measured valve recession indicative of wear remained steady for 550 hours at continuous overload. It was 15 times less than the existing less-stiff valve that ran for a fifth of the time (Figure 3.15). The scanning electron micrograph (SEM) (Figure 3.10c) confirms reduced wear, showing the lack of surface shear and ridge formation. After the initial bedding

Table 3.3 Results of wear by contourograph of seat in the valve of six inlet valves of six cylinders

Inlet valve seat wear (mm)			
	Existing design	Improved design	
		Experiment 1 (using one set)	Experiment 2 (using another set)
Valve No	After 110 hours	After 110 hours	After 550 hours
1	0.1600	0.0145	0.0275
2	0.0810	0.0062	0.0100
3	0.1355	0.0425	0.0550
4	0.1890	0.0080	0.0135
5	0.1035	0.0135	0.0225
6	0.1136	0.0250	0.0275
Average	0.1304	0.0183	0.0260
Std. Dev.	0.0390	0.0140	0.0160

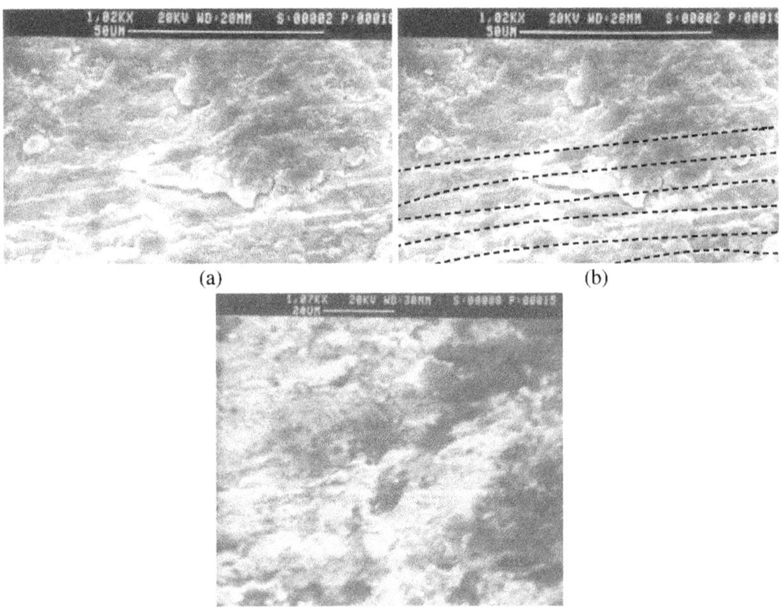

Figure 3.10 SEM photographs of the improved design and existing valves after an endurance experiment. (a) wear of typical valve with ridges along shear; (b) wear of typical valve with ridges along shear lines, highlighted by dotted line; (c) improved valve seat – without any ridges due to shear line.

Table 3.4 Physical properties of high strength material

Valve seat in	Material type	Hardness (HRC)	Young's Modulus (N/mm^2)
Cylinder head	Material A	37–43	180 000
	Material B	40–46	180 000
Valve	Stellite F	40	234 000

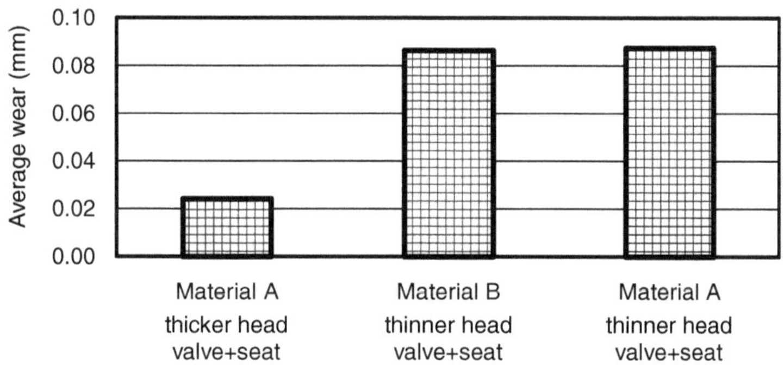

Figure 3.11 Study of seat wears in six valves with a high strength material

in, there is no wear, as opposed to the existing design where run away wear is observed. This is because the reduced initial wear rate maintains the integrity of the original surface due to a reduction in both the shear stress and the relative movement. All the six inlet valves showed similar improvement (Table 3.3).

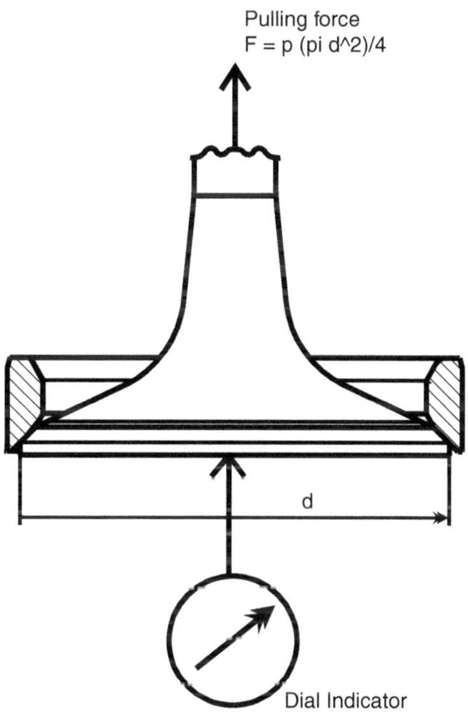

Figure 3.12 Experimental set-up for measurement of valve head deflection under static condition

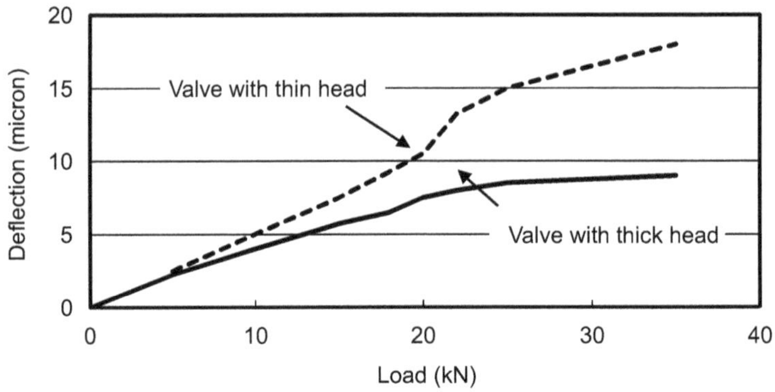

Figure 3.13 Deflection of valve at the centre under static condition, with concentrated load at centre simulating the pressure load (improved and existing design valve)

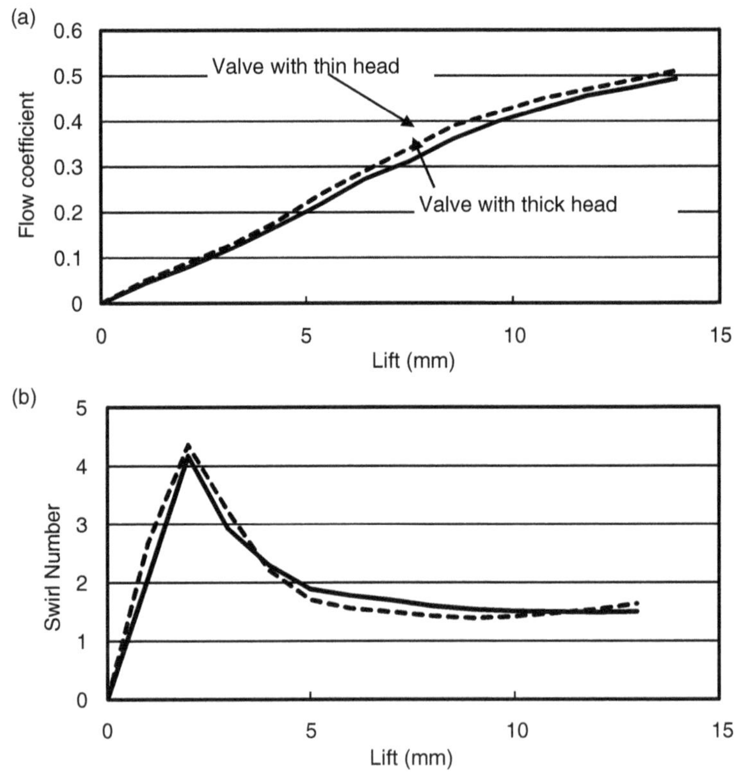

Figure 3.14 Port performance using improved and existing design valves in a steady flow rig

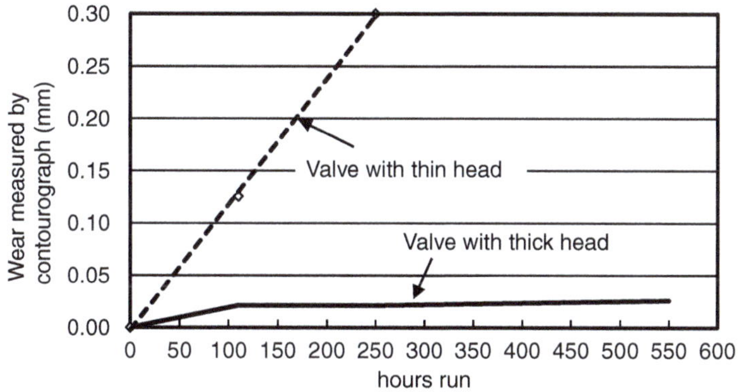

Figure 3.15 Valve seat wear measured using contourograph for existing and improved valve and seat insert

3.7 Summary

The valve wear is a function of the pressure on the seating area and the relative movement of the valve seat and the valve head due to finite elasticity. After the valve has seated, the normal force causes a frictional force between the surfaces to induce surface shear. The friction factor is dependent on the lubrication condition, roughness and hardness of the surfaces. The shear work is converted to surface energy when debris is created by wear. There are other modes of failure or wear, which are not typically considered when estimating the valve life. The inlet valve suffers more often in this mode because of its larger diameter and lack of lubrication. Changing the material of the valve seat in the head is of little use in reducing wear of the valve at the seat. Increasing the stiffness of the valve head decreased the relative movement that caused the wear. Increasing the contact area decreased the shear stress. The actual improvement in wear is many times more than the reduction in relative movement or the normal pressures in the shear zone. This is because the model correlating the wear with relative movement and normal seating pressure indicates the improvement in initial wear rate only and not the vicious wear that follows after the surface roughness has increased due to adhesion and ridge formation in the plastic shear regime.

3.8 Design Rule for Inlet Valve Seat Wear in High bmep Engines

A valve deflection of five micron at the centre of a 60 mm diameter valve is a reasonable starting point for the design of valve. If the product, p'Z, is $<100 \times 10^{-6}$ M J/m^2 (i.e. contact pressure times the relative movement), then the valve wear is contained to a safe level for satisfactory wear life.

References

Abduschelischwili, L.S. and Kotjarow, W. (1965) Stiffness of Inlet and Exhaust Valves of high speed high Bmep diesel Engines. *Energetischer Maschinenbau*, **12**, (in Russian).

Archard, J.F. and Hirst, W. (1956) The Wear of Metals under Unlubricated Conditions. *Proceedings of the Royal Society London, A*, **236**, 397–410.

Dooley, D., Trudeau, T. and Bancroft, D. (2000) Material and design aspects of modern valve seat inserts. Proceedings of the International Symposium on Valve Train System Design and Materials, 14–15 April 2000, Detroit, MI.

Giles, W.S. (1966) Fundamentals of Valve Design and Material Selection. Technical Paper 660471, SAE, Troy, MI.

Hannum, A.K. (1960) Why valves Fail. Technical Paper 600432, SAE, Troy, MI.

Hori, Y., Miyakawa, Y., Asami S. and Kajihara, T. (1989) Si_3N_4 Ceramic valves for internal combustion engines. Technical Paper 890175, SAE, Troy, MI.

Kelley, J.R. and Musil, N.J. (1962) Valve failure finger prints. Paper 620591, SAE, Troy, MI.

Kragelskii, I.V. and Alisin, V.V. (1981) *Friction, Wear, and Lubrication, Tribology Handbook,* Vol. 1. Mir Publishers, Moscow.

Lakshminarayanan, P.A., Nayak, N. S., Aghav, Y.V. and Dani, A.D. (2001) Solving Inlet Valve Seat Wear Problem in high BMEP Engines. Symposium on International Automotive Technology, SIAT 2001–SAE Conference, Technical Paper 2001-01-0024, SAE, Troy, MI.

Mettig, H. (1973) *Die Konstruktion Schnellaufender Vebrennungsmotoren.* Walter de Gruyter, Berlin.

Newton, J.A. and Hoertz, N. (1952) Considerations for valve maintenance. Technical Paper 530210, SAE, Troy, MI.

Newton, J.A., Palmer, J.L. and Reddy, V.C. (1953) Factors affecting diesel exhaust valve life. Technical Paper 190, SAE, Troy, MI.

Oh, J., Jung, S., Kim, Y. *et al.* (1991) The Oxidation effect of the Exhaust valve seat wear. Technical Paper 912507, SAE, Troy, MI.

Rabinowicz, E. (1965) *Friction and wear of materials.* John Wiley & Sons, Inc., New York.

Schaefer, S.K., Larson, J.M., Jenkins L.F. and Wang, Y. (1999) Evolution of Heavy duty engine valves – Materials and Design. Proceedings of the International Symposium on Valve Train System Design and Materials, American Society of Metals, ASM International, Materials Park, OH, pp. 129–139.

Thien, G.E. (1965) Entwicklungsarbeiten an Ventilkanalen von viertakt Diesel motoren. *Osterr. Ingenieurzietschrift,* **8/9**, 291–302.

Updike, S. H. (1989) A comparison of wear mechanics with ceramic and metal valves in firing engines. Technical Paper 890177, SAE, Troy, MI.

Wang, Y.S., Schaefer, S.K. and Bennett, C. (1995) Wear mechanism of Valve seat and insert in heavy duty diesel engine. Technical Paper 952476, SAE, Troy, MI.

Wiles, H.M. (1965) Gas engines valve and seat wear. Technical Paper 650393, SAE, Troy, MI.

4

Wear of the Cam Follower and Rocker Toe[1]

The rate of wear of cam followers in a valve train system is mainly a function of contact stress between the cam and the follower, sliding velocity and hydrodynamic film thickness between the two mating surfaces. The problem of (surface fatigue) wear becomes severe as the contact between the cam and the follower exceeds the plasticity limit of material. It finally leads to an increase in the valve lash and loss of engine performance. The wear is minimized by reducing the coefficient of friction and by minimizing the compressive stress. In this chapter, methods to estimate quantitatively the wear of followers are described. The profile of followers resulting from steady and noncatastrophic wear processes is computed by combining a linear wear relationship and an elastohydrodynamic or boundary lubrication transition model with kinematic analysis. The finite element analysis, a commercially available simulation program, and classical methodology have been effectively used to predict the follower wear. The model was validated on different types of followers that varied widely in size, brake mean effective pressure (bmep) and speed. The predicted wear profiles exhibit satisfactory agreement with experimental observations. A guideline for designing a cam follower for low wear rate is given. Finally, some practical insight into the wear of rocker tip is also provided.

4.1 Introduction

Cams are used in internal combustion engines to provide a specific prescribed motion to valve train system. The forces required to open the intake and exhaust of a conventional valve train system are strongly influenced by the stiffness and damping of the valve spring and valve seat, masses, geometries and frictional behaviours of the contacting elements of valve train components. The motion of the cam is transferred to the follower system by direct contact at high temperatures, with less lubrication and high contact stresses. The contact conditions can affect the performance of the system dramatically by means of severe wear (Neale, 1973).

Chen (1982) studied the wear of the cam follower considering the load, speed and material properties, and suggested that follower wear is a result of a complex set of factors; surface fatigue is the dominant cause for wear of the cam and follower system. Mehenny and Taylor (2001) highlighted the influence of circumferential surface waviness of the cam surface on the lubrication of an automotive cam

[1]All figures in this chapter are reprinted from N.S. Nayak, P.A. Lakshminarayanan, M.K. Gajendra Babu and A.D. Dani, "Predictions of cam follower wear in diesel engines," *Wear*, **260**, 181–192, 2006, with permission from Elsevier.

Critical Component Wear in Heavy Duty Engines, First Edition. P.A. Lakshminarayanan and Nagaraj S. Nayak.

and flat faced reciprocating follower. The study predicted that the circumferential waviness of the cam surface produces large amplitude pressure oscillations that may accelerate pitting of the cam surface at or near the cam nose.

Bell, Davies and Fu (1985) made an exhaustive research on the profiles of finger followers resulting from steady, noncatastrophic wear processes. The wear was estimated by combining a simple wear law in an elastohydrodynamic and boundary lubrication transition model with a new kinematic analysis. The study showed that the wear is influenced by the state of boundary conditions and is highest at positions where the velocity of the contact over the follower is small.

Fries and Rogers (1987) described a computational method for predicting the profiles into which cams and followers wear in service. Similarly, Colgan and Bell (1989) developed a wear model for a conventional cam/tappet of an overhead cam type valve train system. Cheng, Cheng and Yasuda (1994) performed the experimental and analytical research on the contact wear (surface fatigue) of the roller follower. The performance test indicated that the end clearance between the roller and rocker had a significant influence on friction and heat generation on the roller follower surface.

Hugnell and Andersson (1994) developed a simulation model for the wear of the follower based on single point observation, using Archard's wear relation (Archard and Hirst, 1956). The simulation is carried out by calculating the pressures when the surface topography gradually changes. Similarly, Hugnell, Bjorklund and Andersson (1996) studied the followers by finite element analysis and verified the semi-infinite plane assumption in the pressure calculation. The importance of follower rotation on elastohydrodynamic film thickness was shown by Glovnea and Spikes (2001). The variation in speed is significant and it results in a large variation of both oil entrainment and sliding velocity with a corresponding variation of elastohydrodynamic film thickness and friction.

Purmer and van den Berg (1985) predicted the location of maximum wear on the camshaft using kinematic analysis and elastohydrodynamic theory. The wear on the cam lobes is dominant at the locations where hydrodynamic lubrication conditions are critical. The study showed the importance of the frictional torque and wear behaviour on the film lubrication regime. In this chapter, the wear behaviour of followers of both the flat and roller type is studied by modelling the wear coefficient in the boundary lubrication regime. The measured wear satisfactorily correlates with calculated results.

4.2 Wear of Cam Follower Surfaces

Engine performance is greatly influenced by the adjustment of valve lift duration during the cycle, which controls the gas flow rates into and out of the cylinder. The characteristics of valve lift motion are mainly determined by the geometry of contact between the cam and the follower. A proper design of cam profile is very important, as it determines the magnitude of the valve lift with the rotation of the camshaft. The gas flow rates through intake and exhaust valves at the desired moments during the cycle can be maximized by optimizing the cam profile, which differs depending on the valve train system. In this chapter, the overhead valve train system is studied (Figure 4.1).

4.2.1 Wear Mechanism of the Cam Follower

The cam and follower pair under rolling and sliding contact conditions suffers from surface fatigue failures because of the repeated contact load (Nayak *et al.*, 2006). This is characterized by pitting and spalling of the surfaces, and it occurs rather suddenly without any prior visible warning. The resulting pits are relatively large, as shown in Figure 4.2.

In the early stage, a superficial micro crack of irregular shape appears on the surface as a result of contact stress exceeding the fatigue limit. The asperities (small peaks of the rough surface) interact and wear mildly without affecting the contact conditions. Under cyclic loading, the crack propagates quickly

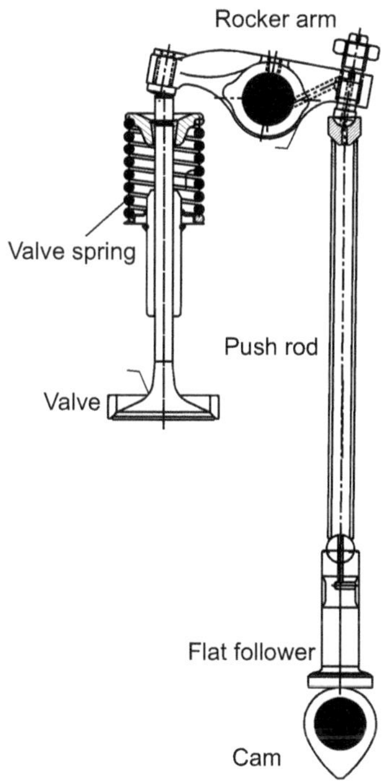

Figure 4.1 Typical diagram representing the overhead valve type valve train system

due to induced compressive stress. The compressive stress increases slowly, with the development of plastic deformed surfaces near the two contacting surfaces. In the later stage of the operation, it leads to disintegration of the surface. The destructive nature of surface (fatigue) crack increases due to high stresses. The sliding action raises the stress and hastens the failure. The surface flakes leave the parent surfaces damaging the soft follower surfaces (Figure 4.3). By selecting the proper cam profile, operating conditions and increasing the material hardness, and by reducing the friction surface, fatigue can be reduced.

Figure 4.2 Typical pitted follower surface

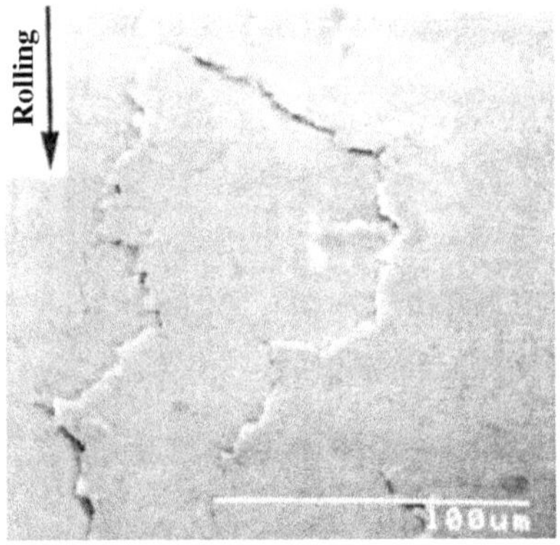

Figure 4.3 Surface flakes of follower surface under cyclic loading (Cheng *et al.*, 1994)

4.3 Typical Modes of Wear

Pitting, polishing and scuffing are the typical modes of the follower wear in diesel engines (Figure 4.4).

4.3.1.1 Pitting

Failure of a surface manifested by breaking-out of small roughly triangular portions of the material surface due to compressive stresses causing fatigue at a point below the surface. The heavily loaded surface will continue to pit with increasing severity in abusive operation.

4.3.1.2 Polishing

General attrition of the contacting surfaces causes polish wear. This mode of follower wear seems to be between pitting and scuffing assisted by some form of chemical action involving the oil. Visually, the surfaces appear to have a mirror finish and only the dimensional change quantifies the wear.

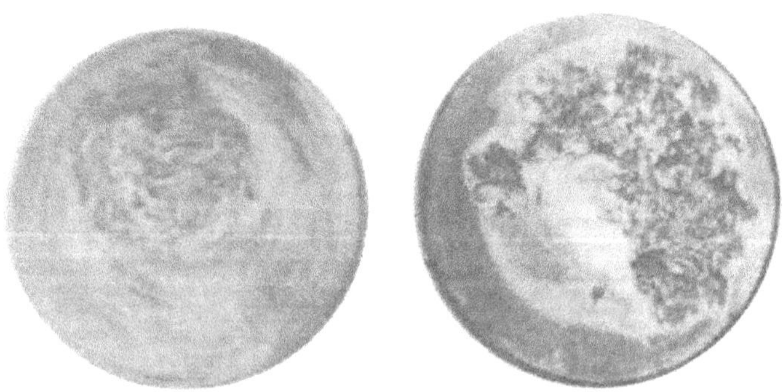

(a) Scuffed (b) Pitted

Figure 4.4 Typical modes of follower failures: (a) scuffed and (b) pitted surfaces (Neale, 1973)

Table 4.1 Engine specifications

Engine	A	B	C
Bore (mm)	118	93	124
Stroke (mm)	135	102	130
No. of cylinders	6	3	1
Aspiration	Turbocharged and After cooled	Naturally aspirated	Naturally aspirated
Follower type	Flat	Flat	Roller
Rating (hp)	325 at 1500 rpm	36 at 2250 rpm	25 at 2250 rpm
Brake mean effective pressure (MPa)	2.2	0.69	0.63
Peak firing pressure (MPa)	16.0	7.0	7.5
Application	Power Generation	Industrial/Tractor/ Generating set	Tractor

4.3.1.3 Scuffing

Scuffing is local welding of two heavily loaded surfaces, particularly when a high degree of relative sliding occurs under poor lubrication conditions, followed by tearing apart of the welded material. It starts at highly stressed zones with poor surface finish and, generally, during the running-in period of new parts.

4.4 Experiments on Cam Follower Wear

Abusive experiments at extreme load conditions and high oil temperatures were carried out on three types of engines (Table 4.1) to study the wear of cam followers.

During the experiments, the tappets showed minor wear, as shown in Figure 4.5. A simple approach to quantify the wear rate is to measure the depth of the wear scar with respect to unworn surface after the experiment.

4.4.1 Follower Measurement

The surface profiles of fresh and endurance-run engine followers were studied on a profiling instrument. Figure 4.6 shows the typical surface roughness of a follower in engine C, indicating little difference

Figure 4.5 Typical photograph of endurance run tappet

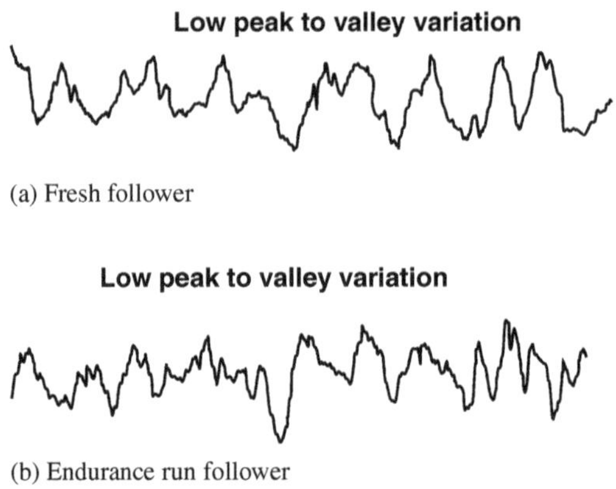

Figure 4.6 Measurement of flat follower, engine type C

between the roughness statistics of the fresh and endurance-run followers. Measurement with respect to the unworn area of the follower enabled wear of about 3.2 micron to be estimated.

4.5 Dynamics of the Valve Train System of the Pushrod Type

4.5.1 Elastohydrodynamic and Transition of Boundary Lubrication

During the valve opening period, large changes occur in the thickness of the fluid film separating the cam and follower surfaces. The positions at which the value of the oil entrainment velocity passes through zero, for which steady state elastohydrodynamic theory predicts total breakdown of the film, are the typical wear zones. Hence, the fluid film thickness in the wear model plays an essential role.

The wear coefficient, K (Figure 4.7), is determined by the nominal separation between the surfaces, h. For thick oil film conditions, no wear is expected at the mating elements and hence K is zero. For very thin oil films, K will approach a limiting value for boundary lubrication conditions, designated here as K_0. The absolute value of K_0 depends on several factors, such as the material hardness, metallurgy, quality of lubricants and operating temperature.

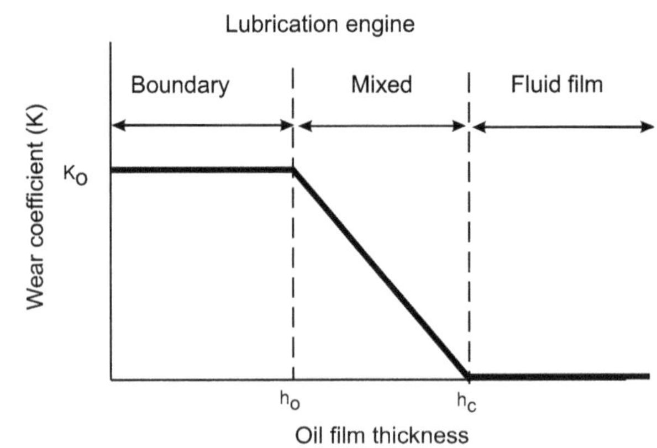

Figure 4.7 Variation of wear coefficient with oil film thickness (Bell *et al.*, 1985)

The lubricating conditions at the transition between boundary and fully hydrodynamic lubrication regimes are important, but are not well understood. The regime is defined by the oil film thickness, h_c, nondimensionalized with respect to the composite roughness, σ_0 (Bell *et al.*, 1985), of the two surfaces given by Equation 4.1:

$$\sigma_0 = \sqrt{\sigma_c^2 + \sigma_f^2} \tag{4.1}$$

where, σ_c, σ_f = r.m.s roughness of the cam and follower surfaces.

The contact between the surfaces takes place when the oil film thickness, λ_c, (nondimensionalized with respect to σ_0) is two to three times the composite roughness of the surface:

$$\lambda_c = h_c/\sigma_0 = 2 \sim 3 \tag{4.2}$$

When the oil film thickness becomes less than σ_0, boundary processes dominate, that is, the critical value of oil film thickness for initial contact between the surfaces, h_c, is given by:

$$\lambda_0 = h_0/\sigma_0 = 0.3 \sim 2 \tag{4.3}$$

The complex relationship between K and h for real surfaces is simplified by setting up the computational wear model where a linear transition of K (h) is modelled over the mixed lubrication regime, as shown in Figure 4.7 (Bell *et al.*, 1985).

As a starting point, a value of 0.5 μm was assumed for σ_0 as being typical for run-in cam and follower surfaces, resulting in values of h_c and h_0 of 1.2 and 0.15 μm, respectively. The nominal oil film thickness, h, is calculated using the formula (Haiqing *et al.*, 1983):

$$H = 2.65 A_1^{0.54} A_2^{0.7} A_3^{-0.13} \tag{4.4}$$

where $H = h/r$; $A_1 = \alpha_0 E'$; $A_2 = (\eta_0 V_e)/2E'R$; $A_3 = L/E'R$; R = equivalent radius of curvature $= \frac{1}{R_c} + \frac{1}{R_f}$; η_0 = dynamic viscosity at the inlet to the contact (Pa-s); and α_0 = pressure/viscosity coefficient at low pressure (Pa^{-1}).

4.5.2 *Cam and Follower Dynamics*

4.5.2.1 Kinematics

A smooth and continuous change in the curve of valve and follower acceleration is a pre-requisite for achieving a good cam profile. The acceleration curve of the cam consists of four portions:

- Ramp portion, a cosine curve.
- Positive acceleration, half the sine wave.
- First section of negative acceleration, quarter sine wave.
- Second section of negative acceleration, a segment of parabola.

The typical cam displacement, velocity and acceleration diagram at the rising portion of symmetric cams is shown in Figure 4.8.

4.5.2.2 Pressure Angle

The pressure angle, α, of a cam is the angle between the line of motion of the follower and normal to the cam surface at the point of contact between the cam and the follower (Figure 4.9).

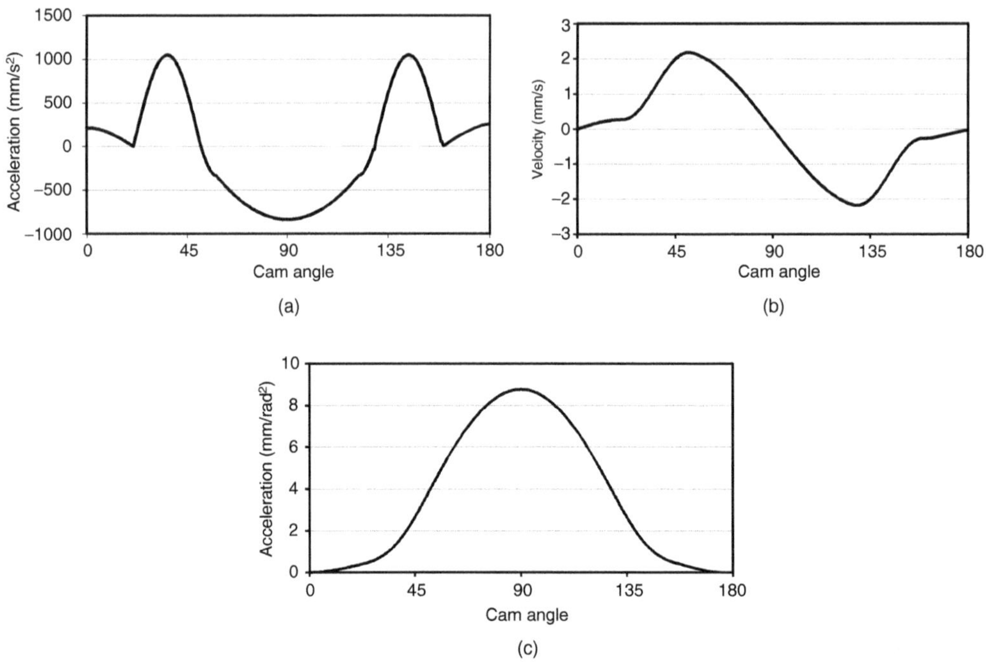

Figure 4.8 Typical diagram representing (a) cam acceleration, (b) cam velocity and (c) cam displacement at the rising portion of symmetric cams (for roller follower)

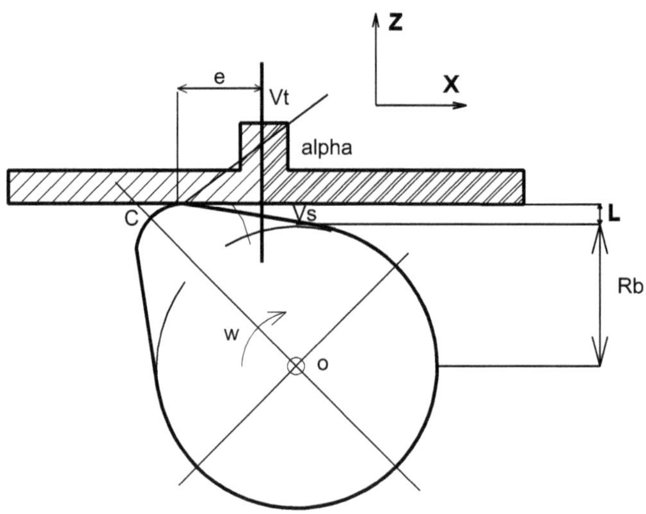

Figure 4.9 Typical diagram of a cam follower system

4.5.2.3 Radius of Curvature

The radius of curviture is a measure of the rapidity with which the curve changes the direction:

$$R = \frac{d^2h}{d\theta^2} + h + r_c \tag{4.5}$$

where, R = radius of curvature (mm); θ = cam angle in radian; and r_c = base circle radius (mm).

4.5.2.4 Dynamics

A flat follower with a given mass is spring loaded against the cam, as shown in Figure 4.9. The normal contact force is computed by cam kinematics approach. The contact force is given by

$$F_c = m\ddot{x} + F_{s0} + kx \tag{4.6}$$

where F_c = contact force (N); m = mass of the follower (kg); $\ddot{x}$ = acceleration of the follower (mm/s^2); F_{s0} = spring pre-load force (N); k = spring rate (N/mm); and x = follower lift (mm).

Coulomb friction conditions exist at the contact path. The presence of a lubricating film at the contact interface influences the dynamics. The effect of the oil film thickness is considered in the wear equation.

4.6 Wear Model

To obtain the contact pressures and sliding velocity, the dynamics of the cam and follower is simulated by a classical approach or, alternatively, by using the finite element method in conjunction with simulation software. The outline of the scheme of modelling and validation is presented in Figure 4.10. These values are used to obtain wear depth by using a judiciously chosen wear coefficient that depends on the lubrication regime and by applying Archard's law (Archard and Hirst, 1956).

4.6.1 Wear Coefficient

A lubricant viscosity, η_0, of 10 cP and pressure viscosity coefficient, α_0, of 15 GPa^{-1} were considered for the SAE 15W-40 multi-grade oil at operating temperatures in the range 100–120 °C.

In the transition model, $K_0 = 8 \times 10^{-11}$ mm^3/mm/N helps the predicted wear agree satisfactorily with the experiments and it is consistent with the measurements of Childs (1980) for iron and steel under boundary lubrication conditions with a good quality oil. For the intermediate or mixed lubrication regime, K is modelled to transition linearly with the oil film thickness (Figure 4.11). The transition between the three regimes takes place at film thickness, $h_0 = 0.15$ μm and hc = 1.2 μm.

4.6.2 Valve Train Dynamics and Stress on the Follower

4.6.2.1 Input to Simulation

The simulation is carried out on naturally aspirated (B and C) and turbocharged (A) engines (Table 4.1). For brevity only the results of engine C are presented here. The elements of valve train system of engine C are shown in Tables 4.2 and 4.3.

The simulation results showed the contact force and pressure on the follower, and forces at valve seat insert zone.

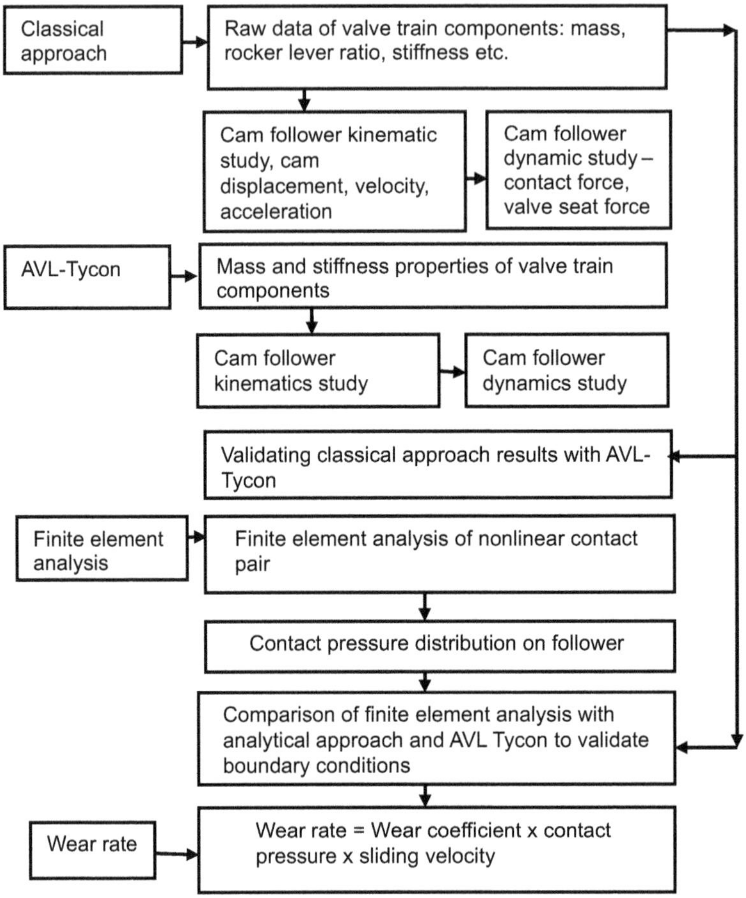

Figure 4.10 Schematic layout of the calculation methodology

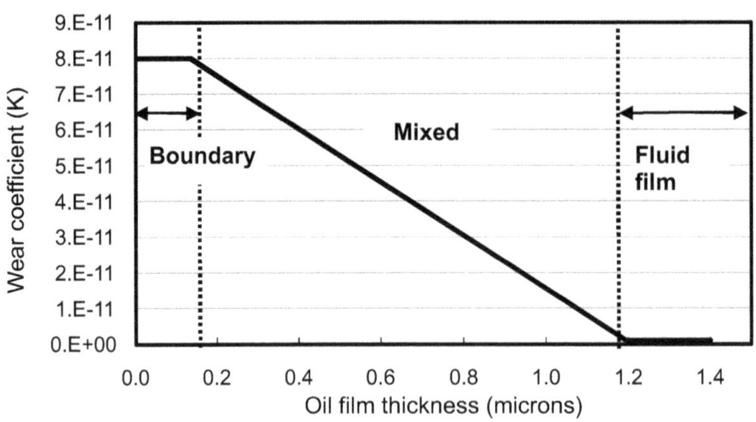

Figure 4.11 Variation of wear coefficient with respect to oil film thickness in mixed lubrication regime

Table 4.2 Individual mass elements and equivalent masses at cam and valve side, engine C

	Rocker lever ratio	1.374
Mass (g)	inlet valve	105.2
	exhaust valve	104.7
	valve cup	33
	valve collette	6
	valve spring	66
	rocker lever assembly	195
	pushrod	115
	tappet	94
	equivalent mass transferred from cam side to valve side	159.2
	total mass on inlet valve side	325.4

Table 4.3 Stiffness details of valve train system, engine C

Valve	length of valve stem (mm)	125
	stem diameter (mm)	8
	Young's modulus (N/mm^2)	206 000
	stiffness (N mm/rad)	6.04×10^{-6}
Valve seat	inlet diameter (mm)	38.4
	exhaust diameter (mm)	36
	Young's modulus (N/mm^2)	206 000
Stiffness	inlet seat (N mm/rad)	5.93×10^5
	exhaust seat (N mm/rad)	5.56×10^5
Pushrod	effective length (mm)	222
	outside diameter (mm)	11
	inside diameter (mm)	8
	Young's modulus (N/mm^2)	206 000
	stiffness (N mm/rad)	2.41×10^{-5}

4.6.2.2 Classical Approach

Hertzian Stress Calculation

The maximum compressive stress at the contact between the cam and the follower is calculated by (Chen, 1982):

$$\sigma_{max} = 0.564\left[\frac{p'\left(\frac{R_c+R_f}{R_cR_f}\right)}{\frac{1-\mu_c^2}{E_c}+\frac{1-\mu_f^2}{E_f}}\right]^{\frac{1}{2}} \tag{4.7}$$

$$P' = \frac{p_{total}}{b \times \cos\alpha} \tag{4.8}$$

where P' = normal load per unit width of the contacting members (N/mm); R_c = radii of curvature of the cam (mm); R_f = radii of curvature of the follower (mm); μ_c = Poison's ratio of cam = 0.3; μ_f = Poison's ratio of follower = 0.3; E_c = modulus of elasticity of the cam (N/mm^2); E_f = modulus of elasticity of the follower (N/mm^2); P_{total} = inertia force + spring force + gaseous force; b = cam width (mm); α = pressure angle (radians).

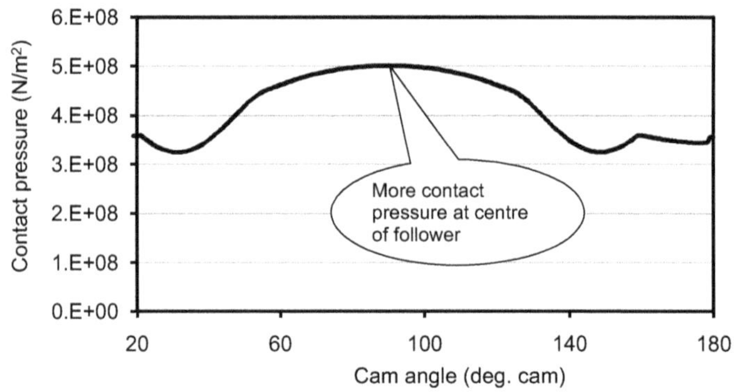

Figure 4.12 Contact stress variation with respect to cam angle (engine type C)

For example, the maximum compressive stress at the contact between the cam and the follower for engine C is shown in Figure 4.12 and Table 4.4.

The permitted maximum compressive stress between the cam and roller follower is less than 1500 N/mm^2 and for a flat follower it should be less than 800 N/mm^2. The results indicated very low contact stresses at the nose portion for both types of the followers.

Tangential Velocity

The cam rotates about a fixed centre; the tangential velocity of sliding of the follower at the contact point is

$$V_t = X\dot{\theta}_{cam} \tag{4.9}$$

where X = distance from centre of cam to contact point and $\dot{\theta}_{cam}$ = rotational speed of the cam

The velocity of sliding, Vs, is given by the component of the tangential velocity, V_t:

$$V_s = V_t \cos \alpha \tag{4.10}$$

The calculated sliding contact velocity of the follower is of the order of 5.6 m/s for engine C (Figure 4.13).

Contact Width

Assuming that perfect cylindrical, elastic, Hertzian contact is maintained between the cam and follower throughout the valve lift cycle, that the cam profile is unchanged by the wear process and that no contact occurs around the base circle of the cam, the finite width of the contact (= 2b) is given by:

$$b = \left[\frac{8LR}{\pi E'}\right]^{0.5} \tag{4.11}$$

Table 4.4 Hertzian stress variation with flat and roller followers

	Hertzian stress (N/mm^2)	
Parameter	Cam with flat follower	Cam with roller follower
At the ramp portion	229.5	385.9
At the nose zone	553.6	500.3

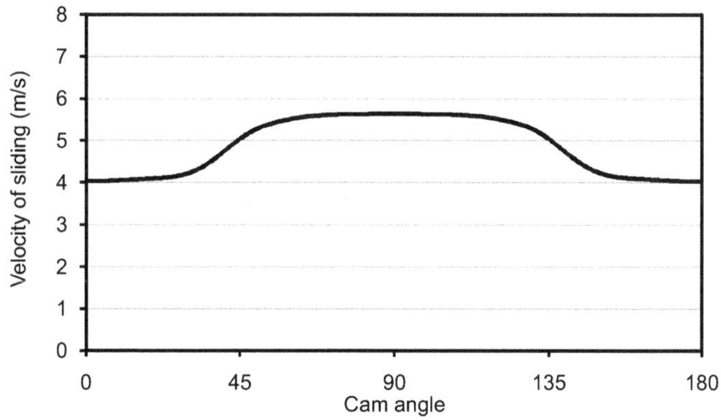

Figure 4.13 Velocity of sliding at the contact between cam and follower (engine type C)

where L is load per unit axial length, R is equivalent radius of curvature and E′ is the reduced elastic modulus.

$$\frac{2}{E'} = \frac{1-\mu_c^2}{E_c} + \frac{1-\mu_f^2}{E_f} \tag{4.12}$$

During the passage of the contact over each position on the follower surface, a significant variation of the contact pressure occurs (Figure 4.14).

Oil Film Thickness
Figure 4.15 illustrates the calculated oil film thickness with respect to cam angle using the Haiqing relationship (Haiqing and Yuxian, 1983) for engine C. A minimum oil film thickness of the order of 0.2 μm near the cam nose zone, well within the permissible minimum oil film thickness, can be observed.

4.6.2.3 Commercially Available Software

Contact Pressure on the Follower by Finite Element Method
The contact pressure distribution on the cam and follower surface is calculated using the finite element method by imposing the boundary conditions of contact force, lubricant damping value and fixity constraint. The problem is defined as nonlinear by imposing the contact pair between the cam and

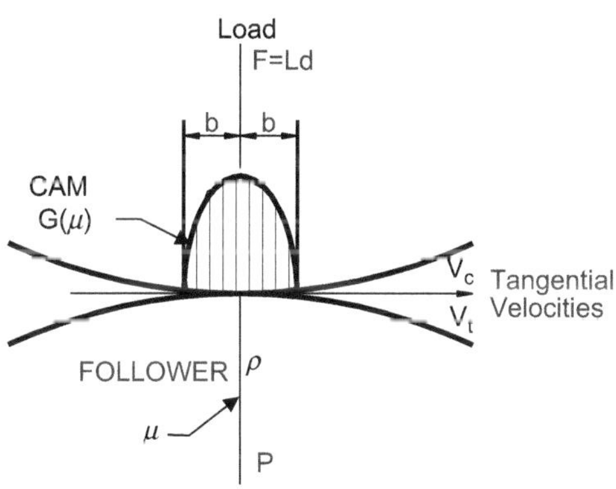

Figure 4.14 Typical contact pressure distribution on the cam follower mating face

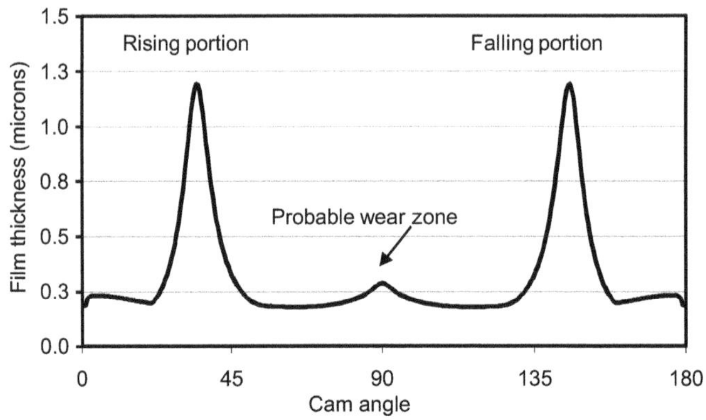

Figure 4.15 Variation of oil film thickness (engine type C)

follower body. The equivalent force applied on the cam predicts the contact pressure on the follower surface and is used as input to the follower wear calculation. Initially, pressure deformation under dynamic conditions and the contact pressure distribution on the follower surface is determined using finite element nonlinear contact pair analysis. The follower surface is treated as target area and the cam surface as contact area. The rigidity boundary condition is applied on the central axis nodes of the cam surface and the lubrication damping effect is considered between the cam and the follower areas. The magnitude of the contact force varies with follower movement on the cam profile; it is predominant at the nose zone of the cam because of the reduced contact area.

Figure 4.16 describes the boundary conditions used in the finite element analysis. The rigid boundary conditions are applied at the centre line of the cam.

The damping effect of lubrication is considered by defining 'combination elements' between the cam and the follower bodies. Figure 4.17 shows the contact pressure plot on the follower, when the follower is

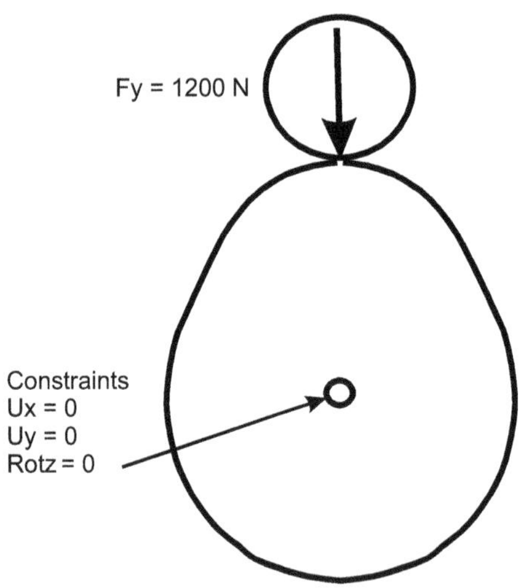

Figure 4.16 Diagram representing boundary conditions of FEM analysis

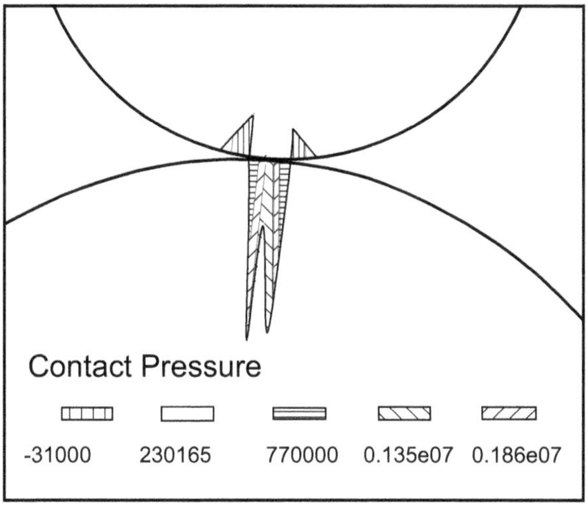

Figure 4.17 Contact pressure plot on the follower (roller type)

on ramp of the cam. A contact pressure value of 285 N/mm^2 is calculated; this is slightly less than the values obtained by the classical approach. When the follower is on the ramp, film thickness is high between them. Due to damping, the contact pressure value is slightly less. In the real case, this manifests as no wear of cam at the ramp or cam rise zone. However, the film thickness is much less at the nose of the cam, leading to rupture of the oil film under high contact forces. Enhanced metal to metal contact of the cam and follower bodies takes place and the surface cracks in fatigue.

Simulation of Valve Train Dynamics

Apart from the classical approach, software available commercially (AVL-Tycon, 2003) can be used for kinematic and dynamic analysis of the valve train system. All the timing drive elements are divided into specific, generic and microelements and the elements contain information about mass, stiffness and damping values. An approximation is used to determine initial positions and velocities of valve train components. The multibody approach is used in dynamic calculation. Due to nonlinear characteristics of the system, the dynamic behaviour is considered in the time domain and a second order predictor corrector method is used while solving the equations. The gaseous pressure at different crank angle positions is calculated from the engine simulation program and is applied at the valve area.

The valve seat forces, valve contact velocities and dynamic behaviour of springs and so on are the outputs. Figure 4.18 shows the variation of oil film thickness and contact pressure between the cam and follower obtained from the simulation. The results match with the output of classical calculations.

4.6.3 Wear Depth

The wear rate on the contact surface of the follower is calculated by using linear wear relation, that is, the wear depth, W, is directly proportional to the wear coefficient, the mean contact pressure and the sliding contact velocity of the follower. The wear coefficient characterizes the significant changes on the topography of the surfaces; under thick oil film conditions, a wear coefficient value of zero was assumed. However, for very thin oil films between mating surfaces the value is limited to boundary lubrication conditions and is dependent on lubricant quality, material properties and

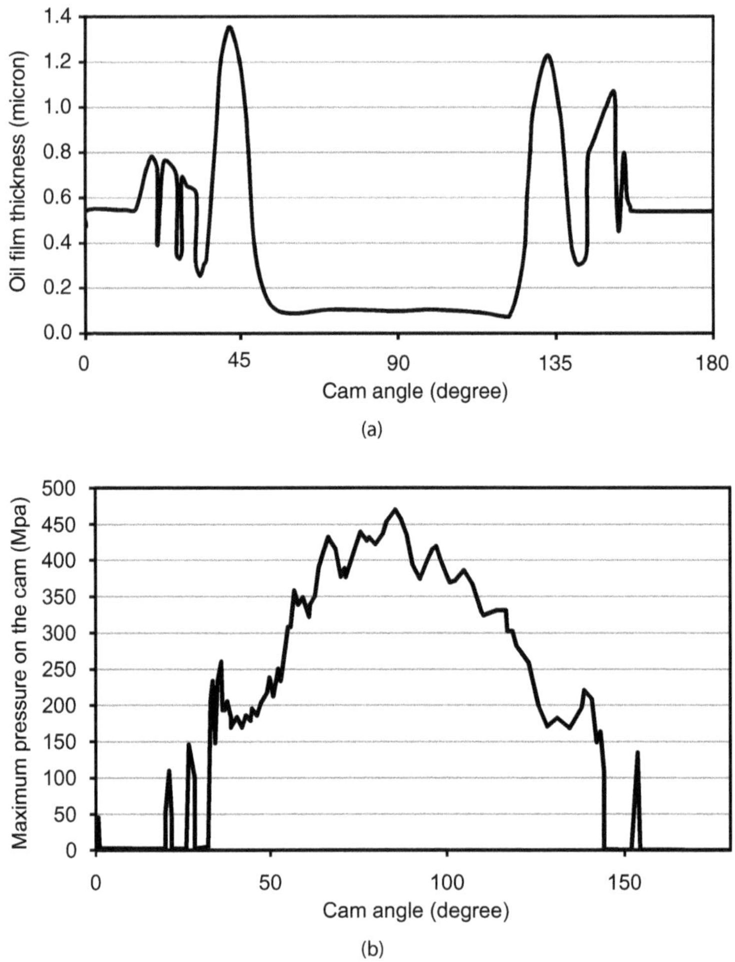

Figure 4.18 Simulation results (engine type C) for (a) variation of oil film thickness and (b) contact pressure between cam and follower

operating conditions:

$$W = KPV_s t \tag{4.13}$$

where K = wear coefficient ($mm^3/mm/N$), P = mean contact pressure (N/mm^2) and V_s = sliding contact velocity of the follower (mm/s).

The contact velocity also varies during the operating cycle. The instantaneous rate of wear, i(x, t), is defined as:

$$i(x,t) = K(h)PGV_s \tag{4.14}$$

where G is a pressure distribution.

For G in Equation 4.14, a semi-elliptical pressure distribution was assumed in the classical approach. The wear coefficient, K (Figure 4.7), is a function of the nominal separation of the surfaces, h, to permit the incorporation of a hydrodynamic and boundary lubrication transition model.

The wear occurring during a single passage of the contact over the follower surface is given by,

$$\Delta W(x) = \int_{t_1}^{t_2} i(x, t')dt' \tag{4.15}$$

where t_1 and t_2 are the time at the start and end of contact at the point on the follower.

4.6.3.1 Wear Profile

From the predicted contact forces, the distribution of mean contact pressure on the follower is calculated. The pressure obtained by the classical approach, finite element analysis and the simulation methods are comparable.

The wear profile is predicted using Equations 4.13 to 4.15 as a function of the contact pressure and the sliding velocity. Figure 4.19 shows the variation of the wear depth of a follower in one engine along its surface. The calculated wear of 3.5 microns matches well with the experimental wear (3.2 microns). Similarly, the measured wear of followers in the three engines under study compared favourably with the results obtained by the three methods described above (Figure 4.20).

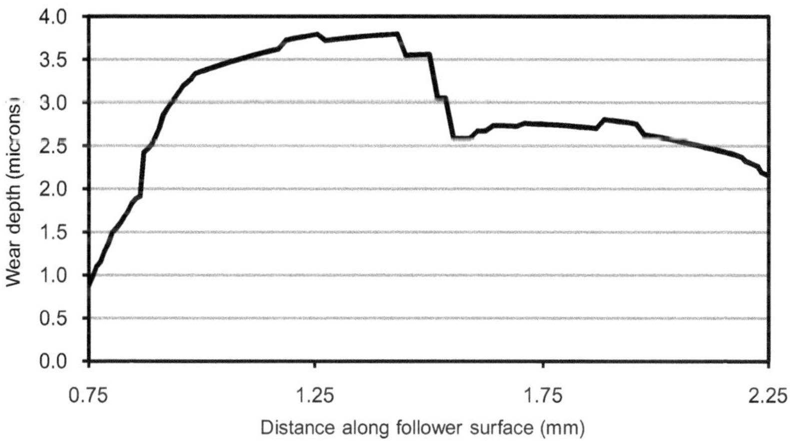

Figure 4.19 Calculated wear depth along the follower surface

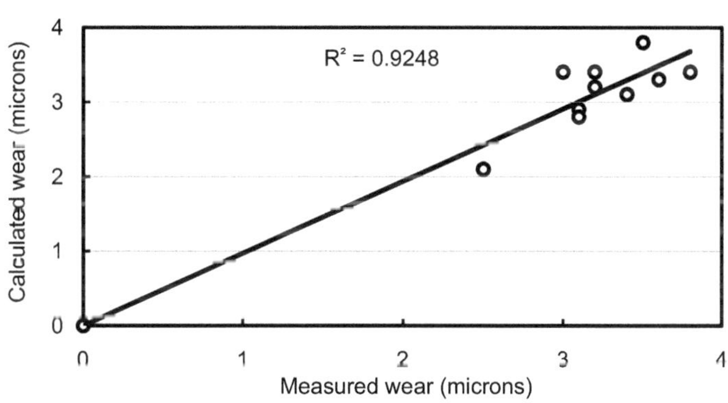

Figure 4.20 Validating the wear model: measured and calculated wear of followers for 500 hours

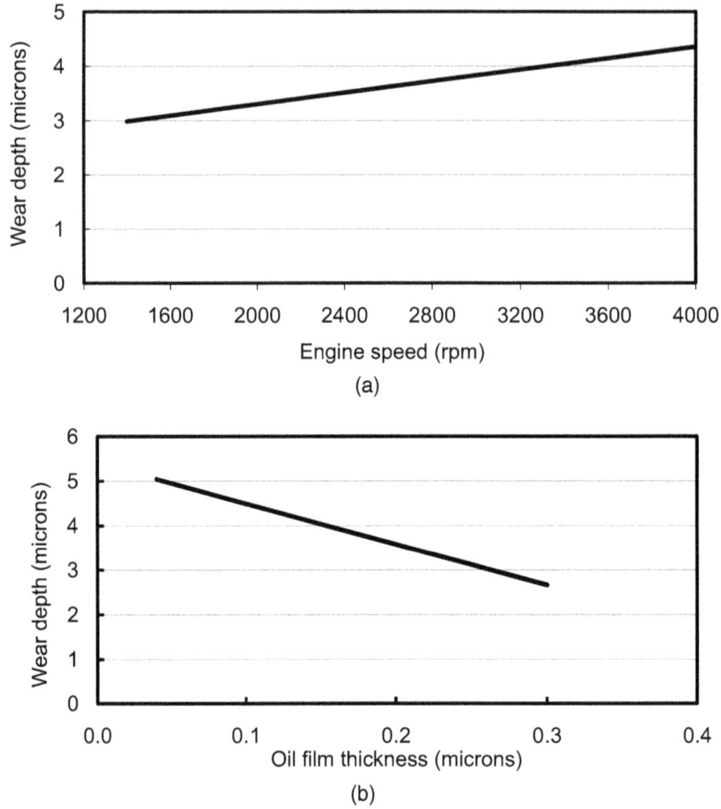

Figure 4.21 Parametric study on follower wear for a period of 500 hours. Variation with (a) engine speed and (b) oil film thickness – increased contact pressure between cam and follower

4.7 Parametric Study

4.7.1 Engine Speed

A parametric study on wear of the follower is carried out for the effect of increase in engine speed from 1400 to 4000 rpm for a constant duration of 500 hours. The total wear at the nose increases linearly with speed due to increase in sliding velocity and the influence of inertia loading on the follower (Figure 4.21a).

4.7.2 Oil Film Thickness

Similarly, the reduced film thickness in the boundary lubrication regime affects the wear process (Figure 4.21b). The breakdown of the film at the cam nose area due to high contact pressure results in metal-to-metal contact of the mating surfaces that undergo plastic deformation simultaneously.

4.8 Wear of the Cast Iron Rocker Toe

The contact between the rocker toe and the valve tip is lubricated by the oil mist in the rocker chamber. In many engines a part of the oil to the rocker shaft is bled to flow over the contact area. Since the (boundary) lubrication regime is hydrostatic, wear sets in if the geometry of the contact is not in commensurate with the hardness of the two mating materials. Usually, the valve tip is extremely

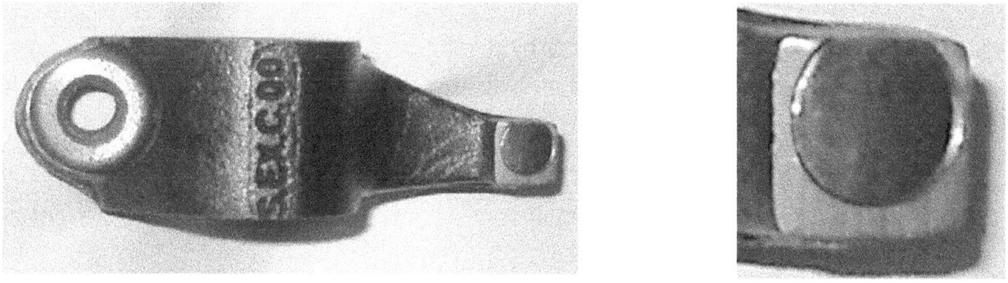

Figure 4.22 Wear of rocker lever tip. Left: Tip wear of cast iron rocker lever with chilled toe, without a valve cap. Right: Enlarged view of dimple shaped wear of the toe (Hardness of the valve tip, Rc 50; hardness of rocker toe, Rc 23)

hard and the rocker toe is softer. If the rocker lever is of steel it is flame hardened or carburized and hardened. If it is of cast iron it is chilled to achieve some hardness. To always maintain the contact and for ease of manufacturing, the rocker toe is usually cylindrical and wider than the diameter of the valve tip. When the hardness of the rocker toe is less than the hardness of the valve tip (e.g. chilled cast iron rocker toe), edge loading by the valve on the rocker toe leads to a severe wear in the form of a dimple (Figure 4.22). This type of wear causes increase in valve lash and, consequently, higher valve seating velocity, which may lead to pitting of the valve seat. To avoid this type of wear, a valve cap wider than the rocker toe is usually provided on the valve tip. Further, the wider cap increases the area of contact during bedding in (Figure 4.23a). The situations with and without the wider cap are shown

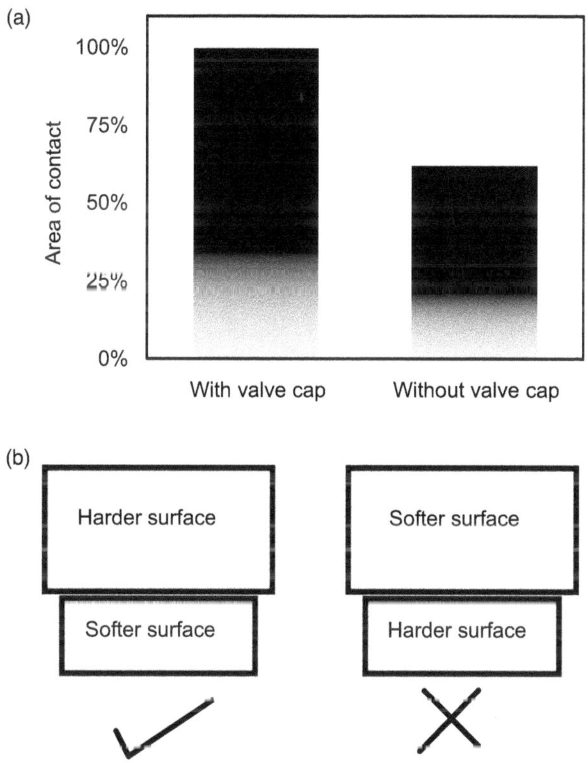

Figure 4.23 Enhanced area of contact of cast iron rocker toe with valve cap after a brief period of running in

schematically in Figure 4.23b. The valve cap is harder than the rocker toe; hence edge loading and subsequent wear are eliminated. To avoid a valve cap, there are alternative but expensive designs for the rocker lever where the rocker tip is extremely hard and cylindrical in shape with a gentle crown perpendicular to the direction of relative movement to take care of any minute misalignment in the angular relationship between the valve axis, the rocker shaft axis and the axis of the super-finished cylindrical toe. On the other hand, the chilled toe of a well-designed cast iron rocker lever beds in after a few hours against a harder and wider valve cap, to achieve super-finished surface quality *in situ* and, hence, improves its strength against wear.

4.9 Summary

The phenomenological model presented here is useful for estimating the wear of followers on the basis of Hertzian compressive force between the cam and follower and velocity of sliding. The current model estimates the initial wear rate and not the ferocious wear that happens after the surface roughness has increased significantly due to adhesion in the plastic shear regime. The improvement in actual wear could be achieved by reducing friction between the cam and follower with proper lubrication and surface finish. The model was successfully validated on roller and flat followers in a wide variety of engines, that is, 0.625–2.2 MPa bmep engines.

A wear coefficient value of 8×10^{-11} mm^3/mm/N is found to give satisfactory agreement with measured results and is the most proven value for the wear calculation of ferrous materials under boundary lubrication conditions with good quality lubricant.

The analysis results indicated that the follower wear increases with high compressive stresses between the cam and follower. For the engine with roller followers 1500 N/mm^2 is the limiting contact pressure, for flat followers it is 800 N/mm^2.

The minimum film thickness should not be less than 0.2 micron in the boundary lubrication regime. This is achieved by cam design, lubrication and the machining process of the cam and the follower.

For parts like the rocker toe working against the valve tip in boundary lubrication regime, a softer surface should have less contact area than the harder surface. This configuration helps to avoid edge loading and consequent wear. Further, super-finish of the engaging surfaces is achieved during the running in period and resistance to wear increases.

References

Archard, J.F. and Hirst, W. (1956) The Wear of Metals under Unlubricated Conditions. *Proceedings of the Royal Society London, A*, **236**, 397–410.

AVL (2003) AVL TYCON Reference Manual. AVL, Graz, Austria.

Bell, J.C., Davies, P.J. and Fu, W.B. (1985) Prediction of automotive valve train wear patterns with simple mathematical models, Mechanisms and surface distress. Proceedings of 12th Leeds-Lyon Symposium on Tribology, Lyon, France, pp. 323–335.

Chen, F.Y. (1982) A text book on Mechanics and Design of cam mechanisms. Pergamon Press.

Cheng, W., Cheng, H.S. and Yasuda, Y. (1994) Wear and life prediction of cam roller follower. Technical Paper 940822, SAE, Troy, MI, pp. 468–479.

Childs, T.H.C. (1980) The sliding wear mechanisms of metals, mainly, steels. *Tribology International*, December 1980, pp. 143.

Colgan, T. and Bell, J.C. (1989) A predictive model for wear in automotive valve train systems. Technical Paper 892145, SAE, Troy, MI, pp. 1221–1236.

Fries, R.H. and Rogers, C.A. (1987) Prediction of cam wear profiles. Proceedings of the 15th Leeds-Lyon symposium on Tribology, UK, pp. 101–109.

Glovnea R.P. and Spikes, H.A. (2001) The influence of cam follower motion on elastohydrodynamic film thickness, in *Tribology Research: From Model Experiment to Industrial Problem – A Century of Efforts in Mechanics, Materials*

Science and Physico-Chemistry. Proceedings of the 27th Leeds-Lyon Symposium on Tribology, Elsevier, Amsterdam, The Netherlands, pp. 485–493.

Haiqing, X.Z. and Yuxian, H. (1983) The computation of unsteadily loaded EHL film thickness and other lubrication parameters of cam-tappet pairs of IC engines and analysis of their performance. Proceedings of the 10th Leeds-Lyon Symposium on Tribology, Lyon, France.

Hugnell, A. and Andersson, S. (1994) Simulating follower wear in a cam follower contact. *Wear*, **179**, 117–122.

Hugnell, A., Bjorklund S. and Andersson, S. (1996) Simulation of the mild wear in a cam-follower contact with follower rotation. *Wear*, **199**, 202–210.

Mehenny D.S. and Taylor, C.M. (2001) Analysis of the influence of cam surface waviness on the lubrication of an automotive cam and flat faced follower, in *Tribology Research: From Model Experiment to Industrial Problem – A Century of Efforts in Mechanics, Materials Science and Physico-Chemistry.* Proceedings of the 27th Leeds-Lyon Symposium on Tribology, Elsevier, Amsterdam, The Netherlands, pp. 863–872.

Nayak, N.S., Lakshminarayanan, P.A., Gajendra Babu, M.K. and Dani, A.D. (2006) Predictions of cam follower wear in diesel engines, *Wear*, **260**, 181–192.

Neale, M.J. (1973) *Tribology Handbook.* Butterworths, London.

Purmer P.D. and van den Berg, W. (1985) Measurement of camshaft wear – Wear and kinematics of overhead camshafts. Technical Paper 850442, SAE, Troy, MI, pp. 232–241.

Part III

Liner, Piston and Piston Rings

5

Liner Wear: Wear of Roughness Peaks in Sparse Contact[1]

The Kragelskii wear model for surfaces considers elastic loading of the normal roughness peaks in sparse contact with the shear load along the direction of sliding. Repeated stressing of the peaks causes fatigue and detaches superficial material in the form of wear debris. The wear of cylinder liners of a wide range of surface finishes and materials of different strengths is predicted by extending the model to take into account surfaces that are not 'normally' rough. The roughness peaks are modelled as tiny hemispheres. The radius of the hemisphere is defined as the geometric mean of the radii in the directions transverse and longitudinal to machining. The transverse radius can be estimated using the actual roughness trace. An alternative and quicker method is presented here to derive this radius as a function of the number of roughness peaks per unit length, the peak roughness, R_{pk}, and the fraction of the bearing surface MR1 at R_{pk}. The radius along the machining direction typical of the machining process.

The wear rate of a normal finish is known to be a strong function of the maximum deviation of the roughness trace from the mean line, R_p. The valley portion of the roughness, R_{vk}, does not take part in wear as actively as the roughness parameter, R_k, lying between R_{vk} and R_{pk}. A general surface roughness, which is not normally peaked, is characterized by the 'plateauness ratio', R_k/R_{vk}. An equivalent roughness, R_{p-eq}, for such a general surface is found to bear a unique relation with the plateauness ratio.

To study surfaces of general roughness, the equivalent R_{p-eq} and the radius of the hemispherical roughness peak were substituted in the wear model known for surfaces with normal roughness. Ten different liner surfaces were studied using the present model. In addition, experiments were conducted on a 105 mm bore diesel engine using normally honed liners to apply the wear model. A satisfactory correlation was obtained between the predicted wear rates and those measured with the model.

5.1 Introduction

Wear of the liner and rings in an engine is a process of destruction of surfaces that rub, resulting in reduced dimensions of the parts in a direction perpendicular to the wear surfaces. In the case of the well-behaved pair of liner and piston ring, the liner wear is predominantly in the top ring reversal zone where the

[1]All figures in this chapter are reprinted with permission from P.A. Lakshminarayanan, N.S. Nayak and A.D. Dani, "Prediction of liner wear by extending the Kragelskii model for wear of roughness peaks in sparse contact," *Proceedings of the Institution of Mechanical Engineers, Part J: Journal of Engineering Tribology*, **216**, 5, 327–342.

Critical Component Wear in Heavy Duty Engines, First Edition. P.A. Lakshminarayanan and Nagaraj S. Nayak.

hydrodynamic lubrication is destroyed, so enabling contact of roughness peaks. The rate of wear of the sliding pair depends on the properties of the materials, the surface finish and the running conditions. Under constant rubbing conditions, three phases of wear exist, namely running-in wear, the steady state wear and severe wear.

Ting and Mayer (1974) developed an analytical method for determining the wear pattern of the bore for a reciprocating engine considering the hydrodynamic lubrication between ring and the cylinder wall, geometry of the ring and load due to piston thrust.

The effect of the surface roughness on the dynamic wear behaviour of lubricated sliding system was illustrated by Hu, Li and Tonder (1991). The optimum roughness of the wear corresponding to the minimum wear rate, stability of wear system and its critical load capacity were predicted using a dynamic system model. Yang and Winer (1991) proposed a thermo-mechanical wear theory to correlate the relationship between the frictional force and the wear of the sliding component.

A wear model was developed by Wu and Cheng (1991) for sliding contacts under partial elastohydrodynamic lubrication, *EHL*. The wear mechanisms of the asperity contacts were studied by thermal desorption at low temperature and by oxidation of contact at elevated temperatures. A transient micro-*EHL* model for wear was used by Chang, Webster and Jackson (1994) to analyse the effect of 3D surface topography on the oil film thickness under *EHL*. Poon and Sayles (1994) studied the plastic contact model of a smooth ball on a directionally structured anisotropic rough surface and validated the model numerically.

In this chapter, the roughness peaks are approximated to small hemispheres. Then, the phenomenological wear model of Kragelskii and Alisin (1981) for surfaces with normally peaked roughness is discussed. For surfaces without peaked roughness, an equivalent maximum roughness is derived from the statistical roughness parameters. By introducing the equivalent roughness in the Kragelskii model, the wear of engine liners is analysed. The calculations are validated using the experimental results of Hill *et al.* (1995), Hegemier and Stewart (1993) and normally honed liners. Archard and Hirst (1956), on the other hand, have presented a simple model of wear as a function load, stroke, speed, pressure and apparent area of contact. Here, the proportionality constant has to be tuned for every type of surface and material. Interestingly, the constant varies from as low as 0.02×10^{-3} for bakelite against bakelite to 45×10^{-3} for wear of low carbon steel against low carbon steel. Since the phenomenon at the surface due to the finish is not fully understood, it is not easy to predict the wear rates from first principles using Archard's work.

5.2 Surface Texture of Liners and Rings

5.2.1 Surface Finish

In heavy duty diesel engines, surface finish of honed liners and rings plays an important role in meeting the demands for low oil consumption, long life and low emission of particulate matter.

5.2.2 Honing of Liners

After achieving a suitable bore geometry and diameter tolerance, free cutting is ensured during honing to avoid glazing. A single stage process for normal honing achieves peaked surface roughness by using an abrasive stone of a unique size. In the case of plateau honing, a coarse abrasive is used initially followed by a few strokes with a finer abrasive stone to remove the surface peaks and hence increase the bearing area. The use of a plateau-type surface finish is usually recommended to reduce running-in time and to provide better control of oil consumption, blow-by and wear resistance. Figure 5.1 shows the surface textures of a typical plateau surface and a peaked surface finish (Hill *et al.* 1995).

5.2.3 Surface Finish Parameters

The surface finish of the liner bore is specified to ensure proper behaviour of the liners. The roughness is one of the characteristics of the surface finish. The other property is the unevenness of the total surface,

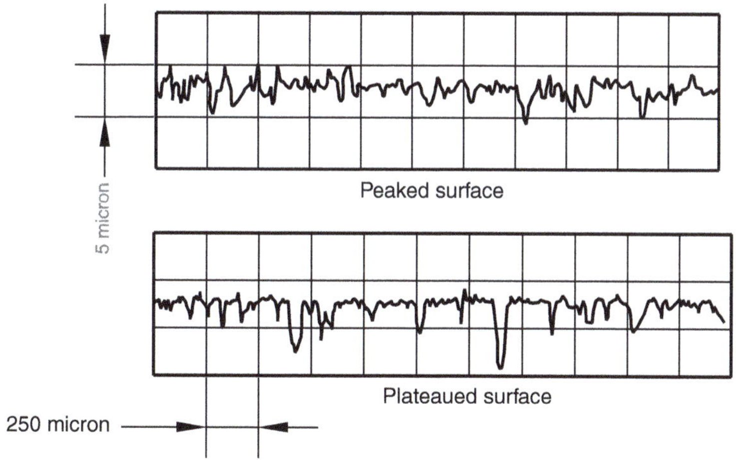

Figure 5.1 Typical surface textures of normally honed and plateau honed liner surfaces

which is typical of the machining process, stiffness of the part and the method of clamping the job during manufacture. The surface roughness parameters are measured using a surface analyser by scratching the surface transverse to machining direction with an appropriate stylus. In Figure 5.2, various roughness parameters for a typical surface are shown using a typical bearing area curve and a roughness trace (Hegemier and Stewart, 1993).

Average Roughness, R_a

The centre line average, R_a, is the arithmetic average of the departure of the profile above and below a mean line throughout the prescribed sampling length. The trace of the centre line of roughness can be estimated from the bearing area curve by finding a roughness depth above and below which the hatched areas are equal, as shown in Figure 5.3a. The sum of absolute values of the two areas gives R_a. The method has been validated for 13 cases of different types of finish, ranging from smooth to rough and plateaued to nonplateaued types. A satisfactory correlation (correlation coefficient = 0.98) is shown in Figure 5.3b.

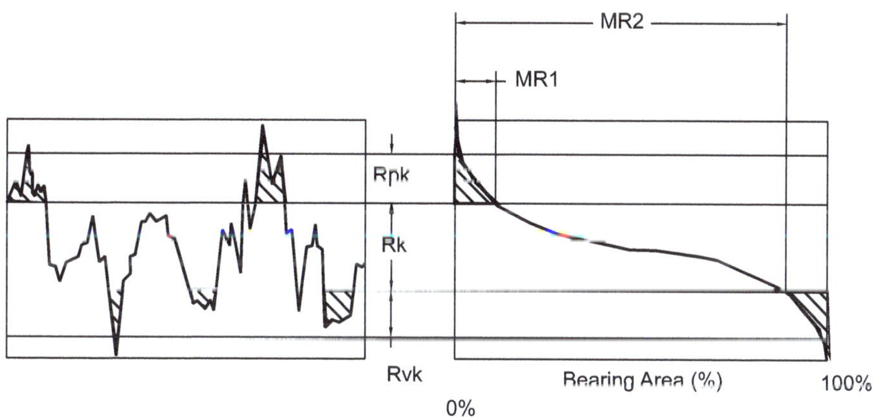

Figure 5.2 Surface roughness parameters

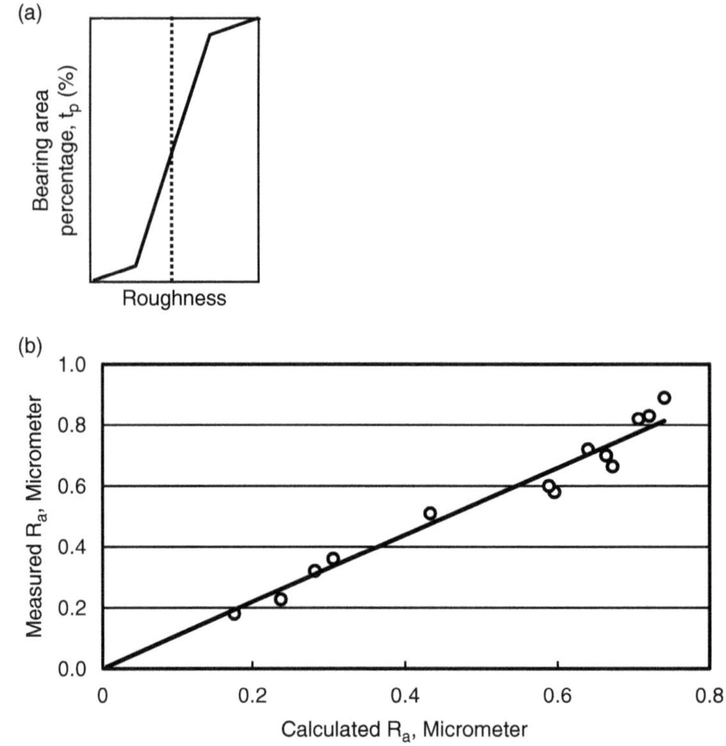

Figure 5.3 Calculation of R_a for the bearing area curve: (a) centre line calculation; (b) validation

5.2.4 Bearing Area Curve

The bearing area curve is the plot of the ratio of air to material (bearing ratio, t_p) at any level, starting at the highest peak, as a function of level. The curve is obtained from the surface profile trace as follows (Bhushan, 1999; Kragelskii and Alisin 1981; Halling 1971). Lines are drawn parallel to nominal surface and the fraction of the line (t_p) which lies within the profile is measured. Mathematically, it is the cumulative probability density function of the surface profile's height and can be calculated by integrating the profile trace.

The curve can be roughly divided into three zones, namely the highest area corresponding to peak region, the middle zone and the lowest denoting the valley of a typical roughness trace. The middle zone contributes to the wear life and the valley to the oil consumption. These three zones are represented by R_{pk}, R_k and R_{vk} according to DIN 4776 (ISO, 1995). The t_p at a roughness depth R_{pk} is represented by MR1. The typical bearing area curves (Casellato, Avezou and Gazzard, 1990) of a normally honed and a plateau honed surface are shown in Figure 5.4.

Roughness Valley, R_{vk}

R_{vk} represents the depth of the valley in the bearing graph. The type of honing is characterized by the value of the plateauness, defined as the ratio of R_k to R_{vk}. When the ratio is less than 1.3, the surface is plateau honed; the higher the ratio the less plateaued the surface is.

Roughness Peak, R_{pk}

This is the height of the peaks taking an active part in the wear. It appears plateauness is not dependent on R_{pk}.

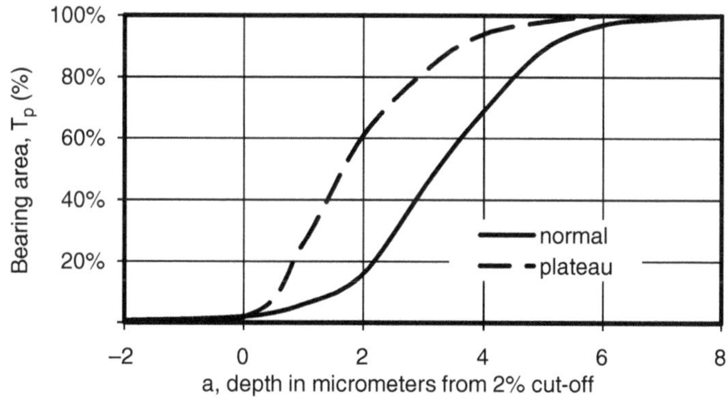

Figure 5.4 Bearing area curves of normally honed and plateau honed surfaces

Medial Roughness, R_k

This parameter gives a good indication of the overall roughness of the liner. R_k is the total roughness excluding the peaks and valleys, namely R_{pk} and R_{vk}. It is derived from the bearing area curve of a surface and is closely related to R_a. For a nonplateaued surface, R_k is nearly equal to π R_a.

Total Roughness, R_z

The maximum roughness is the highest value of the departure of the profile above and below the reference line throughout the sampling length. For brevity, to avoid extreme values that are statistically rare, the difference of the average of five peaks and the average of five valleys is considered. The value of R_z can be approximately calculated from statistical roughness parameters, as shown in Equation 5.1:

$$R_z \approx R_{pk} + R_k + R_{vk} \tag{5.1}$$

Maximum Roughness, R_p

The maximum deviation of roughness from the mean line is defined as R_p.

5.2.5 Representation of Bearing Area Curve of Normally Honed Surface or Surfaces with Peaked Roughness

The bearing area curve completely characterizes the surface finish to estimate the oil bearing capacity of the bore and to predict the wear performance, and hence life, of the cylinder bore. In general, the initial portion of the bearing area curve can be described by two parameters, namely, ν and b (Kragelskii and Alisin, 1981). The parameter t_m is the value of t_p corresponding to the percentage bearing area at the mean line of the roughness graph. The circle in Figure 5.5 corresponds to t_m. The maximum deviation of roughness, R_p, from the mean line and the total roughness, R_z, determine the parameters ν and b:

$$\nu = 2\,t_m\left(\frac{R_\nu}{R_a}\right) - 1 \tag{5.2}$$

$$b = \left(\frac{R_z}{R_p}\right)^{\nu} t_m \tag{5.3}$$

where t_m = bearing area, t_p, at a depth corresponding to the mean line of the roughness trace (Figures 5.4 and 5.5); a = depth of roughness (micrometer).

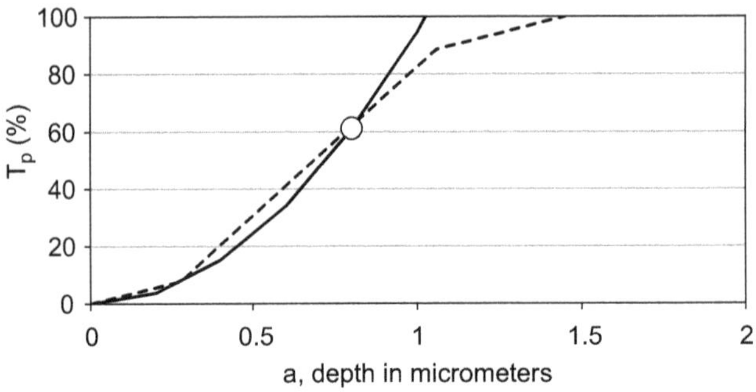

Figure 5.5 Representing the bearing area curve except the valley by an equation of the form, $t_p = t_m\,(a/R_p)^v$

The initial portion of the bearing area curve is given by the following equation, as shown by the full curve in Figure 5.5:

$$t_p = t_m \left(\frac{a}{R_p} \right)^v \tag{5.4}$$

By using an equivalent R_p, the equation can be extended to a plateau honed surface or other surfaces not normally honed. The dotted line curve signifies the generated bearing area curve using an equivalent R_p. The equivalent roughness parameter, R_p, is derived as a function of the plateauness, as explained later in the chapter.

5.3 Wear of Liner Surfaces

Liner wear is a mode of material behaviour occurring at the contact surface when it slides on the ring surface. As seen in Figures 5.9 and 5.10, the sliding of an asperity against the mating piston ring gives rise to a frontal wave of the deformed material subjected to a compressive stress. Cyclic or repeated distortion of the surface at a microscopic level leads to fatigue failure of the roughness peaks.

5.3.1 Asperities

The average asperity is modelled as part of a hemisphere. The statistically stationary value of radius of the hemisphere is the geometric mean of the radius in the direction transverse to machining, r_{tr}, and the radius along machining direction, r_{ln}:

$$r = \sqrt{r_{tr} r_{ln}} \tag{5.5}$$

5.3.2 Radius of the Asperity in the Transverse Direction

The roughness peak can be approximated as a segment of a hemisphere. The radius of the peak is dependent on the roughness parameter, R_{pk}, the corresponding bearing area, MR1, and the number of scratches, f, honed per unit length (Figure 5.6). The average width of the chord of the hemisphere, c, scales with the length, MR1/f. The diameter of the spherical peak could be calculated as follows with an overall scaling factor, k. The value of k is 170.

$$\text{Diameter of the spherical peak, } d_i = \frac{kc^2}{R_{pk}} \tag{5.6}$$

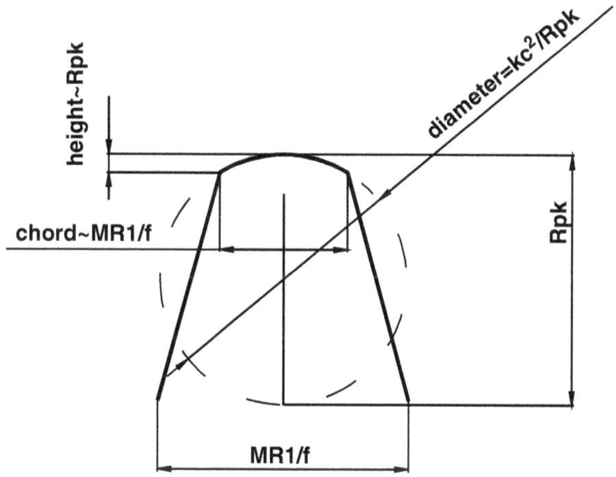

Figure 5.6 Radius of the asperity is obtained by scaling the chord with MR1/f and height with R_{pk}

$$\text{Radius of the spherical peak, } r_{tr} = \frac{d_i}{2} \tag{5.7}$$

where the chord width or width of crest, $c = \frac{MR1}{f}$.

5.3.3 Radius in the Longitudinal Direction

In the wear calculation, the radius of curvature for a liner surface is taken as 900 μm, and 270 μm in the longitudinal direction for piston rings (Kragelskii and Alisin, 1981). Radii usually obtained in the longitudinal direction for different machining processes before honing are given in Table 5.1.

5.3.4 Sparse Contact

5.3.4.1 Areas of Contact

Apparent Area of Contact

This is the geometric area of the top compression ring that is in contact with liner surfaces. The apparent area of contact with the liner can be calculated as follows:

$$A_a = \pi D w \tag{5.8}$$

where D is the liner bore and w the ring height.

The total gaseous pressure, p_g, acts behind the ring over an area less than the axial height by a factor k_n (Figure 5.7).

Table 5.1 Longitudinal radius as a function of the machining method (Kragelskii and Alisin, 1981)

Machining method	Longitudinal radius (μm)
Grinding	100–300
Turning	100–500
Polishing	500–1000

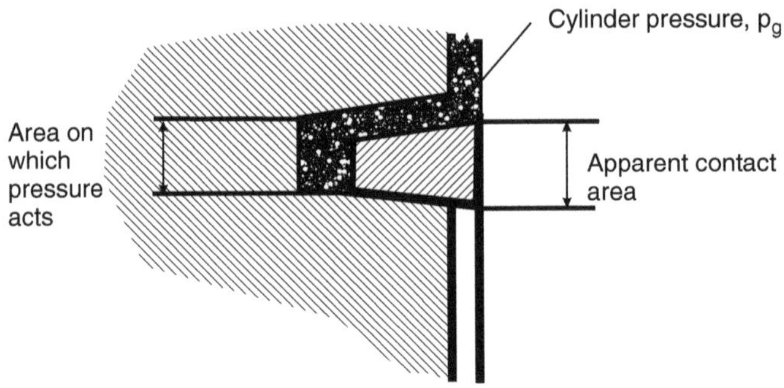

Figure 5.7 Gas pressure exerted on the rear side of the ring over an area less than the apparent area of contact

The normal load, F, on the contact area is given by the gas pressure, p_g, and the projected area, A_a, on which gas pressure acts:

$$F = k_a A_a p_g \tag{5.9}$$

For example, the value of k_a is one for rectangular rings and it is less than one for a keystone ring depending upon the taper angle.

Contour Area of Contact

When two rough surfaces contact, except for the patches of surface as shown by the blue transfer test, there are air gaps everywhere else. The area of the patches is known as the contour areas of contact, A_c (Figure 5.8). The factor by which A_c is less than the nominal area of contact or apparent area of contact, A_a, depends on the waviness of the surface.

Real Area of Contact

On the other hand, when the contour area is looked at in close quarter, the patch is not continuous; it is a group of disjointed pin-tip sized areas. The total of such areas is called the real area of contact, A_r.

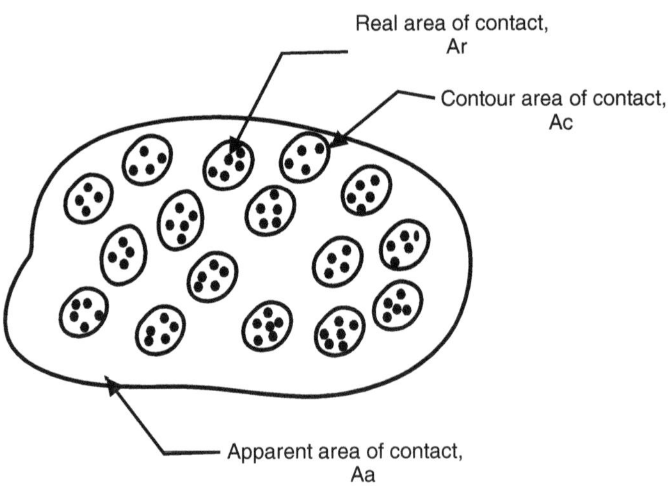

Figure 5.8 Areas of apparent contact, contour contact and real contact

5.3.5 Contact Pressures

The apparent, contour and real pressures of the contact are obtained by dividing the normal load by the corresponding contact areas namely, A_a, A_c and A_r (Equations 5.8, 5.11 and 5.13).

The contour contact area is strictly dependent on the sparse contact of asperities that occurs in the elastic or plastic regime of the materials. Sparse contact occurs usually under elastic conditions when wear takes place. During such wear, the contour contact pressures satisfy the following condition, as a function of a complex surface roughness parameter (Δ), elastic modulus of the material, E, and Poisson's ratio, μ (Kragelskii and Alisin, 1981):

$$p_c = 6 \times 10^{-3}\Delta^{1/2}\left[\frac{E}{1-\mu^2}\right] \tag{5.10}$$

where $\Delta = \frac{R_z}{rb^{1/v}}$.

The contour contact area is given by:

$$A_c = \frac{F}{p_c} \tag{5.11}$$

The liner and the top ring approach each other in the top ring reversal zone under elastic conditions, to allow the atoms and molecules to interact. The normal load per unit area of the real contact is the real pressure, p_r. Under fatigue the pressure tends to be a function of the contact pressure, p_c, the effective radius of asperities, r, roughness parameter, R_a and the Young's modulus of material, E (Kragelskii and Alisin, 1981). The equation is modified for real pressure by changing the index of θ from 2.0 to 1.7 to seek wide applicability for a variety of surfaces with normal roughness:

$$p_r = 0.585\left(\frac{R_a}{r\theta^{1.7}}\right)p_c^{0.04} \tag{5.12}$$

where $\theta = \frac{1-\mu^2}{E}$.

The contour real area is:

$$A_r = \frac{F}{p_r} \tag{5.13}$$

5.3.6 Friction

The metal-to-metal contact at the top ring reversal zone causes wear of the liner under fatigue loading apparently in the elastic regime. The force of external friction is equal to the resistance to ploughing of the soft liner surface by the penetrating asperities of the ring and the molecular forces. The friction force supplies the energy required to detach the asperities, and hence the wear of the mating surfaces. The coefficient of friction depends on the surface roughness parameters and mechanical properties of bore material.

In sparse contact, the molecular coefficient of friction, f_{ml}, is given by (Kragelskii, 1965):

$$f_{ml} = \frac{1.4\tau_0(1-\mu^2)^{2/3}}{E^{2/3}p_c^{1/3}}\left(\frac{r}{R_c}\right)^{1/3} + \beta \tag{5.14}$$

where β – coefficient of strengthening of molecular bond. The coeffficient is 0.025 for cast iron surfaces, 0.013 for steel surfaces, 0.120 for chromium and 0.068 for tin.

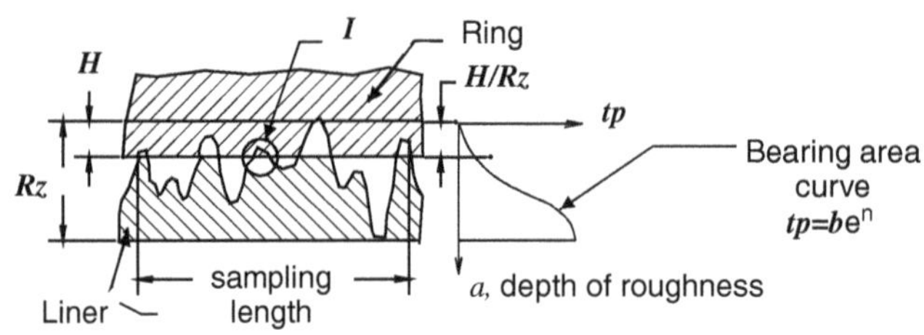

Figure 5.9 Sparse contact model

5.3.7 Approach

The penetration and elastic deformation of the liner surface asperities have the effect of bringing the liner and ring surfaces closer under running conditions. In Figure 5.9, the penetration of the hard asperity in the soft surface is shown. In Figure 5.10, the approach of the spherical peak, *H*, due to deformation is shown, where *r* is the radius of the spherical asperity. The approach determines the character of friction and wear of mating surfaces (Kragelskii and Alisin, 1981; Kragelskii 1965):

$$H = R_{p-eq}\left(\frac{p_c}{\alpha t_m p_r}\right)^{1/v} \tag{5.15}$$

where p_c = contact contour pressure (N/mm^2), p_r = real pressure (N/mm^2), α = elastic deflection factor = 0.5 for metallic surfaces.

The hemispherical asperity makes an indentation of diameter, *d*, and penetrates to a depth equal to the approach. From geometric considerations, an indentation diameter of the deformed asperity is approximated to:

$$d \approx \sqrt{2rH} \tag{5.16}$$

where r = the radius of asperity.

5.3.8 Detachment of Asperities

When gaseous force is applied perpendicular to the direction of sliding of the ring, the compressive force normal to the contact displaces the asperities tangentially. The displacement depends on the approach. Fatigue and detachment are caused with increased slipping in the contact region. The asperity is deformed in the direction of shear and separated.

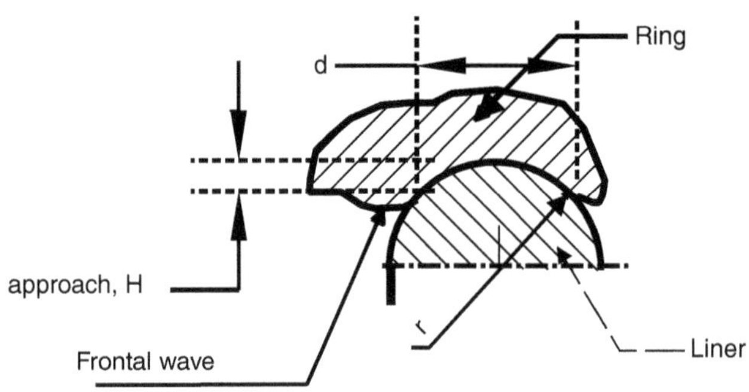

Figure 5.10 View 'I' in Figure 5.9 showing the roughness peak approximated to a hemisphere

5.4 Wear Model

5.4.1 Normally Honed Liner with Peaked Roughness

5.4.1.1 Wear Zone of Liner and Ring

Most of the wear takes place at the end of the compression stroke when the cylinder pressure pressing the ring against the liner is high. In addition, during this period the lubrication transcends to the hydrostatic type (boundary lubrication regime) with a very thin film of oil supporting the ring against the liner. The repeated contact of hard ring surfaces on smooth elastically deformed cylinder wall surfaces causes fatigue failure of the roughness peaks, and hence leads to wear of the liner surfaces. The liner area in contact with the ring is finite and is of the order of the apparent area of contact given in Equation 5.8. The liner wear is usually measured using a profiler by the change in dimension in the specific zone. On the other hand, the ring wear is determined by an increase in end clearance or by the defect in ring mass or by plotting the axial profile with respect to reference.

The relative distance travelled by the two wear surfaces of the ring and the liner per engine cycle is twice the ring height. During a period, t hours, of running, the distance, L, covered by the wear zone can be calculated as a function of engine speed, n:

$$L = \frac{2wnt}{m} \tag{5.17}$$

where m = two for four-stroke engines, w = ring height (mm), t – engine run time (minutes) and n = engine speed (rpm).

5.4.2 Normal Surface Roughness

A rigid liner having a rough surface is subjected to wear; contacting a smooth elastically deformed ring is considered in Figure 5.9. The peaks of roughness are approximated to small equal hemispherical segments with heights so arranged that the real surface and the model have identical bearing area curves. The mutual influence of the asperities is ignored, as the contact is sparse. The area of contact can be approximated to the ratio of the normal force to the contour pressure (Equation 5.11). Assuming only the portions of the penetrated asperities under deformation take part in wear, the wear material separated per unit area, U_v, can be written as follows (Kragelskii, 1965):

$$U_v = \int_0^H A_r \, dH \tag{5.18}$$

The peaks of the rough surface undergo repeated deformation leading to physical and chemical changes on the surface of the liner and, consequently, breaking away of wear particles due to fatigue. The effective contact stress, σ_{eff}, at which the separation of wear particles occurs is proportional to the specific frictional shear stress, τ (Kragelskii and Alisin, 1981). The value of the specific stress is given by

$$\sigma_{eff} = k\tau = k f_{ml} \, p_r \tag{5.19}$$

where, k = factor characterizing contact stress condition = five for brittle materials like the liner and three for highly ductile materials.

5.4.3 Fatigue Loading of Asperities

Elastic interactions at the contact between the deformed asperities lead to changes in the shape and height of the irregularities. A frictional bond is formed at the contact due to fatigue and is subjected to random variations in contour contact zone because of different asperity heights. The asperities will take a number

Table 5.2 Typical values of the empirically derived constant t_f for different surfaces (Kragelskii and Alisin, 1981)

Material type	σ_0 (N/mm^2)	t_f
Steel, grade 45	687	7.9
Cast iron	255	4.1

of cycles to separate from the parent surface. With increasing stress and number of load cycles, N, the fatigue rupture of the peak decreases similar to Veler's law (Kragelskii and Alisin, 1981). The fatigue curve can be empirically described for the number of cycles, N, causing fatigue failure at a given effective stress, σ_{eff}:

$$N \cong 2.25\left(\frac{\sigma_{eff}}{\sigma_0}\right)^{t_f} \tag{5.20}$$

where σ_0 = ultimate tensile stress and t_f = an empirically derived constant = 4.1 for cast iron.
Typical values t_f, for different materials are given in Table 5.2.

5.4.4 *Wear Rate*

The nondimensional wear can be written as the ratio of volume removed per unit area of apparent contact, U, to the total relative distance, L, travelled by the wear surfaces:

$$I = \frac{V}{A_a L} = \frac{U}{L} \tag{5.21}$$

where A_a = apparent area of contact (mm^2) and V = volume of material removed (mm^3).
Usually, wear is taking place over the real area of contact, A_r. Hence, Equation 5.21 can be treated as analogous for a specific wear per cycle, i_H, which is defined as the volume, V, removed per cycle over area A_r, when sliding over a distance equal to the diameter of the contact spot, d:

$$i_H = \frac{V}{A_r d} \tag{5.22}$$

Substituting Equation 5.22 in Equation 5.21:

$$I = i_H \frac{A_r}{A_a} \tag{5.23}$$

The thickness of worn layer corresponding to a single interaction between the irregularities is:

$$V_d = \frac{U_v}{N} \tag{5.24}$$

The integral of the real area over the depth of roughness up to the approach, H, gives the wear rate per unit area, U_v:

$$U_v = \int A_r \, dH = A_c R_{max} \int b\varepsilon^v d\varepsilon = \frac{A_r H}{v+1} \tag{5.25}$$

where ε is the percentage elongation of the peaks under sparse contact.

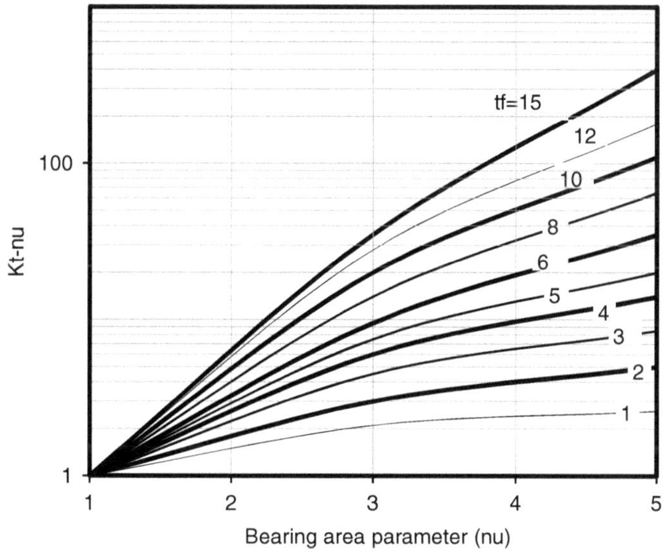

Figure 5.11 Nomogram for determining the fatigue correction factor, k_{tv}

Substituting Equation 5.24 in Equation 5.22 and using Equations 5.25 and 5.16, the nondimensional wear rate, I, is obtained:

$$I = K_1 \left(\frac{p_a}{p_r}\right) \sqrt{\frac{H}{r}} \frac{1}{Nk_{tv}} \tag{5.26}$$

where $K_1$is determined by geometric shape and height of the asperities ≈ 0.2 and k_{tv} = fatigue correction factor (Figure 5.11) (Kragelskii and Alisin, 1981).

Wear of two bearing surfaces is calculated as:

$$W = IL \tag{5.27}$$

5.4.5 *Plateau Honed and Other Liners not Normally Honed*

5.4.5.1 Equivalent R_p (R_{p-eq})

As already defined, in the case of normal peaked roughness, the parameter R_p is the deviation of the surface peaks from the mean line of roughness texture. However, for other surfaces that are not normally honed or are plateau honed, an equivalent R_p exists. The exact value of R_{p-eq} or equivalent R_p depends on the plateauness of the surface, which is defined as the ratio of R_k to R_{vk}. When the plateauness ratio, PR, is greater than 1.3 the surface finish tends to be normally peaked roughness. When the ratio is in the range 0.0–1.3, the equivalent R_p gradually increases from zero to half of R_z. The ratio of R_{p-eq} to $R_z/2$ bears a unique relationship with the plateauness ratio (Figure 5.12) (Lakshminarayanan, Nayak and Dani, 2002):

$$\frac{R_{p-eq}}{R_z/2} = 0.1041\,PR^3 - 0.6852\,PR^2 + 1.4563\,P \tag{5.28}$$

5.4.5.2 Parameters of Bearing Area Curve for a Surface Without Normal Roughness

In a similar fashion, the parameters of the bearing area curve namely, t_m, ν and b are modified for surfaces with a roughness that is not normally peaked.

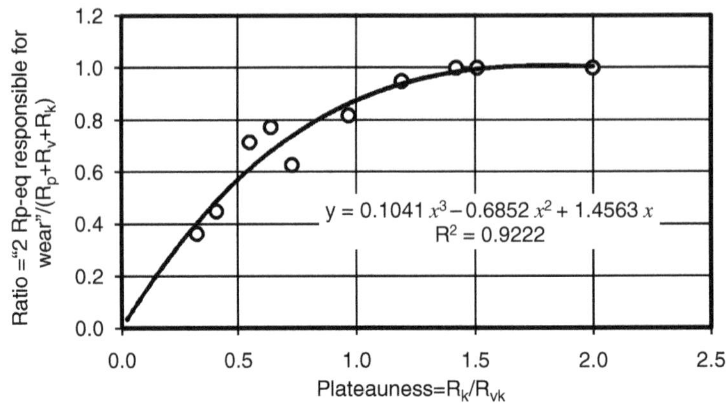

Figure 5.12 Equivalent R_z responsible for wear as a function of plateauness ratio, R_k/R_{vk}

The bearing area curve is in three parts, namely the peak, the central portion and the valley; it can be obtained from the roughness trace, as in Figure 5.2. Each part can be approximated by a straight line represented by coordinates (MR1, R_{pk}), (1 − MR1 + MR2, R_k) and (MR2, R_{vk}), as shown in Figures 5.2 and 5.4. The initial two segments of the bearing area curve are represented by an equation (Equation 5.29) that is in the form of Equation 5.4:

$$t_p = t_{m-eq}(a/R_{p-eq})^{v-eq} \tag{5.29}$$

In Equations 5.2 and 5.3, R_p is substituted by $R_{p\text{-}eq,}$ from Equation 5.28 to reflect the effect of plateauness or smoothness of the surface:

$$v_{eq} = 2\,t_{m-eq}(R_{p-eq}/R_a) - 1 \tag{5.30}$$

$$b_{eq} = (R_z/R_{p-eq})^v t_{m-eq} \tag{5.31}$$

In general, the bearing area at the mean line of the roughness trace is difficult to estimate (or model) for given surface roughness statistics. Instead, the parameter t_{m-eq} is obtained by solving the Equations 5.30 and 5.31 simultaneously, so that the measured bearing area curve fits best to the curve described by Equation 5.29 for the given R_{p-eq}. In other words, the initial portion of the bearing area curve becomes one of the equations in the model. Using the equivalent values in Equation 5.15, the approach, *H*, is given by:

$$H = R_{p-eq}\left[\frac{p_c}{a\,t_{m-eq}p_r}\right]^{1/v} \tag{5.32}$$

The procedure for the calculation of liner wear for a general surface finish is given in Figure 5.13. The calculation is divided into three broad segments. Firstly, the roughness parameters from the measurements are estimated and equivalent parameters for surfaces with nonpeaked roughness are calculated. Then, parameters dependent on operating conditions, liner material and engine size are calculated. In the third segment, the liner wear is calculated for the given roughness characteristics, material and operating conditions. An example is worked out in the Appendix 5.A for a rough plateau honed surface.

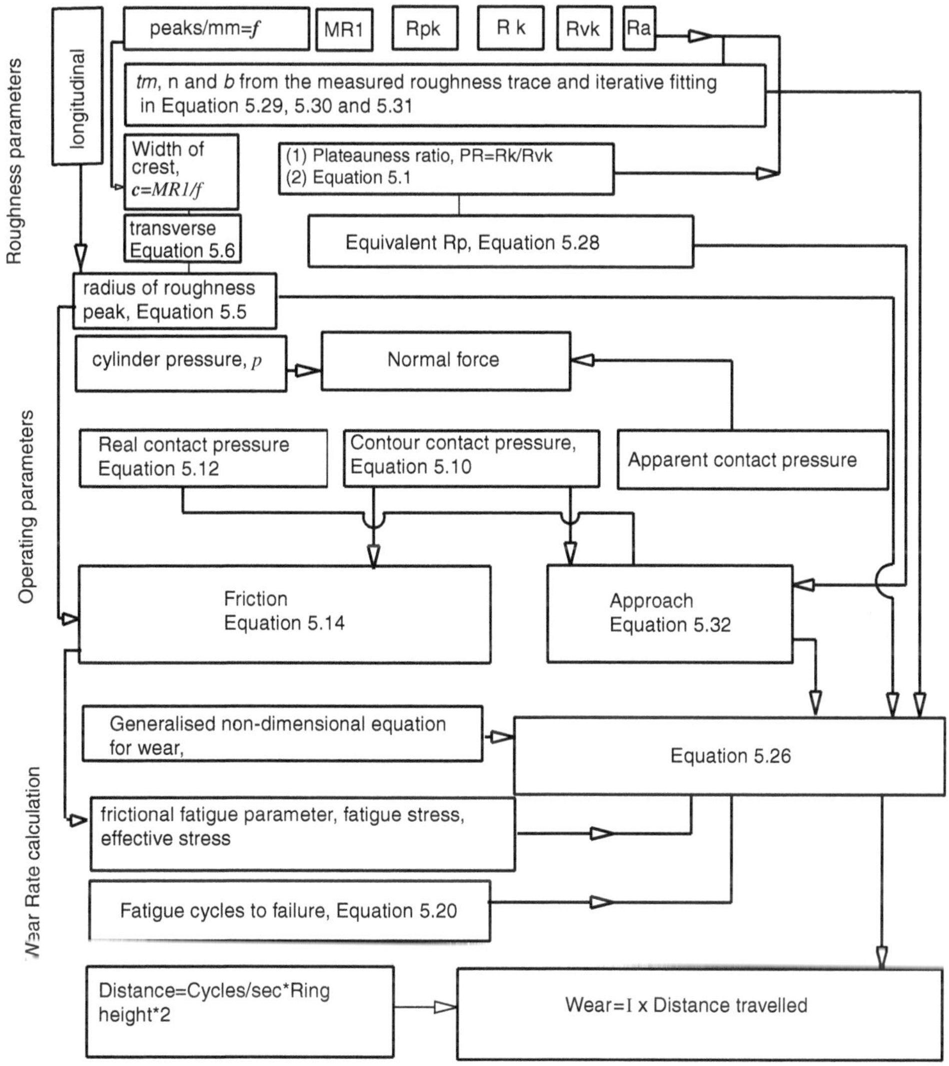

Figure 5.13 Wear calculation procedure for liner surface

5.5 Liner Wear Model for Wear of Roughness Peaks in Sparse Contact

Ten different types of liners reported by Hill *et al.* (1995) and Hegemier and Stewart (1993) and three normally honed liners tested by the authors in a naturally aspirated engine are studied here.

Some of the data are only indirectly available in the literature mentioned above. The cylinder pressure, for example, was derived from the break mean effective pressure (bmep) of the engine and assumed nominal compression ratios of 17 and 18 were used for turbocharged and naturally aspirated engines. The boost pressure after the turbocharger was taken as one seventh of the break mean effective pressure. The peak cylinder pressure was calculated as 1.7 times the calculated compression pressure for naturally aspirated and 1.35 times the compression pressure for turbocharged engines (Obert, 1973). Table 5.3 gives

Table 5.3 Engine specifications and operating parameters

Engines	A (Hegemier and Stewart, 1993)	B (Hill *et al.*, 1995)	C*	Units
Aspiration	Turbocharged	NA	NA	
Swept volume	7.3	1.9	4.16	litre
Number of cylinders	6.0	4.0	3.0	
Stroke/bore	1.2	1.2	1.14	
Swept volume/cylinder	1217	475	1040	cc/cylinder
Diameter of the bore	10.7	7.8	10.5	cm
Speed	2400	2500	2000	rpm
Brake power	179.0	75.0	37.1	kW
bmep	1.23	0.47	0..3	MPa
Compression ratio	17.0	18.0	17.0	
Boost pressure ratio	1.8	1.01	1.01	
Compression pressure	9.5	5.78	5.33	MPa
Peak cylinder pressure	13.19	10	7.5	MPa

*Authors engine data; NA: naturally aspirated

the estimated cylinder pressures and details directly and indirectly available for the two engines in the references. The values of MR2 were arrived at using the volume of valleys given in this reference. These surfaces are plateau or normally honed, and rough and smooth. The initial roughness parameters MR1, R_a, R_{pk}, R_{vk} and R_k obtained from a roughness trace are given in Table 5.4a. The corresponding values at the end of the experimental duration are given in Table 5.4b.

The model involves an integral marching with time as given by Equation 5.25. The values of various roughness parameters deciding the instantaneous real contact area, A_r, are determined as the wear process proceeds. The liner wear contributes to the reduction of R_{pk} and R_k progressively. The duration of wear is considered in several time steps and the integrals are evaluated by using the mean values for each time step. The time step could be 100 hours, as the wear rate is usually very small:

$$\delta U_v = \frac{S}{2}(A_{r1} + A_{r2})\delta H \tag{5.33}$$

where S = step interval.

In the intermediate steps, the current value of R_a is obtained from the bearing area curve, as shown in Figures 5.4 and 5.5. Similarly, the mean values for the instantaneous values of other parameters can be substituted into other equations describing the wear model.

Figure 5.14 shows the strong correlation and successful prediction of wear after 200 hours for each different liner surface type using the extended Kragelskii wear model. The data for the volume of liner wear at the top ring reversal point in 200 hours are available in Hill *et al.* (1995). For smooth plateaued and rough plateaued surfaces it amounts to 0.0107 and 0.0297 mm^3 respectively. The predicted wear volume values using Equation 5.21 were 0.019 and 0.03 mm^3 for smooth and rough plateau surfaces respectively (Figure 5.15.)

5.5.1 Parametric Studies

Using the extended model, some parametric studies were made for the base data given in the Table 5.5, to calculate the wear in 200 hours. With an increase in hardness, the fatigue strength of the material increases and, hence, so does the number of cycles to the failure of the roughness peaks by microcutting. Figure 5.16a shows the decrease in wear rate with increase in hardness. The improvement in wear rate tapers off above a hardness of 350 N/mm^2 as experienced in normal practice.

Table 5.4 Roughness parameters R_a, R_{pk}, R_{vk} and R_k of different liners (Hill *et al.*, 1995; Hegemier and Stewart, 1993)

(a) Start of experiment

Engine	Surfaces	R_a (μm)	R_k (μm)	R_{vk} (μm)	R_{pk} (μm)
B	Low peaked	0.24	0.78	0.39	0.28
	Low plateaued	0.30	0.67	1.05	0.22
	Medium peaked	0.58	1.70	1.20	0.75
	Medium plateaued	0.39	0.98	1.01	0.21
	High peaked	0.82	2.43	1.61	0.88
	High plateaued	0.90	1.56	2.82	0.37
A	Smooth nonplateaued	—	—	—	—
	Smooth plateau	—	—	—	—
	Rough nonplateaued	—	—	—	—
	Rough plateaued	—	—	—	—
C	Normally honed liner	0.96	2.76	2.10	0.63
	Normally honed liner	0.68	2.82	2.19	0.68
	Normally honed liner	0.93	2.80	2.23	0.65

(b) End of experiment

Engine	Surfaces	R_a,μm	R_k,μm	R_{vk},μm	R_{pk},μm
B	Low peaked	0.12	0.28	0.30	0.07
	Low plateaued	0.14	0.25	0.53	0.07
	Medium peaked	0.43	0.90	1.12	0.12
	Medium plateaued	0.32	0.61	0.95	0.14
	High peaked	0.62	1.25	1.67	0.18
	High plateaued	0.88	1.13	3.10	0.31
A	Smooth nonplateaued	0.23	0.71	0.98	0.22
	Smooth plateau	0.55	0.70	1.10	0.39
	Rough nonplateaued	0.66	1.99	4.88	0.69
	Rough plateaued	0.70	2.30	2.00	0.60
C	Normally honed liner	0.83	0.84	3.45	0.31
	Normally honed liner	0.44	0.60	1.53	0.16
	Normally honed liner	0.82	0.71	2.73	0.13

The higher the PR, the lower is the plateauness and the higher is the radius of the hemispherical roughness peak. Hence, the real area of contact is less. The increase in contact area decreases the applied stress on the roughness peaks and their fatigue life increases. Consequently, the wear rate increases with an increase in PR (Figure 5.16b). The reduction in plateauness ratio improves the value of the equivalent maximum roughness, and hence the wear rate. The plateauness ratio indirectly refers to the ability of the surface to hold oil in the tiny pockets, and hence to resist scuffing at microscopic levels.

On the other hand, the increase in the roughness parameter indicative of the peaks of the surface, namely, R_{pk}, leads to an increase in the real pressure, and hence the life of the roughness peaks. The resultant increase in wear rate with R_{pk} is seen in Figure 5.16c. Similarly, the longitudinal radius decided by the machining process determines the real load on the microscopic roughness peaks. The effect is shown in Figure 5.16d.

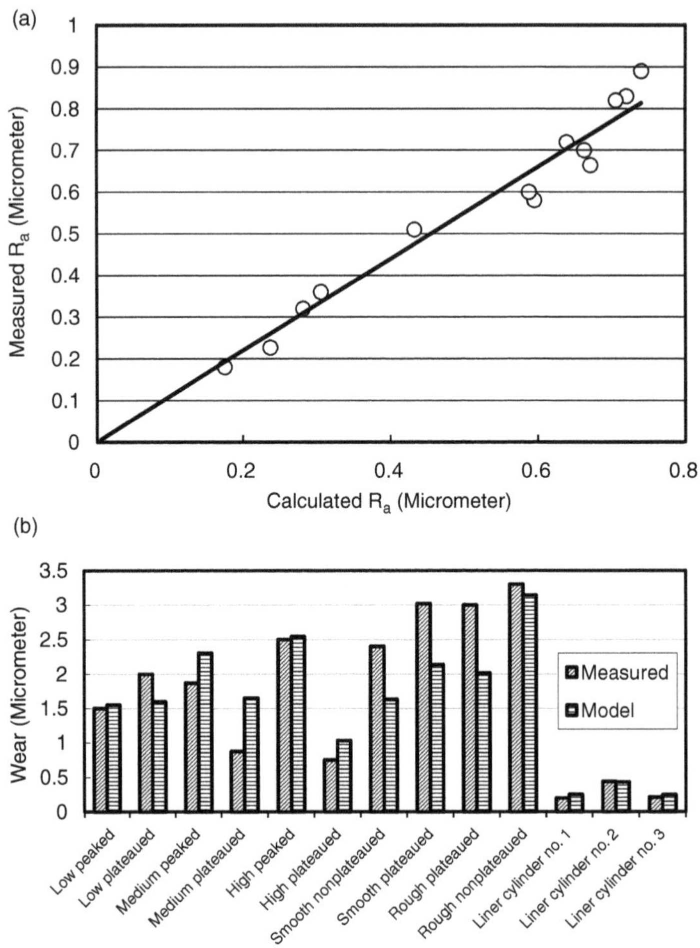

Figure 5.14 (a) Correlation of wear model on different surface roughness after 200 hours. (b) Comparison of measured and calculated wear values of different surfaces normalized for 200 hours (the asterisk denotes normally honed liners from the author's laboratory)

5.5.2 Comparison with Archard's Model

Archard and Hirst (1956) assumed that two asperities of the rubbing surfaces come into contact to form a junction with a constant proportionality, k_w, for the formation of an adhesive fragment. Each fragment is assumed to be a hemisphere of diameter, d, and equal to the junction diameter (Rabinowicz, 1965). The wear pattern along the cylinder wall is calculated by the law of adhesive wear, which states that the loss of material from the parent surface is (a) directly proportional to load and the sliding distance, and (b) inversely proportional to the hardness of the softer surface being worn away. For the reciprocating engine, the bore wear rate at the top ring reversal zone is given by:

$$h_w = \frac{4 k_w F s n}{3 p A_a} \qquad (5.34)$$

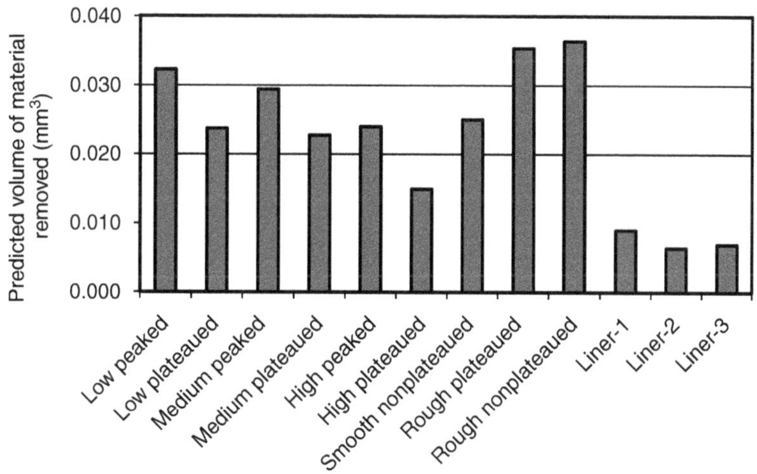

Figure 5.15 Material lost after 200 hours

where k_w = wear coefficient, F = normal load (N), s = stroke (mm), n = speed (rpm), p = hardness of the surfacc (N/mm^2) and A_a – ppparent area of contact (mm^2).

One of the features of this formula is that junction diameter, surface roughness parameters, bearing area parameters and number of cycles causing fatigue failure of asperities do not enter at all into the final expression for the wear volume. A factor 3 empirically appears in the denominator as a shape factor applicable to the circular junctions and hemispherical fragments. The wear rate depends on the factor, k_w, and it is independent of the hardness and load. Usually, the wear constant for this combination lies between 45×10^{-3} and 0.05×10^{-3}. Interestingly, these constants over-predict the wear by up to 20 times. This is probably because the roughness statistics are not considered in the model. It is possible to tune the constant to anticipate the experimental trend, oblivious of the physics or the previous work on cast iron. For example, to match with the experimental data presented in Table 5.6, the proportionality constant may have to be dropped to 0.85×10^{-3}, which is probably the value for low carbon steel on copper (Rabinowicz, 1965) (Table 5.7).

5.6 Discussions on Wear of Liner Roughness Peaks due to Sparse Contact

Wear is measured at the top ring reversal point where it is highest due to loss of hydrodynamic lubrication. The model implicitly assumes that the grade or quality of oil has less influence on the wear rate. It is dependent on the type of the liner surface and the cylinder pressure, which applies the load on the liner

Table 5.5 Base roughness data for parametric study

Surface	Rough nonplateaued
R_a (μm)	0.70
R_{pk} (μm)	0.60
R_{vk} (μm)	2.00
R_k (μm)	2.30
Plateauness ratio	1.15
Hardness (N/mm^2)	255.0

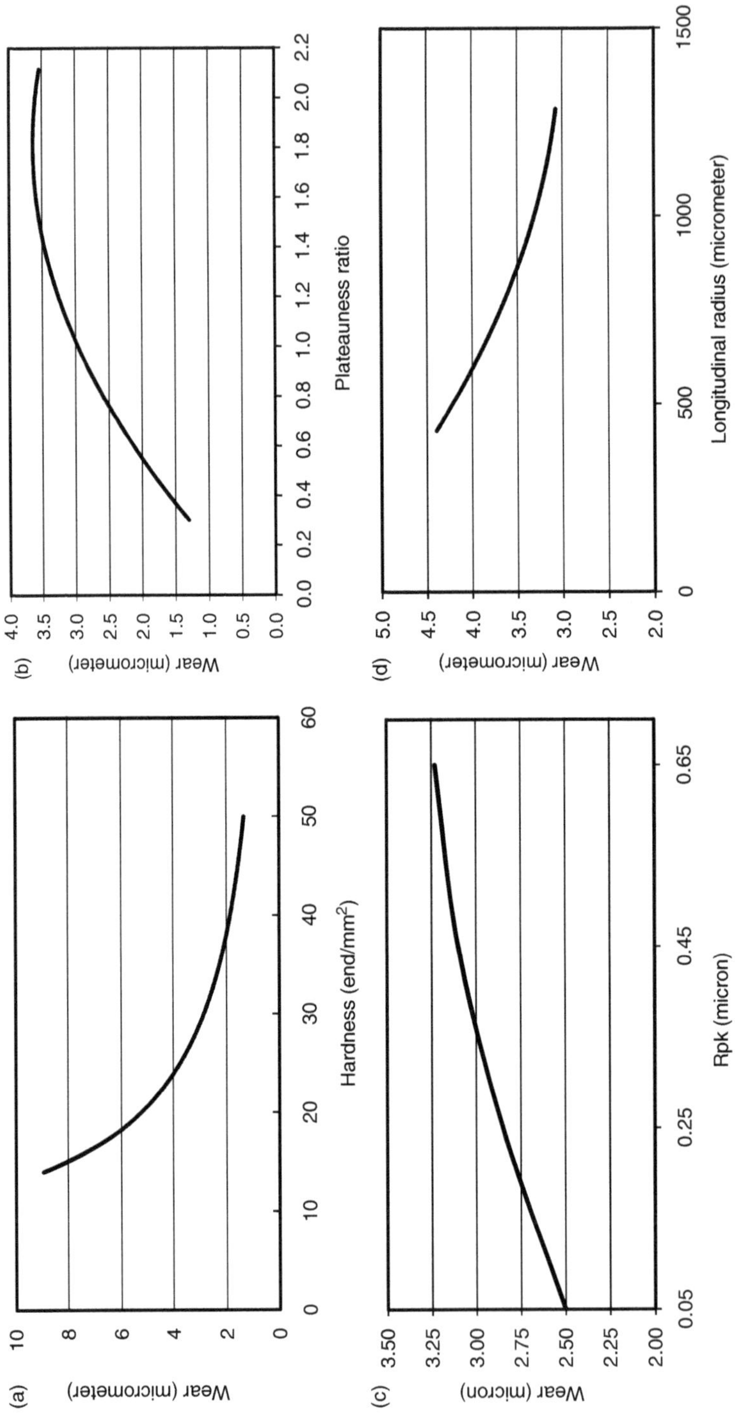

Figure 5.16 Theoretical study of the effect of (a) hardness, (b) plateauness ratio, (c) R_{pk} and (d) R_{ln} on wear

Table 5.6 Wear values of different liner surfaces, Archard's model (the proportionality constant is 0.85×10^{-3})

Engine	Hours run	Surfaces	Wear (μm)	
			Calculated	Measured
B	200	Low peaked	1.883	1.50
		Low plateaued	1.883	2.00
		Medium peaked	1.883	1.88
		Medium plateaued	1.883	0.88
		High peaked	1.883	2.50
		High plateaued	1.883	0.75
A	200	Smooth nonplateaued	2.467	2.40
		Smooth plateau	2.467	3.02
		Rough nonplateaued	2.467	3.00
		Rough plateaued	2.467	3.30
C	135	Normally honed liner	1.177	0.20
		Normally honed liner	1.177	0.44
		Normally honed liner	1.177	0.21

through the piston ring. The use of Veler's law in the model for the number of cycles to failure indicates that there are a finite number of cycles that will be endured by the roughness peaks for any given load. In other words, the life of the wear part, namely the liner surface, is finite. This is different from the case of the fatigue strength of ductile material, below which a part will last for eternity. The harder the material, the longer is the fatigue life. Higher engine pressures cause increased loading of the roughness peaks, and hence increased wear due to fatigue of the roughness peaks.

A highly plateaued surface has a wear rate in a turbocharged engine similar to that in a naturally aspirated engine using rough nonplateaued liners. Here, though the apparent contact pressure is higher in a turbocharged engine, the bearing area increases proportional to the normal load to maintain the same real contact pressure. A rough nonplateaued surface wears at a higher rate than a rough plateaued surface, as in the former case the peaks represented by R_{pk} are subject to higher real pressure. The improvement in machining process and methods of clamping will reduce the longitudinal radius of asperities to below 1000 μm, so reducing wear. Plateauness is characterized by a low plateauness ratio, R_k/R_{vk}, and low value of R_a. The former reduces the apparent value of the maximum roughness while at the same time maintaining a large volume of oil in the pockets of the valleys. If a large volume of valleys is having a direct bearing on oil consumption (Durga *et al.*, 1998), it enables cooling of the contact zone of the rings heated by friction at the top dead centre reversal point. Hence, reduction in the apparent maximum and average roughness values causes a drop in wear rate without causing scuffing.

Table 5.7 Wear constant, k_w, of various sliding combinations (Rabinowicz, 1965)

Combination	Wear constant (k_w)
Low carbon steel on low carbon steel	45×10^{-3}
Stainless steel on stainless steel	21×10^{-3}
Low carbon steel on copper	0.5×10^{-3}
Bakelite on bakelite	0.02×10^{-3}

In general, a smooth plateau surface exhibits lower wear rates than a rough plateaued surface. A smoother surface characterized by lower R_a leads to corresponding lower value of R_k and R_{vk} leading to increased friction. Therefore, a smooth plateau surface wears at a higher rate than a highly plateaued surface. From the experiments and the model, the hierarchy of surface roughness type for resistance to wear, in the order of increasing resistance, is given below:

- Rough nonplateaued
- Rough plateaued
- High peaked and normally honed
- Medium peaked and normally honed
- Low peaked or low plateau or medium plateaued
- High plateau surface

5.7 Summary

The roughness peaks of two mating surfaces in the real area of contact are under higher pressure than the apparent pressure estimated based on the nominal contact area. This is due to the contact of roughness peaks and waviness. Though the real pressure is within the elastic limits, the repeated application of such a high load on the material at the microscopic level leads to fatigue failure of the roughness peaks.

The study results concluded that the shape of the bearing area curve has an overwhelming effect on the real area of contact and the penetration or approach of the two surfaces. In order to comprehend the complexity of the bearing area, a characteristic curve is fitted to the initial portion of the curve. The constants of the curve, namely, t_m, ν and b, are used in the model to predict the wear rate. Many of the statistical wear parameters of the bearing area curve are needed to fully describe its influence on the wear behaviour of the surfaces. They are the parameters corresponding to maximum roughness, total roughness, the peak, the valley and the middle portion of the bearing curve.

Similarly, the plateauness of the surface is given by the ratio of the two roughness parameters, namely, R_k and R_{vk}, describing indirectly the peaks and the valleys of the roughness trace. The ratio determines the equivalent R_{pk} that can be substituted in the wear rate formula for normal surfaces with peaked roughness, in an effort to enhance the applicability to a wide range of surface finishes. Thus, the real and contour contact pressure is estimated by using the equivalent parameters in the established equations for normal surfaces.

The wear itself is shown as the distance travelled by the surfaces in a direction perpendicular to the main relative velocity of the two mating surfaces. The net wear due to fatigue failure of the microscopic peaks is directly dependent on the relative distance travelled by the two surfaces.

The data on different types of liner surfaces distinguished by smoothness, plateauness and hardness have been studied in the current work. The experimental results could be satisfactorily predicted using the extended Kragelskii model for wear. Model constants used are independent of the context and materials and emphasize the phenomenological aspects by considering the roughness statistics and material properties, unlike Archard's model.

The model shows that all types of liner surfaces have a finite wear life. However, surface quality, load per unit area and the hardness of the material determine the life of an engine liner that is not abused, for example, by admission of sand, over-heating or over-loading. In addition, the liners in high bmep turbocharged engines display higher wear rate than in naturally aspirated engines for a given surface quality and material strength. The model indicates that the wear will be low when the hardness is higher, as is seen in some modern engine liners. A surface with hardness above 350–400 N/mm^2 and highly plateaued, indicated by an R_a value of about 0.7–0.8 micrometre with a plateauness ratio of 70%, offers the best compromise for a long wear life and oil consumption.

Appendix 5.A Sample Calculation of the Wear of a Rough Plateau Honed Liner

Equation	Parameter		Unit
	Bore	107	mm
	Maximum cylinder pressure, p	12.85	N/mm^2
	Hardness, HB	255.0	N/mm^2
	R_a	0.66	μm
	R_k	1.99	μm
	R_{pk}	0.69	μm
	R_{vk}	4.88	μm
5.1	R_z	7.60	μm
	MR1	4.1	%
	Number of peaks/μm, f	0.012	μm^{-1}
	r_{ln}	900	μm
5.7	Width of roughness crest, c	3.42	μm
5.6	r_{tr}	2.11	μm
5.5	r	567.1	μm
5.9	Normal load, F	12959.7	N
5.8	Apparent contact area, A_a	1008.5	mm^2
5.28	$2R_{p-eq}$ responsible for wear/R_z	0.484	
	R_{p-eq}	1.84	μm
5.30	v_{eq}	2.10	
5.29	b_{eq}	10.95	
	By iteration using bearing area curve		
	t_{m-eq}	0.56	
5.10	Contour contact pressure, p_c	50.72	N/mm^2
5.11	Contour contact area, A_c	195.1	mm^2
5.12	Real contact pressure, p_r	382.6	N/mm^2
5.13	Real contact area, A_r	F/p_r	
		25.9	mm^2
5.15	Approach, H	0.93	μm
	σ_0	255.0	N/mm^2
5.19	σ_{eff}	186.4	N/mm^2
	t_f	4.1	
5.20	Number of cycles to rupture under fatigue	3.64	
5.26	Nondimensional wear rate, I	2.33×10^{-5}	
5.17	Distance covered in 200 hours, L	86400	mm
5.27	Wear in 200 hours	2.02	μm
5.25	Volume of wear debris	0.022	mm^3

References

Archard, J.F. and Hirst, W. (1956) The Wear of Metals under Unlubricated Conditions. *Proceedings of the Royal Society London, A*, **236**, 397–410.

Bhushan, B. (1999) *Principles and applications of tribology*. John Wiley & Sons, Ltd, p. 123.

Casellato, R., Avezou, J.C. and Gazzard, S.T. (1990) Recent developments in cylinder bore and liner technology. Paper presented at a symposium organized by India Pistons Limited (Chennai, India) in collaboration with Turner and Newhall (UK), India.

Chang, L., Webster, M.N. and Jackson, A. (1994) A line Contact Micro EHL Model with three dimensional surface topography. *Journal of Tribology*, **116**, 21–28.

Durga, V., Rao, N., Boyer, B.A. *et al.* (1998) Influence of surface characteristics and oil viscosity on friction behaviour of rubbing surfaces in reciprocating engines. Proceedings of the 1998 Fall Technical Conference, Vol. 31–2, Paper No. 98-ICE-131, ASME-ICE, pp. 23–35.

Halling, J. (1971) *Principles of Lubrication.* The Macmillan Press Ltd.

Hegemier T. and Stewart, M. (1993) Some effects of liner finish on diesel engine operating characteristics. Technical Paper 930716, SAE, Troy, MI.

Hill S.H., Kantola, T.C., Brown J.R. and Hamelink J.C. (1995) An experimental study of the effect of cylinder bore finish on engine oil consumption. Technical Paper 950938, SAE, Troy, MI.

Hu, Y.Z., Li N. and Tonder, K. (1991) A dynamic system model for lubricated sliding wear and running-in. *Journal of Tribology*, **113**, 499–505.

ISO (1995) DIN 4776 Determination of surface roughness parameters $R_{k,}$ $R_{pk,}$ $R_{vk,}$ MR1 and MR2 serving to describe the material component of the roughness profile. Superseded by ISO 13565.2 GPS – Surface texture: Profile method – Surfaces having stratified functional properties – Parts 1 and 2.

Kragelskii, I.V. (1965) *Friction and wear.* Butterworths Publication.

Kragelskii, I.V. and Alisin, V.V. (1981) *Friction, wear, and Lubrication, Tribology Handbook*, Vol. 1, Mir Publishers, Moscow.

Lakshminarayanan, P.A., Nayak, N.S., and Dani, A.D. (2002) Prediction of liner wear by extending the Kragelskii model for wear of roughness peaks in sparse contact. *Proceedings of the Institution of Mechanical Engineers, Part J: Journal of Engineering Tribology*, **216** (5), 327–342.

Obert, E.F. (1973) *Internal Combustion Engines and Air Pollution.* Intext Educational Publishers.

Poon, C.Y. and Sayles, R.S. (1994) Numerical contact model of a smooth ball on an Anisotropic rough surface. *Journal of Tribology*, **116**, 194–201.

Rabinowicz, E. (1965) *Friction and wear of materials.* John Wiley & Sons, Inc., New York.

Ting, L.L., and Mayer, Jr, J.E. (1974) Piston ring lubrication and cylinder bore wear analysis, Part 1 – Theory. *Journal of Lubrication Technology*, **96**, 305–314.

Wu, S. and Cheng, H.S. (1991) A sliding wear model for partial EHL contacts. *Journal of Tribology*, **113**, 134–141.

Yang, J. and Winer, W.O. (1991) A comparison between the thermo-mechanical wear model and some experimental observation. *Journal of Tribology*, **113**, 262–269.

6

Generalized Boundary Conditions for Designing Diesel Pistons[1]

Diesel pistons are designed to dissipate heat adequately to limit the temperatures and avoid undesirable distortion and scuffing with the cylinder liner. The temperature distribution, leaving sufficient working clearances in running conditions, decides the outer profile of a piston at room temperature. The finite element method enables calculation of piston expansion, and hence the form to be machined at room conditions. However, the results are only as accurate as the boundary conditions imposed on the piston model, irrespective of mesh size and shape, and the method of solution.

In this chapter, the heat transfer coefficients at the boundary surfaces were judiciously chosen from the literature and practice. These are generalized to apply to pistons that vary widely in size, bmep and speed. The average gas temperature and the peak cylinder pressure acting on the piston, obtained from the simulation of the thermodynamic cycle, are fed to the finite element models in two or three dimensions for four different pistons to demonstrate the applicability of the boundary conditions.

The model for thermal expansion simplifies itself to a steady state problem because of large thermal inertia at the relatively high frequency of firing. The transient deformation is mainly due to the cylinder pressure at its maximum. The predicted temperature distribution in the metal piston is satisfactorily compared with the experimental values obtained using relaxation in the hardness of the hypoeutectic aluminium. Piston ovality, waviness of the groove and profile of the piston were obtained using the calculated deformation. Favourable comparison of the estimated piston shape with the proven profiles indicates wide applicability of the generalized boundary conditions.

6.1 Introduction

In the early days, the electrical analogue method or calculation by relaxation technique was used to determine the temperature distribution in piston (Gunjegaonkar, Lakshminarayanan and Chandorkar, 1984). The axisymmetric information obtained using these techniques was helpful in determining the shape of the piston at room conditions. However, either the calculations or the development of the electrodes and application of boundary conditions were too elaborate and difficult for repeated parametric analysis.

---

[1] All figures in this chapter are reproduced with permission from P.A. Lakshminarayanan, N.S. Nayak, S. Talwalkar, T. Babar and A.D. Dani, "Generalized boundary conditions for designing diesel pistons," SAE Paper 2001-28-0009, Symposium on International Automotive Technology, (SIAT 2001) SAE Conference, 2001.

Critical Component Wear in Heavy Duty Engines, First Edition. P.A. Lakshminarayanan and Nagaraj S. Nayak.

Fleming and Percy (1974) described the design, development and implementation of finite element analysis of engine components. Several methods of analysis that could be undertaken during the preliminary stage were proposed. Reipert and Buchta (1981) described details of the interactive piston development program for predicting oil consumption and blow-by. Munro, Griffiths and Ingham (1973) studied the two dimensional axisymmetric finite element model (FEM) of pistons with various cooling gallery forms and extended the work to a complex three dimensional technique. Munro and Griffiths (1979) suggested generalization of a fully integrated computer aided design and computer aided manufacturing system to enable accuracy in calculation. Moebus and Schoeckle (1979) analysed the stress in the nodular cast iron monoblock pistons in four-stroke diesel engines using the three dimensional finite element method. Attention was given to thermal expansion of pistons and its effect on the running clearance between the piston and the liner. Wacker and Sander (1982) estimated the heat loss to the coolant in a monoblock spheroidal graphite cast iron piston with cooling gallery using finite element analysis. Adams (1979) and Saugerud (1980) used the FEM technique for estimating the fatigue life of a piston on the basis of thermal and structural aspects.

6.2 Temperature Distribution and Form of the Piston

6.2.1 Top Land

There is a pronounced temperature drop along the top land indicating substantial heat transfer directly from the lands of the piston to the cylinder wall through the oil layer. However, with reduced top land height, the entire heat is transferred across a smaller area to increase the operating temperature of the ring grooves. A satisfactory operation is not possible when the top groove exceeds 200 °C, as the oil is baked to form carbon in the absence of oxygen. Hence, sufficient top land height of the order of 0.08–0.1 times the cylinder diameter is necessary in diesel pistons to avoid over-heating at the top ring zone.

Sufficient top land clearance avoids scuffing of thermally grown piston with the cylinder. Further, it aids in easily breaking off any build up of a thick layer of carbon at high temperature. On the contrary, large clearances result in higher fuel consumption due to the increase in dead volume. The reduction of the land clearance improves engine efficiency, allows greater heat flow and accumulates less carbon deposits. However, when the clearance is very tight, the thin hard carbon layer polishes the liner at critical areas during service, and hence increases oil consumption. The optimum top land clearance for aluminium pistons would be about 0.1 mm per 10 mm of cylinder diameter.

6.2.2 Skirt

Lack of cooling and massive pin bosses with higher volume to surface ratio, heat up and expand the skirt more along the pin plane than the thrust plane. The required cold clearance assumes an oval shape at any cross-section perpendicular to the piston axis, as the thickness of the wall gradually grows along the circumference from the thrust plane to the pin plane, as shown in Figure 6.1. By the same

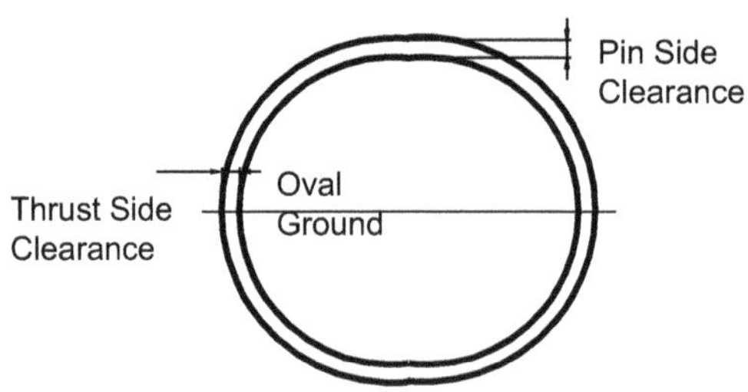

Figure 6.1 Oval ground skirt

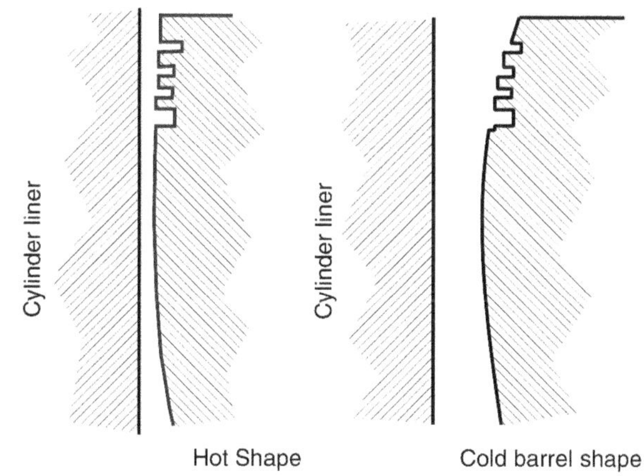

Figure 6.2 Barrelled skirt of a piston

argument, the top portion of the piston expands more as the temperature is higher than the lower portion. Therefore, the cold clearance is higher at the top. The combination of the two effects gives the piston an elliptical barrel shape. A piston machined to such a form at room temperature enables closer skirt clearance at operating temperature (Figure 6.2), avoiding seizure and controlling noise and oil consumption.

A residual barrel shape is left even in the hot conditions. This is to allow piston tilt without having the contact between the top land and the cylinder to avoid bore polishing. The cylinder pressure further distorts the profile and hence the clearance. Overall, the variable ovality along the axis gives the piston a barrelled form when cold (Figure 6.2).

6.3 Experimental Mapping of Temperature Field in the Piston

The piston is run for more than 100 hours at constant load and speed. The drop in hardness of the piston material is correlated with the temperature at which the zone of the piston was soaked. The accuracy of the technique improves when the temperature is higher than 100 °C.

Figure 6.3 gives the curves of the residual hardness of popular hypoeutectic aluminium alloy (silicon 11%) used in many pistons, with respect to the temperature and the time of soaking.

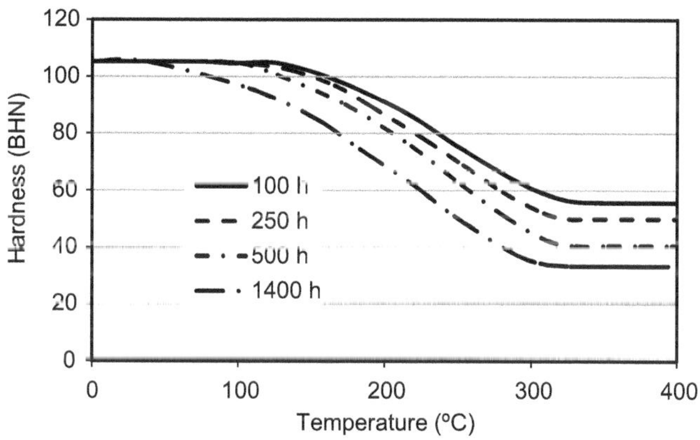

Figure 6.3 Residual hardness values of Mahle-124 alloy

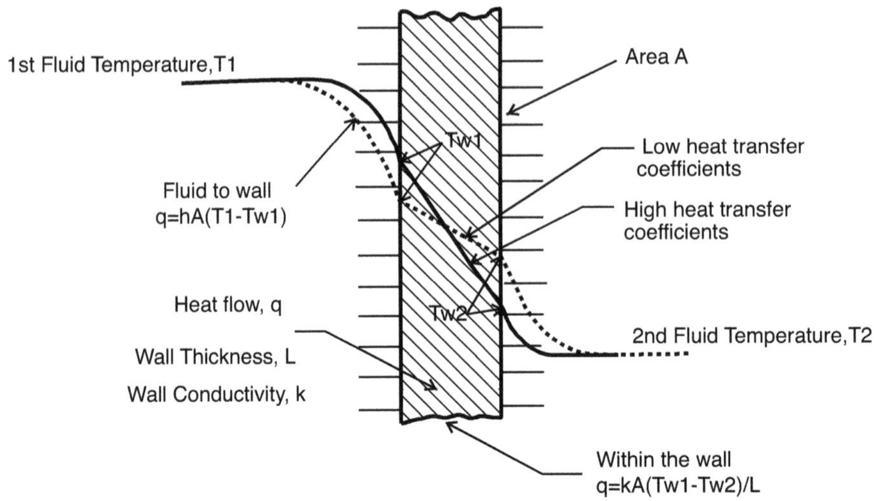

Figure 6.4 Heat transfer in a piston considered as a slab

6.4 Heat Transfer in Pistons

6.4.1 Metal Slab

The simplest model of a piston is probably the single dimensional wall with oil and hot gases flowing on either side at steady state (Figure 6.4). In the wall zone, the heat conduction, Q, follows the Fourier equation:

$$Q = \frac{kA(T_{w1} - T_{w2})}{L} \tag{6.1}$$

where T_{w1} and T_{w2} = the temperatures of the walls (surfaces), k = thermal conductivity (kW/mm K), A = cross-sectional area (mm^2) and L = thickness of plate (mm).

Similarly, on either side of the wall, Newton's heat convection equation holds:

$$Q = h_1 A(T_1 - T_{w1}) = h_2 A(T_{w1} - T_2) \tag{6.2}$$

Here h_1 and h_2 are the heat transfer coefficients of the flow at the wall and T_1 and T_2 are the bulk fluid temperatures.

The temperature distribution within the wall depends strongly on the boundary conditions, namely the values of h_1, h_2 and the temperatures of the bulk fluids. The solution is exact, if the assumed boundary conditions are correct.

The same conclusion is applicable for two or three dimensional steady and unsteady state conduction problems in the piston and cylinder liner. The boundary conditions are in the form of specified temperatures and heat transfer coefficients. The solution is obtained by numerical analysis using finite difference or finite element approach.

6.5 Calculation of Piston Shape

The steady state temperature distribution in the piston is described by the Laplace equation in cylindrical coordinate system (r, y, φ):

$$\frac{\partial^2 T}{\partial y^2} + \frac{\partial^2 T}{\partial r^2} + \frac{1}{r}\frac{\partial r}{\partial r} + \frac{1}{r^2}\frac{\partial^2 T}{\partial \varphi^2} = 0 \tag{6.3a}$$

6.5.1 Popular Methods Used Before Finite Element Analysis

Before the advent of fast computers and the finite element method, the thermal problem of pistons was solved by using an axisymmetric model. For this case, in a cylindrical coordinate system the Laplace equation is reduces to:

$$\frac{\partial^2 T}{\partial y^2} + \frac{\partial^2 T}{\partial r^2} + \frac{1}{r}\frac{\partial T}{\partial r} = 0 \tag{6.3b}$$

The equation was solved by means of the relaxation method (Kreith, 1965) using the given boundary conditions such as heat transfer coefficients and temperatures of the fluids at the surfaces or known temperatures at certain discrete points in the piston.

Alternatively, the cross-section of the axisymmetric piston was simulated using the principle of analogy between the fields of electricity and temperature, as both obey the Laplace equation.

6.5.1.1 Relaxation Method

The cross-section of the piston is partitioned into a network at intervals, Δy and Δr in a grid (Figure 6.5). When, the Laplace equation is satisfied at each node, the sum of the heat conducted through the four grid lines ending at this point equals zero. At the boundaries, the heat flow is calculated using the heat transfer coefficients and the temperature of the bulk fluids, namely gas and oil. The whole temperature field in the piston is determined iteratively using this scheme.

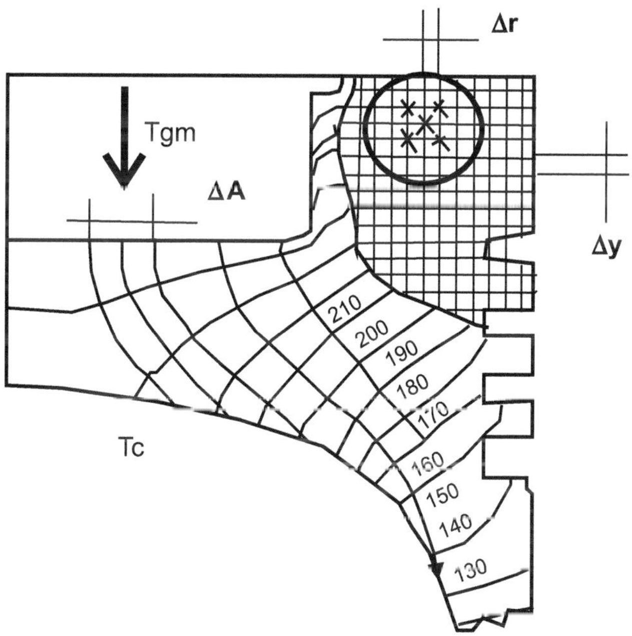

Figure 6.5 Relaxation method (T_{gm} is the bulk mean gas temperature and T_c the temperature of the cooling medium)

6.5.1.2 Analogy of Electric Current and Heat Flow

The piston boundary is divided into several small lengths, defined physically by separate copper electrodes, assuming that the temperature along each length will vary very little. Then for thermal system at the boundaries:

$$-k\Delta A\frac{\partial T}{\partial x} = h\Delta A(T_\infty - T) \tag{6.4}$$

The electrical system analogous to the thermal system can be written similarly:

$$-\frac{1}{\rho}\Delta A'\frac{\partial V}{\partial x'} = h'\Delta A'(V_\infty - V) \tag{6.5}$$

V is analogous to T, that is, $V = \beta T$, and the model dimensions l' and $\Delta A'$ scale with l, that is, $l' = \lambda l$ and $\Delta A'$ can be written as $\Delta A' = \lambda^2 \Delta A$, and Equation 6.5 is simplified to:

$$-\frac{1}{\rho}\lambda\Delta A\frac{\partial T}{\partial x} = h'\Delta A'(T_\infty - T) \tag{6.6}$$

Dividing Equation 6.4 by Equation 6.5 the parameter for similarity between the thermal and electric fields is obtained:

$$\frac{k\rho}{\lambda} = \frac{h}{h'}\frac{\Delta A}{\Delta A'}$$

The specific thermal resistance, R, for the piston becomes:

$$R = \frac{k\rho\lambda}{h\Delta A'} \tag{6.7}$$

Figure 6.6 shows the schematic of the electrical analogy technique. In a tank, a pool of electrolyte bounded by electrodes in a shape geometrically similar to the cross-section of a piston simulates the temperature field in the piston. A wedge shaped pool gives the solution to the axisymmetric problem.

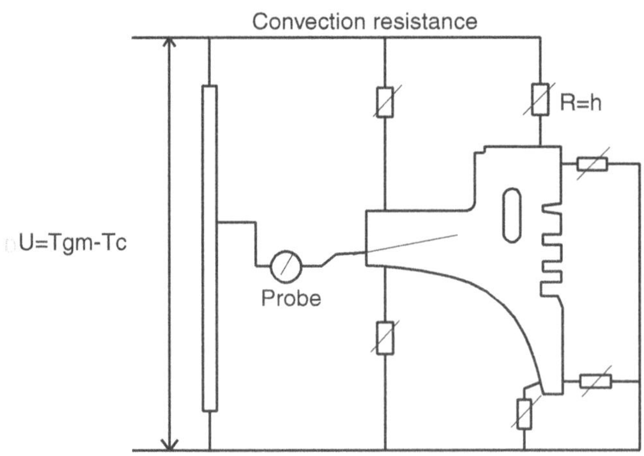

Figure 6.6 Electrical analogy

At the appropriate electrodes, the electric potentials proportional to the fluid temperatures are applied through resistances similar to convection coefficients. The analogous electric field scales with the relative temperature at any point.

6.5.2 Calculation by Finite Element Method

The thermal deformation of the piston shape is calculated by imposing the boundary conditions of convection coefficients and the temperatures of the gases and the oil on a finite element model.

6.5.2.1 Boundary Conditions

Local Convection Coefficient on the Piston Crown and Gas Temperature

The thermal inertia of the piston material is too high to follow the cyclic fluctuations in the gas temperature at high frequency. Hence, the problem of thermal expansion simplifies to a steady state heat conduction with cycle-averaged boundary conditions on the piston crown. The peak cylinder pressure is imposed to consider the second order transient deformation of the piston at the top dead centre.

The mean gas temperature is calculated by simulation of the thermodynamic cycle, including the heat transfer from the gases to the walls using the correlation for instantaneous heat transfer (Woschni and Fieger, 1979). The cycle average of the gas temperature, T_g, and the peak cylinder pressure form the boundary conditions at the piston crown. The mean heat transfer coefficient, h_m, at the crown is obtained by cycle averaging the instantaneous heat transfer coefficient. The experiments on aluminium pistons (Gunjegaonkar, 1984) showed that the variation of the local heat transfer coefficient with the cylinder radius is parabolic, with a weighting factor, w = 1. The heat transfer rate is higher at the centre, due to higher velocity of gases in this zone:

$$\frac{h}{h_m} = 1 + w\left[0.5 - \left(\frac{r}{r_c}\right)^2\right] \tag{6.8}$$

where w = weight, r = radius (mm) and r_c = critical radius (mm).

Convection Coefficients and Bulk Fluid Temperatures in Other Zones

In many of the pistons studied, the heat transfer coefficients in other typical zones were obtained by iteration, for the best match between the calculated temperature fields in the pistons with the measured fields. The best-fitted heat transfer coefficients were found, on average, independent of the type of the piston (Table 6.1). The cooling of the under crown is enhanced by a factor (Table 6.2) depending on the type of cooling.

The temperature at the end of the convection link to the liner wall is considered as 30 °C above the coolant temperature around the cylinder liner. This boundary condition is modified when the coolant is air, by considering the additional convection resistance on the coolant side.

Table 6.1 Heat transfer coefficients, in other zones in a piston

Zone	Value (W/m^{2}K)
Piston top land, upper and back face of top ring groove to liner wall	700
Piston crown	309
Lower face of top ring groove to liner wall	11 700
Lower face of second and third ring groove to liner wall	4 100
All other surfaces on piston outer diameter and ring grooves	700
Piston crown under face to lubricating oil and gudgeon pin	814

Table 6.2 Heat transfer enhancement factor depending on the type of piston cooling

Type of piston under crown cooling	Factor
Splash	1.0
Spray cooling	1.5
Cooling gallery	2.0–3.0

The boundary temperature at the link in the gudgeon pin zone is taken as 30 °C above the oil temperature. The boundary of the piston under crown beyond the convection resistance is at the temperature of the oil measured after the oil cooler.

6.5.2.2 Properties of the Piston Material

The properties of the material input to the finite element calculations are given in Table 6.3.

6.5.2.3 Finite Element Analysis

The finite element analysis involves subdividing the continuum of the piston into elements of finite size. The correct shapes for the elements are chosen to calculate their properties and relate these to the total potential energy within the system. In the present work, two cases of axisymmetric and one case of three dimensional models were analysed to justify the validity of generalized boundary conditions for piston design.

Axisymmetric Model
Two dimensional axisymmetric study of the thermal deformation in a piston easily provides substantial information for the design of piston. In this analysis, two piston geometries are developed in the planes of the thrust and pin. The deformation is studied as if the geometries belong to two independent pistons. For the calculation of the oval shape, the deformations in the two cardinal planes are interpolated as a function of the local thickness at the circumferential angle.

Three Dimensional Model
The three dimensional model of a piston predicts more accurately the thermal field and the deformation than an axisymmetric model. The data preparation and mesh sizing are tedious tasks and need a powerful computing system and a large computer memory. A quarter section of the model is a reasonable compromise and gives satisfactory results notwithstanding the usually offset piston cavity.

Coupled Field Analysis
During the analysis of various pistons, the modulus of elasticity and other properties are assumed invariant with temperature. Therefore, the coupling between the structural and thermal fields is linear, and hence the solutions can be superposed.

Table 6.3 Physical properties of the piston material (aluminium alloy (hypoeutectic))

Parameter	Value	Unit
Young's modulus of elasticity	70 000	N/mm^2
Density of material	2.8	g/cm^3
Poisson's ratio	0.3	
Thermal conductivity	228	W/m K
Specific heat	937	J/kg K
Coefficient of thermal expansion	23×10^{-6}	mm/mm K

Table 6.4 Engine details

Engine	A	B	C
Bore (mm)	175	100	105
Stroke (mm)	220	120	120
Aspiration	TA	NA	TA
Number of cylinders	4	6	4
Engine cooling	Water	Air	Water
Rated speed (rpm)	1000	2500	1500
Rating (kW/cylinder)	96.0	20.5	18.5

NA: Naturally aspirated; TA: Turbo after-cooled.

6.5.3 Experimental Validation

Three types of pistons of widely different bore sizes and brake mean effective pressures were considered for the analysis (Table 6.4). The engines were simple naturally aspirated type or turbocharged and after-cooled type.

Piston A was gallery cooled and pistons B and C were cooled using oil jets from nozzles located in the crankcase. However, for brevity, in each case, details are given selectively to demonstrate the theme of the work.

Pistons A and B were analysed using axisymmetric models. The contours of the temperature and barrel shape of pistons A and B were studied. The proven barrel shapes of piston A and B were found to match the calculated values satisfactorily.

Detailed results from the three dimensional analysis of piston C are described and compared with the experimental values. The oval and barrel shapes of piston C from the FEM are compared with the successful shapes proven in mass production.

6.5.3.1 Piston A

The swept volume of the engine is five litres per cylinder and the brake mean effective pressure is 1.7 MPa. The piston is provided with a cooling gallery. The temperature distribution in the pin plane at rated load and speed is shown in Figure 6.7. From the analysis of the axisymmetric model, the deformations of the piston in the thrust and pin planes are plotted in Figures 6.8 and 6.9. The profiles designed and experimentally optimized for successful performance closely follow those calculated, considering some details as follows: the design clearances on the piston pin plane below the ring pack are kept higher than required by the thermal expansion to allow for the 'drainage' of oil scraped by the oil ring; the top land clearance is designed to be larger to avoid carbon build-up and consequent bore polishing.

6.5.3.2 Piston B

Piston B is used in an air-cooled naturally aspirated diesel engine. The air cooling results in higher temperatures of the cylinder wall and piston surfaces; therefore, the under crown is cooled using nozzles fitted to the crankcase.

Axisymmetric models of the piston considering the cross-sections in the thrust and pin plane were analysed as in the case of piston A. The proven piston profiles in the two planes of thrust and pin axis are shown against those calculated by the finite element method in Figures 6.10 and 6.11. The correlation between the profiles developed painstakingly by experiments and those calculated are satisfactory.

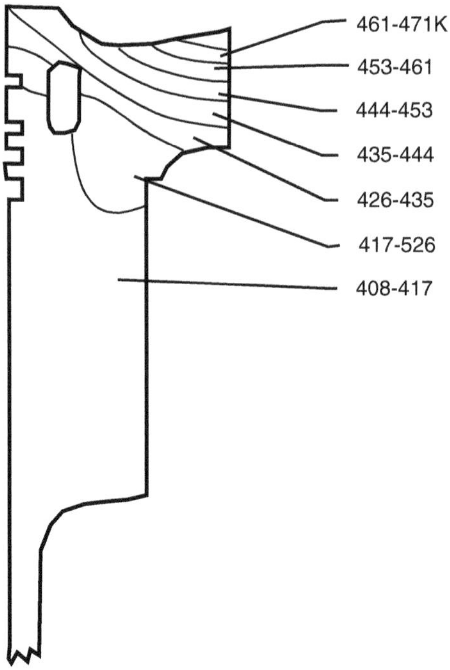

Figure 6.7 Calculated thermal fields (Piston A)

6.5.3.3 Piston C

Piston C is used in a four-cylinder, 4.16 litre capacity turbocharged and after-cooled engine. The peak cylinder pressure was 11 MPa. At the end of an endurance test at rated speed and rated load for 500 hours, the piston was sectioned in the thrust and pin planes and the soaking temperature was mapped by measuring the hardness at the points shown in Figure 6.12.

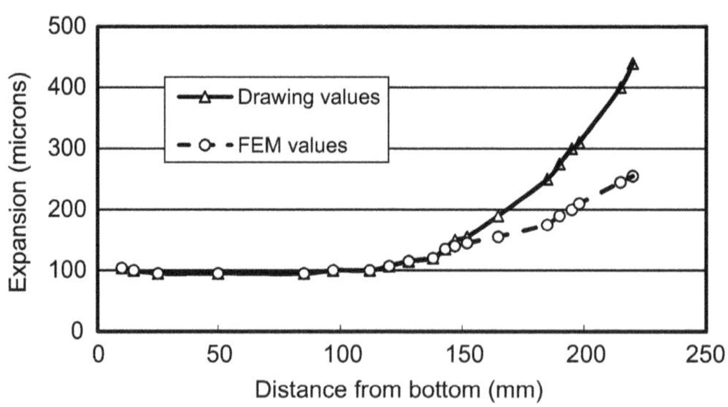

Figure 6.8 Comparison of calculated deformation and the proven shape in the thrust plane (Piston A)

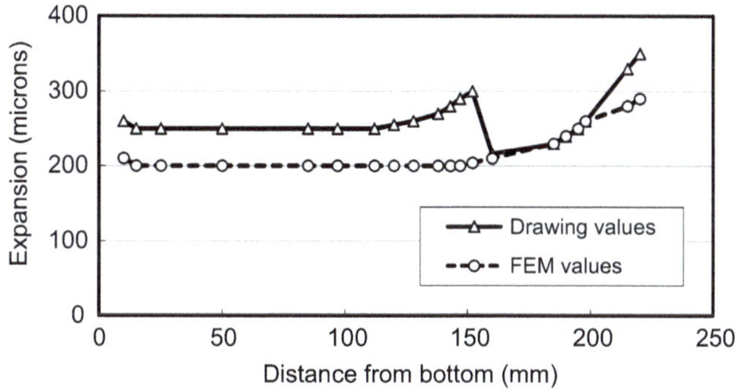

Figure 6.9 Comparison of calculated deformation and the proven shape in the pin plane (Piston A)

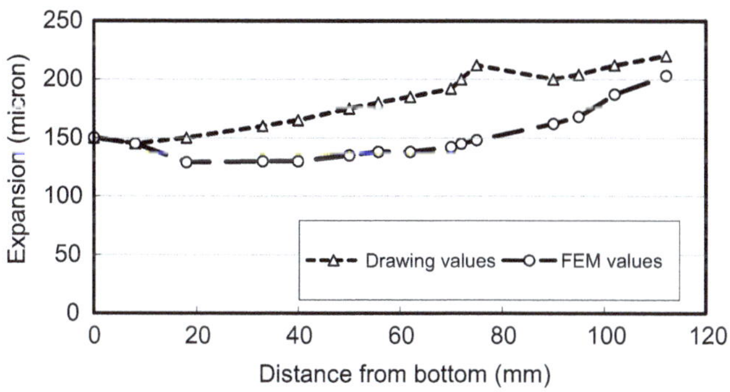

Figure 6.10 Comparison of calculated deformation and the proven shape in the thrust plane (Piston B)

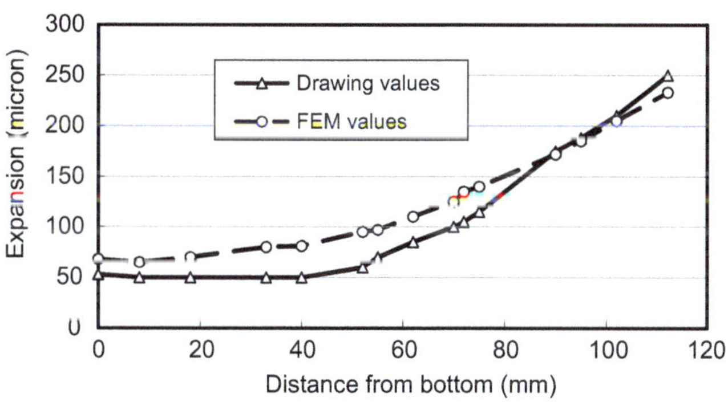

Figure 6.11 Comparison of calculated deformation and the proven shape in the pin plane (Piston B)

6.5.3.4 Three Dimensional Model of Piston C

A quarter model of the piston C was analysed by applying the boundary conditions for thermal deformation. The temperature distribution obtained from the finite element analysis is shown in Figure 6.13.

Temperatures
The correlation between the temperatures measured and those calculated are observed to be satisfactory (Figure 6.14), supporting the recommended generalized boundary conditions.

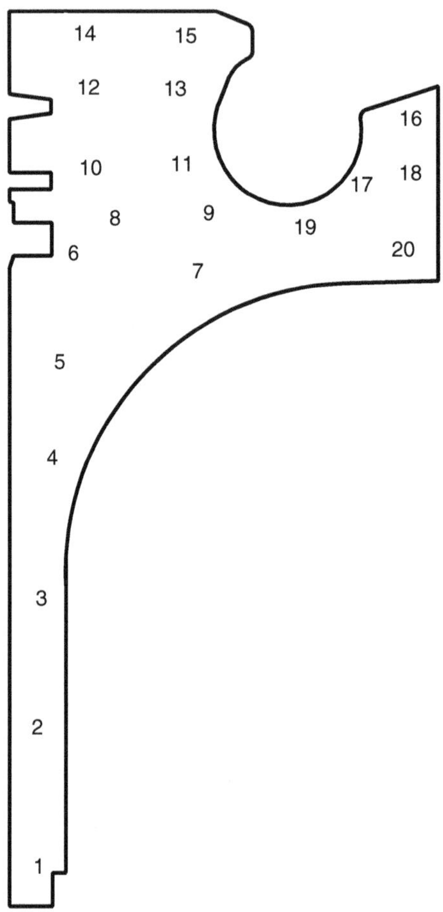

Figure 6.12 Locations where hardness was measured to estimate the running temperatures

Ovality
The required ovality of the piston is the thermal deformation along the radial direction. At any cross-section perpendicular to the piston axis, the ovality is obtained by summing along the radius the components of the deformation in the x and y directions. The deformation in the remaining three quarters of the piston is plotted by considering mirror symmetry of the quarter model.

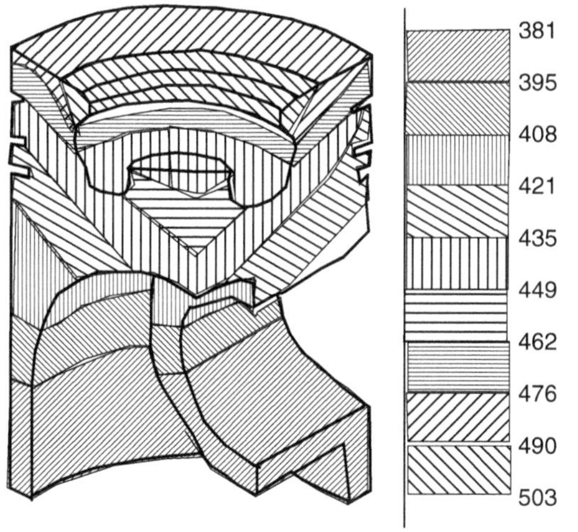

Figure 6.13 Contours of temperature (K) for the quarter model of piston C

For example, the plot of an oval shape at 22 mm distance from piston bottom is shown in Figure 6.15. The oval shape proven during development of the piston for mass production is shown by a broken line. In this piston, the excess clearance on the pin side given for oil drainage, as already explained in case of piston A, is relatively less, and hence is not obvious in the comparison.

Clearance on Diameter Along the Piston Height

Figures 6.16 and 6.17 show the expansion of piston in the thrust and pin planes obtained using the FEM for the prescribed boundary conditions. The successful barrel shape incorporated in the piston is almost identical to that calculated, however, with a hot running clearance varying from 25 to 20 microns over the entire stretch. An extra clearance in the cold condition at the top land is given to avoid the land contacting the liner by tilt.

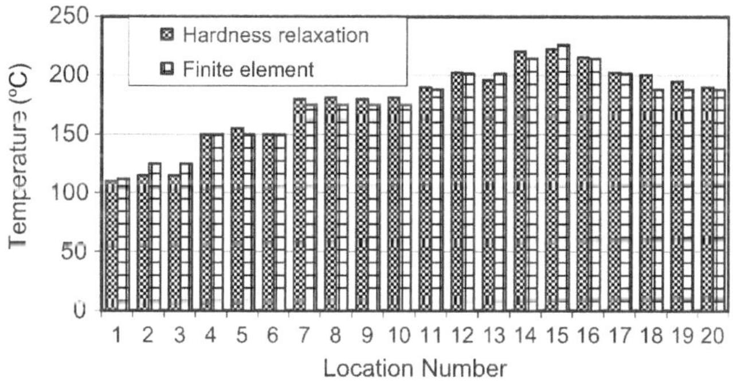

Figure 6.14 Comparison of temperature values using hardness relaxation and FEM techniques (Piston C)

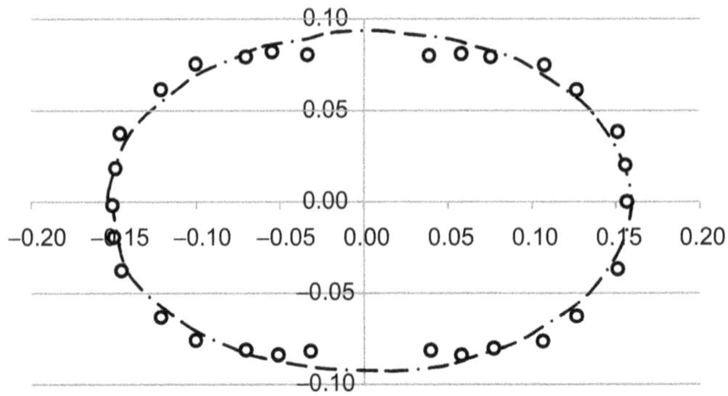

Figure 6.15 Plot of ovality (in mm) at a distance of 22 mm from bottom; data represented by circles are FEM values and the broken line is the ovality proven in production (Piston C)

Waviness of the Top Ring Groove in Hot Running Condition

The bottom face of the top ring groove becomes wavy under hot running conditions, due to inhomogeneous distribution of temperature around the groove. The high points of waviness are observed in well run-in pistons as shiny patches. The waviness and, more importantly, the wave number should be sufficiently small to retain the oil film under the gas pressure. As the operating temperature of the gases increases, the waviness worsens, even if it was satisfactory while machining at room temperature. Figure 6.18 shows the calculated waviness and is nearly within the norms of 10 micron and wave number not more than two, that is, the number of peaks and valleys at the groove zone.

6.6 Summary

The finite element method enables in-depth study of pistons, compared to the relaxation method and electrical analogue practiced in earlier times. The calculation of piston deformation in running conditions using proper boundary conditions gives valuable clues to design the complex form of pistons. In this work, the boundary conditions were applied to pistons that varied widely in size, brake mean effective pressure and type of cooling.

The temperature distribution in a cross-section of a piston was obtained experimentally by measuring the loss in hardness of the piston material for sustained soaking at the operating temperature. The temperature map from finite element analysis compared favourably with the experimental data (Lakshminarayanan *et al.*, 2001).

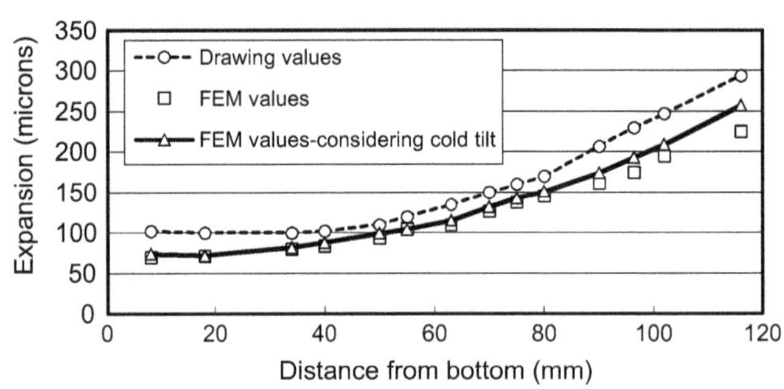

Figure 6.16 Thrust side clearances on the diameter along the piston height (Piston C)

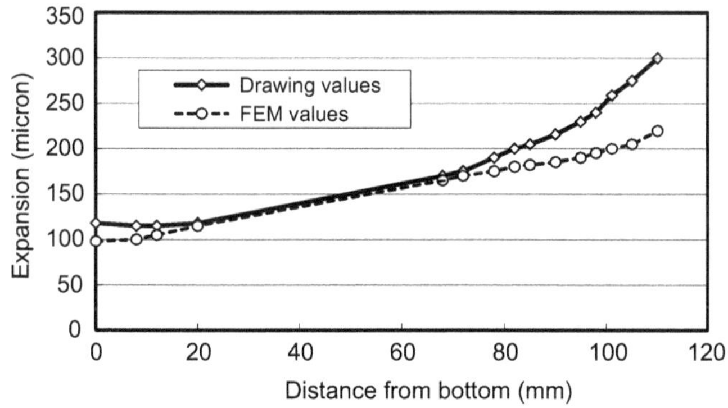

Figure 6.17 Pin side clearances on the diameter along the piston height (Piston C)

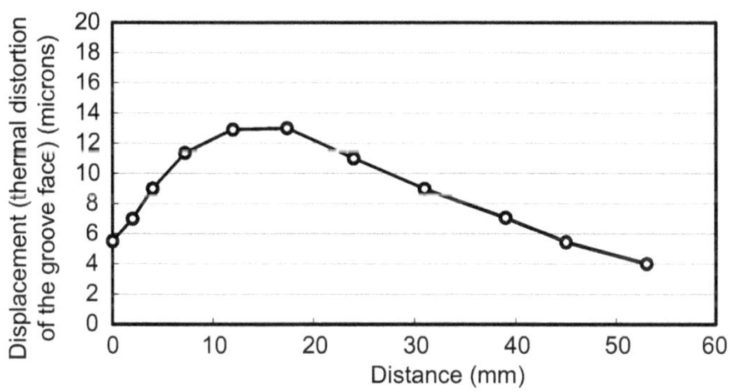

Figure 6.18 Waviness plot of bottom surface of piston ring groove under thermal deformation (Piston C)

The three dimensional model gave results superior to two dimensional models at the expense of time consumed for modelling and computing. The ovality and the barrel profiles of the piston were studied from the thermal deformation and matched with the piston profiles proven in mass produced engines. The favourable results substantiate the recommendation of generalized boundary conditions, which is useful in the study of a piston in an unknown engine.

References

Adams, W.R. (1979) Use of analytical techniques in diesel engine piston design. 13th International Congress on Combustion Engines, Vienna, Austria.

Fleming, J.M. and Percy, M.J. (1974) Diesel engine component design using the finite element method and interactive graphics. Technical Paper 740337, SAE, Troy, MI.

Gunjegaonkar, D.S., Lakshminarayanan P.A. and Chandorkar, S.B. (1984) Electrical analogue for piston isotherms. National Conference on Internal Combustion Engines, Rajkot, India.

Kreith, F. (1965) *Principles of heat transfer*. International Textbook Co.

Lakshminarayanan, P.A., Nayak, N.S., Talwalkar, S. *et al.* (2001) Generalised Boundary Conditions for Designing Diesel Pistons. SIAT-SAE Symposium on International Automotive Technology, Technical Paper 2001-28-0009, SAE, Troy, MI.

Moebus, H.,and Schoeckle, S. (1979) Design and stress analysis of nodular cast iron monoblock pistons for four stroke diesel engines. 13th International Congress on Combustion Engines, Vienna, Austria.
Munro R, Griffiths W.J. and Ingham A.P. (1973) Open gallery pistons for highly rated high speed diesel engines. 10th International Congress on Combustion Engines, 5–9 April 1973, Washington, DC.
Munro, R. and Griffiths, W.J. (1979) The application of predictive techniques in the design and development of medium speed diesel engine pistons. 13th International Congress on Combustion Engines, Vienna, Austria.
Reipert, P. and Buchta, R. (1981) New design methods for pistons. Technical Paper 810933, SAE, Troy, MI.
Saugerud, O.T. (1980) Life prediction of thermally loaded piston engine components. 14th International Congress on Combustion Engines, Helsinki, Finland.
Wacker, E. and Sander, W. (1982) Piston design for high combustion pressures and reduced heat rejection to coolant. Technical Paper 820505, SAE, Troy, MI.
Woschni, G. and Fieger, J. (1979) Determination of local heat transfer coefficients at the piston of a high speed diesel engine by evaluation of measured temperature distribution. Technical Paper 790834, SAE, Troy, MI.

7

Bore Polishing Wear in Diesel Engine Cylinders[1]

Excessive wear of the cylinder liner leads to higher blow-by past the rings and oil consumption. In this chapter, a methodology is developed to estimate the wear of the liner due to bore polishing due to excessive carbon build up on the top land. The wear rate is proportional to the gaseous load, contact stress and sliding action of the piston surfaces. The wear model has been validated on large engines running on heavy fuel at 2.2 MPa bmep. To avoid bore polishing, the piston profile and liner geometry are modified and, most importantly, a scraper ring is designed on the top of the liner. As a result, the engine oil consumption rate improved substantially and is maintained throughout the life of the engine.

7.1 Introduction

Impacts by the piston on the cylinder by transverse movement in the running clearances lead to liner vibration, cavitation and engine noise. The transverse movement of piston also plays a major role in determining the wear phenomenon of the liner surfaces. Laws, Parker and Turner (1973) have made a theoretical analysis of the transverse, rotational motion of the piston, and the effect of frictional forces at the gudgeon pin and the piston periphery. The predicted results are satisfactorily compared with the measured values. Bishop and Leavitt (1975) studied the analytical models of the piston movement, gas flow through the ring assembly and the piston ring dynamics, and developed a gas exchange simulation program to predict the optimum engine performance. Wilson and Fawcett (1974) studied the motion of the slider crank mechanism under gaseous loads by changing the geometry and mass parameters, and predicted the optimum piston transverse motion. Similarly, Wong *et al.* (1994) developed a numerical model for rotational and lateral dynamics of the piston under mixed lubrication conditions. On the other hand, Ting and Mayer (1974) developed an analytical method for determining the bore wear pattern of a reciprocating piston engine by hydrodynamic lubrication action and side thrust loads. Haddad and Tian (1995) and Haddad (1995) carried out a theoretical study on the diesel piston to predict the optimum parameters for low noise and better engine mechanical efficiency.

---

[1] All figures in this chapter are reprinted with permission from N.S. Nayak, P.A. Lakshminarayanan, M.K. Gajendra Babu and A.D. Dani, "Bore polishing wear in diesel engine cylinders," Paper Number ESDA2004-58526, ASME 7th Biennial Conference on Engineering Systems Design and Analysis (ESDA) 2004.

Critical Component Wear in Heavy Duty Engines, First Edition. P.A. Lakshminarayanan and Nagaraj S. Nayak.

Berbezier *et al.* (1989) studied the mechanism of bore polishing by suspended solid particles in the lubricant and showed increasing the wear rate with larger carbon particles. Al Khalidi and Eyre (1987) described bore polishing of the liner due to hard carbon deposits and concluded that a smooth finished area promotes lubricant starvation and delamination of the surfaces. Similarly, research work on high sulfur fuel and spurious quality oil caused excessive deposits on the piston top land, and led to heavy oil consumption (McGeehan, 1980, 1983; Ishizuki, Parker and Turner, 1981; Weiss, Busenthuer and Hardenberg, 1987). The research showed that top land cut back was the choice to prevent liner wear by polishing.

In this chapter, the bore polishing of highly loaded engine liners is studied. Under abusive running conditions, the deposits on the top land become harder due to the ramming action; adhesive or abrasive action on the bore surfaces ensues. The finite element analysis technique was used to determine thermal (and structural) deformation of the piston and bore, and for calculating the contact pressures at different zones of the liner. The transverse motion of piston was studied to calculate the contact forces and sliding velocity of the piston. A linear wear relation was used to predict the wear profile on the contact surface of the liner. Satisfactory correlation was achieved between the calculated and measured results.

7.2 Wear Phenomenon for Liner Surfaces

7.2.1 Bore Polishing

Bore polishing is characterized by the formation of glazed areas with the disappearance of the cross-hatch honing pattern. It is influenced by insufficient top land clearance, unfavourable piston height to bore ratio, combustion and lubricant conditions. Catastrophic failure occurs when the polished areas of the bore can no longer be effectively lubricated and scuffing is initiated. The unetched microstructure of a polished area appears as shown in Figure 7.1.

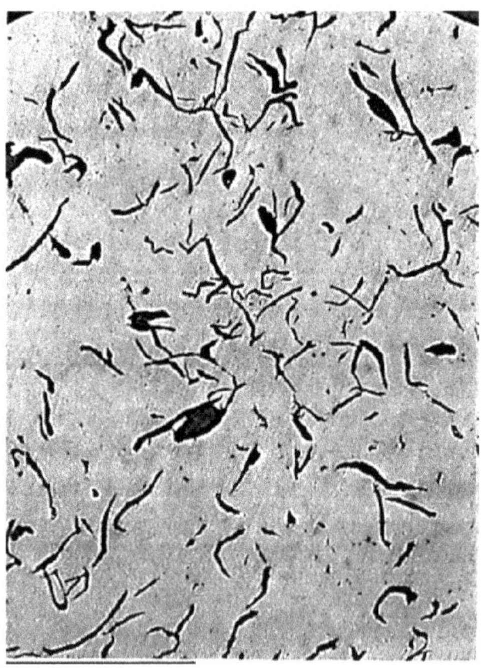

Figure 7.1 Microstructure of the polished liner surface

7.3 Bore Polishing Mechanism

During combustion, the contact pressure on the liner and the piston land is by virtue of the gas forces; the contact area increases when the clearance between them reduces. When the engine load is increased, new pairs of smaller peaks will come into contact and undergo elastic deformation, increasing the contour area of contact and affecting the neighbouring peaks. When the load exceeds a critical value, the deformed asperities undergo plastic deformation. Under continuous running, the asperities of the liner surface detach to create shiny mirror like spots.

7.3.1 Carbon Deposit Build Up on the Piston Top Land

The carbon deposit on the piston top land comes from lubricating oil or fuel. It is very difficult to distinguish the fuel soot and the carbon contribution from the lubricating oil. In most cases, the deposits are formed by fuel soot caught by the oil that adheres to the top land; carbon builds up layer by layer.

7.3.2 Quality of Fuel and Oil

Base load electrical generating sets are not viable for continuous running with high speed diesel engines except for emergency running. Heavy fuel oil or furnace oil is the cost effective fuel for this kind of application with minimum modifications to the engine components. However, adequate care has to be taken care when using heavy furnace oil (Chapter 15).

The sulfur content in diesel fuel and the alkalinity of the oil have a considerable effect on the wear of diesel engine cylinder liners and the piston assembly. A sulfur content higher than 0.5% by weight will cause increased corrosive wear, the effect being more pronounced at lower coolant temperature. The increased sulfur content causes sludge formation in oil; it becomes hard and reduces the clearance at the top land portion. High sulfur is neutralized by specifying alkalinity for the engine oil up to 30 TBN (total base number) and by maintaining the engine cooling water at a high temperature.

The contrast in specification between heavy furnace oil and high speed diesel is shown in Table 7.1. Critical parts such as the piston, ring, cylinder liner and bearing surfaces are subject to adhesive or abrasive wear. The carbon build up at the top land is very large due to carbonization of heavier fractions of fuel. The deposits add to the natural growth of the metallic crown of the piston. This growth causes contact of the piston top land at the mid-stroke of the liner, and hence peculiar wear leading to bore polishing.

7.3.3 Piston Growth by Finite Element Method

The piston is of the composite type with gallery cooling; the top portion is steel and skirt zone is aluminium alloy. A quarter model of the piston was made to predict the temperatures and deformations at different zones. The thermal deformation of the piston shape was calculated by imposing the boundary

Table 7.1 Properties of heavy furnace oil fuel and high speed diesel

Parameters	Heavy furnace oil	High speed diesel
Viscosity at 50 °C (cSt)	176.0	3.3
Density at 15 °C (kg/m^3)	961.7	837.3
Sulfur content (% wt)	3.33	0.049
Water content (% vol.)	0.2	<0.1
Carbon residue (% wt)	10.8	0.0
Ash (% wt)	0.02	<0.01
Flash point (°C)	>70	50
Pour point (°C)	6	<−6

conditions of convection coefficients, and the temperatures of the gases and the oil (Lakshminarayanan *et al.*, 2001). The mean gas temperature was calculated by a thermodynamic cycle simulation program and the heat transfer coefficient, h_m, at the crown by cycle-averaging the instantaneous heat transfer coefficient. The thermal inertia of the piston is too large to follow the high frequency fluctuations in the gas temperatures; therefore, the problem of thermal expansion was simplified to steady state heat conduction with generalized boundary conditions on the piston crown (Chapter 6). The peak cylinder pressure was imposed on the crown and at the top dead centre zone to simulate the pressure deformation of the piston. In the analysis, coupled field technique was used to analyse the structural and thermal deformation of the piston. The deformation output was used in piston transverse motion calculation.

7.3.4 Piston Secondary Movement

When the piston reciprocates in the cylinder bore, at certain times during the cycle it moves transversely across the clearance space. The transverse motion is caused by changes in the direction of the acceleration of the piston and changes in the angularity of the connecting rod mechanism. The transverse motion of the piston across the clearance space between the liners depends upon the geometry, speed and mass distribution of the system. During the exhaust and compression periods, the top of the piston moves towards anti-thrust side, while the bottom of the piston swings slightly towards the thrust side; the opposite is true during the expansion and suction periods. Thus, the transverse motion of the piston assembly causes the piston to contact with liner surfaces and induced impacts cause vibration of the liner surfaces. The governing equations for the case of one edge of the piston in contact with liner are explained below (Figure 7.2.)

For the piston:

$$m_{piston}\ddot{x}_c = F^x_{con-rod} + F^x_{contact} + m_{piston}\, g - P_{gas} \tag{7.1}$$

$$m_{piston}\ddot{y}_c = F^y_{contact} + F^y_{con-rod} \tag{7.2}$$

$$I_{piston}\ddot{\theta}_4 = F^y_{contact}\, h\cos\lambda - F^x_{contact}\, h\sin\lambda - F_{gas}(y_c - eccen) + F^x_{con-rod}\, s_4 \sin\theta_4 - F^y_{con-rod}\, s_4\cos\theta_4 \tag{7.3}$$

For the connecting rod:

$$m_{con-rod}\ddot{x}_d = F^x_{con-rod} + F^x_{crank} + m_{con-rod}\, g \tag{7.4}$$

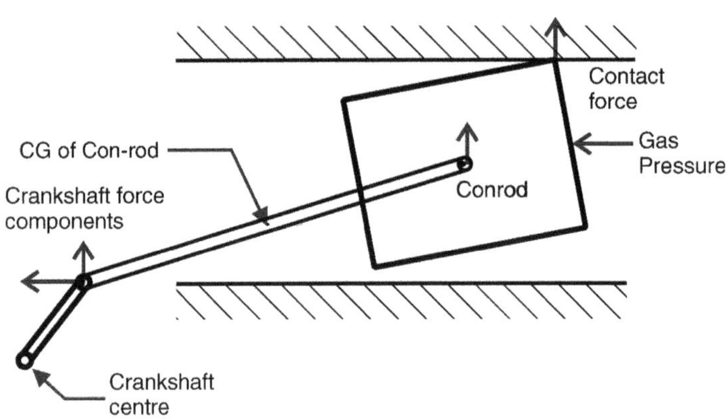

Figure 7.2 Free body diagram; one edge of the piston in contact with liner surfaces

$$m_{con-rod}\ddot{y}_d = F^y_{con-rod} + F^y_{crank} \tag{7.5}$$

$$I_{con-rod}\ddot{\theta}_3 = F^x_{crank}s_3\sin\theta_3 - F^y_{crank}s_3\cos\theta_3 + F^y_{con-rod}(l_3 - s_3)\cos\theta_3 - F^x_{con-rod}(l_3 - s_3)\sin\theta_3 \tag{7.6}$$

where eccen = eccentricity (crank offset to cylinder axis), h = the distance from the piston top to contact point, I = moment of inertia (kg m^2), F = force (N), F_{gas} = gas force = gas pressure × piston area, l_3 = connecting rod length (m), p = pressure (N/m^2), s_3 = distance between crankpin and connecting rod CG (m), s_4 = distance between piston pin and piston CG (m), θ_3 = connecting rod movement angle, θ_4 = piston tilting angle (degrees) and λ = angle component of the piston; subscripts: x = x component, y = y component and $\ddot{}$ = double differentiation of the component.

Thus, by solving the governing equations, the contact force can be calculated (Wilson and Fawcett, 1974).

7.3.5 Simulation Program

Commercially available software (AVL, 2003) was used to predict the piston secondary motion and lubrication of the pistons. The piston stiffness matrix was obtained by an iterative procedure, and for given deformations over the piston height the nonlinear stiffness behaviour was approximated by a third order polynomial function. The equilibrium forces between piston, liner and the pin bearings, and hence the piston motion, were determined for a given piston profile and gas pressure. The oil films and kinetic energy of the piston were also calculated using this tool.

7.4 Wear Model

Bore polishing of the liner is a mode of material wear occurring at the contact surface when the top land has grown due to the temperature and hard carbon deposits slide on the liner surface. As seen in Figure 7.3, the sliding of an asperity against the mating carbon particles gives rise to a frontal wave of the deformed material subjected to compressive stress. The cyclic or repeated distortion of the surface at the microscopic level leads to the fatigue failure of the roughness that produces a mirror finish.

7.4.1 Contact Pressures

The contour contact area is strictly dependent on contact of the asperities that occur in the elastic or plastic regime of the materials. The liner and the hard carbon layer on the piston approach each other when very

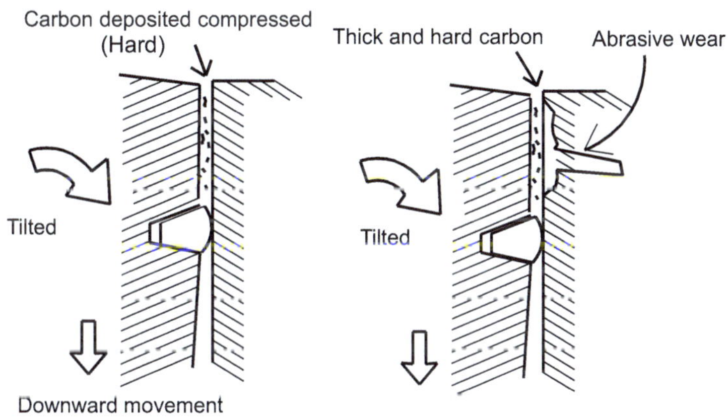

Figure 7.3 Bore polishing mode of liner wear

tight clearance exist between them, allowing the atoms and molecules to interact. When gaseous force is applied, the compressive force normal to the contact displaces the asperities tangentially. Fatigue and later detachment are caused with increased slipping in the contact region.

In the present analysis, the contact pressure was predicted using the Finite Element Contact Pair Static analysis proposed by Tomanik and Nigro (2001), with the consideration of contact force between the piston and the liner and the rigidity boundary conditions at the liner resting face.

7.4.2 Wear Rate

The wear of rubbing surfaces can be expressed as a linear wear. From fatigue theory of different types of wear, the linear rate of wear is proportional to the contact pressure, pc. The nondimensional wear rate is:

$$M = k\,p_c^{m1} \tag{7.7}$$

where $m_1 = 1$–3 ($= 1$ for well run-in surfaces); $k =$ wear coefficient that characterizes the wear resistance of the material, geometry of the surfaces in contact and lubrication; $p_c =$ contact pressure.

The wear between two rubbing surfaces is calculated as:

$$W = M\,L; L = V\,t \tag{7.8}$$

where W = wear (microns), M = nondimensional wear, L = sliding distance (mm) = V.t, V = sliding velocity (mm/s) and t = number of hours run, converted to seconds.

7.5 Calculation Methodology and Study of Bore Polishing Wear

High oil consumption was noticed in the field on a large bore, low speed, highly loaded engine after a relatively short running period of 4000 hours (Table 7.2.) The liner samples showed shiny marks on the liner top and mid portions (Figure 7.4). To understand and solve the polishing problem the calculations were carried out in the following steps, as shown in Figure 7.5.

7.5.1 Finite Element Analysis

Initially, the piston was studied for thermal and pressure deformation to determine the expansion under hot running conditions. Similarly, cylinder bore expansion was studied to predict the bore profile. In addition, the radial elasticity of the piston was calculated along the axial height and at different angular positions for stiffness matrix generation; the data obtained were used in the simulation (AVL, 2003) for the contact force calculation.

Table 7.2 Engine specifications

Engine type and aspiration	Inline, Turbo After cooled
Bore (mm)	280
Stroke (mm)	350
Connecting rod length (mm)	570
Number of cylinders	9
Rating (hp)	3600 at 750 rpm
Brake mean effective pressure (MPa)	2.23
Peak firing pressure (MPa)	14
Application	Power generation and marine

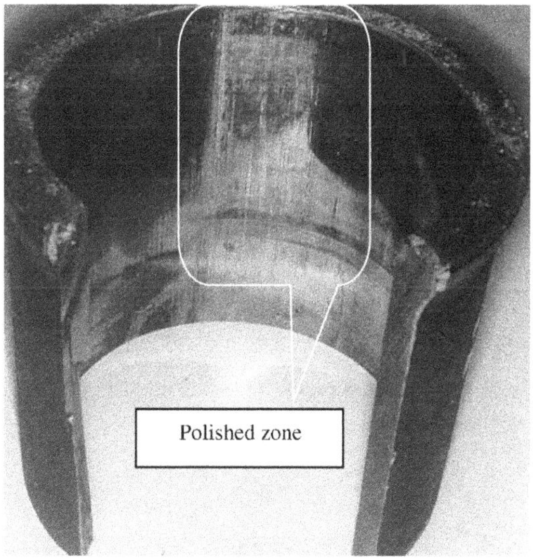

Figure 7.4 Photograph of bore polished liner

From the calculated contact force, the contact pressure distribution on the bore surface was determined using finite element nonlinear contact pair analysis. In the simulation, the bore surface was treated as the target area and the piston surface as the contact area. The rigidity boundary condition was applied on the liner resting face and the radial fixity on the piston pin bore zone. The applied contact force on the piston predicts the contact pressure on the liner, and was used as input in the liner wear calculation. The magnitude of the contact force varies with transverse movement of the piston, and is predominant at the liner top because of the gas pressure loads.

7.5.2 *Simulation*

In this simulation, the piston was considered as an elastic body. The liner, connecting rod and crankshaft were considered as rigid bodies. The gaseous pressure at different crank angle positions was calculated from the engine simulation program and compared with measured data. The piston geometry, mass and inertia parameters were input to simulation to calculate the piston secondary movement (Table 7.3). The analysis was carried out for two different cases, namely (a) running design, that is, piston with carbon build up at the top land, and (b) for the modified piston assembly and liner with a scraper ring, which discourages the carbon build up.

The simulation gave the contact force of the piston on the liner thrust and anti-thrust sides, sliding velocity and kinetic energy loss of the piston due to impact.

The wear rate on the contact surface of the liner was calculated by using a simple linear wear relation, that is, the wear depth is directly proportional to wear coefficient, the mean contact pressure and the sliding velocity of the piston. The wear coefficient characterizes the significant changes on the topography of the surfaces; under thick oil film conditions no wear is expected, a wear coefficient value of zero was assumed; for very thin oil films, the value is dependent on the boundary lubrication, lubricant quality, material properties and operating conditions (Figure 4.5, Chapter 4). A suitable value for the wear coefficient was used in the simulation.

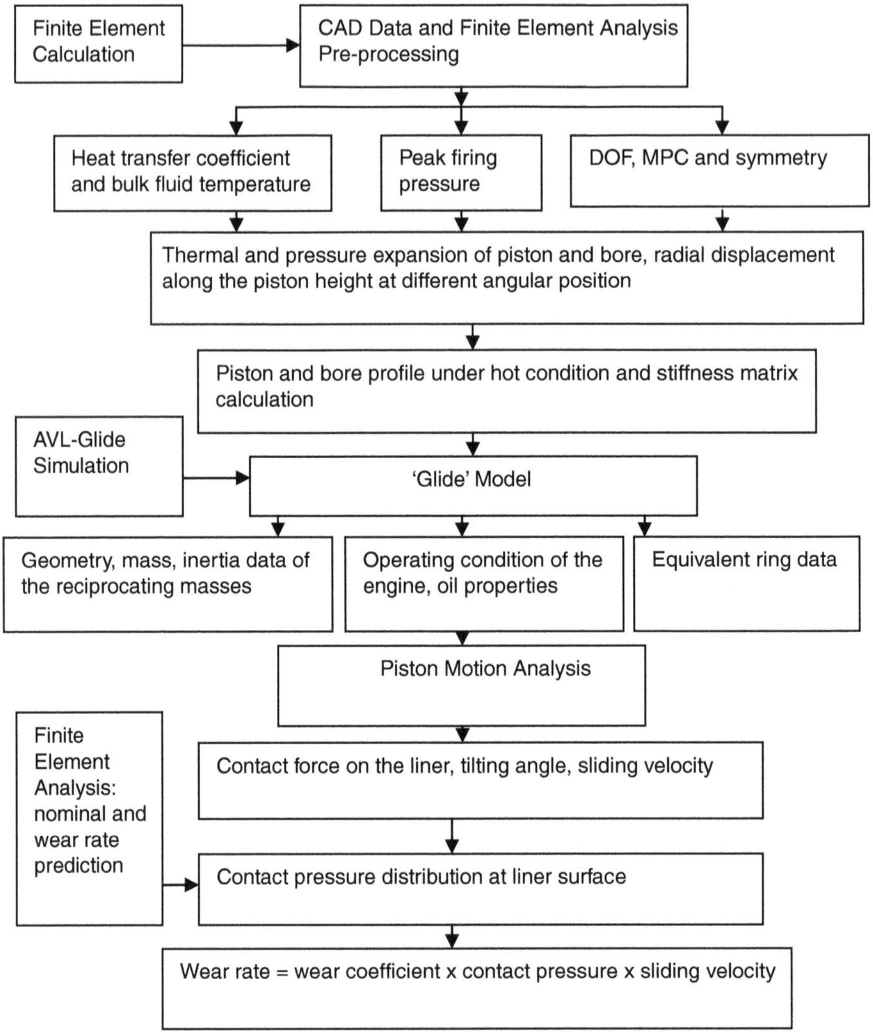

Figure 7.5 Schematic layout of the calculation methodology

7.6 Case Study on Bore Polishing Wear in Diesel Engine Cylinders

7.6.1 Visual Observations

Following observations were made on seven worn out liners and pistons from the engine mentioned in Table 7.1. A thick carbon deposit layer on the piston top land is shown in Figure 7.6. The surface of the deposit was smooth. The worn out liner surface was shiny without any cross-hatch angles, as shown in Figure 7.4. The wear was severe at the top ring travel zone and mid portion of liner, which confirmed the piston top land contact with the liner surface at the top and middle portions due to insufficient operating clearance.

The detailed visual study on the worn out liners indicated that the wear was due to contact by the top land of the piston with the liner throughout the stroke. The simulation of piston motion after

Table 7.3 Geometry, mass and inertia properties – piston assembly and connecting rod

Piston total height (mm)	330
Distance from piston pin centre to top (mm)	209
Piston type	Composite, i.e., steel crown, aluminium skirt
Piston weight (kg)	42.22
Piston CG location from pin centre (mm)	118.72
Piston moment of inertia with reference to CG (kg mm^2)	489 790
Connecting rod mass (kg)	56.75
Connecting rod CG location from crank centre (mm)	212.93
Connecting rod moment of inertia with reference to CG (kg mm^2)	4 364 200
Top compression ring weight (kg)	0.35
Distance from pin centre to ring CG (mm)	170

allowing for the thermal expansion and the layer of carbon on the top land gives input to the wear rate calculations.

7.6.2 Liner Measurements

Fresh and worn out engine liners were studied for roundness and straightness on a form tester (Figure 7.7). The measured roundness data indicate that the fresh liner showed an average $P + V$ (peak + valley) value of 6.35 μm while for the worn out liner the value was 30 μm.

7.6.3 Results of Finite Element Analysis

In the piston assembly, the piston top land and liner diametrical cold clearance was 1.16 mm, which is the optimum clearance when the engine ran with low sulfur high speed diesel fuel. However, when heavy furnace oil fuel was used it created carbon deposits 0.35 mm thick on the piston top land. The location of

Figure 7.6 Field engine piston – hard carbon at the top land

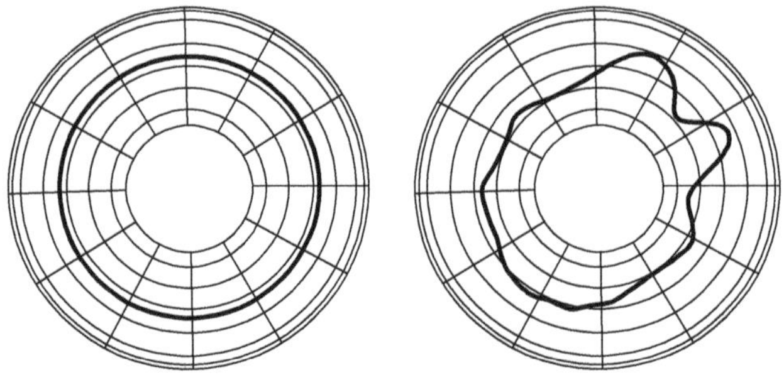

(a) Fresh liner – no waviness

(b) Field liner – uneven shape, 170 mm from liner top. Wavy wear out according to spray impingement at the edge of the piston crown

Figure 7.7 Liner measurements – roundness data

(a)

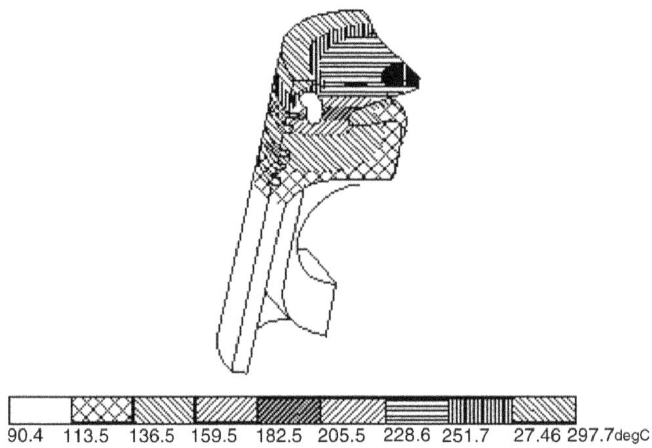

(b)

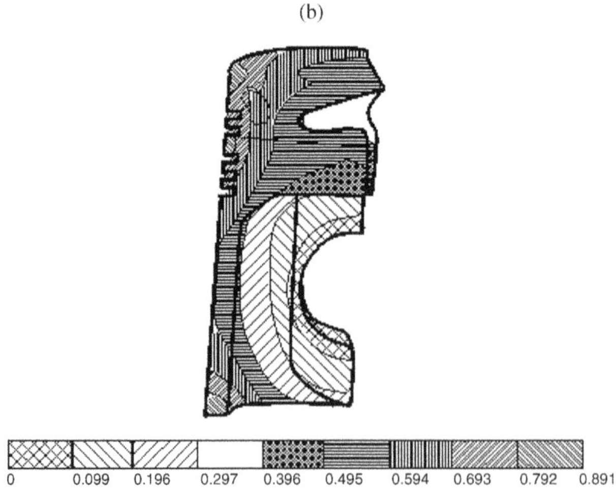

Figure 7.8 Typical FEM results: (a) temperature contour and (b) expansion (thermal + pressure + inertia) plots

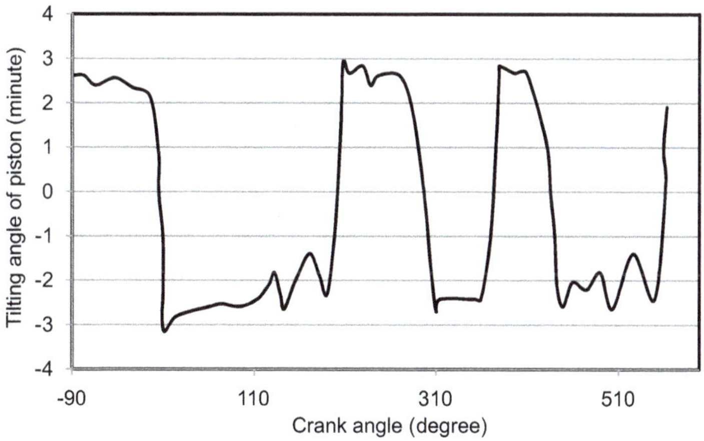

Figure 7.9 Tilting angle of piston with reference to crank angle

the concentrated deposits was spray impingement spots on the piston top where temperatures are high. The piston profile under hot conditions was calculated (Figure 7.8). At the maximum temperature of 298 °C at the bowl rim area, the expansion at the top land zone is 0.89 mm (deformation due to temperature, pressure, inertia forces). This showed that the clearance is completely eaten away under running conditions by the carbon build up and that the contact with the liner by the top land is imminent.

7.6.4 *Piston Motion*

The rotation of the piston, represented as tilting angle with respect to time (crank angle) can be seen in Figure 7.9. The positive angles indicate that the top of the piston is tilted towards the anti-thrust side. The tilt is positive during the upward strokes and negative during the downward strokes.

The loss of clearance between the piston top land and the bore during running conditions leads to rupture of the hydrodynamic lubricating film. The hard carbon layer on the land destroys the soft liner surfaces by its adhesive or abrasive action. Figure 7.10 shows the variation of the contact force against

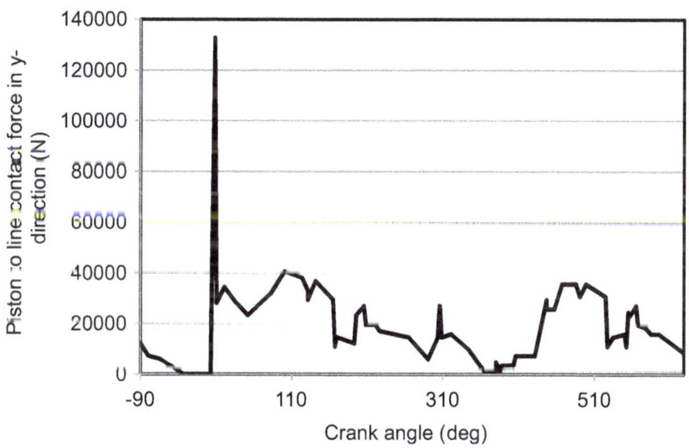

Figure 7.10 Contact force distribution (thrust side)

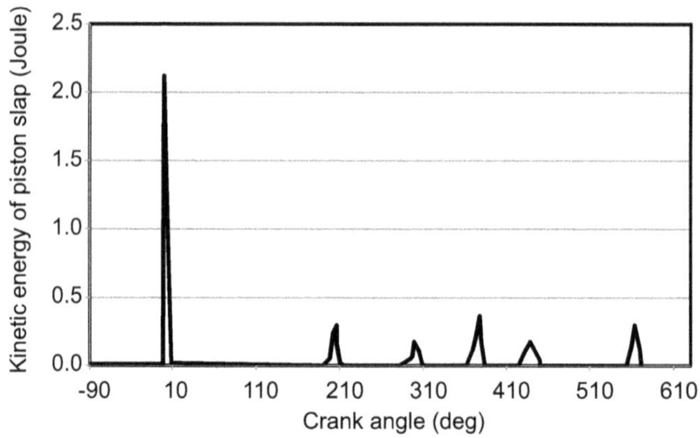

Figure 7.11 Cyclic plot of kinetic energy of piston

crank angle. It is predominant during the beginning of the expansion stroke due to the sudden rise in the cylinder pressure and the reduced axial velocity at the end of the compression stroke. The force on the anti-thrust side of the liner is substantially lower than that on the thrust side. Hence, higher wear is observed on the thrust side than on the anti-thrust side.

The contact forces are high between 80° and 140° CA in the mid-stroke on the liner where the piston velocity is high; consequently, larger wear patches are seen in this zone of the liner.

Along the direction of secondary movement, the piston remains stationary while sliding on the cylinder liner. At top dead centre positions, the piston begins to swing from the anti-thrust side to the thrust side due to the increase in gas pressure. The large force of the connecting rod causes heavy impacts on the

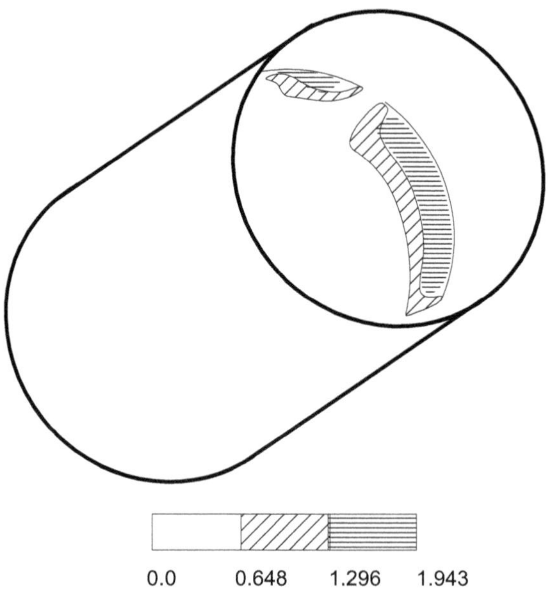

Figure 7.12 Contact pressure distributions on the liner (at middle position by finite element nonlinear analysis

Table 7.4 Contact pressures calculated by finite element analysis

Axial distance from the liner top (mm)	Contact pressure (MPa)
15.0	15.50
30.5	5.85
67.2	3.63
186.3	5.23
247.0	5.83
318.0	4.18
357.6	4.84
365.0	1.77

liner (Figure 7.11). In addition, at the end of the expansion small peaks are noticed, due to inertia of the piston. The same is true when the suction stroke begins; however, the impact is less, corresponding to lower pressure in the cylinder. Figure 7.11 shows the impact locations and kinetic energy of the piston during the secondary movement. The latter can be used to calculate engine vibrations due to the piston slap.

7.6.5 Wear Profile

From the predicted contact forces, the distribution of pressure on the liner was calculated using finite element nonlinear analysis (Figure 7.12). The maximum contact pressure somewhere in the middle of the upward stroke is 5.83 MPa. The distribution is not uniform around the periphery of the piston crown (Figure 7.7) because of the nonuniform collection of carbon deposits, more near the (hot and sooty) spray tips, and due to varying clearance about the periphery of the piston top land. In Table 7.4 the calculated contact pressures at different axial locations on the liner are given.

Figure 7.13 illustrates the sliding velocity of the piston at pin centre; minimum sliding velocity can be observed at the liner top.

Seven liners worn severely by bore polishing were measured for wear and used to validate the model. The wear profile was predicted using simple wear relation, from Equations 7.7 and 7.8:

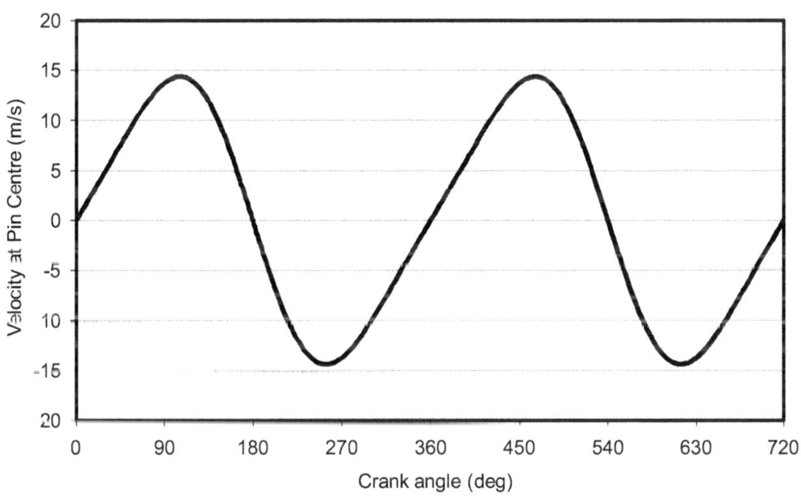

Figure 7.13 Sliding velocity of the piston

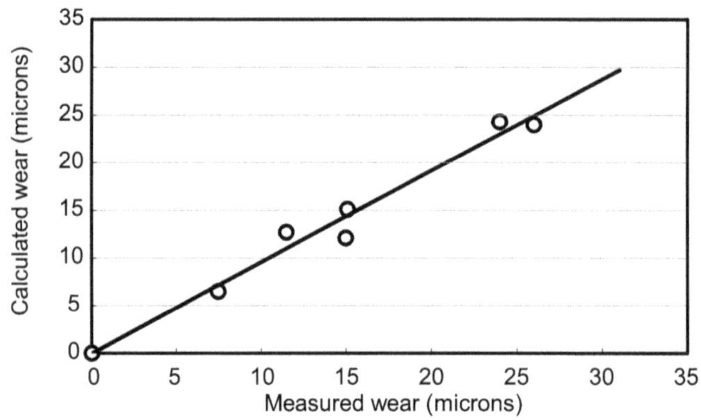

Figure 7.14 Measured and calculated liner wear after 6000 hours (optimized wear coefficient: 1.4×10^{-11} mm^3/mm/N)

$$\text{Wear} \sim \text{the contact pressure} \times \text{the sliding velocity}$$

In the wear simulation, a coefficient value of 1.4×10^{-11} mm^3/mm/N was used to calculate the wear using the sliding velocity and the contact force. Figure 7.14 shows the good comparison of the experiments with the estimated wear at different points on the liner. In Figure 7.15, the measured wear profile along

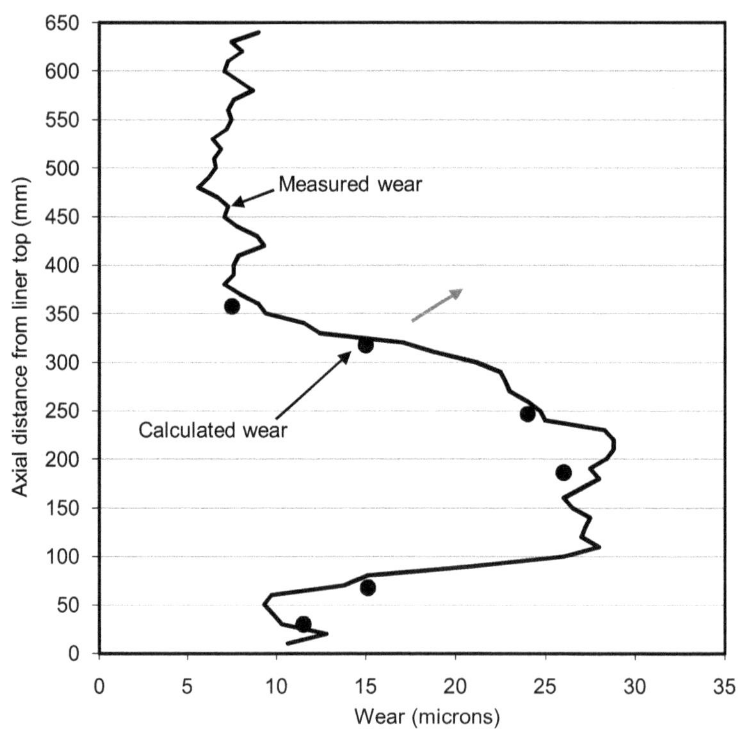

Figure 7.15 Wear profile along the liner axial position after 6000 hours

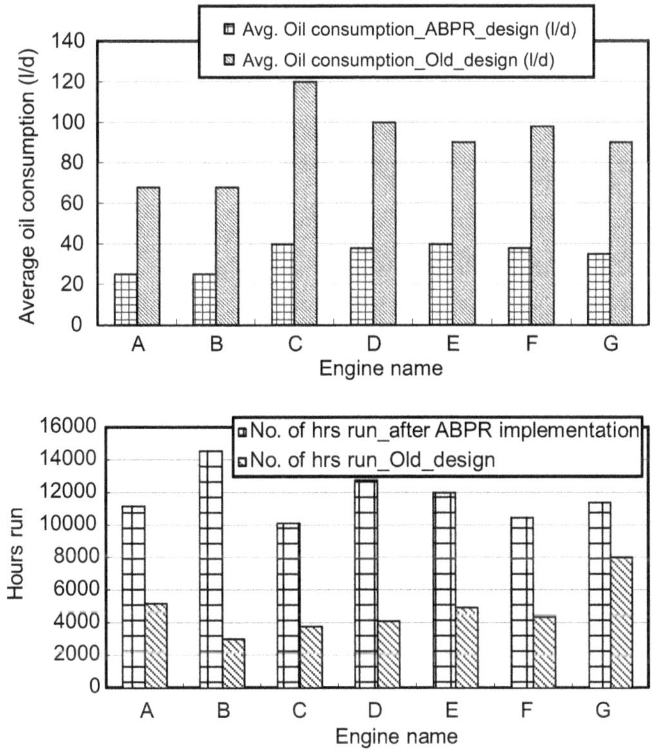

Figure 7.16 (a) Engine oil consumption and (b) engine life study with bore polishing and without bore polishing (when a scraper ring is used)

the liner axial position is superimposed on the calculated profile. Higher wear is observed in the range of 100–350 mm.

7.6.6 Engine Oil Consumption

The carbon layer built up on the top land polished the liner (Figure 7.4). The honing pattern disappeared, and hence the oil retention on the liner surface as well. Further, the bore lost its circularity due to nonuniform wear around the periphery and along the length. For this reason, the piston ring assembly lost 'light tightness' in the liner that would control the oil consumption. The engines with polished liners showed higher oil consumption (of 120 litres/day) than normal (30 litres/day) (Figures 7.16a and 7.16b).

7.6.7 Methods Used to Reduce Liner Wear

Polishing wear is predominately due to the loss of top land clearance under running conditions. To provide a sufficient clearance at the top land and to prevent the contact with the liner when it tilts a cut back on the piston top land is employed. With this concept substantial improvement of the engine oil consumption and the liner wear could be obtained (Ishizuki, Sato and Takaso, 1981). However, it leads to deterioration in the engine performance and smoke. Therefore, a scraper ring ('anti-bore polishing' ring of lamellar

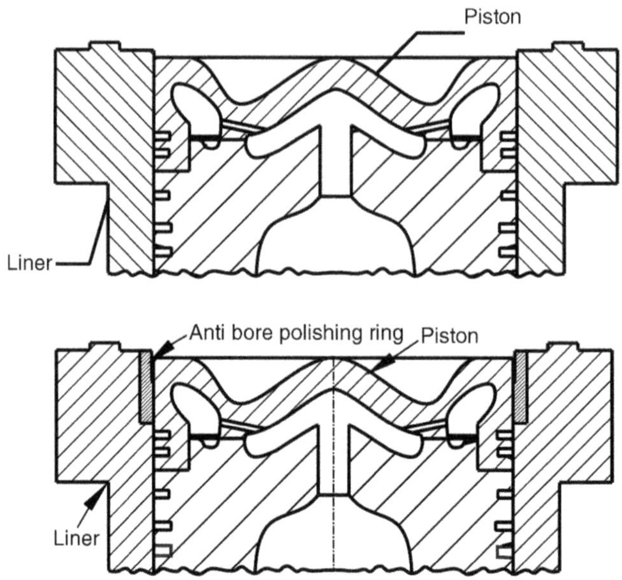

Figure 7.17 Liner designs (a) without and (b) with anti-bore polishing ring (scraper ring)

graphite cast iron) is designed at the top of the liner that allows sufficient clearance in the mid-stroke to avoid bore polishing wear and at the same time scrape away any build up of carbon at the top land (Figure 7.17b). In addition, it keeps the dead volumes at the top dead centre low so that the performance is not lost.

When the piston top land cold clearance is cut back by 90%, the contact at the mid portions of the liner completely disappears, even with some carbon build up (Figure 7.18.) The oil consumption behaviour substantially improves; it was 30 litres/day for the entire operating life of 20 000 hours, while without the scraper ring the life was only 3000 hours (Figures 7.16 and 7.17).

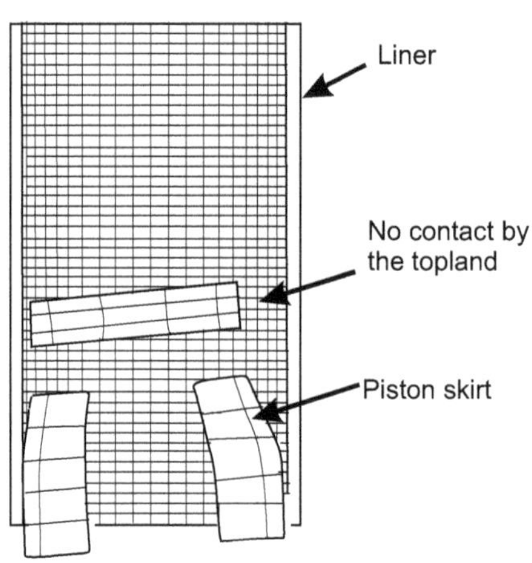

Figure 7.18 Modified design – no contact at the mid portion of the liner

7.7 Summary

Cylinder liner polishing by carbon was found to be a major cause of the increase in engine oil consumption and the mode of wear was studied in detail in relation to the transverse motion of the piston and the carbon deposit build up on the piston top land.

The cut back piston can be used to prevent carbon polishing wear. However, it adversely affects engine performance. A scraper ring inserted on the top of the liner area avoids bore polishing, higher engine oil consumption and loss of performance. Engine operating life is improved more than five times.

The wear model using the finite element method and simple linear wear relationship (Nayak *et al.*, 2004) predicted the wear behaviour of the liner correctly for both the standard piston and the cut back piston.

References

Al Khalidi, G.F. and Eyre, T.S. (1987) Bore polishing – Identification and simulation, *Tribology International*, **20** (1), 1987, 18–24.

AVL (2003) AVL GLIDE Reference Manual. AVL, Graz, Austria.

Berbezier, Martin, J.M., Kapsa, P. and Mansot, J.L. (1989) Higher resolution analysis of the mechanism of bore polishing. Proceedings of the 16th Leeds–Lyon Symposium on Tribology, 5–8 September 1989, Lyon, France.

Bishop, G.R. and Leavitt, A.H. (1975) Performance simulation of a diesel piston and ring system. Technical Paper 750768, SAE, Troy, MI.

Haddad, S.D. (1995) Theoretical treatment of piston motion in IC piston engine for the prediction of piston slap excitation. *Mechanism and Machine Theory*, **30** (2), 253–269.

Haddad, S.D. and Tjan, K.T. (1995) An analytical study of offset piston and crankshaft Designs and the effect of oil film on piston slap excitation in a diesel engine. *Mechanism and Machine Theory*, **30** (2), 271–284.

Ishizuki, Y., Sato, F. and Takaso, K. (1981) Effect of cylinder liner wear on oil consumption in heavy duty diesel engines. Technical Paper 810931, SAE, Troy, MI.

Lakshminarayanan, P.A., Nayak, N.S., Talwalkar, S. *et al.* (2001) Generalised boundary conditions for diesel engine pistons. Technical Paper 2001-28-0009, SAE, Troy, MI.

Laws, A.M., Parker, D.A. and Turner, B. (1973) Piston movement in the diesel engine. Paper No. 33, CIMAC Conference, Washington, DC.

McGeehan, J.A., A single-cylinder high bmep engine for evaluating lubricant effects on piston deposits, ring wear, oil consumption, and bore polishing. Technical Paper 800437, SAE, Troy, MI.

McGeehan, J.A. (1983) Effect of the piston deposits, fuel sulfur, and lubricant viscosity on diesel engine oil consumption and cylinder bore polishing. Technical Paper 831721, SAE, Troy, MI.

Nayak, N.S., Lakshminarayanan, P.A., Gajendra Babu, M.K. and Dani, A.D. (2004) Bore Polishing Wear in Diesel Engine Cylinders. Proceedings of ESDA04, 7th Biennial Conference on Engineering Systems Design and Analysis, 19–22 July 2004, Manchester, UK (Paper No. ESDA 2004–58526).

Ting, L.L. and Mayer Jr, J.E. (1974) Piston ring lubrication and cylinder bore wear analysis: Part 1 – Theory. *Journal of Lubrication Technology*, **96**, 305–314.

Tomanik, E. and Nigro, F.E.B. (2001) Piston ring pack and Cylinder wear modelling. Technical Paper 2001-01-0572, SAE, Troy, MI.

Weiss, F.K.J., Busenthuer, B.B. and Hardenberg, H.O. (1987) Diesel fuel sulfur and cylinder liner wear of a heavy-duty diesel engine. Technical Paper 872148, SAE, Troy, MI.

Wilson, R. and Fawcett, J.N. (1974) Dynamics of the slider-crank mechanism with clearance in the sliding bearing. *Mechanism and Machine Theory*, **9**, 61–80.

Wong, V.W., Tian, T., Lang, H. and Ryan J.P. (1994) A numerical model of piston secondary motion and piston slap in partially flooded elastohydrodynamic skirt lubrication. Technical Paper 940696, SAE, Troy, MI.

8

Abrasive Wear of Piston Grooves in Highly Loaded Diesel Engines[1]

The rate of wear of piston ring grooves is mainly a function of the peak cylinder pressure, hardness, surface roughness, depth of penetration and the hard particles like soot produced by combustion or quartz (sand) entering through the air filter. The problem becomes acute as the blow-by past the rings and oil consumption increase with the wear of the grooves and the rings. A method is presented in this chapter to estimate the wear of piston grooves quantitatively. The wear rate is correlated directly with the gaseous load, amount of hard particles past the piston lands and radius of abrasive particles, and inversely with the hardness of the groove surface.

8.1 Introduction

Wear is injurious, as it reduces the operating efficiency; it increases the power losses, oil consumption and, ultimately, ends in failure of the engine parts. Any change in load, speed or environment can cause a dramatic change in the wear rates of the contacting surfaces. In some cases, increasing hardness will reduce the wear. However, greater improvements can be obtained by changing the operating conditions than by changing the material.

Abrasive wear occurs when there is friction between a metal stressed under a harder abrasive body, leading to surface destruction by cutting, scratching and plastic deformation. A semi-quantitative model by Suh, Sin and Saka (1978) explains the dependence of abrasive wear rate on the size and relative sharpness of the grit. Davis and Eyre (1990) studied abrasion and delamination wear in the piston, rings and cylinder liner of an engine, and demonstrated that friction modified oils give better wear resistance between two contacting bodies. Mashloosh and Eyre (1985, 1986) studied the effect of the angle of attack by the abrasive particle and the physical properties of the metal on wear. The type of debris produced by abrasion clearly indicated the mechanism of deformation and fracture occurring in the wear process. It was concluded that the wear is directly proportional to load and inversely related to hardness, but not dependent on the sliding distance because of the degradation of the abrasive surface. Rabinowicz and Mutis (1965) looked at the abrasive particle size that results in maximum abrasive action. Hirano and Levy (1978) correlated the abrasive wear of metals with hardness and yield strength. Sarkar (1976), on the

[1]All figures in this chapter are reprinted with permission from N.S. Nayak, P.A. Lakshminarayanan, M.K. Gajendra Babu and A.D. Dani, "Abrasive wear of piston grooves of highly loaded diesel engines," Paper Number ESDA2004-58520, ASME 7th Biennial Conference on Engineering Systems Design and Analysis (ESDA) 2004.

Critical Component Wear in Heavy Duty Engines, First Edition. P.A. Lakshminarayanan and Nagaraj S. Nayak.

other hand, has presented a simple model for abrasive wear of metals as a function of total normal load, density of metal being abraded, friction and maximum possible strain energy of the metal. The deformation work varies from metal to metal and it is about 1% of stored strain energy. Golothan (1978) described piston ring groove wear of diesel engines and indicated the main causes for groove wear to be abrasion and corrosion. The study revealed that abrasion wear is caused by dust in the intake air and hard deposits derived from combustion of fuel or lubricating oil.

In this chapter, the wear of piston grooves where abrasive particles enter between the two sliding surfaces (ring and piston), and allow the softer surfaces to undergo plastic deformation, is studied. Under abnormal conditions, the deformed material will chip off from the parent surface, as a function of cyclic loading by hard carbon particles.

8.2 Wear Phenomenon in Piston Grooves

8.2.1 Abrasive Wear

The mechanism of abrasive wear is the detachment of material from the original surface by the unidirectional sliding action of hard discrete particles of another material. In general, parts such as the piston, ring, cylinder liner and bearing surfaces undergo this mode of wear either due to abrasive particles entering through the air and oil filters or to loose hard carbon produced during incomplete combustion of oil and the heavier fraction of the fuel. The number of particles passing through the filters is a function of the filtration efficiency for a given type of filter and the concentration of abrasive particles at the inlet to the filter. The latter is a function of environment. Newer types of filters have been introduced to improve filtration efficiency, at the same time reducing the pressure drop across the filter.

8.2.2 Wear Mechanism

8.2.2.1 Three-Body Abrasive Wear in Piston Grooves

The loose hard particles entering the sliding interface act as grit and the process of metal removal by these has been termed three-body abrasive wear. An examples of this type is abrasion due to trapped hard carbon particles in a piston groove (Figure 8.1).

When an abrasive grain slides against the target material at a constant velocity under the action of load, the surface layer deforms first because the applied load is below the yield limit of the surface (Figure 8.2a); subsequently the softer layers below the first layer deform. Thus, the elastic deformation spreads uniformly over a large volume, without forming any microchips. Under abnormal conditions of high

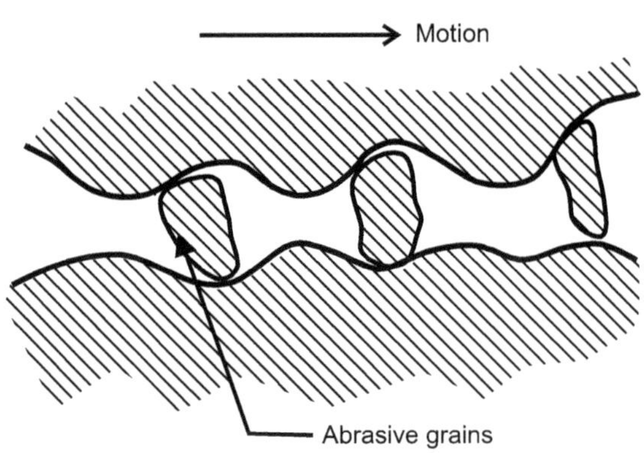

Figure 8.1 Diagram representing three-body abrasive wear

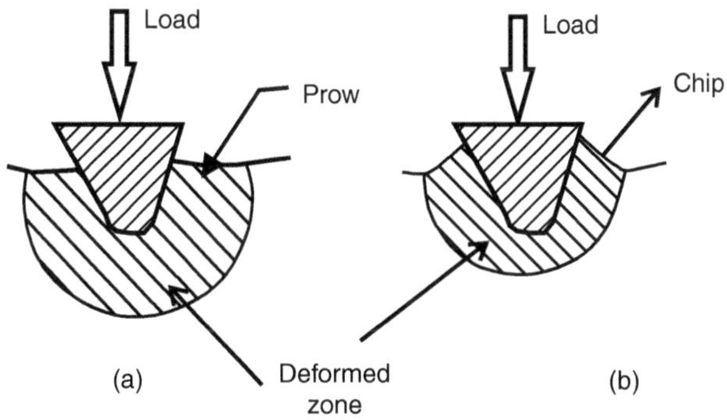

Figure 8.2 A schematic illustration of the localization of deformation underneath the abrasive grain: (a) spreading of elastically deformed zone and (b) localized plastically deformed zone

applied load, the deformed material softens as the shear strain due to friction exceeds the plasticity limit of the material. A layer of material just underneath the abrasive grain deforms first and delaminates superficially (Sundararajan, 1985). The increased slipping and shaking within the layers lead to localized deformation near the surface and the formation of microchips (Figure 8.2b). Finally, the chip detaches from the parent surface, thus the surface wears.

Wear of Piston Grooves

The carbon deposit on the top land comes from incomplete combustion of lubricating-oil and the fuel, which is trapped between ring and groove surfaces. Alternatively, the dust in air taken in can be trapped here. Under cyclic loading, the compact harder particles wrap the groove surface, and impart contact pressure between the materials. The relative sliding motion ploughs the softer surface, Figure 8.3.

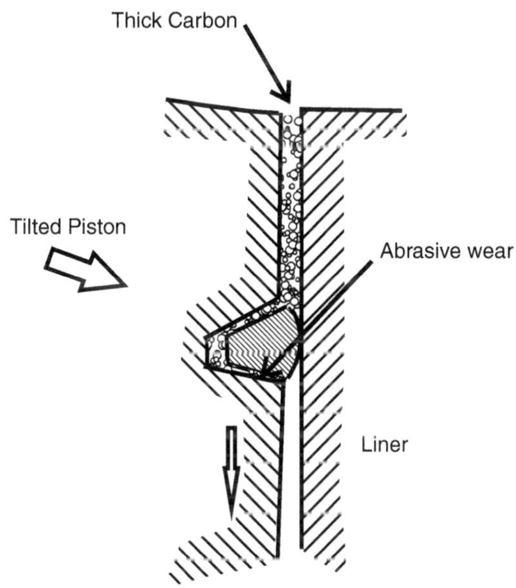

Figure 8.3 Schematic diagram of the abrasive mode of piston groove wear

Table 8.1 Hardness of materials

Material	Hardness (N/mm^2)
Hard carbon particles – fuel and oil composition	800
Steel alloys	650
Aluminium alloys	220

8.3 Wear Model

8.3.1 Real Contact Pressure

The pressure at the top ring groove zone is nearly equal to the combustion chamber pressure, since the lubricant occupies the clearance between the ring, piston groove and the liner surfaces, allowing no gas to pass past the contact faces. The gas leakage is restricted through an orifice formed by the radial gap between the ring and cylinder liner. The inter ring-groove pressure between the rings is obtained using quasi-steady approach with orifices that are in series.

The gaseous load on the piston top land area allows the hard carbon particles to slide on the softer piston groove area. The relative velocity of the abrasive particles produces long parallel scratches on the groove surfaces. The normal load per unit area of the real contact is the real pressure, p_r, and, in the plastic deformation condition, it is proportional to the material hardness, H (Table 8.1):

$$p_r \approx H^{1/k} \tag{8.1}$$

where k = constant for hardness, which varies from material to material (= 8 for wear of hard steel surfaces with hard abrasive particles, = 2 for wear of soft aluminium surfaces with hard abrasive particles) and H = hardness.

8.3.2 Approach

Under the cyclic loading of the combustion pressure, the hard carbon particles detach from the piston top land surface, pass the groove zone to abrade the groove surfaces and cause the surfaces to undergo plastic deformation.

When the average normal stress in the areas of contact between the asperities reaches the value of the Brinell hardness of the ring groove material, then abrasive grain penetrates the surfaces of the softer groove material. The magnitude of approach determines the friction and wear of the contacting surface:

$$\frac{h}{r} \approx 5.4(1-\mu^2)^2\left(\frac{HB}{E}\right)^2 \tag{8.2}$$

Where μ = Poisson's ratio, HB = Brinell hardness of the groove surface (N/mm^2), E = Young's modulus (N/mm^2), H = approach (micron) and R = radius of the deformed asperities (micron).

8.3.3 Wear Rate

The shape of the embedded abrasive grains on the groove surface can be approximated to small hemispherical segments, as shown in Figure 8.4. The nondimensional linear wear rate, I, can be written

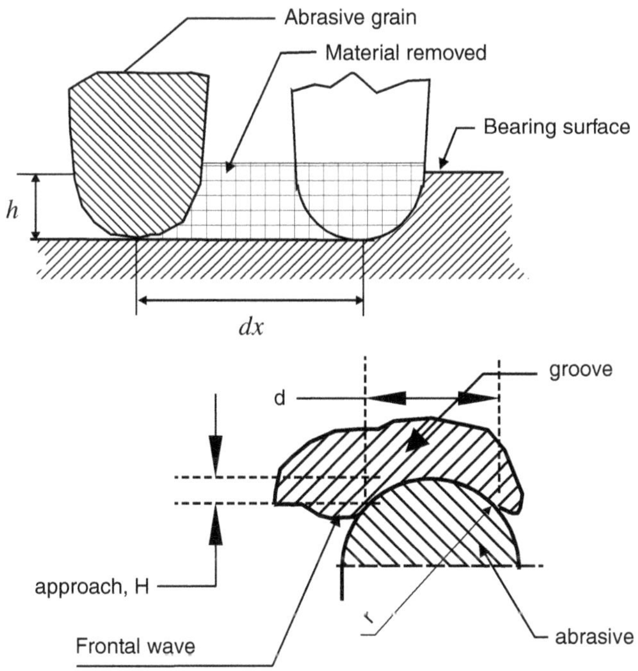

Figure 8.4 Model for abrasive wear

as a ratio of volume, U, removed per unit area of apparent contact to the total distance, L, travelled by the wear surfaces:

$$I = \frac{V}{(A_a L)} = \frac{U}{L} \tag{8.3}$$

where A_a = apparent area of contact and V = volume of material removed.

Since wear takes place over the real area of contact, A_r, Equation 8.3 can be written as analogous for specific wear per cycle:

$$i_h = \frac{Vd}{A_r d} \cong \frac{h}{d} \tag{8.4}$$

where h = depth of penetration and d = diameter of the deformed asperity.

Substituting Equation 8.4 in Equation 8.3:

$$I = i_h \frac{A_r}{A_a} = i_h \frac{P_a}{P_r} \tag{8.5}$$

The diameter of the deformed asperity undergoing plastic deformation is calculated as:

$$d \approx \sqrt{2rh} \tag{8.6}$$

$$\frac{h}{d} \approx 0.7\sqrt{\frac{h}{r}} \tag{8.7}$$

Table 8.2 Values of μ_{ml} and t_f

Material	t_f	μ_{ml}
Aluminium alloys	7.0	0.12
Steel	9.5	0.14

From Veler's fatigue law, the number of cycles to cause fatigue rupture of roughness peaks is calculated as:

$$f = \left(\frac{\sigma_{aff}}{\sigma_0}\right)^{t_f} \tag{8.8}$$

$$\sigma_{eff} = K_3 \mu_{ml} p_r \tag{8.9}$$

where f = number of cycles causing fatigue failure of soft surfaces σ_0 = ultimate tensile stress (N/mm^2), σ_{eff} = effective stress (N/mm^2), K_3 = factor characterizing the contact stress condition, μ_{ml} = molecular coefficient of friction and t_f = frictional fatigue parameter.

The molecular coefficient of friction and frictional fatigue parameter are different for aluminium and steel alloys (Table 8.2) (Kragelskii and Alisin, 1981).

For multiple contacts, the simplified forms of non-dimensional wear rate, I, is expressed by the equation below (for ingress of n abrasive particles).

$$I = K_1 \frac{n\ R\ p_a}{f\ R_a\ p_r} \sqrt{\frac{h}{r}} \tag{8.10}$$

where K_1 = determined by geometric shape of abrasive grain ≈ 0.15–0.2 for hard carbon particles, N = number of abrasive particles, R = radius of abrasive particles past the piston groove surfaces (micron) and R_a = average roughness parameter (micron).

The wear of soft piston groove surfaces under the action of abrasive grain, W, is:

$$W = I\,L \tag{8.11}$$

where W = wear (microns), I = nondimensional wear and L = sliding distance traversed by the abrasive particles (microns).

The distance traversed by the abrasive particles is of the order of the diametrical gap between the liner and piston under hot conditions:

$$L = K_2 N\, t\, d_c \tag{8.12}$$

where K_2 = constant derived from abrasive particles entrapment at the ring groove surfaces ≈ 125, N = engine speed (rpm), T = engine run time (minutes) and d_c = diametrical clearance between piston and liner under hot conditions.

8.4 Experimental Validation

8.4.1 Validation of the Model

8.4.1.1 Carbon Deposit Build Up on Piston Top Land

The carbon deposit on the piston top land comes from the lubricating oil and fuel, and it is very difficult to distinguish between the fuel carbon and the soot contribution from the oil. The deposits are formed by fuel soot caught by the oil adhering to the top land and the thick deposits significantly increase.

Table 8.3 Engine specifications

Engine details	
Number of cylinders	9
Application	Generating set/Marine
Bore × Stroke (mm)	280 × 350
Compression ratio	11
Rated power (kW)	295
Rated speed (rpm)	750
Brake mean effective pressure (MPa)	2.22
Peak cylinder pressure at top land (MPa)	14.0
First and second groove material	Steel
3rd, 4th and 5th groove material	Aluminium alloy

As an example, a piston used in an engine (Table 8.3) running on heavy furnace oil is considered. Due to incomplete combustion of the heavier fraction of the furnace oil, carbon is formed excessively near the piston lands. Piston, ring grooves and ring surfaces suffer abrasive wear. The wear increases oil consumption and blow-by.

8.4.2 *Wear Measurement*

A surface analyser was used to measure the average roughness of the piston grooves. The measured values were used in the wear relationship to predict the wear depth. The piston top groove showed an average roughness of 1.5 μm after wear and the fresh pistons were 0.8 μm rough. In addition, roundness of the piston grooves was measured (Figure 8.5). An average P + V (peak + valley) of 6 m for the fresh piston and 15 μm for the worn piston were observed.

The abrasive particles' shape, size and quantity play a complex role in the wear rate calculation. The deposits of abrasive particles typically on a piston top groove of a large engine are shown in the top photograph of Figure 8.6. The particles are a mixture of hard carbon, silica particles and silica coated with hard carbon. A sample of the particles collected from the top groove is washed and shown in the bottom photograph of Figure 8.6. The particles are hard and transparent like plain glass. The edges of the particles have been partly rounded due to three-body wear. The quantity and size of carbon particle deposition on

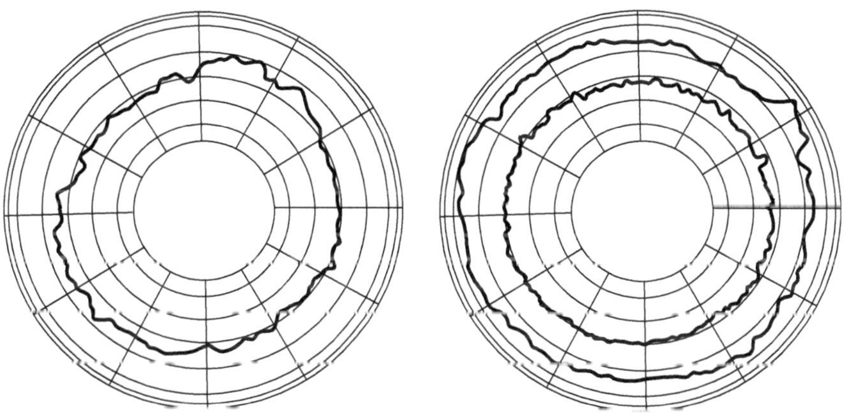

Figure 8.5 Roundness measurement of the piston top groove. Left. Piston top groove – severely worn surface. Right: Fresh piston – smooth surface

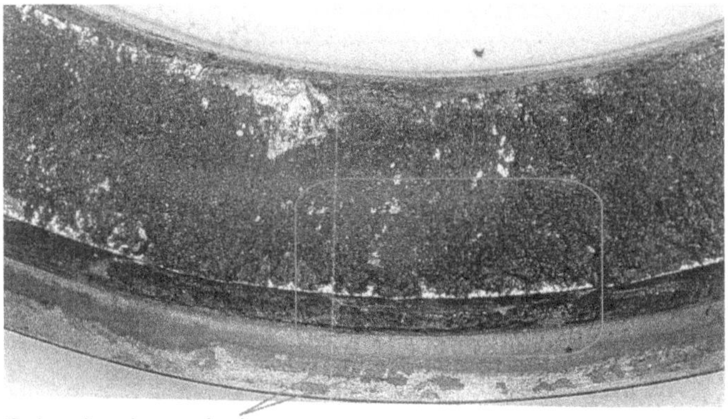

Carbon deposit at top land groove

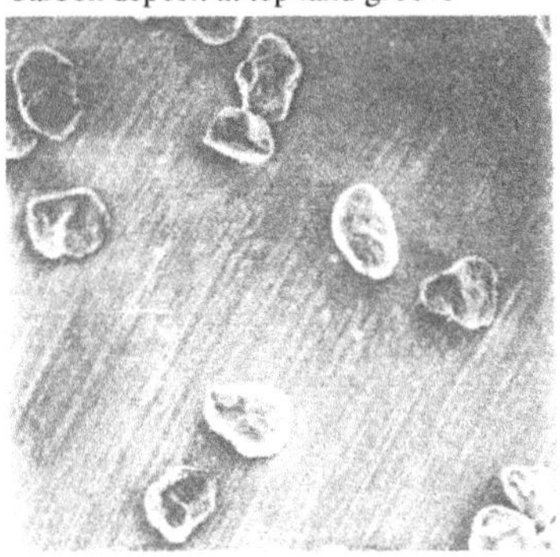

Abrasive particles – shape

Figure 8.6 Abrasive particles – top land area and shape of particles

the groove surface is difficult to estimate. A simple study of these parameters has been carried out using particle image analysis system. If the relative number of grains collected on the first groove zone is taken as 100, and the radius of a grain is, on average 19 μm, the number of grains and their sizes in other grooves are given in Figures 8.7 and 8.8.

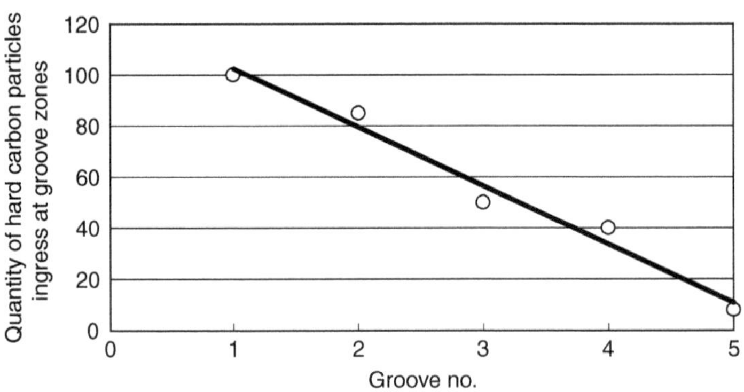

Figure 8.7 Relative quantity of hard carbon particles trapped in the piston groove zones

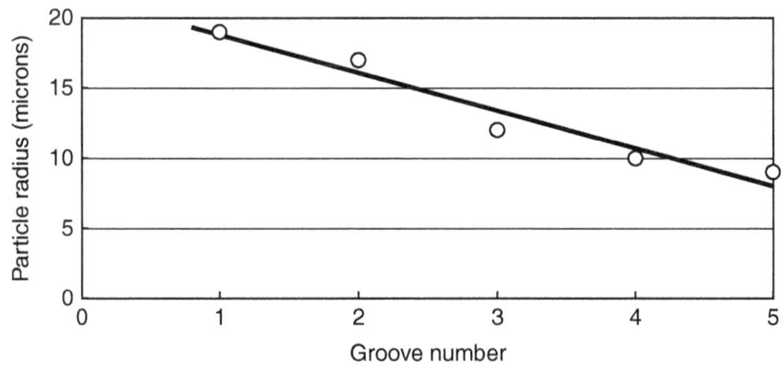

Figure 8.8 Average particle sizes at different groove zones

Table 8.4 Sample Calculation for Wear of the Top Ring Piston Groove

Equation	Parameter		Unit
	Max cylinder pressure at top ring, p_{cyl}	13.24	N/mm^2
8.1	Real contact pressure, p_r	559.2	N/mm^2
8.2	Approach, h/r	3.281×10^{-5}	
	Relative quantity of hard carbon particles trapped at top groove	100	
8.8	Number of cycles causing fatigue failure	1.26E6	
8.10	Nondimensional wear rate, I	1.48×10^{-11}	
	Distance covered, L	1.03×10^{10}	μm
8.11	Wear, W	0.15	μm

The flow of particles to lower lands of the piston decreases as the inter-ring pressure and flow velocity progressively drop due to the controlled butt gap of the ring. By finite element analysis, the minimum hot diametrical clearance between piston and liner at the top land zone was calculated and was of the order of 0.89 mm at rated load conditions; the clearance is less at other piston lands.

The depth of penetration of the softer surface is calculated from Young's modulus, hardness, quantity of particles entrapment, size and shape of the particles and peak firing pressures. The sliding distance of abrasive grains was determined using a constant, K_2, which is derived from the relationship between hard carbon particle depositions in different groove zones. The constant represents the total quantity of hard particles entrapped in the recesses. Table 8.4 describes the calculations estimating the wear of the top groove surface using the wear correlations (Equations 8.10 and 8.11). The calculated wear imitates the measured data (Figure 8.9) corroborating the model for abrasive wear.

8.5 Estimation of Wear Using Sarkar's Model

Sarkar (1976) postulated that hard asperities indent into the groove surface and that the load acting along the normal direction removes metal particles from the original surface. Thus, the metal removed by abrasion accumulates the maximum amount of stored energy due to mechanical working. The wear mass is given by:

$$M = \frac{\alpha \mu \rho F g}{S_m} \tag{8.13}$$

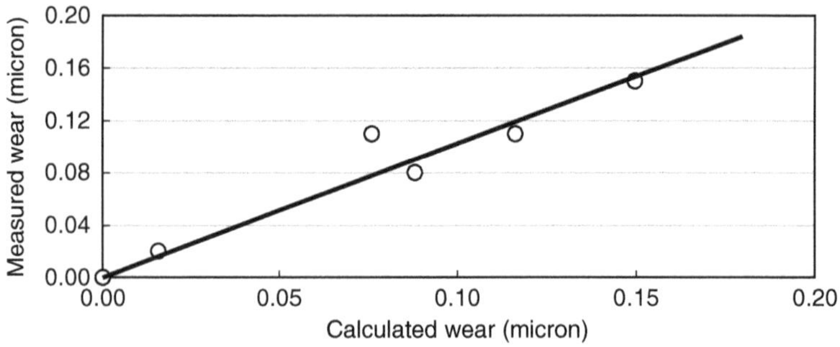

Figure 8.9 Validation of the piston groove wear model, that is, relationships between the measured and calculated wear values of piston grooves after 4000 hours

where M = mass of wear debris (g), α = fraction of deformation work, μ = coefficient of friction, ρ = density of material being abraded (kg/mm^3), F = total normal load (N), G = acceleration due to gravity (mm/s^2) and S_m = maximum possible strain energy per unit volume of soft metal.

The constant, α, for deformation work is usually about 1% of the stored strain energy. Figure 8.10 shows the comparisons between Sarkar's wear model values with measured and predicted values after 4000 hours.

Satisfactory correlation was achieved for the first and second grooves; however, the predictions by Sarkar's model were way off from the actual wear. This may be due to surface roughness parameters and abrasive grain sizes, which are not fully dealt with in Sarkar's abrasive wear relationship.

8.5.1 Parametric Study

The parametric study on the number of abrasive particles on the piston groove surface and particle size was carried out to see the significance of these on wear phenomenon after 4000 hours. The increase in the quantity and the size of carbon particles at different grooves increases the apparent contact area and cyclic (fatigue) stresses on the roughness peaks, and thus chip-off from the parent surface. Figure 8.11 shows the response to the two parameters.

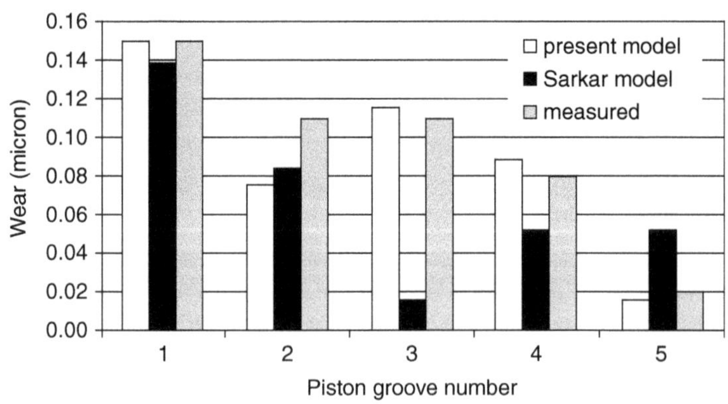

Figure 8.10 Comparison of Sarkar's model with measured wear values after 4000 hours

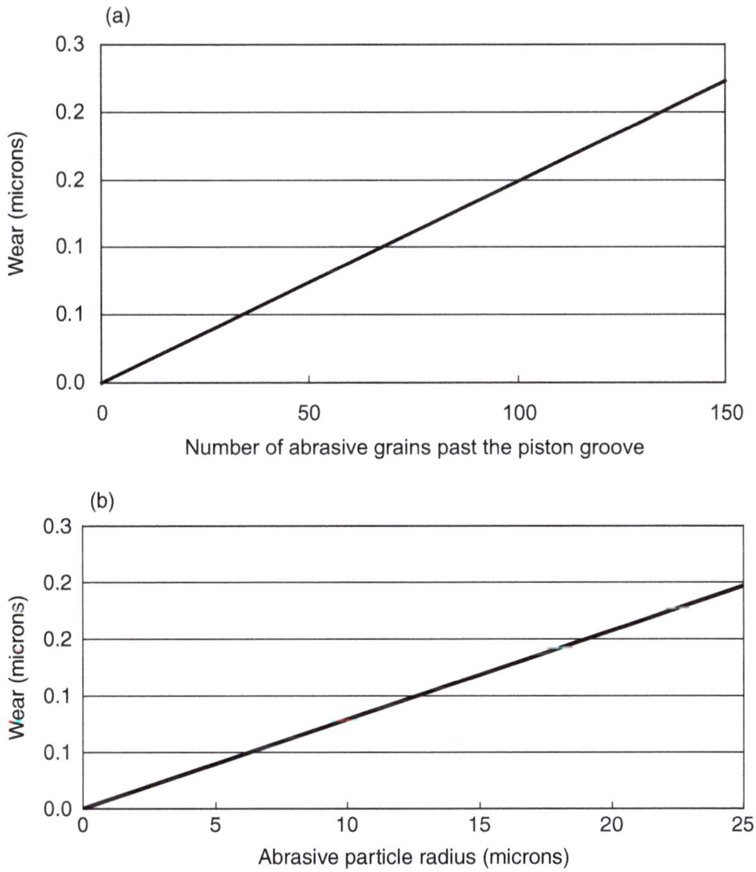

Figure 8.11 Parametric study on (a) the effect of quantity and (b) the size of abrasive particles on piston groove wear after 4000 hours

8.6 Summary

Abrasive wear is complex and is dependent on various factors, such as peak cylinder pressure, surface roughness parameters, abrasive grain size and material properties. The abrasive contaminants damage the groove surface by breaking the lubricating oil film. Hard particles contacting the soft groove surface chip away material in a plastic shear regime.

A mathematical wear model for the three-body abrasive mode is proposed to estimate the abrasive wear rate. The hard abrasive particles behave as rigid bodies that dislodge the softer material from the parent surfaces. The model assumes that chip formation and detachment of debris take place simultaneously. The initial wear rate is estimated when ploughing and cutting of the groove surface by abrasive particles would cause debris. Since the model could take into account physical and geometrical aspects of abrasive wear, it could mimic the true wear rate of different materials in contact with varying concentrations of abrasives. Satisfactory results were achieved for bores of different sizes and brake mean effective pressure.

The phenomenological model is useful for estimating the wear of ring grooves caused by the sliding abrasive particles trapped between the ring and groove surfaces.

References

Davis, F.A. and Eyre, T.S. (1990) The effect of friction modifier on piston ring and cylinder bore friction and wear. *Tribology International*, **23** (3), 163–171.

Golothan, D.W. (1978) Pistons and Cylinders. *Industrial Lubrication and Tribology*, **30** (3), 72–77. doi:10.1108/eb053152.

Hirano, H.H. and Levy, A.V. (1978) An investigation of two-body abrasive wear, Fundamentals of tribology. Proceedings of the international conference on tribology, June 1978, MIT, Cambridge, MA, pp. 519–542.

Kragelskii, I.V. and Alisin, V.V. (1981) *Friction, wear, and Lubrication*, Tribology Handbook, Vol. 1, Mir Publishers, Moscow.

Mashloosh, K.M. and Eyre, T.S. (1985) Abrasive wear and its application to digger teeth. *Tribology International*, **18** (5), 259–266.

Mashloosh K.M. and Eyre, T.S. (1986) *Effect of attack angle in abrasive wear. Metals and Materials.* Springer, pp. 426–430.

Rabinowicz, E., and Mutis, A. (1965) Effect of abrasive particle size on wear. *Wear*, **8**, 381–390.

Sarkar, A. D. (1986) *Wear of metals.* Pergamon Press, Oxford.

Suh, N.P., Sin, H.C. and Saka, N. (1978) Fundamental aspects of abrasive wear, Fundamentals of tribology. Proceedings of the international conference on tribology, June 1978, MIT, Cambridge, MA, pp. 493–518.

Sundararajan, G. (1985) A new model for two-body abrasive wear based on the localisation of plastic deformation. *Wear*, **117**, 1–35.

9

Abrasive Wear of Liners and Piston Rings

Today, the effect of dust on the early wearing out of the critical parts is well understood. This has become more important with more and more engines being turbocharged, which results in fourfold higher blow-by than in naturally aspirated engines for the same wearing out of the liner and rings. Air filters of high efficiency that filter even very fine quartz are used. However efficient the air filter may be, though, a minute fraction of the dust still enters the engine. This chapter considers the wear caused by sand ingress. Also proposed is a concept of critical dosage of sand which spells the end of the life of the parts being worn.

9.1 Introduction

The causes of wear of the piston, rings and the cylinder are abrasion, corrosion or a combination of both. An abnormal piston wear pattern could also be the result of bore distortion due to wrong assembly methods or the design itself. The symptoms of excessive wear of an engine are an increase in rate of blow-by and oil consumption.

9.2 Wear of Liner and Ring Surfaces

After running-in an engine for a short time, a normally honed liner will exhibit a surface profile similar to a plateau honed liner. The initial wear of normally honed liners is higher. The initial period is shown by the curve AB in Figure 9.1. The curve is for a 'normally' honed liner. The initial period and wear rate are shorter for a plateau honed liner. After this period, the surface is relatively stable in terms of wear, curve BC. During this second period, the plateau honed surface continues to possess relatively large and smooth plateaus, which provide a large bearing area and also deep valleys that help to retain the oil for lubrication and provide relief pockets for wear particles. A smooth and stable plateau honed liner offers substantial improvement in oil consumption, ring wear and liner wear at the ring reversal zone (Hegemier and Stewart 1993). In the final phase of the wear rate curve, CD, the engine wear increases dramatically on account of corrosion.

Critical Component Wear in Heavy Duty Engines, First Edition. P.A. Lakshminarayanan and Nagaraj S. Nayak.

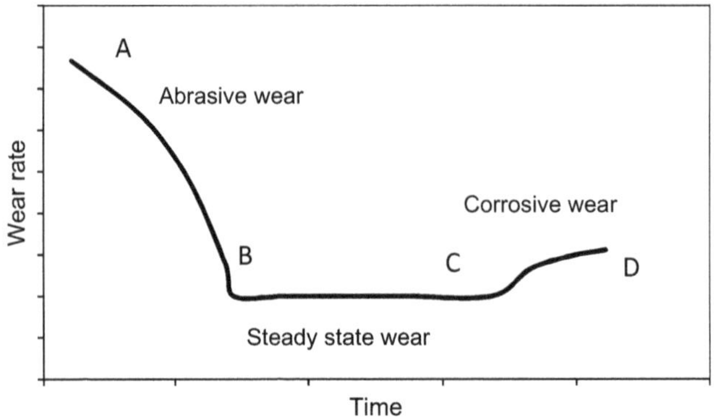

Figure 9.1 Bath tub nature of wear rate

The general wear pattern for a cylinder of an engine is given in Figure 9.2 (Bolis, Johnson and Daavetilla, 1977). It shows that most liner wear occurs at the top and bottom of the stroke. This is because the highest friction for a piston and ring, and cylinder assembly occurs at the top and bottom of the liner due to the combined effects of gaseous and inertia loads, and low piston velocity.

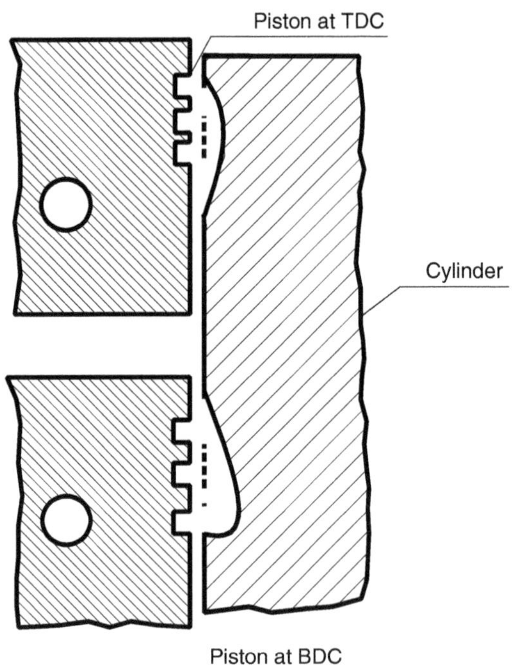

Figure 9.2 Liner wear pattern

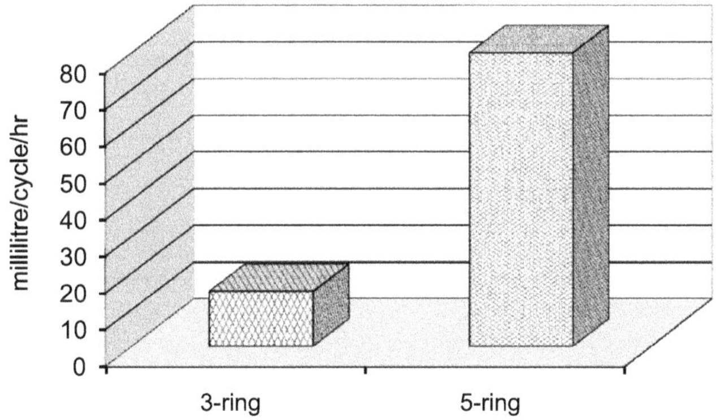

Figure 9.3 Oil consumption of engines with three rings and five rings (with one ring below the piston pin) per piston

9.3 Design Parameters

9.3.1 Piston and Rings Assembly

Of all the parts of the engine, the piston and rings are usually the most critical and undergo rapid wear under abusive operation or due to poor design. The skirt of the piston is expected to glide on the liner. In the case of a piston assembly with five rings where one oil ring is below the piston pin, the lubrication of the skirt is impeded. In addition, it has the downstream effect of wearing out the rings above the piston pin by oil starvation. Research shows that the oil consumption with a three ring assembly is superior with all the rings above the piston pin (Figure 9.3).

Similarly, the percentages of complaints with three rings piston are lower than with five rings system (Figure 9.4), all other factors remaining the same.

9.3.2 Abrasive Wear

The statistics of engine failures due to wear of the liner, piston and ring surfaces while under warranty and post-warranty (Durella, 1978; Wallace, 1978) is given in Table 9.1. Abrasive wear is the most important factor affecting the life of the engine.

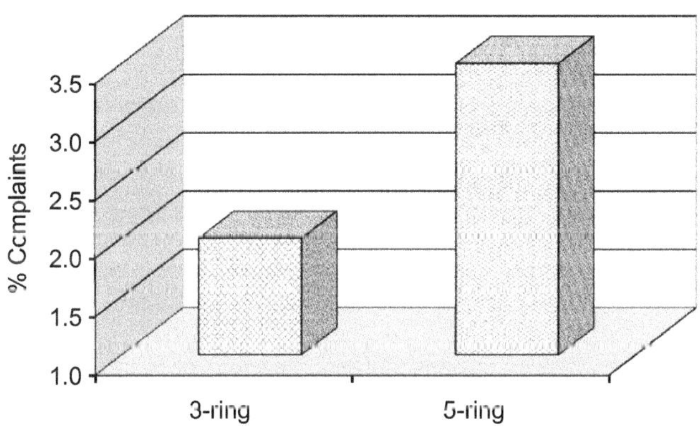

Figure 9.4 Statistical studies on field complaints for three ring and five ring piston engines

Table 9.1 Warranty claims distribution

Type of failure	Percentage returns from field
Abrasive wear	44.0
Scuffing and scoring	10.5
New rings installed in worn groove	10.5
Improper ring installation	9.5
Other causes	2.5
Unknown	15.0

9.3.3 *Sources of Abrasives*

Abrasives that produce wear can enter the engine in three ways (Golothan, 1978; Mahle, 1990).

- Induction through improper air cleaner system.
- Dirt entrapment during assembly.
- Improper engine combustion and heavier fractions of engine fuel.

Airborne abrasives drawn into the engine through the intake system hold the greatest potential for premature wear. Built-in-dirt consists of core sand, metal machining fragments, honing grit and dirt present in the assembly, which play a minor role in reducing the life of the engine.

9.4 Study of Abrasive Wear on Off-highway Engines

9.4.1 *Abrasive Wear of Rings*

The faces and sides of abrasively worn rings have excessive butt gap and a dull grey lapped appearance. Hence, abrasive wear is indicated by heavy blow-by. The cross-sections of a new ring set and those of a worn out are shown in Figure 9.5. In Table 9.2, the average closed gaps of the sample rings are given. The contourograph in Figure 9.6 shows the lip formation at the bottom flank of a typical worn out top ring. Figure 9.7 shows the wear-out of the corresponding piston grooves similar to the reports by Mahle (1990), Nural (1992) and Schilling (1972). The closed gap of the top compression rings is more than the gap at the oil control rings. This clearly shows that the dust ingress is through the air intake system.

9.4.2 *Abrasive Wear of Piston Pin and Liners*

Figure 9.8 shows the characteristic wear of the piston pin due to abrasives. The friction is much less at the oil groove in the small end bearing of the connecting rod as the oil clearance is large. In addition, it is flushed with copious amount of oil. Thus, the zone near the oil groove remains unworn. If the pip in the wear graph at the oil groove is taken as reference (not worn), the area engaging with the connecting rod bearing is worn by 15 microns and the areas moving in (aluminium) piston bosses are worn by about eight microns.

The relative velocity with which the pin moves in the connecting rod bearing is higher than in the piston boss. Also, the contact pressure in the connecting rod bearing is higher, as the area over which the load acts is nearly half of the piston pin bosses. Therefore, the connecting rod bearing area of the pin wears at about twice the rate of the pin in the piston bosses.

Table 9.2 Statistics of ring wear out (number of hours run: 400)

Ring type	Average closed gap (mm)		Samples measured
	New	Worn out	
Top compression	0.47	0.98	6
Taper faced second compression	0.45	0.52	6
Oil control	0.33	0.61	6

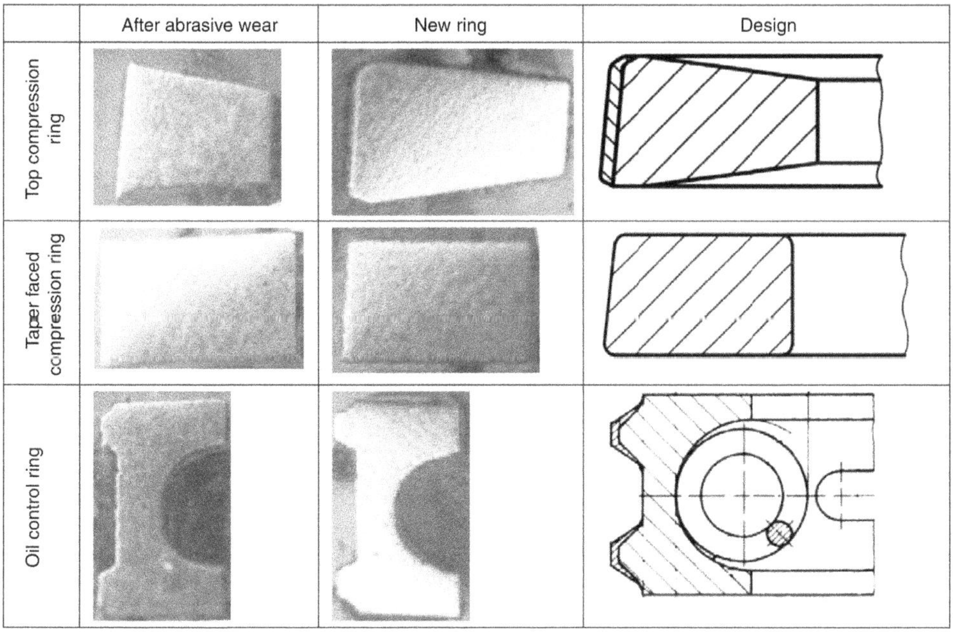

Figure 9.5 Photographs of sections of new and abrasively worn out rings

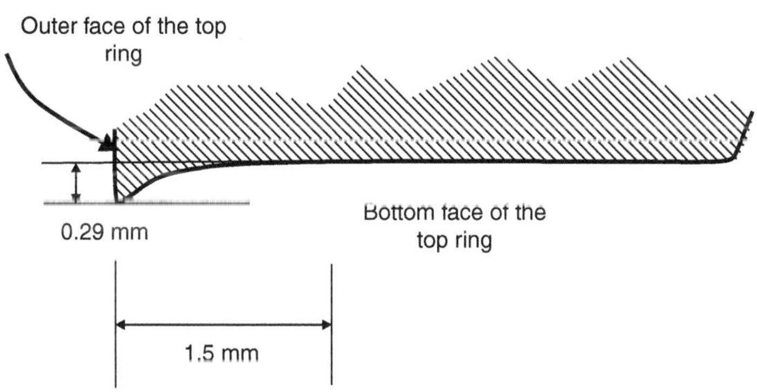

Figure 9.6 Contourograph – top ring

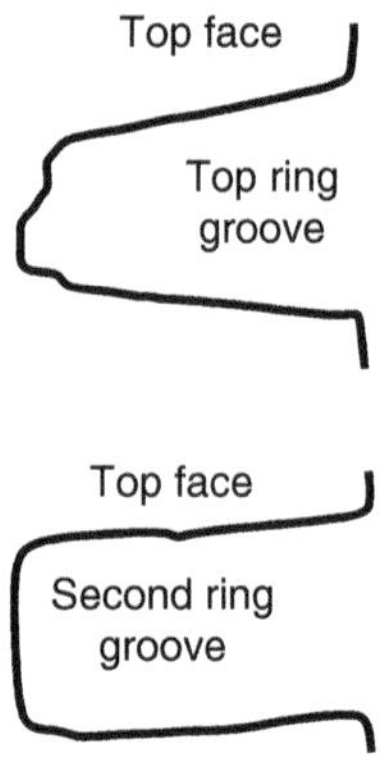

Figure 9.7 Wear out of piston groove surfaces

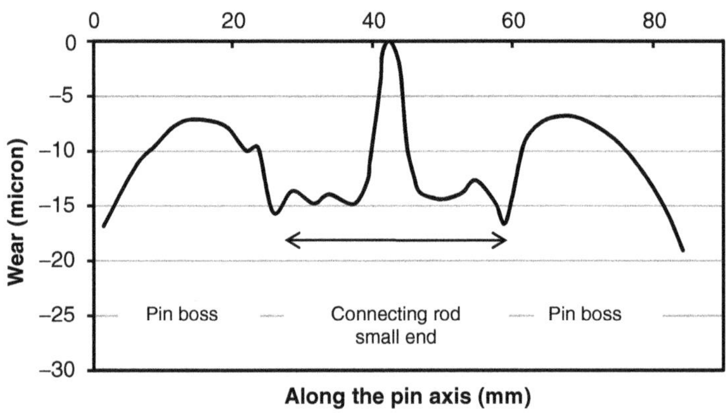

Figure 9.8 Piston pin wear due to inadequately filtered air

The dull lapped finish of liners in Figure 9.9 shows that the wear is abrasive. It is not shiny as in the case of bore polishing. Intense wear is seen at the top ring reversal zone, which is in the sensitive boundary lubrication regime (Table 9.3, Figure 9.10). In addition, varnish formed on the piston skirt and ring lands can be seen on the pistons in Figure 9.11. This is due to heavy blow-by caused by the increased ring gap and insufficient sealing by liner surface because of nonuniform wear.

9.4.3 Accelerated Abrasive Wear Test on an Engine to Simulate Operation in the Field

A new three-cylinder naturally aspirated engine with manifold configuration as shown in Figures 9.15 and 9.13d was taken from the assembly line and tested for 15 hours at full load and rated speed using an oil bath type air cleaner. The oil bath type air filter was filled with oil up to 50% level to accelerate the test. At the end of the test, a band of vertical scratches was observed, predominantly in the central liner. The width of the band was about 30 mm and the depth was 5 μm, as shown in Figure 9.12. The scored band is due to

Table 9.3 Liner wear at different locations (400 hours average run time, under-warranty period)

	Wear (micron)	
	Along x	Along y
Top of liner	22.3	30.0
Middle of liner	8.9	9.5
Bottom of liner	9.7	10.4

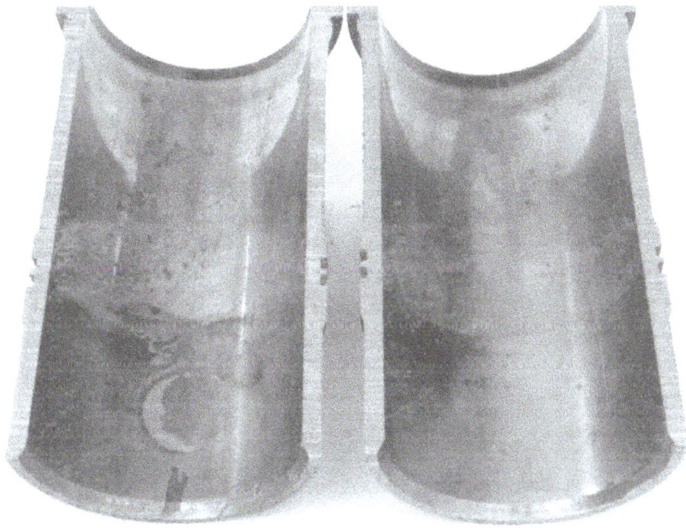

Figure 9.9 Cylinder wall showing dull lapped appearance from abrasive wear

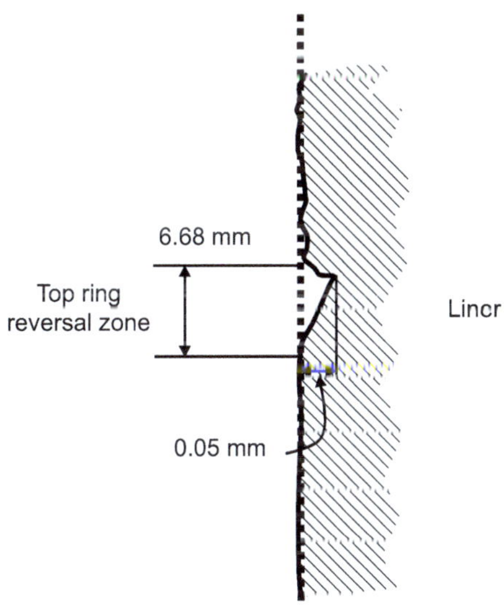

Figure 9.10 Contourograph of liner worn at ring reversal zone

Figure 9.11 Formation of varnish on the piston skirt and land areas

dust ingress through the inefficient air cleaner system. The sand particles collected from the manifold are shown in Figure 9.13. The study highlighted that the central liner of the three-cylinder engine is scored relatively more, as is observed in engines in the field that use inefficient air cleaners. It is due to the winnowing effect (Section 9.5).

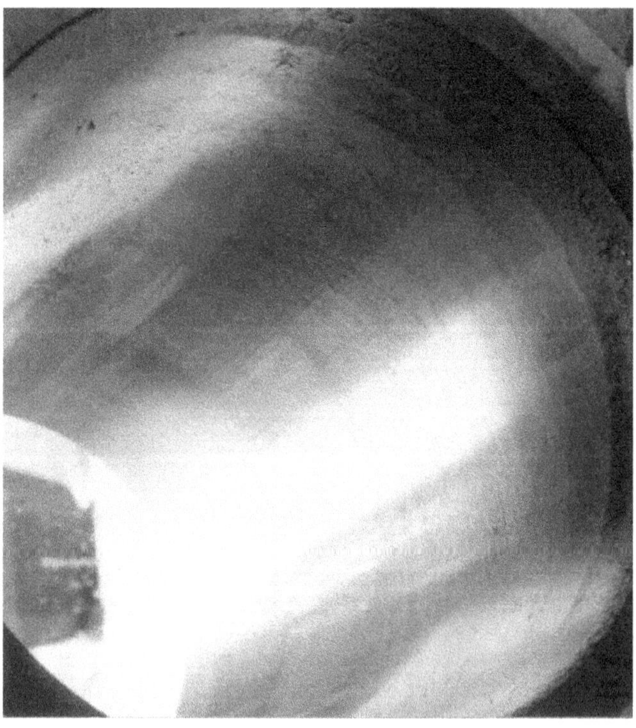

Figure 9.12 Liner wear during accelerated test using old type air cleaner – abrasive action

(a) (b)

Figure 9.13 Dust in the inlet manifold due to an inefficient oil bath type air cleaner. (a) Dust particles collected; (b) dust on the surfaces in the inlet manifold

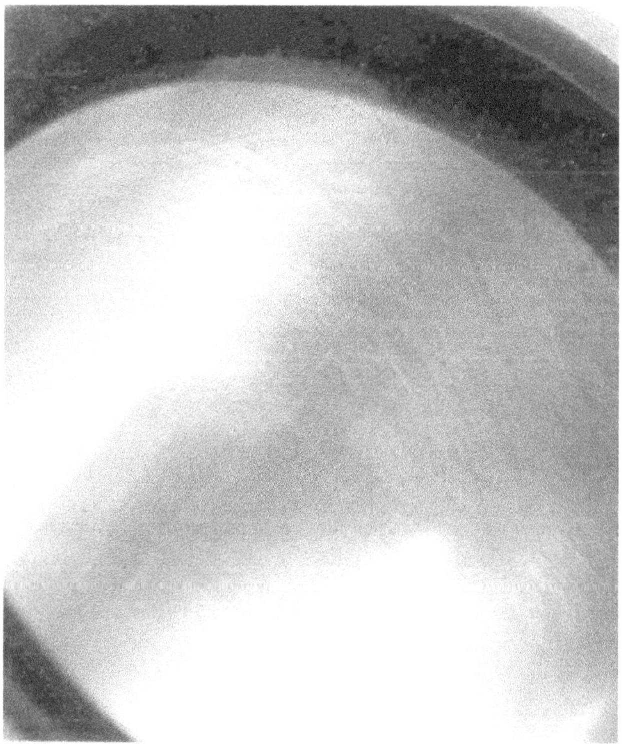

Figure 9.14 No wear marks on the liner are noticed after the accelerated test using a paper type air cleaner; the honing pattern is intact

Later, the engine was reassembled with fresh liners, pistons and rings and an efficient dry type (paper) air cleaner. The test was repeated for 100 hours at full load and rated speed. There were no scratches on any of the liners, as the air was filtered efficiently (Figure 9.14).

9.5 Winnowing Effect

Denser dust particles bear a high ratio of inertia force to drag force in the air flow. Hence, denser dust continues to flow in the original direction and deviate from the streamlines of the air flow. In a

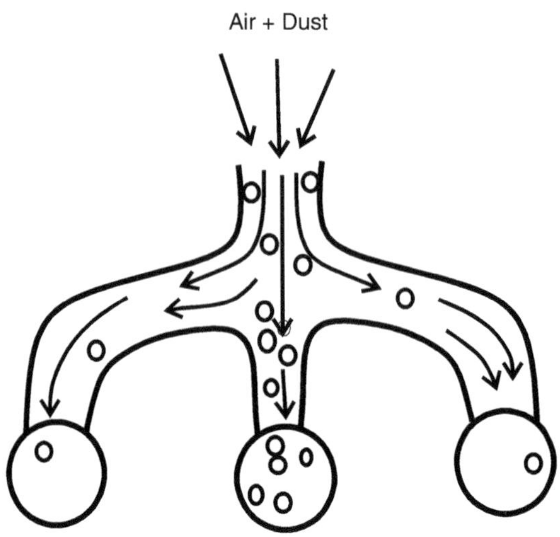

Figure 9.15 Diagram representing winnowing effect – more wear at the middle cylinder

three-cylinder engine with the particular layout of the inlet manifold given in Figure 9.13, the heavier dust from the air filter short-circuits to the central liner and avoids following the air flow to the neighbouring cylinders. This winnowing effect is shown schematically in Figure 9.15. Consequently, in a multi-cylinder engine there are a few cylinders into which hard particles enter preferentially and these cylinders wear out faster than the rest.

9.6 Scanning Electron Microscopy of Abrasive Wear

Figure 9.16a shows the Scanning Electron Micrograph of the surface texture of the abrasively worn liner at 100× magnification. The honing pattern has completely disappeared and the graphite flakes are clearly visible on the surface of the cast iron liner. The micrograph also shows the scratch lines along the piston travel. The size of the dust could be read as 5–10 micron in Figure 9.16b at 600× magnification. It could be concluded that dust in the range 1–15 micron causes harmful wear of liner, piston and rings.

9.7 Critical Dosage of Sand and Life of Piston–Ring–Liner Assembly

An experiment was conducted on a 5.7 litre engine at no load and 2000 rpm to evaluate the life of the engine by admitting fine silica (quartz or sand) dust having a size of about 10 micron (ISO fine dust), which is injurious. The rate of top ring wear out was established by correlation with the rate of admission of fine dust. For every one gramme of dust entering the engine 0.01 gramme of the piston ring was worn out. After 40 g of silica ingress the blow-by increases abruptly, indicating the end of the life of the piston–ring–liner assembly. The blow-by and silica addition were normalized with the swept volume of the engine and are plotted in Figure 9.17. It can be seen that, for a diesel engine, the limiting dosage of sand sized below ten microns is seven grammes per litre of swept volume. This important parameter can be used for calculating the life of an engine operating in dusty areas, such as mines or desert, or when using a given air intake system.

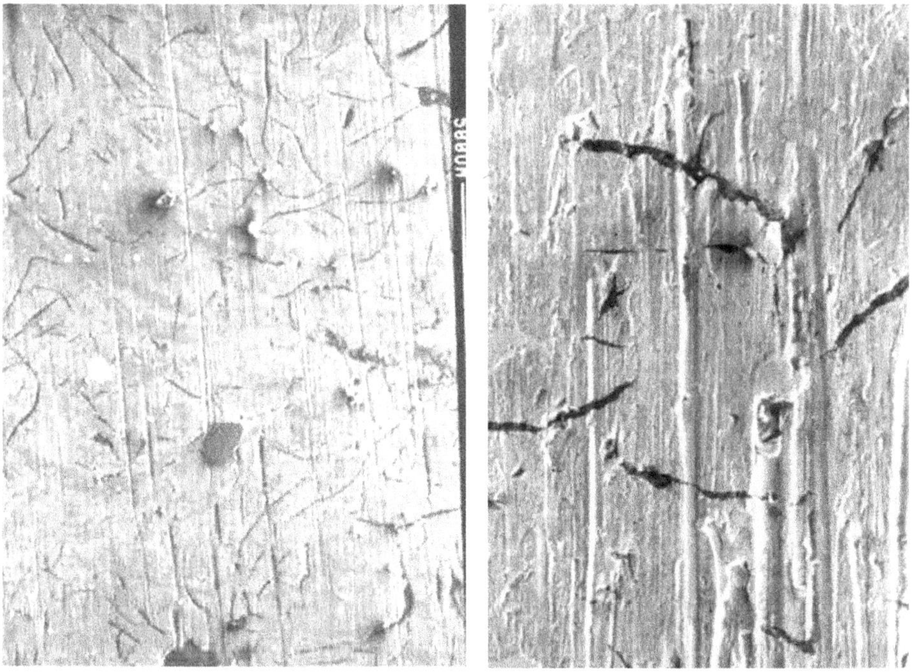

(a) (b)

Figure 9.16 SEM photographs at (a) 100× and (b) 600× magnification of a liner – lapped finish – the abrasive mode of wear of liner surfaces

9.7.1 Simulation of Engine Life

A simulation was carried out using the data available in Schilling (1972) and the experiments mentioned above to estimate the wear life of a three litre naturally aspirated engine operating in a dusty environment. The concentration of dust in the air was 10 mg/m^3. The dust particles less than 10 micron in size cause

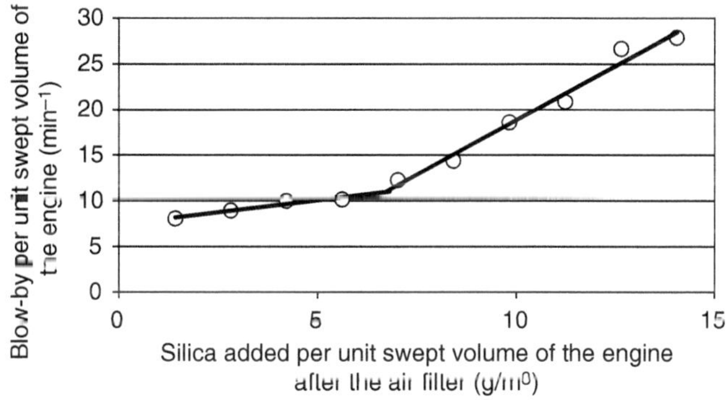

Figure 9.17 Increase in blow-by past the piston rings and the turbocharger due to addition of silica at the inlet after the air cleaner in a 5.7 litre engine

Table 9.4 Input data for simulation of wear for different conditions and types of filters

Dust concentration in industrial area (mg/m^3)	10
Dust fraction with diameter less than 10 micron (%)	25
Air flow per cylinder of off-road engine at 2000 rpm (m^3/h)	50
Ring diameter (mm)	100
Ring width (mm)	3
Increase in closed gap at condemnable limit (mm)	0.50
Density of iron (mg/mm^3)	7.50

Table 9.5 Simulation of wear for different conditions and types of filters of a 2.6 litre three-cylinder engine

	Oil bath type			Paper type
Condition of air filter	Without pre-cleaner and without oil	With pre-cleaner and without oil	With pre-cleaner and with oil in the bath	With pre-filter
Efficiency (%)	47	92	97	99.9
Fraction of dust transmitted to the engine (%)	53	8	3	0.1
Dust ingress into engine in operation (mg/h)	265	40	15	0.5
Top ring wear rate (mg/h)	1	0.15	0.056	0.0019
Increase in radius of ring (mm/h)	0.000141	0.000021	0.000008	0.00000027
Increase in closed gap of ring (mm/h)	0.000889	0.000134	0.000050	0.00000168
Life, h (life ends when the closed ring gap doubles from 0.5 mm to 1.0 mm with double the initial blowby)	563	3727	9937	298125
Observed life, h	400	1800	6000	>8000

wear; this formed about 25% of the total (ISO medium) dust by weight. The other data given in Table 9.4 were used to calculate the life. The calculation steps are shown in Table 9.5 for an efficient dry type (paper) filter and an inefficient oil bath type air filter. The average engine life was drastically reduced when the pre-filter was missing and there was no oil in the oil bath type filter. If the engine works continuously in areas where the dust concentration is 250 mg/m^3, the wear rates will be 25 times higher.

The engines with warranty complaints and a poor wear life of 400 hours had lost both oil and the prefilter in a very short time. The wear life of 1800 hours corresponds to engines with pre-filter in good working condition and oil bath filter with no oil during the failure period. An efficient air intake system enhances the life of an engine and allows the engine to age normally (Chapter 2).

9.8 Summary

Abrasive wear is the most important factor responsible for diminishing the life of an engine. A dull lapped finish of the surfaces betrays abrasive wear. High blow-by and oil consumption are usually the result of abrasive wear of the liners and piston rings. 1–15 micron sized dust in the air is responsible for higher wear than other sizes. When the dust ingress is from the air intake system, the top ring wears out more than the oil ring at the bottom and, correspondingly, the ring grooves in the piston. Similarly, the top ring reversal zone on the liner shows higher wear.

Dry type air cleaners filter the dust efficiently and allow the engine to last for the expected life. Oil bath type filter are far less efficient and the engine life suffers. Other problems, such as broken pre-filters, high oil carry over in the bath type filters and leaking air intake systems, aggravate wear.

References

Bolis, D.A., Johnson, J.H. and Daavetilla, D.A. (1977) The Effect of Oil and Coolant Temperatures on Diesel Engine Wear. Technical Paper 770086, SAE, Troy, MI.

Durella, M.J. (1978) Engine Bearings. SAE SP235, SAE, Troy, MI.

Golothan, D.W. (1978) Pistons and Cylinders. *Industrial Lubrication and Tribology*, **30** (3), 72–77. doi:10.1108/eb053152.

Hegemier, T. and Stewart, M. (1993) Some effects of liner finish on Diesel Engine Operating Characteristics. Technical Paper 930716, SAE, Troy, MI.

Mahle (1990) Piston Handbook. Mahle, Stuttgart, Germany.

Nural (1992) Piston Handbook. Alcan Aluminiumwerk Nuernberg GmbH, Nuernberg, Germany.

Schilling, A. (1972) *A text book on Automobile Engine Lubrication*, Volume II. Scientific Publications Ltd, UK.

Wallace, D.G. (1978) Piston rings, Pistons and cylinders. SAE SP235, SAE, Troy, MI.

10

Corrosive Wear

Sulfur in fuel oxidizes during combustion and becomes sulfuric acid after reaction with the water that condenses on the surfaces where the temperature is below the local dew point. It is highly corrosive to bearings and piston rings. The temperature of the coolant should be sufficiently high to avoid condensation of water of combustion on the cooler walls and, at the same time, should be low enough to provide sufficient oil film to assure lubrication of all the critical wear parts. Also discussed in this chapter is the wear of aluminium parts in contact with the coolant in specific cases where acidity is higher than five ppm.

10.1 Introduction

The causes of wear of the piston, rings and the cylinder are abrasion and corrosion or a combination of both. Usually, the symptoms of excessive wear of engine parts are an abrupt increase in blow-by and oil consumption.

10.2 Operating Parameters

10.2.1 Corrosive Wear

Corrosive wear is observed predominantly in the third stage of the wear rate diagram of an engine, that is, along the curve CD shown in Figure 9.1. The temperature and pressure in the cylinder and chemical composition of the combustion gases play a vital role in the corrosive wear of piston rings and liner surfaces. If the environment in the cylinder is such that the dew point of water is lower than the temperatures of the cylinder wall and the piston, sulfuric acid will not form and corrosive wear will not occur on ring and liner surfaces (David, Johnson and Daavetilla, 1977). Higher corrosive wear rates are usually attributed to very low coolant temperature, that is, coolant temperature less than 60 °C (Schilling, 1972). The probable causes for corrosive wear are:

- Effect of fuel quality.
- Water jacket temperature.
- Stop and go driving.
- Coolant leaking into the cylinders.

Critical Component Wear in Heavy Duty Engines, First Edition. P.A. Lakshminarayanan and Nagaraj S. Nayak.

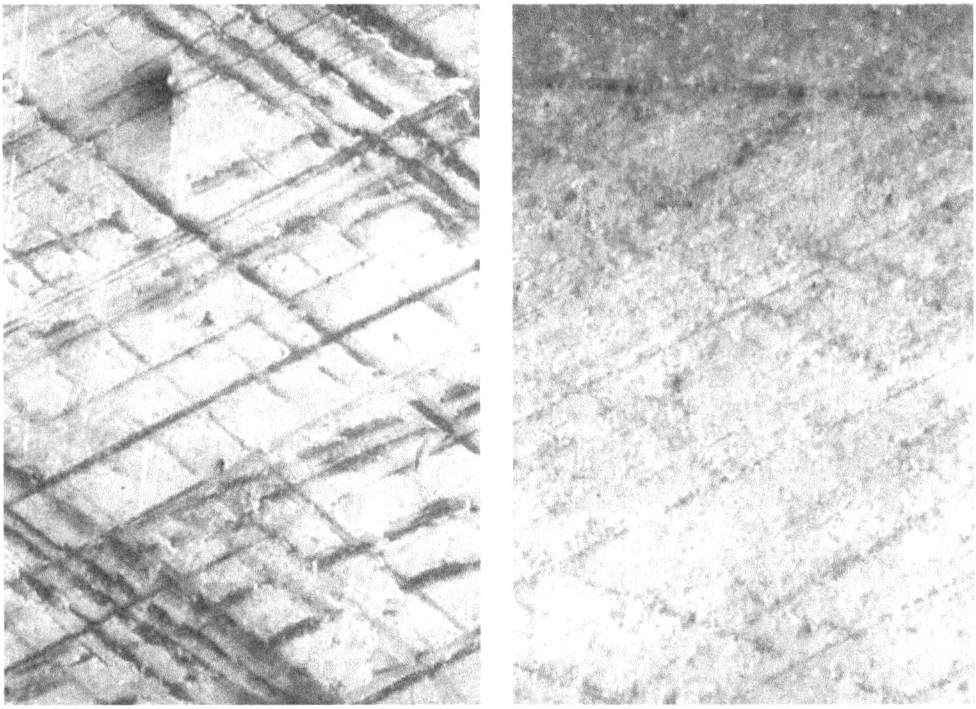

(a) (b)

Figure 10.1 SEM photographs of liner surface at the top ring reversal zone: (a) with thermostat – liner cross-hatch angles are intact; (b) without thermostat – liner wear can be observed

When corrosive wear is severe, a mottled, greyish, pitted appearance on the liner surface is seen. A thermostat in the engine, if set to limit the coolant temperature in the range of 70–95 °C, would reduce the corrosive wear substantially by avoiding acid formation.

10.3 Corrosive Wear Study on Off-road Application Engines

10.3.1 Accelerated Corrosive Wear Test

To study the effect of a thermostat, an accelerated engine test was carried out by cooling the engine with cold water at 35 °C and switching-off the engine for one hour after every three hours of running. The engine running at half load and cooling using cold water enabled liner temperatures lower than the dew point. In addition, to accelerate the corrosive wear test, lubricating oil of low total base number (= 5.6) was used. All necessary precautions to separate abrasive and corrosive wear were taken when the test engine was assembled. A paper type air filter having the best efficiency, 99.99%, was used while studying corrosive wear, thus eliminating the effect of dust in the surroundings. The engine was run for 300 hours without a thermostat.

Similarly, the experiment was repeated for 300 hours after changing the liners, pistons, rings and oil, and with a proper thermostat assembly. The performance of the engine was stable throughout both the tests (Table 10.1). The test duration was too small to detect liner wear by using a simple bore gauge. The differences due to corrosive wear would be pronounced later and affect the life of the liner and rings. However, the incipient corrosive wear of the liner surface could be easily seen in the most stressed top ring

Table 10.1 Test results of accelerated test for corrosive wear after running for 300 hours

Description		Speed (rpm)	Power (hp)	sfc (g/hp-h)	Water temperature (°C)	Smoke (BSU)	Blow-by (litre/min)
Endurance test without thermostat	Final	2000	31.73	188.3	80.0	1.7	9.6
	Initial	2000	31.7	181.9	47.0	2.0	9.6
Endurance test with thermostat	Final	2000	32.0	176.6	85.0	1.8	9.0
	Initial	2000	32.0	177.5	85.0	1.9	9.3

reversal zone under a scanning electron microscope at 100× magnification (Figure 10.1) or by careful interpretation of the surface profile (Figure 10.2).

In addition, the bearing curves (t_p %, Chapter 5) of the liners at ring reversal points, the piston ring gap and tangential load of all rings were measured before and after the endurance tests.

The bearing curves at the top ring reversal zone of the typical liner without a thermostat are shown in Figure 10.3 and with a thermostat in Figure 10.4. The honing marks at the top ring reversal point are still intact when a thermostat was used. The wear rate at the top-ring-reversal without a thermostat is twice that

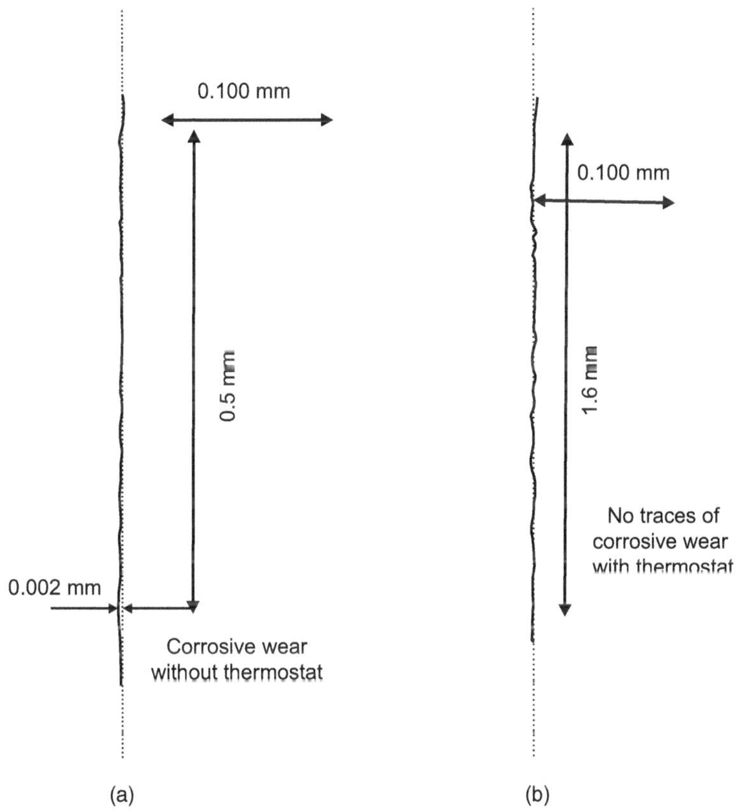

Figure 10.2 Contourograph of liner surface – at top ring reversal zone: (a) without thermostat; (b) with thermostat

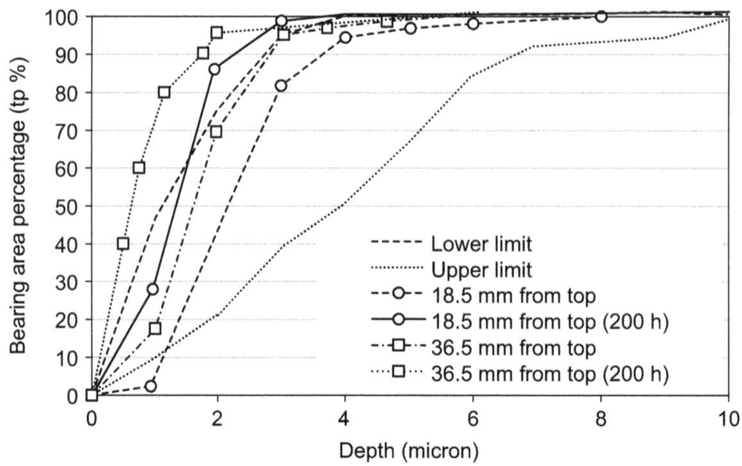

Figure 10.3 Bearing area of liners before and after the endurance test – without thermostat

with a thermostat. The lower coolant temperature (without a thermostat) at low loads causes water to form on the liner surfaces. The water dissolves sulfur dioxide and forms corrosive sulfuric acid. There is an optimum water temperature range for minimum wear of 80–90 °C, as shown in Figure 10.5 (Baumgartner, Inhelder and Mikulicio, 1987). At high temperatures, the thickness of the oil film on the wear parts reduces and boundary lubrication is beginning to cause wear due to metal-to-metal contact.

The accelerated engine test results showed that the general conditions of the pistons and the liner are superior when a thermostat is used. Figure 10.6 shows the condition of the main bearings, liner and pistons for engins with and without thermostat (over-cooling liners). The quality of the bearings is better with a thermostat because acid is not formed on the liner walls. Similarly, reduction in the fuel consumption is observed at low loads when a thermostat is used. This is by virtue of lower friction at raised oil temperatures and of less wear of the parts (Figure 10.7).

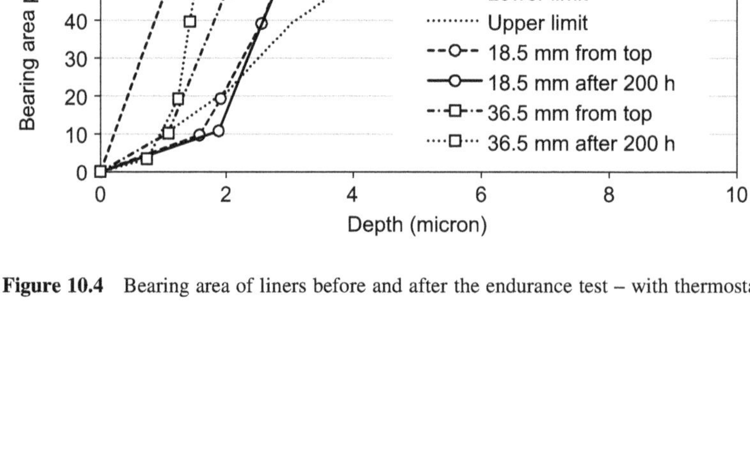

Figure 10.4 Bearing area of liners before and after the endurance test – with thermostat

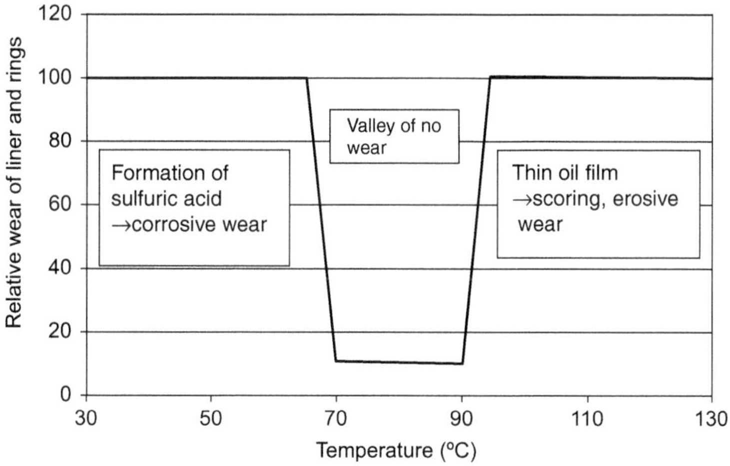

Figure 10.5 Relative wear of liner and rings against water temperature

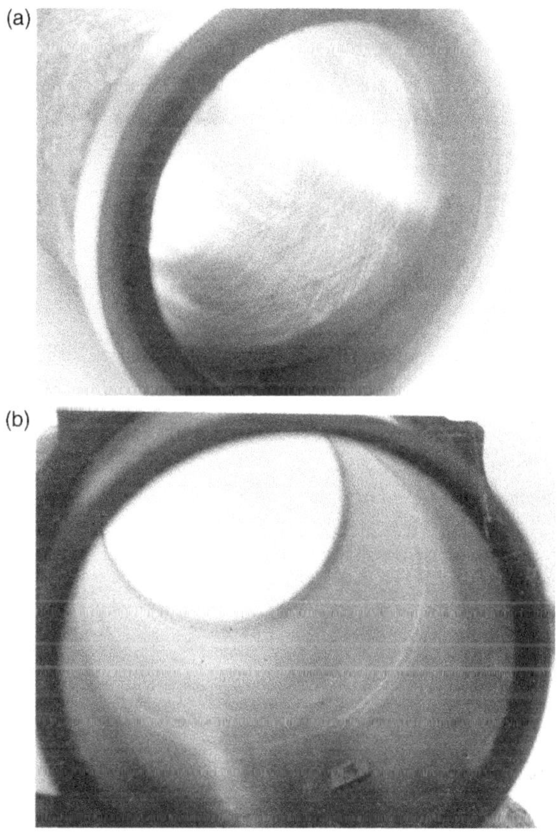

Figure 10.6 Liner, piston and main bearings condition after endurance test with and without thermostat: (a) liner with thermostat; (b) liner without thermostat; (c) piston with thermostat; (d) piston without thermostat; (e) main bearings with thermostat; (f) main bearings without thermostat

(c)

(d)

(e)

(f)

Figure 10.6 (*Continued*)

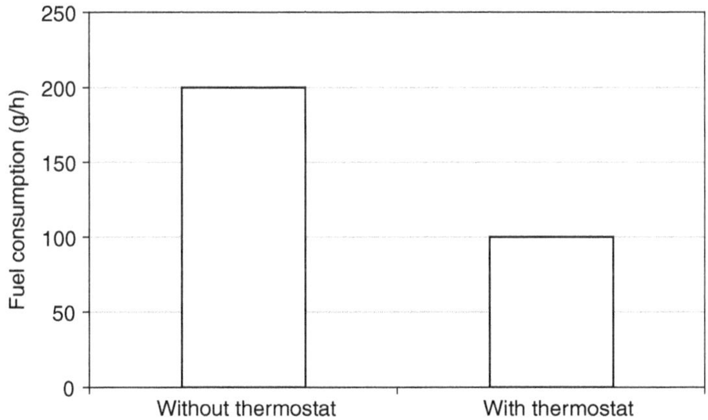

Figure 10.7 Diesel fuel consumption at low speed (1200 rpm) and at no load with and without thermostat

10.4 Wear Related to Coolants in an Engine

10.4.1 Under-cooling of Liners by Design

While too cold a liner causes acidic corrosion, too hot a liner resulting from less cooling leads to a breakdown of the lubrication of the piston ring, and early wear of the liner and rings sets in. Wet liners are popular in the design of heavy duty engines. Since the outer metal surface of the liner is in contact with the coolant, these liners are cooled well. However, to reduce the weight of the engine, dry liners are often designed. When they are interference fitted, the heat transfer is good and the temperature is maintained within safe limits. However, there are some manufacturers who prefer slip fit liners for ease of servicing (Mettig, 1973), leaving an air space of 50 micron between the liner and the parent bore in the crankcase, which is equivalent to a cast iron wall of 60 mm, and the temperature of the liner can go up to 250 °C (Figure 10.8). In addition to heat transfer problems, such designs seriously affect the strength and wear life (Figure 10.5) of the liner, and frequent failures are common because of combined high frequency fatigue caused by the cylinder pressure and low cycle fatigue by temperature cycling (Mizutani *et al.*, 1996). Oil is allowed to flow in the air gap in some designs to gain strength by damping (high frequency) vibrations and to improve heat transfer capability. These designs are serious compromises which can be avoided by interference fitting the liner in the parent bore.

10.4.2 Coolant Related Wear

The engine coolant is a mixture of water and additives to protect against the hardness of water and corrosion. The typical requirements of the coolant and additives are given in Tables 10.2, 10.3, and 10.4. Glycol is added to the coolant up to 50% to depress the freezing temperature (−34 °C at one atmosphere) of the coolant for temperate and colder climates. It also raises the boiling point slightly (107 °C at 1 atmosphere).

10.4.2.1 Electrolytic Corrosion by External Electricity due to Poor Earthing

With an increase in demand for the efficiency of vehicles, engine designs are made lighter with plastics in contact with oil and aluminium with coolant. The corrosion of aluminium parts is slow, but can be disastrous. Corrosion can take place due to electrolysis when the potential difference between dissimilar metals, for example cast iron and aluminium, exists. A survey of heavy vehicles with corrosion problems, for unearthed electricity after prolonged use and no maintenance is shown in Figure 10.9. It can be seen

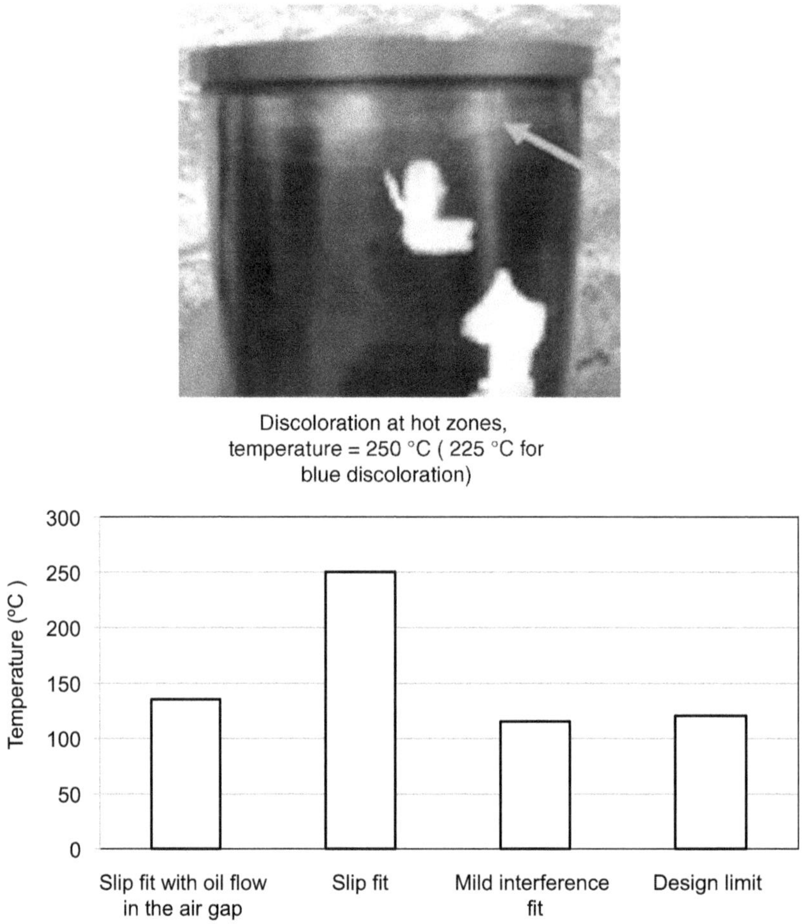

Figure 10.8 Effect of under-cooling by design

Table 10.2 Basic requirements of coolant additives

Test	Specifications
Appearance	—
Density at 20 °C (g/cm^3)	1.112 min.
Boiling point (°C)	155.0 min.
Foaming property (ml)	4.0 max.
30 vol-% aqueous solution	
water content by KFT (%)	5.0 max.
pH value (normal temperature)	
(a) pH Neat	—
(b) pH at 30 vol-% aqueous solution	7.0–11.0
Reserve alkalinity ml	13.5
Freezing point °C	
(a) 50 vol-% aqueous solution (glycol)	−34.0 max.
(b) 30 vol-% aqueous solution (glycol)	−14.5 max.

Table 10.3 Corrosion property of cast aluminium at heat transfer surface

Property	Specification
Change of mass (mg/cm^2), 25 vol-% test solution (135 ± 2 °C, 168 ± 2 hours) Note: no change in colour is allowed	–0.15

Table 10.4 Metal corrosion property: 30 vol-% in corrosive water tested at 88 ± 2 °C, 336 ± 2 hours

Property	Metal test piece	Change in mass (mg/cm^2)
Metal corrosion	Aluminium	± 0.30
	Cast iron	± 0.15
	Steel	± 0.15
	Brass	± 0.15
	Solder	± 0.30
	Copper	± 0.15
	Appearance	There should be no visually noticeable corrosion on the test piece, excluding the part contacting with spacer, but change in colour is permissible.
Foaming during test		No foam flooding out of the cooler.
Properties of solution after test	pH value	6.5–11.0
	Change of pH value	± 1.0
	Rate of change of reserve alkalinity (%)	10
	Liquid phase	No significant change of colour. No significant change of liquid such as separation generation of gel.
	Amount of precipitation (vol-%)	0.5 max.

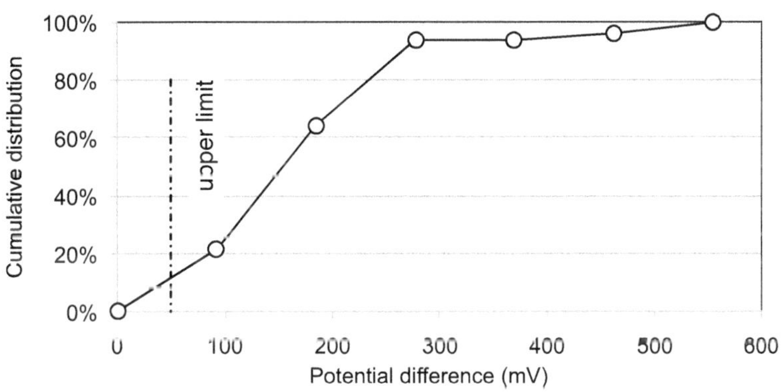

Figure 10.9 Cumulative distribution of potential difference in prolonged use without maintenance, to cause early corrosion of dissimilar metals

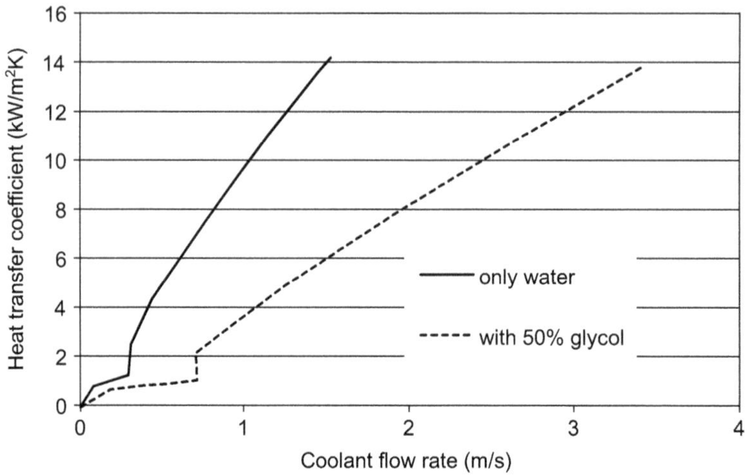

Figure 10.10 Effect of ethylene glycol on coolant side heat transfer coefficient for a radiator tube

that hardly 5% of these vehicles have a potential difference between the engine and the chassis less than the safe limit of 50 mV.

10.4.2.2 Heat Transfer Coefficient with Glycol

Glycol is poor at heat transfer (Figure 10.10). For example, the heat transfer coefficient of coolant mixed (50:50) with glycol is 4 kW/m^2 K, compared to 11 kW/m^2 K for water for a nominal coolant flow velocity of 1 m/s (Kanefsky, Nelson and Ranger, 1999). Therefore, adding glycol to water is not recommended for tropical or warm countries, where the heat exchanger equipment must handle the heat load at a higher ambient temperature. If the burden of glycol is to be carried for high temperature application, the cooling package is correspondingly increased in size, initial cost and fan power. Insufficient cooling raises the coolant temperature, and hence the oil temperature. Higher temperatures decrease the oil film thickness; the critical parts of an engine are pushed to the boundary lubrication regime and subjected to increased wear.

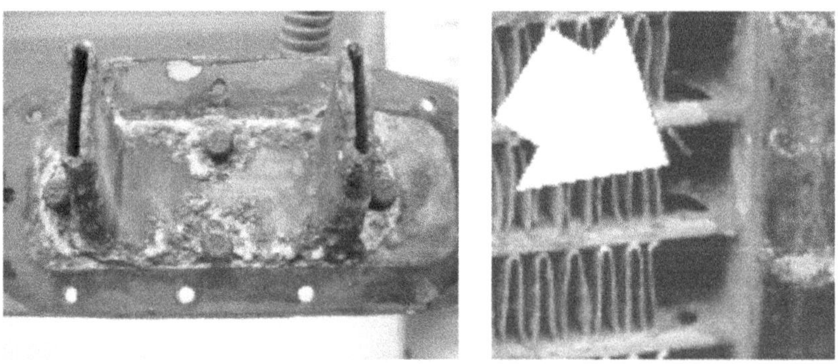

(a) (b)

Figure 10.11 Corrosion caused because of the high acidity of the water. (a) aluminium cover plate; (b) aluminium radiator tubes

10.4.2.3 Corrosive Wear of Aluminium Parts by Acidic Coolant

In the case of organic additives, the acidity of water due to chlorides and sulfates should be limited to 5 ppm. Inorganic additives can handle up to 50 ppm of acidity. Two examples, namely aluminium radiators and covers, are shown in Figure 10.11, where the acidity exceeded 250 and 300 ppm, respectively. Such a situation can arise if the water is initially acidic or the coolant was topped up frequently with poor quality water without adding alkaline additives to neutralize the acids introduced with the top-up water.

10.5 Summary

Experiments on corrosive wear showed that the wear is severe when the coolant temperature is below 60 °C. It can be minimized and eliminated by the selection of low sulfur fuel, lubricating oil with proper corrosion inhibiting additives and, most importantly, by a thermostat assembly. There is no wear in the optimum coolant temperature range of 70–95 °C. At higher coolant temperatures, oil film thins out and boundary lubrication begins.

References

Baumgartner, P., Inhelder, E. and Mikulicio, N. (1987) The development of new Sulzer medium speed engine ZA40S and further development steps on A20 and AT25 engine types. 17th International Congress on Combustion Engines, Volume 2. CIMAC, Frankfurt, Germany

David, A.B., Johnson, J.H. and Daavetilla, D.A. (1977) The Effect of Oil and Coolant Temperatures on Diesel Engine Wear., Technical Paper 770086, SAE, Troy, MI.

Kanefsky, P., Nelson, V.A. and Ranger, M. (1999) *A Systems Engineering Approach to Engine Cooling Design.* Society of Automotive Engineers, Warrendale, PA, ISBN: 076800540X.

Mettig, H. (1973) *Die Konstruktion schnellaufender Verbrennungsmotoren.* de Gruyter, Berlin, Germany.

Mizutani, K., Murata, K., Suzawa, T. and Niitsu, Y. (1996) Analysis of Dry Cylinder Liner Behaviour during Engine Operation. Technical Paper 960059, SAE, Troy, MI.

Schilling, A. (1992) *A text book on Automobile Engine Lubrication*, Volume II. Scientific Publications Ltd, UK.

11

Tribological Tests to Simulate Wear on Piston Rings

A. Igartua[1], X. Fernandez[1], M. Woydt[2], R. Luther[3] and I. Illarramendi[4]

[1]*Tekniker-IK4, Eibar, Spain*
[2]*Federal Institute for Materials Research and Testing (BAM), Berlin, Germany*
[3]*Fuchs, Petrolub AG, Mannheim, Germany*
[4]*CIE Tarabusi, Igorre, Spain*

Substantial energy content of fuel is lost in friction in engines. The reduction of friction due to the surface treatment of the piston rings and cylinder liner, and by lubricating with alternative engine oils, is discussed. In order to reduce experimental effort, tribological tests are performed for coating selection, simulating the friction, wear and scuffing behaviour between the piston ring and cylinder liner, as well as lubricant evaluation. The creation of test standards in piston ring/cylinder liner simulation is discussed. Final tests are performed in engines by the manufacturer of piston rings CIE Tarabusi to validate the results. PVD coatings, High Velocity Oxi-fuel (HVOF) coatings and nitriding treatments have been applied to piston rings as alternatives to chromium coatings. The results are promising and CrN/TiN coating seems very effective in increasing wear and scuffing resistance. Biodegradable engine oils based on esters and polyglycols reduced friction and reached or exceeded the wear resistance of reference engine oils based on hydrocarbons and showed, in combination with tribo-reactive coatings, an increase of the extreme pressure resistance in tribo-test and a reduction of oil consumption during engine tests.

11.1 Introduction

In engines, approximately 15% of the energy content of the fuel is lost due to mechanical friction. Half of this frictional loss is attributed to the friction and wear between the piston, piston rings and cylinder liner walls. Friction losses caused by the piston ring pack are produced by normal pressure against the cylinder

Critical Component Wear in Heavy Duty Engines, First Edition. P.A. Lakshminarayanan and Nagaraj S. Nayak.

wall; this is needed for tightness (for sealing oil and gases). Currently, the pressure of the compression rings reaches around 25 MPa, but in the case of the top ring this pressure is mainly forced from the back side of the ring by the peak pressure of the combustion gases, which in recent years has been increased to values as high as 20 MPa. Secondly, the amount of mixed/boundary lubrication is really higher than expected decades ago. So a reduction in the friction coefficient of piston ring coatings is of utmost importance.

To properly carry out their function, piston rings must be tight on the periphery in contact with the cylinder wall. To ensure a seal throughout the engine life, wear of the peripheral face should be kept as low as possible. Face wear of the top ring increases its clearance gap, leading to increased gas flow through it; in turn this changes the inter-ring pressure values, the secondary movement of piston rings inside their grooves and the gas flow characteristics. The consequences are increased blow-by and oil consumption and reduced engine power output.

11.2 Friction and Wear Tests

The methods of testing wear are key factors in the development of new materials and surface treatment processes when identifying new piston ring and liner materials.

In current engines the most exhaustive tests by the engine manufacturers or road testing are used to evaluate wear. The results are more realistic. However, the high costs and long duration of these tests make it very difficult to carry out a great number of tests (e.g. with different numbers of candidates) to determine the optimal composition of new materials and coatings, as well as of lubricants for the piston rings. (Liang and Strong, 1998)

To develop a new coating and surface treatment processes for the piston ring, it is necessary to produce different types of material until the most adequate one is found. It is necessary to carry out fundamental tests in order to eliminate those materials which are not meaningful, because testing these materials in engines would be very expensive.

As a first step, therefore, some basic tests, which are transferable to the application, allow the meaningless alternatives to be rejected and all efforts to be focused on the most promising ones. (Onoda, Kuroishi and Motooka, 1998)

The next step is to analyse the friction and wear properties of the selected development when applied to piston rings and to compare the properties with, for example, reference piston rings plated with chromium by means of simulation tests, using a specially designed ‘piston ring-cylinder liner’ test configuration.

Finally, to ensure that the simulation test results are transferrable, some engine tests are carried out with the reference chromium-plated coating in terms of calibration and the selected alternatives in the simulation tests.

11.2.1 Testing Friction and Wear of the Tribo-System Piston Ring and Cylinder Liner Outside of Engines

Industry seeks to use test systems that enable a qualitative or even semi-quantitative correlation between affordable model tests and expensive, as well as time consuming, component testing in engines or field testing. The lifetime of engine oils specifications shortens more and more. The number of these tests increases rapidly and fragments between Asia, Europe and North America, increasing the test matrices dramatically. Today, there is a strong demand for test procedures that can rapidly screen potential lubricants, materials and coatings before system-level life tests are performed. They aim to elucidate the effects of parameters (base oil type, finishing, texturing, additives etc.) on friction and wear.

During the last decade, in the USA costs have increased fivefold to complete all the engine tests required for the principal heavy duty diesel (HDD) engine oil category, while passenger car motor oil (PCMO) costs have almost doubled, as:

- Each new set of specifications is technically more complex than the previous one.
- The frequency of issue is increasing.
- Costs are escalating or spiralling.

In Europe, the recent new engine oil specifications of Renault SAS (RN 0720), DaimlerChrysler AG (DC 229.31), Ford WSS-M2C934-A and Volkswagen AG (VW 504.00) display an interest in passenger car engine oils (PCEO) with reduced metal–organic additives. The homologue ACEA (European Automobile Manufacturers' Association) European oil sequences of the category 'Cx' word 'catalyst compatible oils'. The CJ-4 and GF-5 follow in a reduced form this path.

Hydrocarbon-based engine oil formulations have recently reduced the ash content down to 0.7/0.8 wt-% (MidSAP), some with help of ester addition. Prototype engine oils blended with 30–60 wt-% esters having 0.5 wt-% ash (low SAP) are now under evaluation.

The new engine oil denominations 'ESP' (Environmental System Protection) of Mobil or 'Exhaust Treatment Protection' of Fuchs Petrolub underline this approach.

The evolution of engine oil specifications is detailed in Chapter 13. After a flurry of activity and debate since 2005 around the test limits, the International Lubricants Standardization and Approval Committee (ILSAC) members accepted a compromise offered by the oil marketers, which allows ILSAC GF-5 to head straight to an API licensing (API SN) of the Starbust Certification Mark after fall 2010, ensuring that ILSAC GF-5 oils are available for use in 2011 model year vehicles. Significant points are as follows:

(a) Meet the 0.5% improvement in fuel efficiency over GF-4 (GM V6 VVT LaCrosse engine) requiring higher content of friction modifier or the introduction of new friction modifiers.
(b) New phosphorus volatility (retention) test requirements, for example less volatile zinc dithiophosphate (ZDP) additive.
(c) New emulsion ROBO bench test to replace IIIC used oil low temperature pumpability requirement.
(d) Two new bench tests.
(e) An emulsion test.
(f) A panel rust test to help ensure compatibility of GF-5 with alcohol-containing fuels.
(g) Re-introduce the TEOST 33C turbocharger deposit bench test requiring a significant increase in detergency.
(h) The 30 mg maximum limit for the TEOST 33C test effectively eliminates oils containing high levels of molybdenum.

The European sequences contain nine, eight and six engine tests respectively; these are developed by the CEC (The Coordinating European Council for the Development of Performance Tests for Fuels, Lubricants and Other Fluids), which ensures acceptable test precision and then monitors the test and test methods. In contrast, the tests used for in-house specifications are controlled by the OEMs. These engine tests are related to wear, extreme pressure, fuel economy (friction) and soot, detergent behaviour or engine cleanliness.

In order to complete all the engine tests required for actual US-API PC-9 oil or CJ-4 oil, assuming only one test on each test type, an additive company spends around US$ 500 000 or more, which is the same amount as for the European ACEA E7 HDDO oil. Completing all the engine tests required for actual PCMO ACEA A5/B5-04 oil, assuming only one test on each test type, an additive company spends around US$ 223 000, followed by the specific OEM tests.

In consequence, the industry should seek out and remove test redundancy and repetition. Another way to reduce the costs and shorten development time is to carry out a significant number of wear tests outside engines using samples prepared from production parts or certified engines tribologically stressed in a tribometer. This requires an extensive exchange of OEM wear test data achieved in engine sequence tests (like CEC-L-51-A-98 using MB OM602A or CEC-L-54-T-96, using M-111) to be compared with those from the test rigs (tribometers) in order to identify the 'correct' or transferable test parameters and to ensure that the lubricant performance is correctly evaluated. Here, the input of OEMs and component suppliers is crucial for the calibration of the tribometer test. Also, tribometer-based test procedures need to be consistent and effectively monitored across all test laboratories.

11.3 Test Procedures Assigned to the High Frequency, Linear Oscillating Test Machine

Today, several DIN and ASTM test methods (Woydt and Kelling, 2003; Woydt and Ebrecht, 2003, 2008) prove the high technological level and versatility of SRV test philosophy (SRV is the abbreviation in German for 'Schwingung, Reibung, Verschleiß'; in English this is 'Oscillation, Friction, Wear'). The precision statements in the test methods underline 'best in class', which are annually or biannually verified by international round robin tests. The SRV test principle was initially developed in 1968. After more than 42 years of experience with more than 300 test machines worldwide, several test procedures (Woydt and Elbrecht, 2003) have been developed and standardized; these are summarized in Table 11.1.

In China (SAC TC 280) and Japan, SRV related working groups have been established since 2005 in order to transfer and extend these standards to Asian methodologies (for China: SH/T 0784-2006 and SH/T 0721-2002). Based on all these efforts, more and more OEMs base their technical requirements on tribological tests with SRV. It can be said, that the 'high frequency, linear oscillating' test machine 'SRV®' covers a wide range of tribological testing purposes and needs, and represents the ideal technical platform for the issues of piston ring cylinder liner testing outside of engines in light of active OEM support and standardization working groups guarantying highly valuable and transferable test results.

In the basic test configuration for all models, an upper test specimen is rubbed against a lower specimen (Figure 11.1). According to the standardized test procedures, a few milligrams of lubricant (0.3 ml for D6425/D7421) or grease may be placed into the tribo-contact. After the lubricant has been placed on the test specimen and these have been installed in the test chamber, a force is applied mechanically to the upper specimen in a direction normal to the direction of motion at a given test frequency and stroke.

Table 11.1 Standard test procedures referring to the high frequency oscillating test machine

Type of lubricant	DIN	ASTM
Oils, fluids	51834, part 2 (white print)	⇒ D6425-10
	51834, part 3 (white print)	
	51834, part 4 (under preparation)	
Oil EP	51834, part 5	⇒ D7421-08
	ISO TC28	
Greases	51834, as future part 6	⇐ D5706-10
	51834, as future part 7	⇐ D5707-10
Needle and sinker oils	62193, part 1	
Plastic socket suspension joints	51834, part 9	⇒ D7420-10
Solid bonded films (formerly DIN 65593, yellow print 2001, EP and film life)	51834, part 8	⇒ D7217-10
Anti-fretting action of greases		⇐ D7594-10
Evaluating compatibility of binary mixtures of lubricating greases	D6185-08	
Friction and wear of piston ring cylinder liner	(under preparation)	
Extreme pressure properties of engine oils	(under preparation)	
Fuel economy under aged oils	(under preparation)	
Hydrodynamic friction	(under preparation)	
Reciprocating wear test for ceramic coatings (CEN/TC 184/WG5 Document N148)	EN 1071	
Test apparatus	51834, part 1 (white print)	

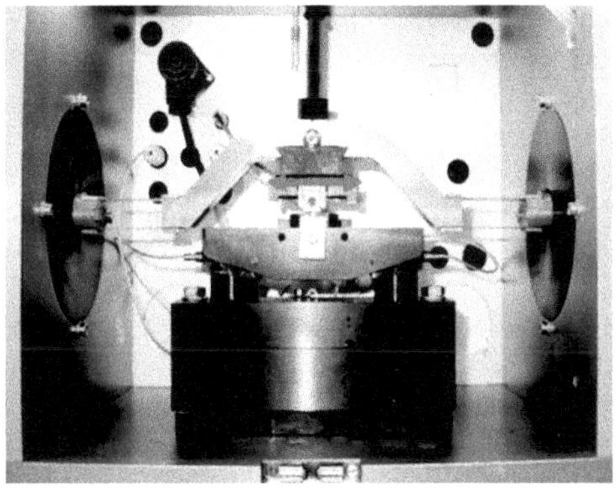

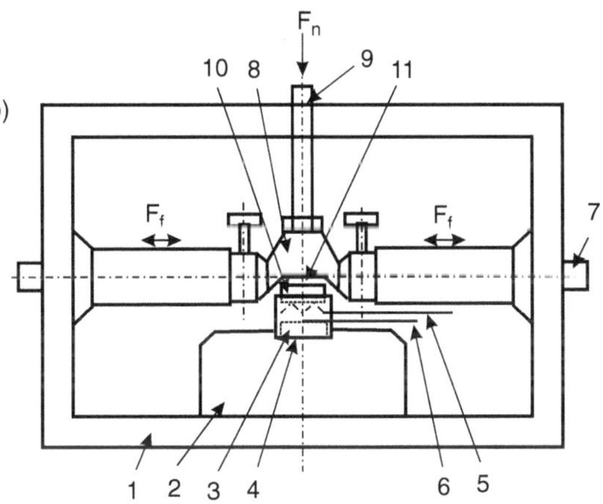

1. Oscillating drive rod
2. Test ball holder
3. Load rod
4. Test ball
5. Test disk
6. Electrical resistive heater
7. Resistive thermometer
8. Test disk holder
9. Piezoelectric measuring device
10. Receiving block

Figure 11.1 (a) Schematic view of the SRV® test chamber and (b) the test arrangement

The friction force is measured continuously by means of a piezoelectric load cell under the lower specimen holder, which is attached to a rigid test block. The wear scar and track dimensions are determined after a given test duration by optical microscopy, three dimensional laser microscopy or stylus profilometry.

The test block and the holder can be heated to +295 °C (optionally up to +1000 °C) and cooled to control the temperature of the lower specimen.

The DIN working group has compiled in detail for internal use the test procedures used at DaimlerChrysler AG, Volkswagen AG, BMW AG, Fiat AUTO, GM, Ford, Toyota, Mahle, Kolbenschmidt, Federal Mogul, Nissan, Tekniker, Tsinghua University, Lanzhou University, Tribology Testing, BAM, AC2T, VTT, NPR, Tarabusi and many others. Some of the test geometries are presented in Figure 11.2. The consensus is clearly visible, that only test specimens prepared from liners and rings are in use.

In the contact area, the amount of oil differs from 0.5 ml over a bath of 30 ml to an oil circuit, which injects the oil from a heated nozzle into the contact. The latter lubrication system is also suited for low activated radio nuclide wear evaluation (nVCT), which enables a real time response on a nano-scale to wear by additives and test parameters. Experience suggests that for tribo-couples with a low wear rate the remaining wear particles in the tribo-contact and oil bath do not affect the ranking. The capability to tilt and rotate the test chamber (from SRV 4 models) ensures that wear particles can leave the tribo-contact.

The test temperatures varied from + 135 to + 200 °C within a mean of around +160 °C. The stroke for friction and wear testing is selected from 2–4 mm and the test frequencies from 30–100 Hz.

11.4 Load, Friction and Wear Tests

11.4.1 EP Test

The test procedures for determining the results under extreme pressure conditions are based on ASTM D5706-10 (greases) and 7421-08 (oil). Cam shaft systems normally do not exceed 1300 MPa of Hertzian contact pressures (but some exceptions persist!).

During a 30 second break-in period the steel plate is vibrated at 50 Hz in a reciprocating horizontal motion with the ball loaded at 50 N. After the break-in period, the load is increased to +100 N using the slow ramp speed rate; this load is maintained for five minutes. The five minute interval includes the loading ramp sequence. Then, the load is increased by + 100 N every two minutes using the slow ramp until a load of 2000 N is reached, or the load limit of the test apparatus is attained, or failure occurs.

11.4.2 Scuffing Test

Meaningful friction and wear results can be determined with the SRV test rig under linear, oscillating sliding motion for mixed lubricated conditions.

When the digital timer reaches 30 seconds (break-in), the load is increased to the chosen load, 300 N, using the slow ramp speed rate. This load is maintained until the end of the test or failure occurs.

11.4.3 Reagents and Materials

11.4.3.1 Test Rings

The liner specimen is prepared from production engines and the mating piston rings used. Preferably, the diameter of the piston ring should be $5 \pm 0.5\%$ smaller than the bore diameter of the liner and the length (along the axis) of the liner test piece is 20 ± 1 mm and 10 ± 1 mm wide (Figure 11.3).

Before running the test, the ring specimen has to be weighed or the profile in the centre of the contact zone has to be measured by means of profilometry or by interferometry, if the surfaces are for not transparent, in order to determine the wear loss after the test at the same place.

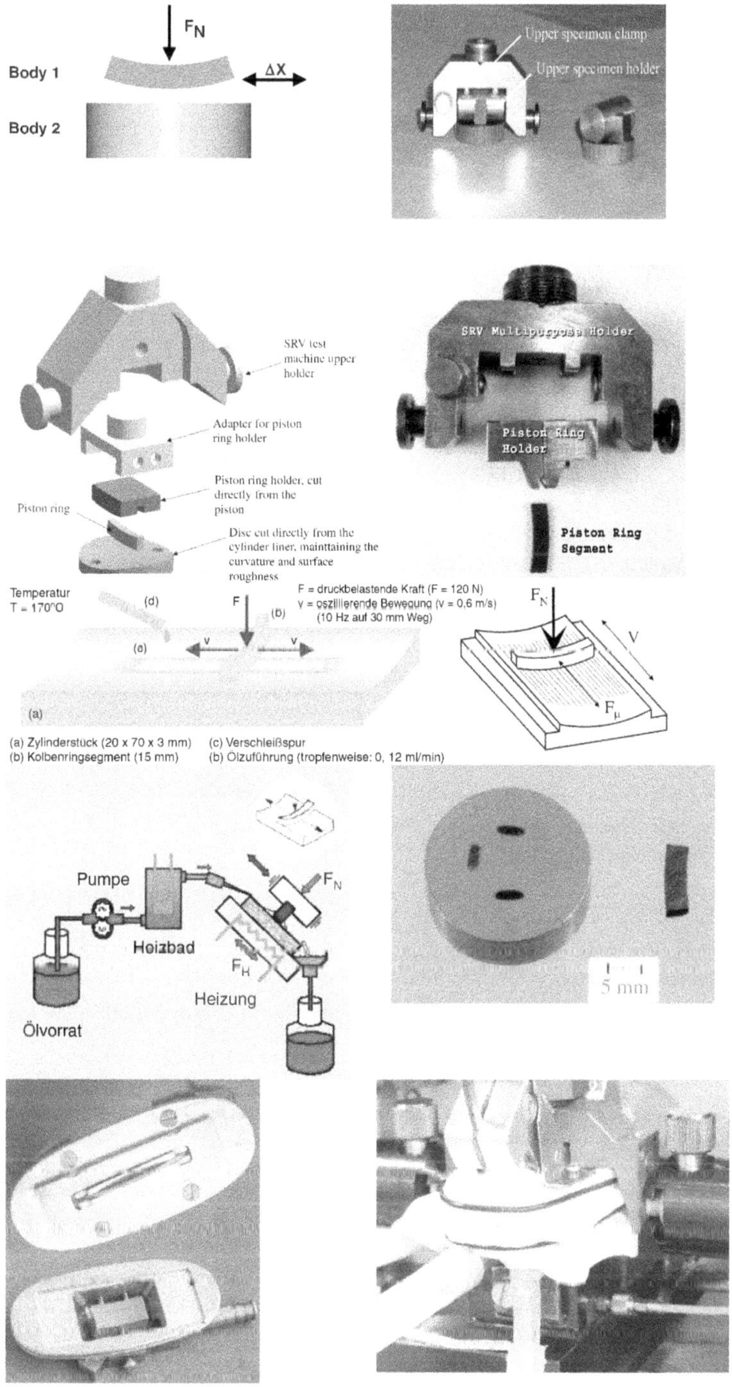

Figure 11.2 Examples for testing piston ring/cylinder liner using the high frequency, linear oscillating test machine SRV

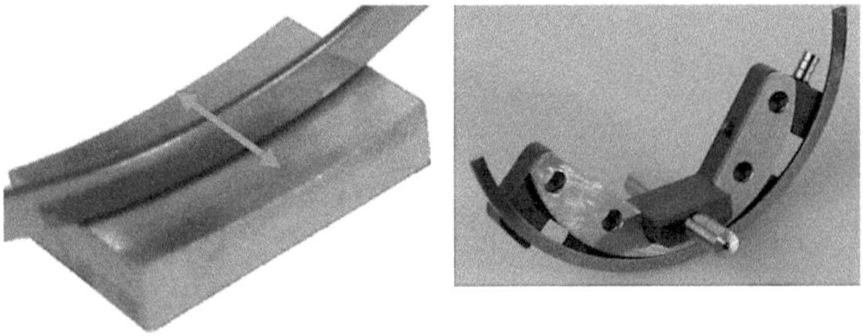

Figure 11.3 Schematic view of test geometric piston ring/liner and ring holder, SRV

11.4.3.2 Test Liners

The liner specimens were cut directly from honed cylinder liners (Figure 11.4) of a certified engine to a size of at least six millimetres along the stroke length, 25 mm in width and with a minimum thickness of five millimetres. The surface opposite to the tribological contact has to be machined flat and the topography of the liner specimen has to be determined by four roughness values (in μm): R_z, C.L.A. (R_a), R_{pk} and R_{vk} (Chapter 5).

Before running the test, the liner specimens have to be weighed. The preferred procedure to determine the wear volume is stylus profilometry (three dimensional confocal microscopy, three dimensional stylus profilometry and/or laser profilometry).

The determination of the wear volume of piston ring and liner specimen represents a key issue, because the parts have a deviation in shape and differ from engine to engine. Table 11.2 presents the outcome of three different methods. Even the laser profilometer operates in a noncontacting mode; the measuring time is 15 times faster than with the stylus profilometer. It is believed that the stylus profilometer determines the most realistic figure. On the other hand, the very short scanning time of two seconds makes the white light profilometer attractive, even if the figure for the wear volume is 20% smaller than those determined by the stylus profilometer.

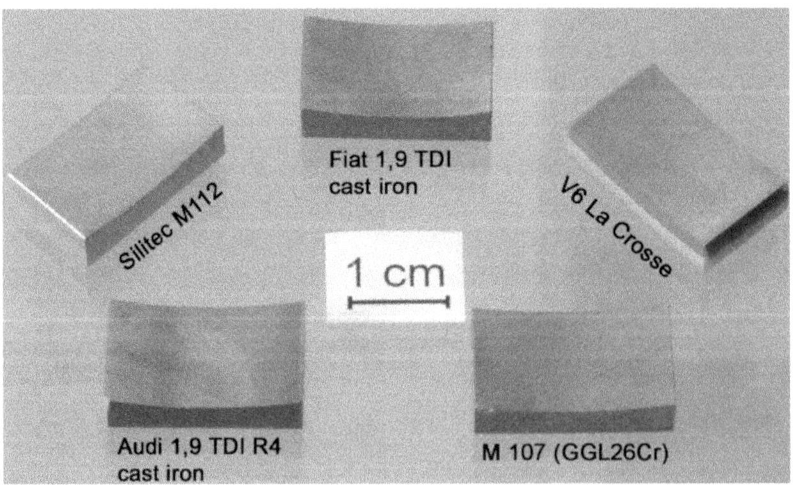

Figure 11.4 SRV test samples prepared from certified engines

Table 11.2 3D wear volume of liner specimen and imaging of a wear track of a grey cast iron liner specimen

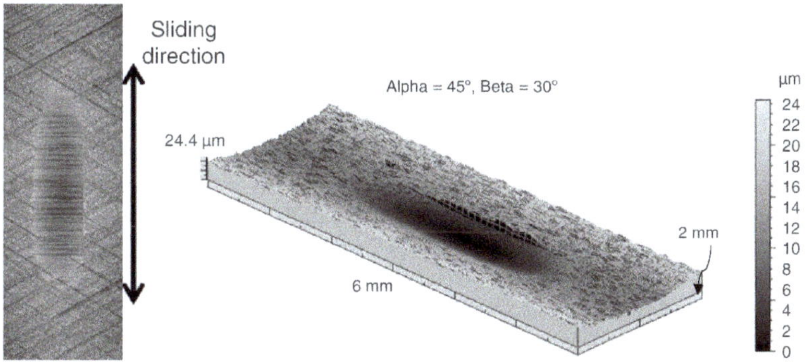

Measuring principle	Scanning time	Wear volume of disk (µm^3)	Wear volume of disk (mm^3)
3D Stylus profilometer	240 min	12 200 000	0.0122
Laser profilometer	16 min	11 317 301	0.0113
White light profilometer	2 s	10 030 000	0.01

Liner specimen: GGL26Cr: Ring specimen: nitrided; T = 135 °C, FN = 100 N, v = 50 Hz, Δx = 2 mm, Castrol FF 0W-30 (VP1)

11.5 Test Results

11.5.1 Selection of Coatings for Piston Rings

Power density and thermal load of diesel engines have dramatically increased in recent years with heavily super-charged diesel passenger car engines expected to reach 90 kW/l and/or brake mean effective pressure (bmep) of up to 2.5 MPa by 2012. As increased thermal loads promote the risk of piston ring scuffing, new piston ring coatings need to be developed to overcome this problem.

For many years, a large number of investigations into piston friction and wear have been carried out in order to obtain greater engine power, greater thermal efficiency, lower emissions and the longest life (Shen, 1997).

For many years chrome plating has been an almost universally used coating for the peripheral face of top rings, due to its excellent anti-wear and sliding properties. In recent years, however, the use and emissions of hexavalent chromium (Cr^{VI+}) have been banned by various regulatory bodies (European Directive 200/53/EC) due to its adverse health and environmental effects.

Chrome coatings have not been able to overcome the scuffing problems, especially those with particulates, so more sophisticated coating procedures have been introduced, the most popular being plasma spraying of metal–ceramic mixtures.

Molybdenum-based ring coatings, such as Tarabusi PL72, have represented for decades a reference for ring coatings in diesel engines, but the cost of molybdenum has spiralled. Therefore, other coating processes and compositions (Landa *et al.*, 2004) have been developed and tested.

In the case of petrol engines, the continuous trend to reduce the top ring height down to 1.0 or 1.2 mm, has favoured the introduction of nitrided steel piston rings, due to the advantage that the ring sides are also nitride coated and, therefore, more wear resistant.

Composite coatings based on an electroless nickel–phosphorus matrix reinforced by silicon carbide (SiC) particles are well known as coatings on AlSi liners. As silicon nitride (Si_3N_4) has been shown to have optimal physical characteristics, such as low density, high toughness and good sliding properties, it was considered as a suitable particulate in coating materials.

Physical Vapour Deposition (PVD) coating has been extensively used to coat cutting tools, increasing considerably the tool life, and recently the use of PVD coatings applied in engine components has been reported (Masuda *et al.*, 1997). In this reference, the multilayer coatings CrN/Cr/TiN/Ti and WC/C have been successfully applied to Tarabusi piston rings.

In the 1980s the high velocity oxi-fuel (HVOF) process was been introduced as an alternative to plasma spraying, due to the advantages of improved adherence to the substrate as a consequence of the high velocity of the spraying, and less decarburation of certain ceramics, due to the lower temperature of the process. Today, thermally deposited liner coatings based on FeC0.3 or FeC0.8 wires are in production in small volume applications, such as PTWA (Plasma Transferred Wire Arc) in the Ford Mustang V8 or Twin Arc Wire Spraying (TWAS) in the Mercedes AMG.

Thermal sprayed Cr3C2-NiCr coatings are being used to improve the wear resistance and aim to decrease the friction coefficient between various sliding components in automotive applications. They are widely used in high temperature wear resistance and corrosion resistant applications in the aerospace and powder engineering industries (Zimmermann and Kreye, 1996; Stein, Schorr and Marder, 1999; Barbezat, Nicoll and Sickinger, 1993).

11.5.1.1 High Pressure-High Velocity Oxi-fuel (HP-HVOF), Physical Vapour Deposited (PVD) and Electrolytic Composite Coatings

The coatings studied were deposited on test samples made from the piston ring substrate (spheroidal graphite cast iron), using the HVOF thermal spray process. The feedstock powder was a commercially available 2075-NiCr (Cr2C3 75% + NiCr20 25%) and WC86-Co10Cr4 powder produced by WOKA (now Sulzer Metco). The 2075-NiCr powder was mechanically treated to obtain a nano powder. Later, the HVOF thermal spray process was used by Thermico to produce conventional and nano-crystalline CrC-NiCr coatings.

Tribological tests were carried out by Tekniker using a 'cylinder-on-disc' basic configuration (Figure 11.5) in order to compare the friction and wear behaviour of the 75Cr3C2-25NiCr coatings with small powder size (called nano coatings), in relation to the standard 75Cr_3C_2-25NiCr//WC86-Co10Cr4 and the reference electrolytic chrome plating.

The different coatings deposited on the samples of the piston ring material (cast iron) were tested using the cylinder liner material as counter material in a lubricated test under the following conditions:

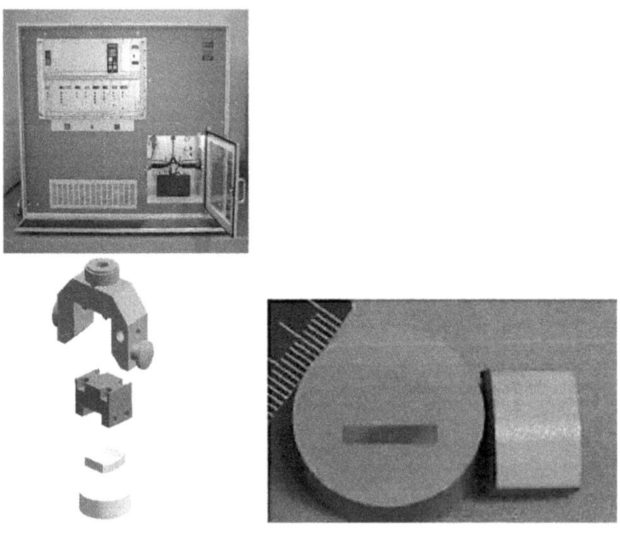

Figure 11.5 Cylinder-on-disc basic configuration, Tekniker

Table 11.3 Cylinder-on-disc test results

	Powder size (μm)	Roughness (R_a, μm)	Average friction coefficient	Total mass loss (mg)
Coatings without grinding				
WC86-Co10Cr4	15–30	4.16	0.375	22.17
$75Cr_3C_2$-25NiCr Standard	10–30	4.87	0.390	9.39
Nano-$75Cr_3C_2$-25NiCr	1.4–30	2.1	0.195	1.03
Nano-$75Cr_3C_2$-25NiCr	10 max.	1.81	0.196	1.03
Nano-$75Cr_3C_2$-25NiCr	5 max.	1.78	0.187	1.9
Coatings after grinding				
WC86-Co10Cr4	15–30	0.5	0.163	2.17
$75Cr_3C_2$-25NiCr Standard	10–30	0.5	0.158	0.45
Nano-$75Cr_3C_2$-25NiCr	1.4–30	0.3	0.160	0.27
Nano-$75Cr_3C_2$-25NiCr	10 max.	0.5	0.167	0.30
Nano-$75Cr_3C_2$-25NiCr	5 max.	0.4	0.160	0.27
Electrolytic chromium (reference)		0.5	0.182	0.45

$F_N = 200$ N, $\nu = 50$ Hz, $\Delta x = 1$ mm, $T = 100\,°C$ and $t = 60$ minutes. The reference oil was Repsol 15W-40. The results are presented in Table 11.3.

As a result of these tests, it can be seen that the surface roughness is, in general, the most critical parameter determining the friction and wear properties under lubricated conditions. The friction and wear properties of the different HVOF coatings improve considerably when the roughness is reduced by means of a grinding process.

The reduction of the powder size in the 75Cr3C2-25NiCr coating produces an important reduction in the coating roughness and, consequently, an improvement in the friction and wear properties.

The differences between the coatings are not significant when the coatings are ground. The 75Cr3C2-25NiCr presented better wear properties than the alternative WC86-Co10Cr4.

A chemical and metallographic analysis of the reference coating (chrome) was performed to compare them with the nano-HVOF, PVD and electrolytic composite coatings applied to the piston rings. All the coatings were produced by Tarabusi (CIE Automotive Group). Table 11.4 summarizes the thickness and the hardness of the layers.

Table 11.4 Chemical and metallographic analysis (Landa *et al.*, 2004)

Coating Process	Coating composition	Piston ring code	Thickness (mm) UNE-EN ISO 1463:1996	Layer hardness (HV) UNE-EN ISO 6507	Roughness (μm)
HP-HVOF	'A' coating	Nano-$75Cr_3C_2$-25NiCr Powder size 10 μm max.	0.15	801	0.11
HP-HVOF	'B' coating	WC CoCr//CrC-NiCr Powder size 30–40 μm	0.19	1039	0.33
HP-HVOF	'D' coating	$75Cr_3C_2$-25NiCr Powder size 45–65 μm	0.13	798	0.23
Electrolytic	NiPCO + Si_3N_4	NiPCO + Si_3N_4	15 μm	1057	0.245
PVD	TiN-Ti/CrN-CrN	CrN/TiN	8.5 μm	1965	0.175
PVD	WC/C	WC/C	2.4 μm	1052	0.065
Electro-deposition	Cr	Cr	0.145	1136	1.21

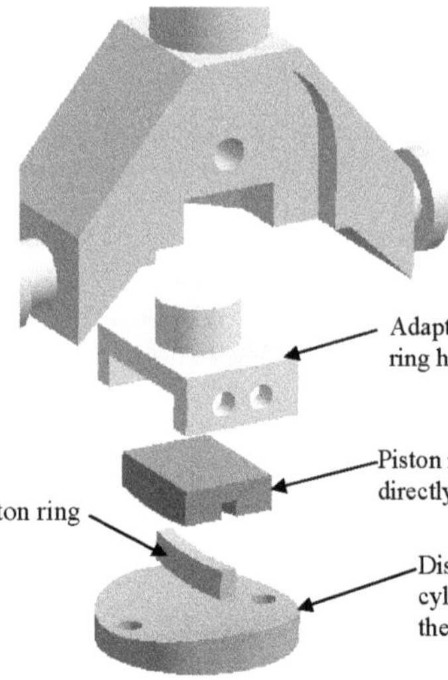

Figure 11.6 Piston ring–cylinder liner configuration, Tekniker (Landa *et al.*, 2004)

11.5.2 Scuffing Tribological Test

The tribometer simulates the sliding reciprocating motion between the piston and piston wall in an operating engine (Figure 11.6) (Höhn and Michaelis, 2001). The piston rings move in the same groove in the piston, as it is the case in the engine.

The piston ring segments performed a reciprocating motion with a stroke of three millimetres and an oscillating frequency of 50 Hz. Normal load was 50 N during a short run-in period (30 s) and 300 N during the test (90 min). During the test, the piston ring segment and the cylinder liner sample were immersed in the oil, whose temperature was constant at $+200\,^{\circ}$C. The reference oil was Repsol 15W-40.

The diamond-like PVD coating 'WC/C' has the lower friction coefficient and wear rate, probably caused by the low roughness.

The newly developed coatings 'A' (Nano 75Cr3C2-25NiCr) and 'D' (75Cr3C2-25NiCr) have better friction and wear properties than the reference chrome-plated piston ring. However, when changing the composition to WCCoCr/CrC-NiCr (coating 'B'), no good results were obtained. When only the powder size is decreased, the piston ring with a lower powder size exhibits better behaviour than a piston ring with a bigger powder size. Abrasive, large particles remain in contact, causing an important increase in the friction and wear. In addition, the NiPCO + Si_3N_4 and TiN-CrN coatings have better friction and wear properties than the reference chrome-plated piston ring (Table 11.5).

Finally, engine tests were carried out by Tarabusi (CIE Automotive Group) to compare the behaviour of the developed coatings with reference piston rings (Table 11.6) coated with chrome and nitrided steel. For this purpose, a turbo diesel engine was selected, where the reference and developed coatings have been tested at full load and high temperature. For assessment purposes, wear levels of the cylinder block (bore), piston (first groove sides near the top ring) and piston ring (face and sides) and fuel consumption were used. Wear and fuel consumption results were judged favourably for both the coatings developed. The lower wear was observed for the CrN/TiN PVD coating followed by the CrN coating, the WC/C, 75Cr3C2-25NiCr coatings and NiP + Si_3N_4 coatings. The lowest fuel consumption was obtained with the nano-HVOF coating and the WC/C PVD coatings.

The engine employed for this task (Figure 11.7) was a four-stroke water cooled direct injection turbo diesel with the following characteristics: bore 90 mm, stroke 90 mm, displacement 2.0 dm^3, power 50 kW at 3000 rpm, and maximum torque 172 Nm at 1800 rpm.

Table 11.5 Piston ring–cylinder test results with different coatings, Tekniker

Deposition process	Coating composition	Friction $f_{average}$	Piston Ring Mass loss (mg)
Electrodeposition	Hard chrome	0.16	0.2
Electrolytic	NiPCO + Si_3N_4	0.15	0.1
PVD	TiN-Ti-CrN-CrN	0.16	0.0
PVD	WC/C	0.13	0.0
HP-HVOF	Nano-$75Cr_3C_2$-25NiCr Powder size −10 μm	0.15	0.1
HP-HVOF	WC-CoCr/CrC-NiCr Powder size -30 + 10 μm	0.21	0.0
HP-HVOF	$75Cr_3C_2$-25NiCr Powder size -45 + 20 μm	0.15	0.2

To compare results, eight different tests were carried out, two using reference coatings (chromium over spheroidal graphite cast iron and gas nitriding of steel), and the rest using the coatings previously described.

The parameters applied in all the tests to assess the functional behaviour were the scuff resistance and wear resistance.

Wear evaluation after engine testing was performed on the piston ring coatings and on the contact surfaces of pistons and blocks (cylinder bores), namely:

- Piston ring faces (measured as closed gap increase).
- Piston ring sides (height decrease).
- Sliding surface of cylinder bores (bore diameter increase in the top dead centre position of the first ring).
- Piston top groove sides (groove height increase).

11.5.3 Hot Endurance Test

The target of the experiment was to prevent the piston ring sticking and to measure the wear of parts involved.

The test conditions were:

- Torque: 167 Nm
- Speed: 3000 rpm
- Water temperature: 110 ± 5 °C
- Oil temperature: 130 °C max.

Table 11.6 Piston ring set is composed of three units: two compression and one oil control

TOP	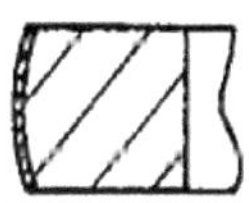	Rectangular, barrel face, coated
2nd	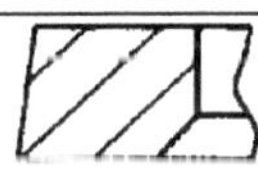	Taper face, reverse torsion
3rd	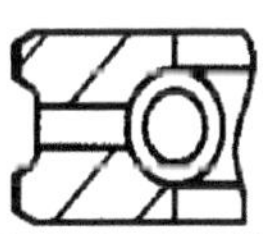	Spring loaded

Figure 11.7 The engine arrangement in the test bed, CIE Tarabusi

Wear test results are displayed in the Figures 11.8, 11.9 and 11.10.

It was demonstrated that wear results show the same trend as measurements based on closed gap increase. Additionally, at the end of hot endurance tests, fuel consumption measurements, using the AVL 733 S device, were made. The values obtained are plotted in Figure 11.10.

Engine tests confirmed the previous results, being possible to select CrN/TiN PVD coatings from the point of view of wear and 75Cr3C2-25NiCr nano coatings from the point of view of fuel consumption. The Nano-75Cr3C2-25NiCr allows a reduction of 50% in coating thickness, when compared to chromium. The diamond-like WC/C coating deposited by PVD represents a good compromise between wear, friction and low fuel consumption. The advantage of the PVD coating is that it is possible to deposit a thin layer of around eight micron at the last step on a piston ring manufacturing process, while keeping the shape of the piston ring unaffected (Landa *et al.*, 2004).

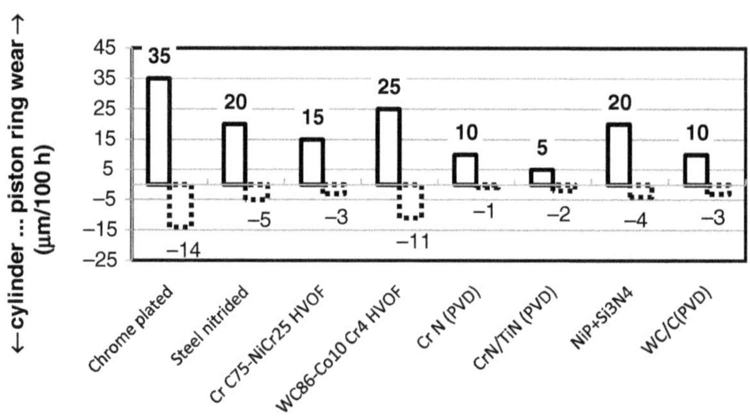

Figure 11.8 Piston ring and cylinder wear under hot endurance test conditions, CIE Tarabusi: (1) closed gap increase (piston ring wear) is plotted along the positive axis (upward); (2) Bore diameter increase in top dead centre position of the first ring plotted along the negative axis (downward)

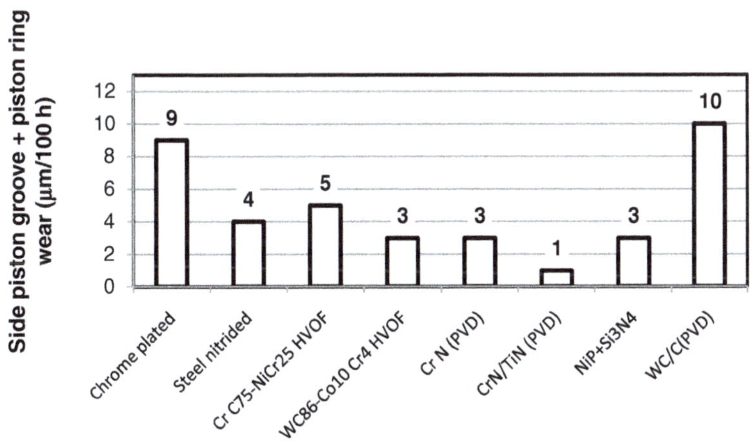

Figure 11.9 Piston ring wear under hot endurance test conditions, CIE Tarabusi

11.5.3.1 Influence of Nitrided Piston Ring in Place of Hard Chrome in Cylinder Liner by PVD

Scuffing Tribological Test

Cylinder liners and test samples were coated using magnetron sputtering. To achieve adhesion on the internal side of the cylinder, it was necessary to tune up the process conditions, so a plasma penetration was developed by GENCOA, UK. Figure 11.11 (right) shows the plasma penetration of an enhanced ion cleaning method. Only when certain conditions were achieved did the plasma penetrate into the cylinder liner. Enhanced ion cleaning enabled coating adhesion to pass the HRc indentation adhesion test.

The coating process could be performed from both ends of the cylinder liner. Most of the wear occurs on the hot side of the cylinder at the point of minimum piston ring linear speed. Therefore, it was decided to coat the cylinder from a single side only. In this way there is a coating gradient from 10 µm on the near side of the plasma to 1 µm on the far side. Cast iron piston rings were supplied by Tarabusi (CIE Automotive) in Spain and nitrided by means of a plasma deposition process by ELTRO, UK. The cylinder liner was supplied by CROMODURO, Spain.

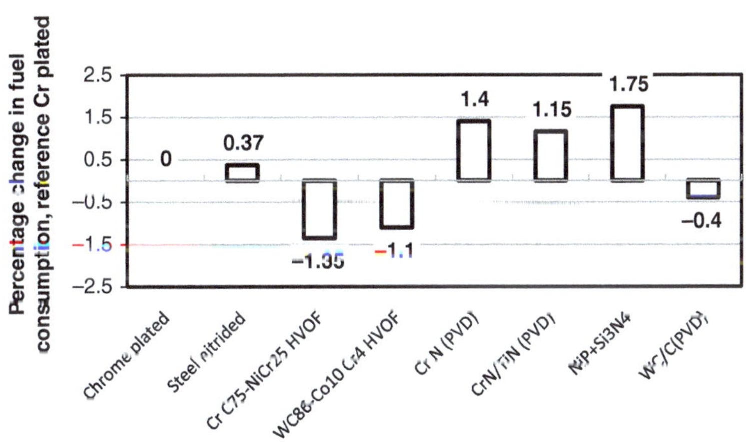

Figure 11.10 Fuel consumption measurements, CIE Tarabusi

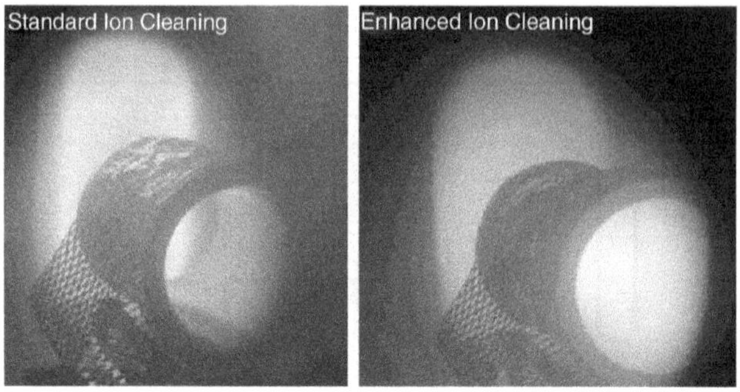

Figure 11.11 Photo shots of cylinder liners in front of Cr target during ion cleaning; the standard ion cleaning (left) did not produce plasma penetration whilst an enhanced ion cleaning (right) produced the desired penetration (Igartua *et al.*, 2004)

Once cylinder liners were coated, samples for different tests were cut out. Figure 11.12 shows a cylinder after samples were cut out for tribological tests performed by Tekniker. The tribological test assembly with ring holder can be seen in Figure 11.6.

The evolution of the coefficients of friction for different tested coating treatments is presented in Figure 11.13. Counterpart material was taken from cast iron piston rings, supplied by Tarabusi. Testing uncoated cylinder liner material against as-received piston rings produced very little ring mass loss (0.56 mg), but significant mass loss in the cylinder material (12.69 mg) and overall wear of 0.209 mm. However, when chrome plating was applied to the cylinder liner, ring mass loss increased to 69.14 mg and overall wear to 0.548 mm (Table 11.7). Subsequent trials therefore used cast iron piston rings that had been pulse plasma nitrided by ELTRO (UK). This nitriding treatment considerably reduced the mass loss from both surfaces. The reference oil was 15W-40 motor oil from REPSOL

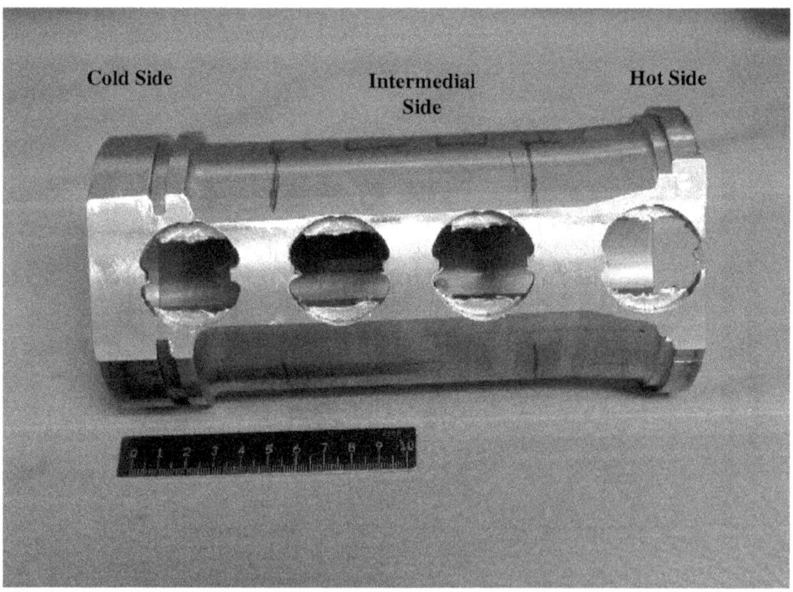

Figure 11.12 Photo of cylinder liner with the scheme to prepare test samples, TEKNIKE

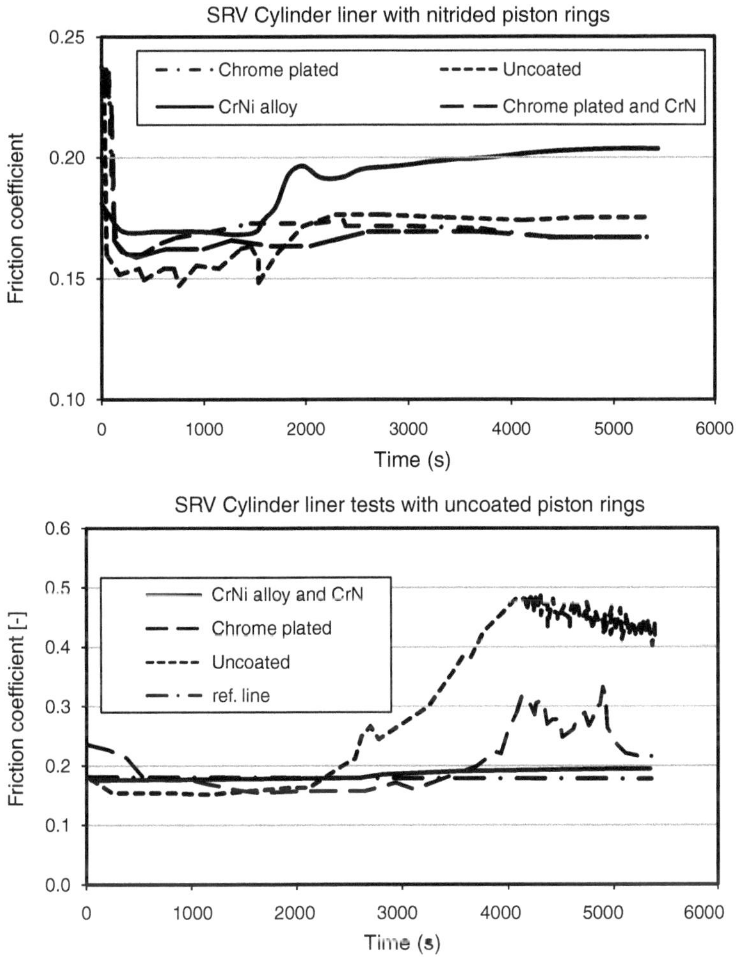

Figure 11.13 Friction coefficient for uncoated, chrome plated coated and PVD coatings (CrN and CrNi alloy), Tekniker

Table 11.7 Piston ring–cylinder test results with different coatings (Igartua *et al.*, 2004)

Cylinder liner	Piston ring	$f_{average}$	Disc mass loss (mg)	Ring mass loss (mg)	Wear (mm)
Chromium plated	Uncoated	0.186	—	69.14	0.548
Uncoated		0.267	12.69	0.56	0.209
CrN		0.172	—	1.96	0.110
CrNi alloy		0.183	1.0	0.51	0.147
Chromium plated	Nitrided	0.167	—	8.04	0.156
Uncoated		0.167	2.1	0.18	0.120
CrN		0.168	—	2.06	0.139
CrNi alloy		0.189	1.29	0.133	0.143

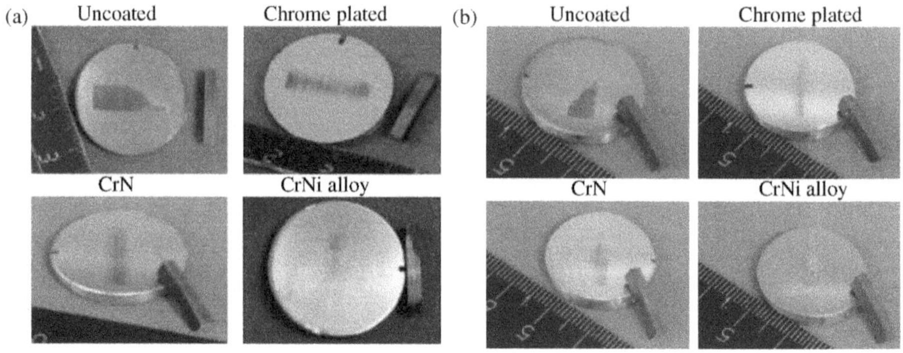

Figure 11.14 Wear tracks on cylinder liners after 1.5 hours using SRV tribo-tests (a) when using uncoated piston rings; (b) when using nitrided piston rings, Tekniker

When using untreated cast iron piston rings, the uncoated cylinder liner presented the worse wear performance, followed by the chrome-plated cylinder liner samples. Both PVD coatings (CrN and CrNi alloy) performed well in this test with uncoated piston rings. Both CrN and chrome plated performed well in this test sliding against nitrided piston rings. Images of the wear scars for samples can be seen in Figure 11.14, where the uncoated sample clearly shows the biggest wear scar and where the chrome plated and CrN samples show a similar wear scar. To achieve a minimum wear and friction in the piston ring cylinder liner, it seems here that the type of material of the piston ring has more influence than the cylinder liner material itself. Nitriding piston rings decreased the friction and wear independent of the liner material.

Although PVD coatings offered an alternative to traditional chromium plating due to environmental impact considerations, the latest advances in the chromium plating processes have matched the values of process 'cleanliness'. When testing PVD and chromium plating coatings under laboratory conditions, it was possible to obtain better results for PVD coatings on friction and wear with uncoated piston rings, but when the different cylinder liners were tested with the nitrided piston rings the friction and wear behaviour were quite similar.

Using standard engine working conditions, the chromium-plated cylinder liners performed better than the CrN coated ones. When testing PVD coatings under scuffing conditions, the performance of chromium coatings decreased, then again PVD coatings represent a good alternative for modern turbocharged engines, where higher combustion pressures are involved (Igartua *et al.*, 2004).

11.6 Selection of Lubricants

Oil is necessary at the piston ring and liner interface to reduce friction and wear through hydrodynamic or mixed lubrication, as well as to prevent seizure and to cool. In addition to the lubricating aspect, the oil acts as a 'heat transfer fluid' which moves away heat (combustion, frictional heat) from the piston and the ring–liner interface. The oil is further needed in the ring groove to prevent the ring from sticking in the groove and to clean the groove.

Oil is supplied to the piston and piston rings from the crankcase, directly or indirectly. The oil supply method usually depends on the size of the engine and on the required amount of oil. Smaller, high speed engines use splash lubrication, as the amount of oil supplied by this way is usually sufficient.

Larger engines that require a higher amount of oil need to have the oil supplied up into the piston. One solution for oil supply to the piston is that the oil is fed from the main bearing or bearings to the crankshaft and further on to the connecting rod and the piston.

To limit the oil flow volume, which is directly linked to the fuel consumption, as the oil pump is directly driven by the crank shaft, alternative esters with viscosity indices above 200 (Yoshio, Akira and Tatsuya, 2008) or with higher volumetric heat capacities (Woydt, 2007), like polyalkylene glycols, are under development. Another option is the use of oil pumps with a controlled oil flow volume.

11.7 High Performance Bio-lubricants and Tribo-reactive Materials for Clean Automotive Applications

In a new approach, environmentally friendly lubricants will be formulated by substituting the additives by the base oil's molecular oxygen (polarity). In addition, the functionality of the EP/AW additives will be transferred to the metallurgy of the surfaces by the introduction of new types of materials, especially new coatings (Landa *et al.*, 2007, Gili *et al.*, 2010, Igartua *et al.*, 2009). Besides functional targets, the alternative engine oils comply with eco-toxicological (environmental) requirements, as well as the requirements to have a renewable source content and mid/low/no SAPs.

11.7.1 Synthetic Esters

This type of 'chemical' bond covers a broad range of substances (mono-, di-, TMP- and tetra-esters as well as complex'esters). Their polar base components (alcohols and carboxylic acids) may have a natural or synthetic origin (0-90% content of naturally grown components). They can be more or less biodegradable, depending on their structure; their thermal/oxidative stability and cold flow properties can be much better than those of triglycerides. Before ecological aspects became part of lubricant development, ester oils were used in special lubricants for technical reasons, for example as base fluids for aviation turbine oils, hydraulic oils and 'lubricity additives' for fuel economy oils. Ester oils are predestined as base oils for engine oils by combining high viscosity index, lubricity and low evaporation loss, with full compatibility with the world of mineral-based lubricants and fuels. Moreover, suitable synthetic esters can be produced from renewable raw materials, for example by chemical modification of vegetable oils. With regard to the important application criterion of oil viscosity, the combination of dicarboxylic acids (e.g. azelaic acid (C9), sebacic acid (C10) or dodecanedioic acid (C12)) with fatty acids will be an excellent instrument for adjusting viscosity.

11.7.2 Polyalkyleneglycols

The polyglycols (aliphatic polyethers) aim to significantly reduce the viscosity at low temperatures in order to improve the fuel economy in city driving cycles. This is achieved by ensuring that the base oil has a high, intrinsic viscosity index, but can be further improved by viscosity index improvers.

Firstly, these offer lean burn behaviour and reduce polymeric deposits on the intake valves in direct injecting engines, as they are polymer free. Secondly, as the polyglycol-based formulations are ash free, these formulations do not increase the back pressure of the particulate filter over time. Thirdly, theses formulations are metal free, thus do not cause deterioration of the lambda sensor and contamination of the active centres on catalysers.

Polyalkyleneglycols with different backbone compositions, namely polypropylene glycols (PPG 32-2/ B48), polyalkylene glycols (PAG D46-4/D23) and fully hydrocarbon soluble polyglycol (PBG B20), were used.

It is believed that the PPGs are better suited for the dilution of biodiesels and the PAGs for the alcoholic-based biofuels. Polypropyleneglycols (PPG) are (partially) miscible with mineral oils or esters and are ultimately biodegradable. Fully hydrocarbon soluble polybutylene glycol (PBG, B20) may bridge the gap between PPGs and PAGs.

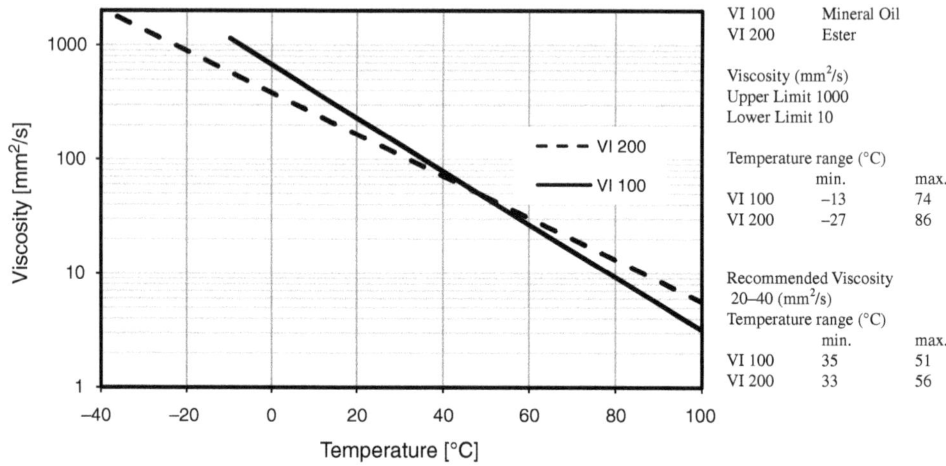

Figure 11.15 Viscosity–temperature behaviour of esters and mineral oils, Fuchs

Therefore, additionally to the use of hydrocarbon/ester blends containing renewables, intensive R&D efforts are being made to develop polyglycols towards this end.

11.7.2.1 Relevant Properties of Ester Oils and Polyglycols

Viscosity–Temperature Behaviour

The use of an ester or PAG instead of a hydrocarbon-based base oil improves the viscosity–temperature behaviour of formulations. This general relationship is displayed in Figure 11.15.

The higher viscosity index of such uncommon base oils results in a better fluidity at low temperatures, but with the recommended high temperature, high shear (HTHS) viscosity. The impact of these high viscosity index formulations on fuel economy in city driving especially is obvious. Recently, low viscosity engine oil formulations with a high viscosity index based on PAO4 (also with PAO6) blended with a high content of polymers appeared in the market. The NOACK volatilities of $>13\%$ of these formulations represent a weakness in terms of oil consumption and thickening over drain.

Esters or polyglycols enable the reduction or elimination of viscosity index improvers from the formulations (Igartua *et al.*, 2009).

Boundary Lubrication

Most vegetable oils, synthetic esters and glycols displayed excellent lubricity in boundary lubrication conditions exceeding the extreme pressure capabilities of mineral base oils. This is well established in gear lubrication. Experimental investigations on twin disc test rigs showed that the friction coefficient of vegetable oils, synthetic esters and glycols (Laukotka, 1985; Godet, Roussel and Thebault, 1984) is half that of mineral oils (Bartz and Wienecke, 2000; Laukotka, 1985) as shown in Figure 11.16.

Figure 11.17 displays extreme pressure loads at $+135\,°C$ with the associated Hertzian contact pressures for different, fully-formulated alternative engines and factory fill engine oils based on hydrocarbons (Woydt, 2010; Igartua *et al.*, 2009). The load carrying capabilities until seizure of the hydrocarbon-based engine oils are lower than those of the polyglycols and hydrocarbon/ester blends (Fuchs), whereas the latter presented the highest values. In the case of diols (EO/PO-polyglycols), the load carrying capabilities of the unadditivated base oils were superior to the corresponding additivated formulations. The calculated Hertzian contact pressures until seizure of all engine oil formulations exceeded the FZG load stage of 14 ($\cong 2.170\,MPa$).

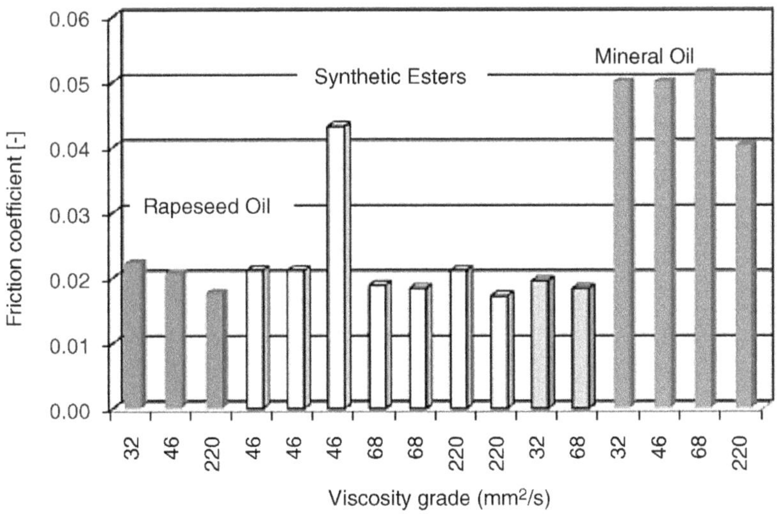

Figure 11.16 Friction coefficients of various base fluids, (Höhn and Michaelis, 1994)

11.7.2.2 Additivation Concepts for Alternative Versus Hydrocarbon-Based PCEOs

The evaporation loss of esters and PAGs as base oils are, at iso-HTHS, in the range of <50% of those of hydrocarbons, which reduces the oil consumption. Those of PAGs can be a further 50% lower than those of ester-based formulations. Due to this it will be probably possible to decrease the emissions and oil consumptions.

The low pour points, especially of esters, means that they do not require pour point depressants (Table 11.8).

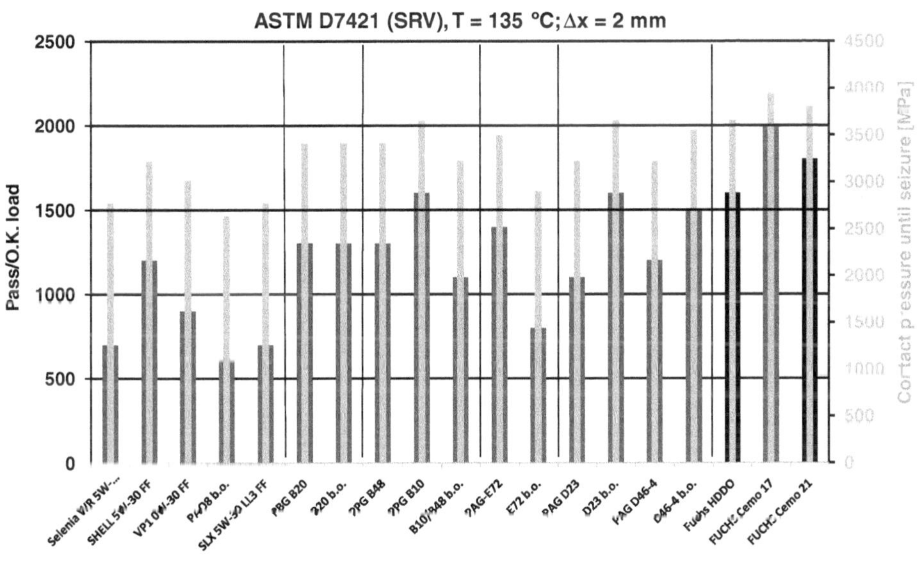

Figure 11.17 Extreme pressure behaviour of alternative engine oils by using ASTM D7421-08 (t <40 min), BAM

Table 11.8 Comparison of ester- and polyglycol-based prototype oils with factory fill oils

Lubricants	NOACK (%)	Ash (%)	VI	Pour point (°C)	η_{40} (mm^2/s)	η_{100} (mm^2/s)	η_{150} (mm^2/s)	HTHS 150°C (mPa)s	OECD 301 B/F (%)	Aquatic toxicity algae daphnia OECD (mg/l)	
										201	202
TOTAL FF 5W-30	12.8	1.2	159	−42	55.1	9.6	4.2	3.0	<50	<100	>100
Selenia WR 5W-40	10.8	1.15	175	−40	89.9	14.5		3.8			
Hydrocarbon/Ester blends											
Fuchs HCE mid SAP (epc-49)	5.2	0.50	184	−45	44.0	8.8	4.26	2.9	78-87	>1000	>100
Fuchs HDDO (ehd-25)	7.0	0.80	143	<−48	57.0	9.3	4.3	3.0	71.3	>100	>500
Fuchs cemo3 HCE	6.9	0.81	182	−57	49.8	9.6		2.9	71.4	>100	>1000
Fuchs 100E	5.5	1.2	144	−39	43.26	8.23		2.95	<60	>100	256
Fuchs cemo17 HCE	6.6	0.80	192	−54	83.1	15.1		4.1	>60	>100	106
Fuchs cemo21 HCE	6.7	0.90	182	−58	50.2	9.7		3.1	>60	>100	400
Polyalkylene glycol (PAG)											
PAG B25	5.5	<0.01	192	−45	31.0	7.01	3.46	3.01	>60.0	>100	
PAG D46-4 + add.	6.2	<0.01	146	−33	49.6	8.44	3.7	3.6	80	>100	301
PAG D23	2.5	<0.01	160	−30	68.5	11.5	5.0	4.81	>60	>100	211
Polypropylene glycol (PPG)											
PPG 32-2 + add.	4.8	<0.1	156	−45	34.3	6.70	3.2	2.78	74.5	>100	694
PPG B48	3.9	0.04	172	−39	48.1	9.05	4.80	3.59	>60	>100	136
Polybutylene glycol (PBG)											
PBG B20	3.2	<0.01	132	−36	66.1	9.90	5.43	3.61	~60	>1000	>1000

The viscosity index of lubricants shows how much viscosity drops when the temperature increases (Figure 11.15). The paraffinic mineral oil raffinates generally used in lubricants today have a viscosity index of about 100–120. Synthetic hydrocarbons can reach up to 145. For a number of lubricant applications, polymeric viscosity index improvers are used to improve viscosity-temperature behaviour. The disadvantage of these polymer compounds is that they can be mechanically sheared and generate the so-called 'shear loss'. As is visible from Table 11.8, the dynamic viscosity at 150 °C of polyglycols is most close to the HTHS and describes a more Newtonian flow behaviour. The stability of polymers over drain is limited and chemical reactions with the molecular fragments can negatively influence the lubricant (e.g. sludge formation, deposits on machine elements, valve intake deposits in direct injecting engines etc.).

Natural fatty oils, synthetic ester and polyglycols have a much better viscosity index than mineral oils, mostly over 150, with some over 200, so that viscosity index improvers could be avoided from the formulation (Table 11.8). As no polymer compounds will be present in the alternative lubricants, the rheology will not be affected when shearing forces act at high temperatures and loads, thus decreasing sludge formation and deposits in the engine and power train.

Ester and polyglycols with good low temperature performance were selected in order to avoid pour point depressant additives.

Most vegetable oils and synthetic esters used in lubricants display better lubricity in boundary lubrication conditions associated with high load carrying capability than mineral oils, because of their polarity generated from the molecular oxygen. The problem of metal–organic sulfur compounds is that they generate ash, and the retention of the conversion rate of catalytic converter systems necessary to reduce emissions will require a drastic reduction in engine oil ash content (Table 11.8). The elimination or minimization of the anti-wear and EP additive substitutes anticipated through the intrinsic properties of the base oil (for example, the polarity of the base oil), and the role of the tribo-reactive materials seem to be the right paths.

Vegetable oils, synthetic esters and polyglycols show a high intrinsic polarity. This also applies to corrosion inhibitors and can result in a competitive reaction on the metal surface. The corrosion protecting additives have also been carefully selected for ecolubes, because as 'surface active substances' they significantly affect the biodegradation and aquatic toxicity.

Antioxidants are still needed, but the reaction steps differ between esters and polyglycols. Special phenolic and aminic materials are suitable antioxidants for the formulation of biodegradable lubricants. Their ecotoxicological properties affect those of the formulations and deteriorate those of the base oil. Mostly, mixtures based on specific phenolic and aminic components have been used successfully in biodegradable lubricants.

Detergents and dispersant (DD) additives are responsible for the sludge carrying capacity and the cleaning potential of circulating engine oils. Ecotoxicological properties are sensitive to these classes of additives. The DD additives are most necessary for long oil change intervals of state-of-the-art engine oils. With DD components the oil can buffer the acid impact from combustion and fuel as well as have 'anti-wear' properties by forming solid films. The polarity of the chemical oxygen in each monomer of polyglycols does not consequently requires 'DD' additives.

11.7.2.3 Interaction of Lubricants with Tribo-Materials

The current high performance lubricants are reaching their technological limits, especially in mixed and boundary lubrication. These fluids are mineral oils or synthetic hydrocarbons with a relatively large amount of additives (up to 25%). In these classic additivated lubricants, chemical reactions of additives form reaction layers on the material surfaces and these layers carry load and provide low friction or anti-wear properties (Andersson, Tamminen and Sandström, 2002). In the new approach, the liquid (base oil) provides cooling, removes wear particles, provides hydrodynamic lubrication and has a polar character to interact with the tribo-reactive material and increase the viscosity index. Biodegradable lubricants without environmentally toxic additives can be combined with tribo-reactive materials (surface coatings on industrial base metals or monolithic materials) to replace friction modifiers, extreme pressure additives and anti-wear agents. The concept of tribo-reactive materials is not new but there has not been much introduction in high volume production yet because little is known about their tribological behaviour in real applications, especially in combination with the polar molecules of POEs (polyolesters) and PAGs as biodegradable lubricants. However, for alternative base oils, the patent applications FR 2 792 326 of Renault SAS, US 6 194 359 and DE 10 2005 011 776 of Daimler AG and the latest papers of Ford (12th Int. Coll. Tribology 2000-plus, in Esslingen, Jan. 2010) claim this concept to be the future-orientated solution, which can be influenced directly by the automotive OEM.

11.8 Tribo-Active Materials

Coatings and surface treatments are used with success in a few industrial applications to reduce wear and/or friction. They are intended to replace extreme pressure and anti-wear additives. The additional cost of the surface treatment has to be balanced with a reduction of the additive cost and improving lifetime, increasing the quality of the product and environmental advantages (Igartua *et al.*, 2009). One group of the tribo-reactive materials is based on substances (oxides) with a high portion of ionic bonding of carbides, nitrides and carbo-nitrides, which form these oxides by tribo-oxidation, like (Ti, Mo) (C, N). The tribo-reactive materials are crystalline with an under-stoichiometry either:

- on the oxygen or anion side (Magnéli type phases with the common principles $(Ti, V)_nO_{2n-1}$ or $(W, Mo\ _nO_{3n-1}$ or
- on the cation or metal side ($Co_{1-y}O$, $Cr_{2-y}O$, $Fe_{3-y}O_4$, $Ni_{0.75-1.25}O$, $Fe_{0.95-1.00}O$, $Cu_{0.4-0.6}O$, Mn_yO).

The under-stoichiometric oxides of titanium, vanadium, molybdenum and tungsten as well as those in the system Ti-Cr-O (Woydt, 2000) form an homologous series of planar defects in a regular spacing (so called crystallographic shear structures). The incorporation of lower valent cations, such as Fe_2O_3, V_2O_5, Cr_2O_3, Nb_2O_5 and so on, enhances the formation of 'crystallographic shear planes' (CSP) by tribo-oxidation. These defects can be described as missing of a planar oxygen layer. These materials exhibit a superior wear resistance with wear rates lower than $<10-7\ mm^3/Nm$ under dry friction up to 800 °C and support 'PV' value under dry friction greater than $> 10\ Watt/mm^2$ (up to $200\ Watt/mm^2$ are known); they are also effective under boundary lubrication (Berger *et al.*, 2009; Woydt, 2000). The surfaces of these tribo-active oxides are polar ($[O]^-$ and $[Me]^+$) due to the greater atomic radius of oxygen, this pronounced electron negativity thus enabling the adsorption of polar molecules on the surface, like PAG and POE.

The interaction of the base fluid with the tribo-reactive surfaces can be applied using the oxide by itself or to enhance to tribo-oxidative formation of these. In PAG the tribo-oxidation will be more pronounced, because the PAGs (ethylene/propylene oxides) selected dissolve water by nature (Storz, Gasthuber and Woydt, 2001).

The preferred oxides are the under-stoichiometric ones due to a higher degree of polarity, which are available as thermal spray powders at pilot plant scale.

The polarity of the PAG and POE leads to a greater attraction between the base oil molecules, so overall the viscosity index improver, the pour point depressants, the friction modifiers, extreme pressure and anti-wear additives can be avoided or their levels reduced.

11.8.1 Thematic 'Piston Rings'

In a reciprocating engine, most of the friction losses are generated in the contact of piston rings and cylinder block or liner (Mufti and Priest, 2009). The friction conditions of piston rings are very different from those of other engine elements, like crank shaft bearings. They are characterized by:

- Speed variations from zero in the dead centres to a maximum of 15–20 m/s in mid-stroke.
- Variable contact pressures over a thermodynamic cycle of an engine (360° crank angle, i.e. one revolution in two-stroke engines, and 720° crank angle, i.e. two revolutions in four-stroke engines).
- A mixture of hydrodynamic and mixed/boundary lubrication.

The difference between success and failure of piston ring operation is given by the lubricant conditions. With moderate loads, speeds, careful design and generous oil supply, a full film may be established, in mid-stroke by a hydrodynamic wedge, in the dead centres by squeeze action. With still higher pressures, the material resilience of the running partners may assist matters in establishing an elastohydrodynamic film that still fully separates the surface asperities. If the loads are still higher or if there is not enough oil (on account of insufficient supply, irregular distribution or burning away by the blow-by gases), so that a sufficient boundary film to separate the rubbing partners and prevent high wear does not exist anymore, scuffing will be the inevitable result.

Scuffing proneness is influenced by two different factors:

Scuffing severity	= Speed × contact pressure × friction coefficient
	= known as P × V factor
Scuffing index	= (Young's modulus × thermal expansion coefficient)/(Specific heat × density)

When selecting piston ring coatings all the influencing factors should be taken into account. Most current piston rings are chromium coated. But it is well known that the deposition processes of chromium (Cr^{VI+}) are carcinogenic. To improve ecology, lifetime and reduce the friction coefficient of piston rings, new under-stoichiometric tribo-reactive coatings will be deposited by physical vapour deposition (PVD) and plasma spray (APS) techniques.

The study is focussed on passenger car motor oils (PCMO) with reduced metal–organic additives. This is necessary in order to reduce the ash build up in the after treatment system and, therefore, improve its efficiency and lifetime. High fuel efficiency and long drain intervals are expected, as well. These oils have to be biodegradable and nontoxic to the aqueous environment according to the directive EC/1999/45 (amended by EC/2006/8), coherent with other international standards. In a modern diesel or gasoline engine, the engine oils have to fulfil quite a number of different functions, such as lubricating and cooling the system, wear protection, soot and particle handling with less deposit tendency and so on.

A study of the biodegradability, toxicity and the tribological properties has been carried out for newly developed prototype engine bio oils (Igartua *et al.*, 2009). Also, some different plasma sprayed tribo-reactive coatings have been deposited on cast iron piston rings and their tribological properties studied. Finally, the behaviour of the new bio oils selected and plasma sprayed tribo-reactive coatings on piston rings have been screened in a real engine.

Both new oils produced by the company Fuchs Europe are hydrocarbon-ester-based (HCE) on the same base fluid composition; they differ mainly in additivation. Due to their high viscosity index, the desired high temperature, high shear viscosity (HTHS) at +150 °C of 2.9 mPa s is reached at significantly lower viscosities at 100 °C, 40 °C and −25 °C. This offers the potential for improved fuel efficiency in city driving. The hydrocarbon-ester blends of Fuchs mostly reached an oxidation resistance of 144 h in the iron catalysed oxidation test (ICOT).

Two new polyglycol-based engine oils formulations (GLYMOT) supplied by BAM apply an ash-, metal- and polymer-free philosophy and respect biological nontoxicity (bio-no-tox) properties. Their thermo-physical properties, HTHS properties, pressure–viscosity coefficients, heat capacity and conductivity, as well as tribological properties, have been measured and are detailed elsewhere (Schmidt, Klingenberg and Woydt, 2006).

Oxidative evaporation losses and the resistance against oxidation of polyalkyleneglycols have been evaluated for water soluble PAGs, oil tolerant PPGs and oil soluble PBGs using the French (iron catalysed oxidation test, which is standardized in GFC Lu 36T 03 (now as TOC AFNOR NF 55-3099). In this test 96 h corresponds to a potential drain interval of more than 30 000 km (Barbezat, Nicoll and Sickinger, 1993) for turbo diesel engines with filter. PPG32-2 and PAG D46-4 with proprietary additives (Phopani) met the >96 h criteria. The matching in correlation between ICOT (TOC) and road testing is detailed elsewhere (Fitamen, Tiquet and Woydt, 2007). Further, polyglycol formulations exceeded a lifetime of 130 h, reaching recently 196 h, underlining a potential for exceeding ACEA A5/B5 or a drain interval of >40 000 km.

11.9 EP Tribological Tests

11.9.1 Piston Ring Cylinder Liner Simulation

Study of these new oils has been carried out by Tekniker using a test configuration (Figure 11.6) with chosen different coated piston rings and the reference grey cast iron cylinder liner alloy. The new lubricants were compared with PCEO reference oil.

The test conditions were: time, up to 41 minutes; load, 50/2 000 N; frequency, 50 Hz; stroke, 3 mm, temperature, +200 °C. The load was increased in fixed intervals until seizure occurred. Higher load corresponded to a better protection at highest load spikes. Figure 11.18 shows the impact of alternative engine oil formulations and tribo-active coatings on scuffing (extreme pressure) resistance (Woydt, Kelling and Hartelt, 2006).

These two developed coatings have better extreme pressure behaviour than the reference chromium plated piston ring. Also, the new oils developed, epc-48, epc-49, PPG-32-2 and PAG D46-4, have similar or better behaviour than the PCEO reference oils. It seemed that the combination tribo-reactive coating/polar lubricant increases the load for seizure, thus improving the extreme pressure properties.

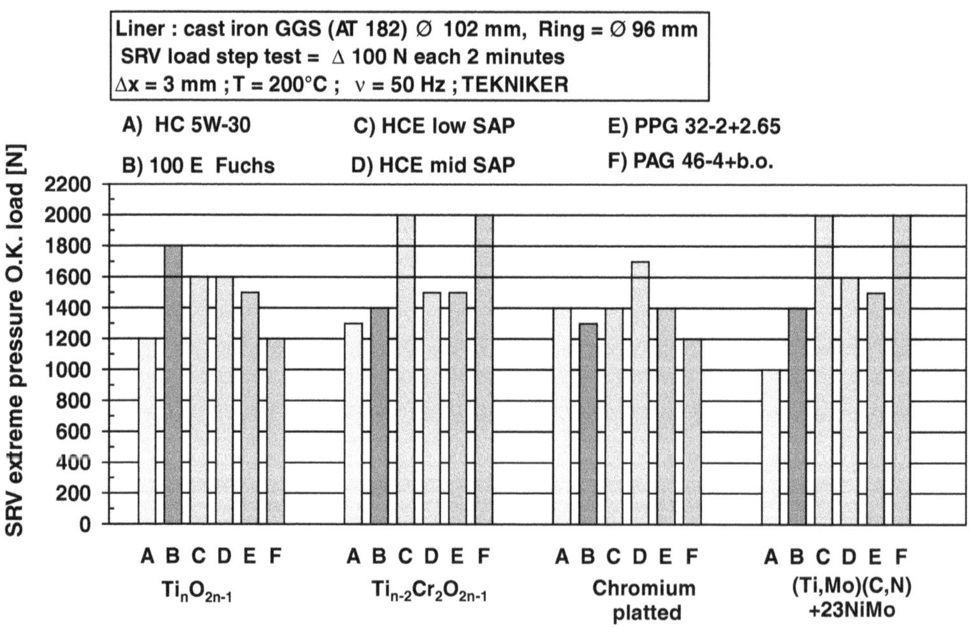

Figure 11.18 Cylinder liner–piston ring scuffing test results, Tekniker (Woydt *et al.*, 2006)

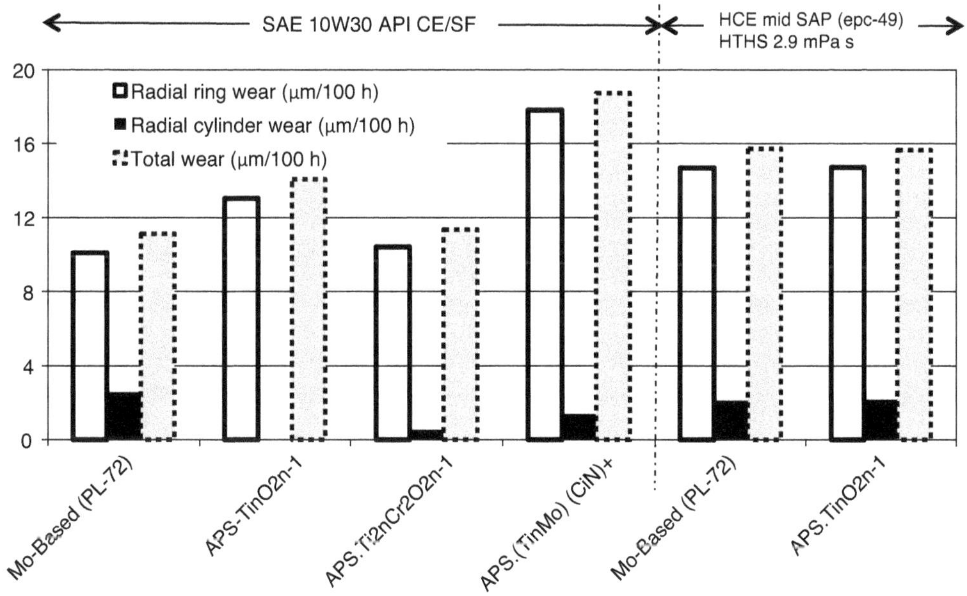

Figure 11.19 Testing of new piston ring coatings in a turbo diesel engine scuffing test, CIE Tarabusi (Igartua *et al.*, 2009)

Engine Tests

The piston ring scuffing tests were performed by Tarabusi (CIE Automotive Group) in a 2.7 l turbocharged four-stroke diesel engine ($T_{oil} = +155\,°C$, $T_{water} = +140\,°C$, 4000 rpm, 100 h) using as reference the molybdenum-based plasma coating PL72 (Figure 11.19). Under the test conditions the engine was operated over 100 h in a mode which favours scuffing. Basically, three tests are needed for each tested assembly:

(i) Scuffing tests to determine the scuffing resistance of the coating at high temperature and pressure with minimum component clearance.
(ii) Hot test, long run at maximum power and high speed to set the reference values for wear.
(iii) Cycle test between low and high thermal conditions to determine the thermal fatigue and adherence resistance of the developed coatings.

The components were measured before and after the test. Figure 11.19 presents the wear rates of different ring coatings in the engine scuffing testing. The main conclusions of the tests can be summarized as follows:

- Good wear and anti-scuffing properties of TiOx ($TinO_{2n-1}$).
- Wear resistance of $TinO_{2n-1}$ is similar to standard molybdenum-based coating for both standard and alternative oils.
- The best tribological properties were found for the $Ti_{n-2}Cr_2O_{2n-1}$ coating, but showed some spall-offs at edges of the ring substrate during the hot scuffing engine tests.
- Wear results with alternative PCEO mid-SAP epc-49 are similar to the standard oil.

Overall, the wear results with mid-SAP oil epc-49 are similar to the reference oil, but the oil consumption in the scuffing test of the epc-49 was lowered by 45%, probably due to the lower NOACK volatility of this HCE formulation.

Finally, the engine tests confirmed the prognostics achieved in SRV testing in relation to scuffing and wear resistance of tribo-active coatings and alternative engine oils.

Acknowledgements

The authors express their gratitude to the European Union for the financial support given to this work through the GROWTH Project No. GRD2-2001-50119, Contract No. G3RD-CT-2002-00796-EREBIO; the Project EFCAP 'Environmentally-Friendly Coatings for Automotive Parts' (EFCAP), 5th Framework Programme of the European Community for Research, Technological Development and Demonstration Activities; Project No. G3ST-CT-2001-50097, Contract No. CRAFT-1999-7045; the Project Clean Engine 'Advanced technologies for highly efficient Clean Engines working with alternative fuels and lubes', under contract TST5-CT2006-031241; and the Project Powerful 'Powertrain for future light duty vehicles', under the 7th Framework Programme (FP7/2007-2013) under grant agreement No. SCP8-GA-2009-234032. Also, the authors acknowledge the Spanish financing agency MICIIN, MICYT, CDTI and the Basque agency SPRI for the financing of the National projects AUMORE, REMTRAL, EQUIMOTOR, EQUIMOTOR Plus, MANUENGI and MOTORGARBI in, some of the cases, support from European FONDOS FEDER.

References

Andersson, P., Tamminen, J. and Sandström, C. (2002) Piston ring tribology. A literature survey. VTT Tiedotteita Research Notes 2178, VVT Finland.

Barbezat, Nicoll, A.R. and Sickinger, A. (1993) Abrasion, erosion scuffing resistance of carbide and oxide ceramic thermal sprayed coatings. *Wear*, **162–164**, 529–527.

Bartz, W.J. and Wienecke, D. (2000) Influence of gear oil formation on fuel economy of passenger cars. Technical Paper 2000-01-2050, SAE, Troy, MI.

Berger, L.M., Stahr, C.C., Saaro, S. *et al.* (2009) Dry sliding up to 7.5 m/s and 800°C of thermally sprayed coatings of the TiO2-Cr2O3 system and (Ti, Mo) (C, N)-Ni(Co). *Wear*, **267**, 954–964.

Directive 2000/53/EC of the European Parliament and of the Council, End-of-Life Vehicles, Official Journal of the European Communities, 21.10.2000, page L269/34.

Fitamen, E., Tiquet, L. and Woydt, M. (2007) Validation of oxidative stability of factory fill and alternative engine oils using the Iron Catalysed Oxidation Test. *Journal of ASTM International*, **4** (8), online.

Gili, F., Igartua, A., Luther, R. and Woydt, M. (2010) The Impact of Biofuels on Engine Oil's Performance. *Proceedings of the 17th International Colloquium on Tribology*, January 2010, Esslingen, Germany, ISBN 3-924813-80-9.

Gili, F., Igartua, A., Luther, R. and Woydt, M. The Impact of engine oils in engine oil performance. *Lubrication Science Journal*, 2010, LS-10-0033-RA-LS.

Godet, A., Roussel, G. and Thebault, M. (1984) Frictional behaviour of experimental synthetic and semi-synthetic gear oils. *Journal of Synthetic Lubrication*, **1** (2), 106–126.

Höhn, B.R. and Michaelis, K. (2001) Frictional behaviour of synthetic gear lubricants, Tribology Research: From Model Experiment to Industrial Problem – A Century of Efforts in Mechanics, Materials Science and Physico-Chemistry. *Proceedings of the 27th Leeds-Lyon Symposium on Tribology, Tribology and Interface Engineering Series*, Vol. 39, pp. 759–768.

Igartua, A., Mendoza, G., Fernández, X. *et al.* (2004) Influence of nitriding piston rings in the replacement of hard chrome in cylinder liners by PVD coatings. *Proceedings of the first workshop on 'COST 532 (European Project) Superior friction and wear control in engines and transmissions'* 86–100, 18–19 October 2004, Ghent, Belgium.

Igartua, A., Fernandez, X., Areitioaurtena, O. *et al.* (2009) Bio-lubricants and tribo-reactive materials for automotive applications, *Tribology International*, **42**, 561–568.

Landa, J., Illarramendi, I., Montalban, J. M. *et al.* (2004) Wear and friction in piston rings. *Proceedings of the first workshop on 'COST 532 (European Project) Superior friction and wear control in engines and transmissions'* 86–100, 18–19 October 2004, Ghent, Belgium.

Landa, J., Illarramendi, I., Kelling, N. *et al.* (2007) Potential of thermal sprayed TinO2n-1-coatings for substituting Molybdenum based ring coatings. *Industrial Lubrication and Tribology*, **5**, 217–229.

Laukotka, E.M. (1985) Lubricants of gears with synthetic lubricants. *Journal of Synthetic Lubrication*, **2** (1), 39–62.

Liang, X. and Strong, G.R. (1998) Recent developments in valve seat insert allows. Technical Paper 982047, SAE, Troy, MI.

Masuda, M., Ujuno, M., Shimoda, K. *et al.* (1997) Development of titanium nitride coating shim for a direct acting OHC engine. Technical Paper 970002, SAE, Troy, MI.

Mufti, R. A. and Priest, M. (2009) Effect of Engine Operating Conditions and Lubricant Rheology on the Distribution of Losses in an Internal Combustion Engine. *Journal of Tribology*, **131** (4), 041101. doi: 10.1115/1.3176988.

Onoda, M., Kuroishi, N. and Motooka, N. (1988) Sintered valve seats inserts for high performance engine. Technical Paper 880668, SAE, Troy, MI.

Schmidt, R., Klingenberg, G. and Woydt, M. (2006) Thermo physical and viscosimetric properties of environmentally acceptable lubricants. *Industrial Lubrication and Tribology*, **58** (4), 210–224. (Can also be seen in BAM Research Report 277 (2006) 'New lubrication concepts for environmentally friendly machines Tribological, thermo physical and viscosimetric properties of lubricants interacting with tribo-active materials', Wirtschaftsverlag NW, Bremerhaven, ISBN 3-86509-528-3.).

Shen, Q. (1997) Development of a material surface engineering to reduce the friction and wear of the piston ring. Technical Paper 970821, SAE, Troy, MI.

Stein, K.J., Schorr, B.S., Marder, A. (1999) Erosion of thermal spray MCr-Cr3C2 cermet coatings. *Wear*, **224**, 153–159.

Storz O., Gasthuber, H. and Woydt, M. (2001) Tribological properties of thermal sprayed Magnéli-type coatings with different stoichiometries (TinO2n-1). *Surface and Coatings Technology*, **140**, 76–81.

Woydt, M. (2000) Tribological characteristics of polycrystalline titanium dioxides with planar defects. *Tribology Letters*, **8** (2–3), 117–130.

Woydt, M. (2007) No/Low SAP and Alternative Engine Oil Development and Testing. *Journal of ASTM International*, **4** (10), online.

Woydt, M. (2010) Polyalkyleneglycols as next generation engine oils ASTM D02. *Symposium on Testing and Use of Environmentally Acceptable Lubricants*, 6 December 2010, Jacksonville, FL, USA.

Woydt, M. and Ebrecht J. (2003) Influence of Test Parameters on Tribological Results – Synthesis from International Round Robin Tests. *TriboTest*, **10** (9), 59–76.

Woydt, M. and Ebrecht, J. (2008) SRV-Testing of the Tribosystem Piston Ring and Cylinder Liner Outside of Engines. *TriboTest*, **14**, 113–126.

Woydt, M. and Kelling, N. (2003) Testing the tribological properties of lubricants and materials for the system piston ring/cylinder liner outside of engines. *Industrial Lubrication and Tribology*, **55** (5), p. 213–222.

Woydt, M., Kelling, N., Hartelt, M. *et al.* (2006) Tribological performance of bio-no-tox engine oils and new tribo-active materials for piston ring/cylinder liner systems. *Proceedings of the 15th International Colloquium on Tribology*, January 2006, Esslingen, Germany, ISBN 3-924813-62-0.

Yoshio, H., Akira, Y. and Tatsuya, E. (2008) Lubricant base oil for internal combustion engine and lubricating oil composition for internal combustion engine. Japan Patent JP2008280381.

Zimmermann, S. and Kreye, H. (1996) Chromium carbide coatings produced with various HVOF Spray Systems. *Proceedings of the 9th National Thermal Spray Conference*, ASM International, Materials Park, OH.

Part IV

Engine Bearings

12

Friction and Wear in Engine Bearings

Elena Fuentes[1] and Kedar Kanase[2]
[1]*Tekniker-IK4 Foundation, Eibar, Spain*
[2]*Kirloskar Oil Engines Limited, Pune, India*

Bearings are an important part of any engine. Engine bearings support the operating loads of the engine at all engine speeds and, along with lubricant, minimize friction. They have a very hard job to fulfil and their importance can never be underestimated. It is important to know the friction and wear behaviour of the engine bearings in their service conditions. This chapter deals with all the aspects related to lubrication, friction and wear of this type of component. As well, interesting information about engine bearing materials and coatings is gathered in it. Wear characterization tests help in selection and application of bearing materials. The test for adhesion quality of overlays, tests with the SRV tribometer for the coefficient of friction and thrust washer configuration test with the Falex tribometer further aid in describing the bearing materials and validating the design. These tests effectively predict the bearing wear mechanisms and correlate well with real engine tests.

12.1 Introduction

The machine elements that support a moving shaft against a stationary housing are called bearings (Bhushan and Gupta, 1991). A bearing can be defined as a device that supports a mechanical element and provides its movement relatively to another element with minimum power loss (Kopeliovich, 2010). The rotating components of internal combustion engines are equipped with sleeve type sliding bearings. Bearings can be classified broadly according to the movement they allow and according to their principle of operation in addition to by the directions of applied loads they can withstand. Rolling contact bearings, in which a load is transmitted by means of rolls (usually balls) to a relatively

Critical Component Wear in Heavy Duty Engines, First Edition. P.A. Lakshminarayanan and Nagaraj S. Nayak.

small area of the ring surface, cannot withstand the loading conditions of these types of engine. Only sliding bearings providing a distribution of the applied load over a relatively wide area can work in internal combustion engines. Due to the fact that the bearings provide a surface that allows the parts to rotate while avoiding the friction of metal against metal, their importance cannot be underestimated. They support the crankshaft and connecting rods and, in pushrod engines, they also support the camshaft. Because of the high loads generated by combustion and the reciprocating mass of the pistons, bearings have a very hard mission to fulfil (CTP, 2011). The science and engineering of bearing technology has evolved through the years to suit the specific and demanding requirements of engines. The engine bearings form the support system of the engine to carry loads, transmit power and act as a critical part of the power train system.

Bearings in engines are lubricated to provide a significant decrease in the friction coefficient, and to remove the heat generated by friction and the foreign particles from the rubbing surfaces. Engine bearings generally work in a hydrodynamic regime of lubrication that implies the presence of a continuous oil film between the bearing and the shaft. A constant supply of oil in sufficient amounts is necessary for good behaviour from engine bearings.

Figure 12.1 presents a scheme of a typical four-stroke reciprocating internal combustion engine, in which different sliding bearings can be observed.

Sliding bearings are usually conformal; that is, the components of a sliding pair fit together fairly closely (Ludema, 2002). In sliding bearings the load is transmitted between moving parts by sliding contact.

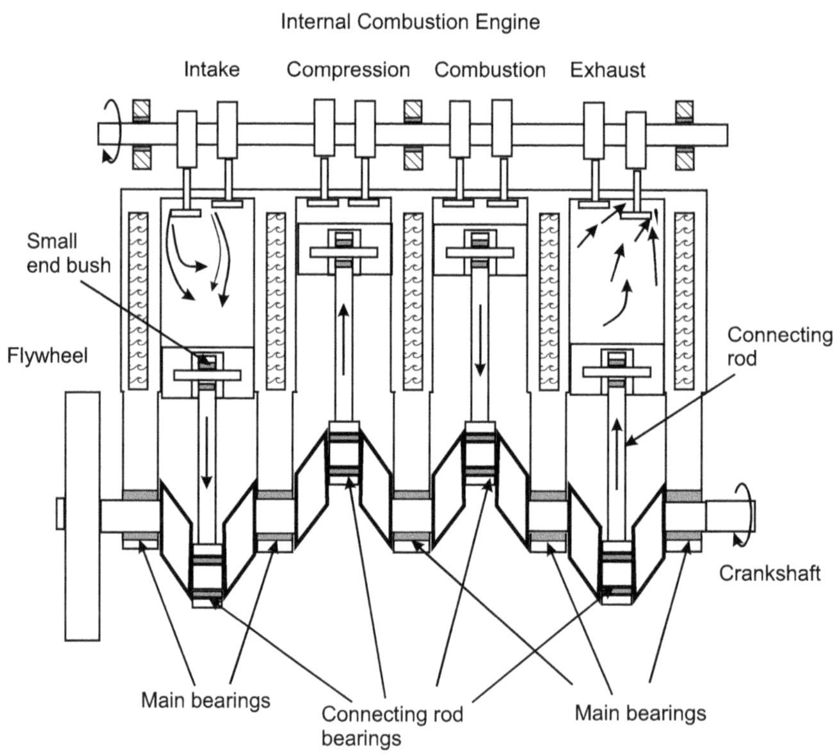

Figure 12.1 Scheme of a typical internal combustion engine

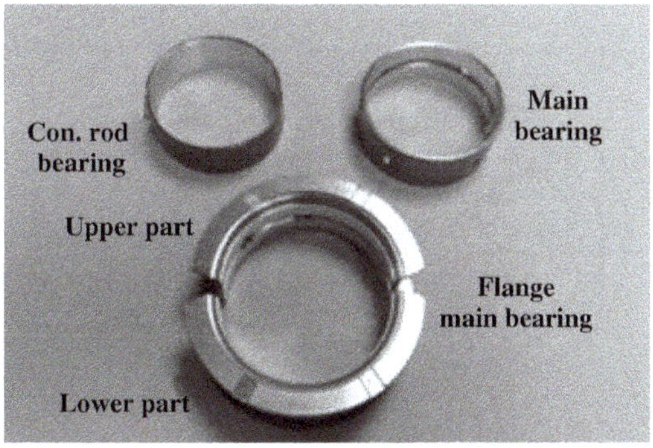

Figure 12.2 Types of engine bearings

The journal bearing is perhaps the most known and most widely used of all bearing types. It consists mainly of a sleeve of bearing material wrapped partially or completely around a rotating shaft or journal, and it is designed to support a radial load. The main types of sliding bearings used in internal combustion engines are (Kopeliovich, 2010) (Figure 12.2):

- **Main crankshaft bearings** support the crankshaft providing its rotation under inertia forces generated by the parts of the shaft and oscillating forces transmitted by the connecting rods. Main bearings are placed in the crankcase. This type of bearing has a hole in it for passing oil to the feed holes in the crankshaft. A main bearing is composed of two parts: upper and lower. The upper part commonly has an oil groove on the inner surface.
- **Connecting rod bearings** support the rotating motion of the crank pin within the connecting rod, which transmits cyclic loads applied to the piston. Connecting rod bearings are placed at the big end of the connecting rod.
- **Small end bushes** provide motion of the piston relative to the connecting rod joined to the piston by the piston pin. End bushes are placed in the small end of the connecting rod. Small end bushes are cyclically loaded by the piston pushed by the alternating pressure of the combustion gases.
- **Camshaft bearings** support the camshaft and provide its rotation.

The main journal bearings, the connecting rod bearings and the small end bushings are the most critical part of the engine system. The main journal bearing and the connecting rod bearings operate in various lubrication regimes based on the conditions of operation. At the starting conditions, when there is very less oil between the shaft and the bearing, the journal tends to climb up the right side of the bearing, as shown in Figure12.3a, but as the oil is pumped in between the shaft and the bearing a wedging action forces the journal in such a way that a minimum oil film thickness is formed at the bottom of the journal on the other side, that is clockwise in Figure 12.3b. This is the fundamental principle of operation of the pressurized converging film separating the shaft and the bearing in hydrodynamic operation. This shows the difference between the dry and lubricated regimes in the bearing operation. Figure 12.4 shows the pressure distribution in the bearing resulting from the hydrodynamic operation from the converging oil film. The distribution is a function of load, speed, oil viscosity, clearance and the diameter of the shaft. The potential wear regions due to these modes of operation are where the oil pressure is the highest or the film thickness is the minimum to cause fatigue failure or adhesive and abrasive wear respectively (Budynas and Nisbett, 2006).

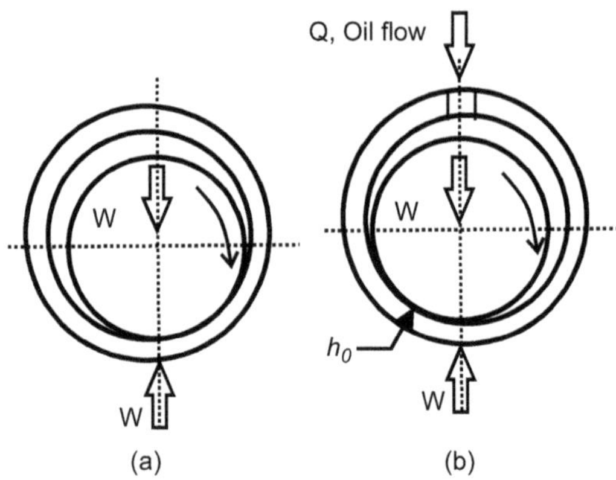

Figure 12.3 Formation of oil film as indicated in a (a) dry and (b) lubricated bearing

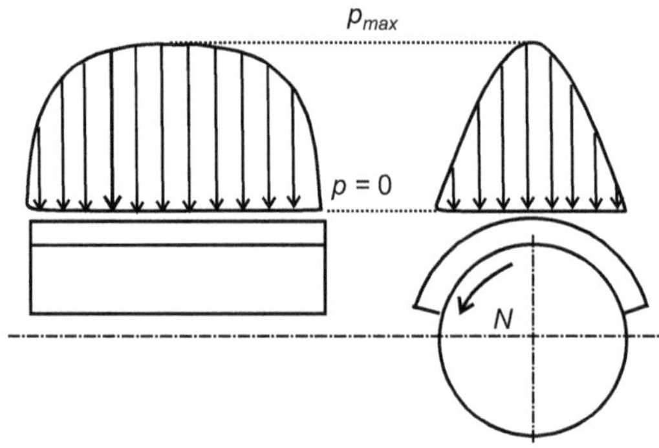

Figure 12.4 Pressure distribution in the bearings

12.2 Engine Bearing Materials

The selection of materials for sliding bearings is a multifunctional optimization problem. In general, the standard requirements are as follows: score resistance, conformability, embeddability, compatibility (seizure resistance), compressive strength, fatigue strength (load capacity), thermal conductivity, wear resistance, corrosion resistance and cost (Bhushan and Gupta, 1991).

Soft metals, in various combinations, are used in high performance bearings, and they are virtually always lubricated. The soft metals are sometimes used in thickness up to six millimetres, but to increase fatigue resistance, they are coated or plated on a steel (or other stronger metal) backing in layers as thin as 25 μm (Ludema, 2002).

The three major groups of soft metal bearing materials are Babbitt or white metal, copper–lead alloys and aluminium-based materials.

12.2.1 Babbitt or White Metal

Lead and tin white metal patented by Isaac Babbitt in 1839 are commonly the first choice for bearing materials; they offer superior compatibility with steel shafts, the ability (due to their softness) to embed foreign particles and an unique ability to adapt to misalignment by lid wiping on initial running-in as enabled by their low melting points (Khonsari and Booser, 2001). These materials are basically either antimony–tin or antimony–lead alloys having excellent conformability, embeddability, compatibility and corrosion resistance, but they fatigue readily.

Tin Babbitt alloys commonly contain about 3–8% copper and 5–8% antimony. Within a soft solid–solution matrix of antimony in tin, small, hard Cu_6Sn_5 copper–tin inter-metallic particles are dispersed. Increasing copper increases the proportion of Cu_6Sn_5 needles in the microstructure. An increase in antimony above 7.5% results in antimony–tin cubes. Hardness and tensile strength increase with greater copper and antimony content, while ductility decreases. Low antimony (3–7%) and low copper (2–4%) content provide maximum resistance to fatigue cracking. Since these low-alloy compositions are relatively soft and weak, a compromise is often made with fatigue resistance and compressive strength. Despite its higher cost, tin Babbitt is often used in preference to lead Babbitt for its excellent corrosion resistance, easy bonding and less tendency towards segregation. SAE 12 (ASTM Grade 2) is widely used in industrial and automotive bearings. SAE 11 and ASTM Grade 3 also find extensive industrial use.

Lead Babbitt commonly contains 9–16% antimony and up to 12% tin to provide hard cuboid crystals of antimony–tin in a eutectic matrix of the three metals. To minimize segregation during casting, up to 0.5% copper is usually added. SAE 15, containing 1% arsenic for finer grain structure, has been used in automotive bearings for its resistance to fatigue and better high temperature properties. This alloy also provides excellent service in large hydroelectric generator thrust bearings.

Sufficient tin is added in SAE grades 13–16 for reasonable corrosion resistance. With 10% tin, SAE grade 14 has frequently been used in railroad, industrial and automotive bearings. Softer SAE 13 is used in similar applications.

For smaller bearings, such as used in automotive engines and small industrial equipment, a bimetal strip is first produced by casting the Babbitt on a continuous steel strip or by electroplating lead Babbitt of 10% tin and 3% copper. For high fatigue strength in automotive bearings, a very thin layer of Babbitt enables much of the stress to be accommodated in the stronger backing material. Relative fatigue resistance was found to be higher, as follows with decreasing thickness of tin Babbitt (Table 12.1).

Three-layer strip bearings are usually used for connecting rod, camshaft and main bearings for heavy duty service in reciprocating engines. These consist of a low carbon steel backing; an intermediate layer about 0.3–0.8 mm thick of copper–lead, leaded bronze, aluminium or electroplated silver; and a thin Babbitt layer 0.025–1 mm thick. The thin Babbitt layer functions primarily during running-in, after which the higher strength intermediate layer supports the load

12.2.2 *Copper–Lead Alloys*

These alloys show higher hardness and have higher fatigue strength than the white metal alloys, but have comparatively less conformability, embeddability, compatibility and corrosion resistance. Copper-based engine bearing materials provide superior fatigue strength (load capacity).

Table 12.1 Fatigue resistance increase with a decrease in Babbitt thickness

Babbitt thickness (mm)	Relative fatigue resistance
1.00	1
0.50	1
0.25	1.5
0.13	3.2
0.08	4.6

Copper-based bearing alloys commonly contain tin (up to 10%) as a strengthening component and a soft component (lead or bismuth) distributed in the copper–tin matrix as a separate phase in the form of small particles. The soft component provides compatibility, conformability and embeddability properties to the material. Maximum content of the soft phase is about 25%. These types of bearing alloy are manufactured by either casting or sintering technology. Sintered materials are less strong than cast materials of the same composition due to possible voids and pores characteristic for sintered alloys. Besides, the grain structure of cast bearing alloys consists of copper dendrites orientated perpendicular to the bearing surface and therefore capable of resisting progressing fatigue cracks.

The copper-based bearing alloys can be mainly split into three main groups:

(i) Copper–lead 70/30% alloys, used for low to moderate duty petrol engines.
(ii) Lead bronze with 24% lead, 1.5% tin and 74.5% copper, used for heavy duty petrol engines.
(iii) Lead bronze with 8% lead, 5% tin and 87% copper, used for heavy duty naturally aspirated and turbocharged diesel engines.

In spite of the presence of a soft phase in most copper-based bearing alloys their anti-friction properties are poor. Therefore, copper (bronze) bearings are rarely used in engines in mono-metal (solid) or bi-metal (steel backed) forms. To improve their poor conformability, embeddability, compatibility and corrosion resistance, a soft layer of either lead–tin–copper or lead–indium alloy is applied. Typical copper bearings are steel backed and over-plated with a thin sliding layer (usually referred to as overlay) of a soft material (tri-metal structure). The overlay improves the anti-friction properties of the copper alloy. Modern copper-based bearing alloys do not contain lead because of its harmful effect on the environment and health. Lead can be replaced by harmless bismuth having excellent anti-friction properties in comparison with lead. Tri-metal bearings are typically CuPbSn or CuBi.

Typical copper-lead bearing alloys are SAE 48, 49, 480 and 481 (CDA alloys C98200, C98400, C98600 and C98800, respectively). All of these alloys have been employed as main bearings and connecting rod bearings of engines. They are, therefore, typically covered with a thin layer of lead–tin, lead–tin–copper or lead–indium alloy to increase the corrosion resistance. In some cases, a nickel barrier about 1 μm thick is placed between the intermediate layer and the overlay to prevent migration of tin from the overlay to the copper in the intermediate layer.

12.2.3 Aluminium-based Materials

These materials provide a good combination of moderate fatigue strength (load capacity), good corrosion resistance, high thermal conductivity and low cost with a moderate level of compatibility, conformability and embeddability properties. Most aluminium-based bearing materials have bi-metal structure. The manufacturing technology of this type of material includes continuous casting followed by bonding the alloy with a steel strip. Aluminium-based bearing materials are less expensive than copper-based ones, contain no (or low) hazardous lead, possess good corrosion resistance, low wear rate and, in most cases, do not require overlays for protection from corrosion; however, the overlay is often provided for start-up protection.

These bearing alloys cover a similar application range to the copper-based alloys and are suitable for medium to heavy duty operating conditions with both petrol and diesel engines. These alloys can be divided into three main groups:

(i) 20% tin, 1% copper and 79% aluminium. These alloys are used unplated for moderate duty petrol engines.

(ii) 6% tin, 1% copper one, 1% and 92% aluminium with a lead–tin overlay (SAE 770 and SAE 780). These alloys are used for solid wall and bi-metal aluminium bearings in medium and heavy duty diesel engines.
(iii) 11% silicon, 1% copper and 88% aluminium (SAE 784) with a lead–tin overlay of 0.025 mm standard thickness. These alloys are used for highly loaded diesels.

A typical tin containing alloy is AlSn6 and the most popular no tin, silicon containing alloys are AlSi4, AlSi11 and AlZnMg. The tin–aluminium alloys are roll-bonded to steel backing using a pure-aluminium-foil bonding layer. The silicon–aluminium alloys are directly cold-roll-bonded on to their steel liners. The high tin content aluminium alloy exhibits desirable bearing properties, hence does not require an overlay. The low tin and silicon alloys are harder, and hence are overlaid.

12.3 Functions of Engine Bearing Layers

Most engine bearings have a layered structure, which provides the optimum combination of the bearing material properties: fatigue strength, compatibility, wear resistance, conformability, embeddability, corrosion resistance and cavitation resistance. The hardness and modulus of elasticity of the oil film and boundary lubricated bearing materials should be as low as possible while providing enough strength to withstand the applied load. In initial operation, the layered structure provides the maximum possible surface softness and elasticity to aid in compensating for misalignment and other geometric errors. In addition, the soft bearing material is better adapted to absorb foreign dirt particles and thus minimize scoring and wear of the bearing and shaft (Khonsari and Booser, 2001).

According to the number of the main functional layers, the bearings can be mono-metal (solid), bi-metal, tri-metal or multi-layer. Most engine bearings are either bi-metal or tri-metal. Typical engine bearing structures are presented in Figure 12.5. The main components are:

(a) *A steel back* usually supports the bearing structure and provides its stiffness and press fit under harsh conditions of increased temperature and cycling loads. Steels such as SAE 1008, SAE 1010 and SAE 1020 are often used for steel back of this type of bearings.
(b) *The intermediate layer* is the layer placed between the steel back and the overlay. This layer is made of a bearing material with moderate or low anti friction properties, but high fatigue strength providing durability to the bearing, working under alternating loads generated in the internal combustion engine.

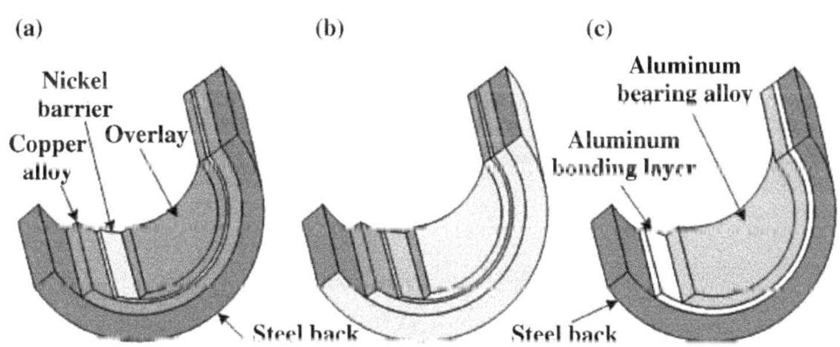

Figure 12.5 Engine bearings structure. (a) Tri-metal bearing; (b) tri-metal bearing with cosmetic tin flash; (c) bi-metal bearing

A certain level of surface properties is necessary to prevent catastrophic failure of the bearing in emergency conditions of boundary lubrication, oil starvation, significant misalignments or shaft deflections. This intermediate layer is referred to as tri-metal and multi-layer bearings and it is made mainly of copper alloys such as CuPbSn (leaded bronzes), CuAl (aluminium bronzes) or CuBiSn. Copper materials are applied on the steel back surface by either a casting or sintering processes. Some aluminium alloys, such as AlSn6, AlSi11, and AlSi4, are used also as intermediate layer materials. These materials are bonded to the steel back by the cold-rolling bonding method. The thickness of these intermediate layers is usually within the range 0.2–0.4 mm.

(c) *An overlay* is a thin surface layer made of a soft material (PbSnCu, SnCu, PbIn, or graphite/molybdenum disulfide polymer-based materials) with very good anti-friction properties (seizure resistance, compatibility, conformability and embeddability). Overlays are referred to as tri-metal bearings and their thickness is usually within the range of 12–25 μm. The small thickness of overlays is needed to achieve the required level of fatigue strength that is dependent on the overlay thickness (the thinner the overlay, the higher its fatigue strength). An overlay is applied on the surface of the intermediate layer after the final machining operation. Overlay properties become crucial, particularly under the conditions of boundary lubrication at low rotation speeds of the engine (engine start and shutdown).

(d) *A nickel diffusion barrier* is deposited between the intermediate layer and the overlay to prevent diffusion of tin from the overlay into the intermediate material (usually copper). Migration of tin into copper causes formation of brittle Cu–Sn inter-metallic compounds (such as Cu_3Sn, Cu_6Sn_5), which decrease the adhesion strength of the overlay to the intermediate layer. Apart from this reduction of the tin content in the overlay, the material of the intermediate layer can deteriorate the corrosion resistance of the lead-based overlay alloy. The thickness of this nickel diffusion barrier is about 1–1.5 μm. Nickel electroplating methods are used for the nickel diffusion barrier deposition. If the overlay is coated by Physical Vapour Deposition (sputter bearings) the nickel barrier is also applied by the vapour deposition method.

(e) *The bonding layer* is generally used to achieve good adhesion between the steel back and aluminium–tin alloy. The presence of tin particles on the surface of an aluminium–tin alloy bonded to a steel back weakens the adhesion strength between both materials. A bonding layer of pure aluminium between both layers allows achieving strong adhesion. The thickness of the aluminium bonding layer in an engine bearing is about 25–50 μm. Another type of bonding layer is a nickel layer deposited on the steel surface prior to bonding with aluminium–tin alloys.

(f) *A nonplated bearing material* is referred to as mono-metal (solid copper or aluminum-AlSi11) and bi-metal (cast or sintered copper alloys such as CuPb10Sn10, CuAl, soft tin containing alloys (AlSn), heat and nonheat treatable tin containing alloys (AlSnSi, AlSnNiMn)) bearings. No overlay is deposited on the surface of the material. Nonplated materials combine fair anti-friction properties with good fatigue strength. Most aluminium–tin bearing alloys (AlSn20Cu1, AlSn12Si3Cu1) do not require overlays. The thickness of nonplated bearing alloys is usually within the range of 0.2–0.4 mm.

(g) *A tin flash* is a very thin (max. 1 μm) layer of tin applied to the overall bearing surface. A tin flash improves cosmetic appearance of the bearing and imparts additional corrosion protection. Tin alloy electroplating methods are used for tin flash deposition.

12.4 Types of Overlays/Coatings in Engine Bearings

Overlays can help to eliminate or minimize metal-to-metal contact between the bearing and the shaft surfaces during the initial period of the bearing operation. Overlay promotes conforming of the bearing surface, resulting in lower wear of the bearing material. Components of coating have excellent anti-friction properties (very high seizure resistance, low coefficient of friction, good embeddability and conformability). However, coating is a sacrificial layer that can be worn fast under high load in metal-against-metal contact. When the overlay is removed by friction, the bearing clearance is increased by the value of the coating thickness. Therefore, coatings should not be too thick (not more than 0.01 mm). Coatings have no effect on bearing fatigue strength and they can be more effective when applied onto high strength bearings to improve their surface properties.

The main and crank bearings of modern engine crankshaft are produced as removable steel shells covered with various bearings alloys. Nowadays, engine bearings are manufactured almost exclusively as thin two or three layer shells, shells with a bearing layer of inhomogeneous structure (so called Rillenlager), or shells of a fine-grained bearing layer with evenly spread soft matrix (so called sputter bearings) (Krzymién and Krzymién, 2002).

At present, produced so called bi-layer shells are made of a steel strip covered with a thin layer of a bearing alloy. Often they are additionally covered with another thin layer of a soft alloy of good frictional properties (tri-metal bearing), separated from the basic bearing material with a nickel intermediate layer, which prevents diffusion of tin from the overlay into bronze alloys and formation of new compounds at the bronze and third layer contact area. Besides, other protecting overlays are used, for example those made of tin or copper, placed on the outer surface of the shell, which prevent the housing seizure, ease the heat transfer and protect the shell against corrosion. Figure 12.6 illustrates the construction of a multi-layer shell (Krzymién and Krzymién, 2002).

Multi-layer bearings combine the good mechanical properties of high strength bearing alloys with the good slide features of soft bearing materials.

As mentioned before, an overlay is a thin (12–25 μm) surface layer made of a soft material that possesses very good anti-friction properties. Overlays are usually referred to tri-metal and multi-layer bearings. There are four main types of overlays used in engine bearings.

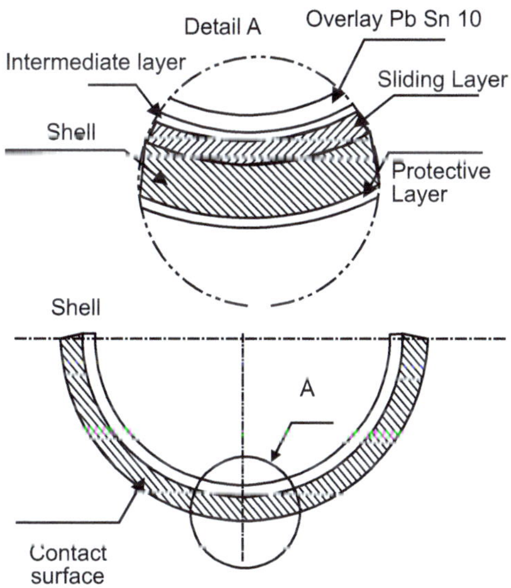

Figure 12.6 Construction of a multi-layer shell

12.4.1 Lead-based Overlays

Lead based alloys (PbSnCu, SnCu) are still commonly used as overlay materials, mainly in copper-based tri-metal bearings. However in many applications the alloys containing hazardous lead are being replaced by newly developed lead-free materials based on, for example, bismuth. Lead-based overlay alloys have excellent compatibility, conformability and embeddability, but relatively low fatigue strength, thus being limited in their applications in highly loaded engines. As well, pure lead has poor corrosion resistance in acidic oils and, for this reason, conventional lead-based overlay materials contain tin (not less than 10%), which inhibits corrosion attack of the alloy. In order to impart the alloys with higher fatigue strength, lead–tin compositions are alloyed with copper (2–8%), indium (9–12.5%) or dispersed alumina particles (0.5–2%). PbSn10Cu2 and PbSn10Cu3 compositions are the most popular lead-based overlay alloys. The hardness of this kind of overlay materials is in the range 8–20 HV, depending on the content of the hardening component. The fatigue strength of these overlays is about 50–70 MPa.

12.4.2 Tin-based Overlays

Tin-based overlays contain copper (3–6%) as a hardening component. These alloys have been developed to replace toxic lead-based overlays. Tin-based overlays possess excellent corrosion resistance and cavitation resistance. Overlay multi-layer systems of SnCu6 alloy deposited on the bi-layer diffusion barrier Ni–SnNi allows a micro-hardness gradient to be created across the overlay thickness due to varying content of copper (soft at the top layer and hard at the bottom layer of the overlay). Such multi-layer bearings are capable of withstanding alternating loads up to 80–90 MPa. Hardness of tin–copper overlays is 20–30 HV. Tin–copper overlays are deposited by electroplating methods.

12.4.3 Sputter Bearing Overlays

Sputter overlays (normally AlSn20 or AlSn40) are deposited by the Physical Vapour Deposition (PVD) method. The sputtering method provides a completely homogeneous distribution of tin within the aluminium matrix. The hardness of aluminium–tin sputtered materials is about 90 HV, which is three times higher than the hardness of aluminium–tin alloy produced by casting technology. Cast copper or high strength aluminium-based bearings are commonly coated by sputter overlays. Load carrying capacity of sputter bearings is the highest of all bearing materials and ranges between 100 and 120 MPa. The disadvantages of sputter bearings are their high production cost (slow deposition process) and poor soft compatibility, conformability and embeddability properties. The improvement of anti-friction properties and cost reduction are achieved by a combination of a sputter bearing shell in the high load position with a common tri-metal bearing shell in the less loaded position. Sputter bearings are mainly used as connecting rod bearings in highly loaded diesel engines with direct fuel injection system.

12.4.4 Polymer-based Overlays

Polymer-based overlay materials are polymeric matrix composites composed of particles of solid lubricants and hard abrasive particles (dispersed phases distributed within a polymer binder (matrix phase). Thermosetting resins (epoxy, phenolics and poliamide-imide) are used as the polymeric matrix. As solid lubricant can be used molybdenum disulfide (MoS_2), graphite, zinc sulfide, tungsten disulfide (WS_2), boron nitride (BN) and polytetrafluroethylene (PTFE). Alumina, silica, silicon carbide and silicon nitride can be used as hard abrasive particles, although some polymeric overlays do not contain hard particles.

Polymer-based materials containing solid lubricants and abrasive particles and thinned with appropriate solvents are applied onto the bearing surface by spraying techniques and then cross-linked at about 150–250 °C.

12.5 Coatings for Engine Bearings

The number of layers, the kind of material used and the way of production influence notably the total cost of bearing manufacture. In the case of tri-metal bearings, the cost of production simply exceeds that for the conventional bi-metal bearings and, for this reason, this kind of bearing is most often applied to the engine crank mechanisms. Rillenlager and sputter bearings are far more expensive; therefore, they are applied only to highly loaded sports car engines.

Examples of materials used for tri-layer slide bearings and information on their manufacturing and chemical composition are presented in Tables 12.2, 12.3 and 12.4.

Materials used for production of tri-metal bearings presented in this chapter are only some examples of a great variety of possible compounds covering the steel support and applied as bearing sliding surface.

Table 12.2 Composition and overlay of bronze bearings

Material		
Support material	Bearing material	Overlay
Steel	Sintered bronze CuPb30	PbIn7
	Sintered bronze CuPb22Sn4	PbIn7
	Sintered bronze CuPb26Sn1.5	PbIn7
	Cast bronze CuPb23Sn2	PbIn7
	Sintered bronze CuPb30	PbSn10Cu2
	Sintered bronze CuPb22Sn4	PbSn10Cu2
	Sintered bronze CuPb26Sn1.5	PbSn10Cu2
	Cast bronze CuPb23Sn2	PbSn10Cu2
	Statically cast bronze CuPb23Sn1	PbIn7
	Statically cast bronze CuPb23Sn1	PbSnCu
	Statically cast bronze CuPb23Sn2 + PbSnCu	PbSn10Cu8

Table 12.3 Typical composition of bearings used in modern engines

	Material			Remarks
Bearing alloy	Support material	Bearing material	Overlay	
Babbitt or white metal	Steel	CuPb22Sn4.5	PbSn	Sintered overlay
		CuPb26Sn1.5	PbSn	
			PbSnCu	
Lead bronze	Steel	CuPb23Sn1.5	PbIn	Cast overlay
			PbSnCu	
			PbSnAl	
			AlSn-sputtered	
		CuPb10Sn10	PbSnCu	
Aluminium	Steel	AlSi4CuMg	PbSnCu	
			PbSnAl	

Table 12.4 Typical composition of engine bearings with intermediate nickel dam

Bearing alloy	Support material	Bearing material	Intermediate layer	Overlay	Remarks
White metal (Babbitt)	Steel	IgSn85CuNi	Ni	PbSn18Cu2	(1)
Lead bronze	Steel	CuPb22Sn2	Ni	PbSn18Cu2	(1)
		CuPb20Sn4			
		CuPb24Sn	Ni	PbSn18Cu2	(1)
		CuPb22Sn4			
		CuPb22Sn2	Ni	PbSn10TiO_2	(1)
		CuPb10Sn10	Ni	PbSn18Cu2	(1)
		CuPb15Sn7			
		CuPb22Sn2	Ni	IgSn85CuNi	cast
		CuPb20Sn4			
		CuPb22Sn2	Ni	PbSn10TiO2	(1) (2)
		CuPb22Sn2	Ni	AlSn22Cu	sputtered
		CuPb22Sn2	Ni	PbSn18Cu2	(1) (2)
		CuPb10Sn10	Ni	CuPb30	(1)
		CuPb22Sn2	Ni	CuPb40	(1)
Aluminium	Steel	AlSn6CuNi	Ni	PbSn18Cu2	(1)
		AlZn4SiPb	Ni	PbSn18Cu2	(1)
		AlZn4,5SiPb	Ni	PbSn18Cu2	(1)
		AlSn6CuNi	Ni	PbSn18Cu2	(1) (2)
		AlZn4SiPb	Ni	PbSn18Cu2	(1) (2)
		AlZn4SiPb		AlSn20Cu	sputtered

(1) Electroplated
(2) Rillenlager

12.6 Relevance of Lubrication Regimes in the Study of Bearing Wear

Engine bearings are always lubricated by means of engine oils that are constantly supplied in a sufficient amount to the bearing surfaces. Lubricated friction is characterized by the presence of a thin film of pressurized oil (squeeze film) between the surfaces of the bearing and the shaft. The classic curve used to determine the operating regimes of a bearing is shown in Figure 12.7. The viscosity of the lubricant, the speed of the shaft and the applied load are important parameters. The three general lubrication regimes are clearly distinguished in the Streibeck curve (Figure 12.7), which demonstrates the relationship between the coefficient of friction and a dimensionless bearing parameter:

$$\text{Sommerfeld number} = \eta \times N/p_{av}$$

where $\acute{\eta}$ = dynamic viscosity of the oil, N = rotation speed and p_{av} = average bearing pressure.

This curve clearly shows the minimum value of friction as the demarcation between full fluid film lubrication and some solid asperity interactions. The ‘Streibeck curve’ is a classic teaching element in Tribology courses.

The stability of the different lubrication regimes can be explained by means of the Stribeck curve.

The combination of low speed, low viscosity and high load will produce boundary lubrication. **Boundary lubrication** is characterized by little fluid in the interface and large metal-to-metal contact between the bearing surface and the shaft. In this case, the friction is very high, resulting in high temperatures and wear.

As the speed and viscosity increase, or the load decreases, the surfaces will begin to be separated, and an oil film begins to be formed. The film is still very thin, but acts to support more and more of the applied load.

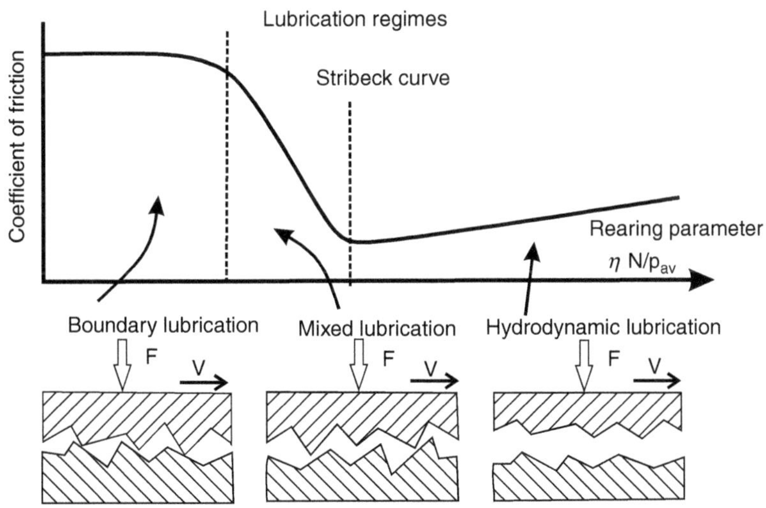

Figure 12.7 Schematic presentation of friction, Stribeck curve

Mixed lubrication is the result, and is easily seen on the Stribeck curve as a sharp fall in friction coefficient. **Mixed lubrication is unstable**. A fall in friction is a result of decreasing surface contact and less viscous lubrication. The surfaces will continue to be separated as the speed or viscosity increase until there is a full fluid film and no surface contact. The friction coefficient will reach its minimum and there is a transition to hydrodynamic lubrication. At this point, the load on the interface is completely supported by the oil film. In hydrodynamic lubrication, there is low friction and no wear, since there is a full fluid film and no metal-to-metal contact. The increase in the bearing parameter due to temperature rise (lower viscosity) in hydrodynamic lubrication regime causes the coefficient of friction to drop, with a consequent decrease in the temperature. The system corrects itself. Thus, **hydrodynamic lubrication is stable**.

Bearings under normal conditions of operation work under fluid friction, which is in the hydrodynamic lubrication regime. During starting and stopping they operate in the boundary or mixed film region. Also, the bearings operate in this regime at maximum torque conditions and the asperity contacts become predominant.

However, these modes of operation change with time or remain for prolonged duration. In addition, the transition from one operating regime to another may be gradual. The bearings, therefore, have to exhibit good emergency running properties as well as resilience to sustain varying conditions. Considering the above, the friction and wear characteristics of bearings must be in line with the mode of lubrication and the transient operating conditions. The roughness of the shaft and bearings, material conditions, form tolerances, assembly conditions and geometry play important and critical roles, along with the lubricating oil, in the tribological performance of the pair.

The study of the phenomenon of friction and wear of the tribological pair focuses on the thickness (h_{min}) of the lubrication film (or protective layer, Figure 12.8) with respect to the surface roughness of the bearing (e_b) and shaft (e_s), given by the ratio, λ:

$$\lambda = \frac{\textit{thickness of lubricating film or protective layer, } h_{min}}{\sum \textit{surface roughness } (R_a)}$$

$$\sum \textit{surface roughness } (R_a) = \sqrt{e_s^2 + e_b^2}$$

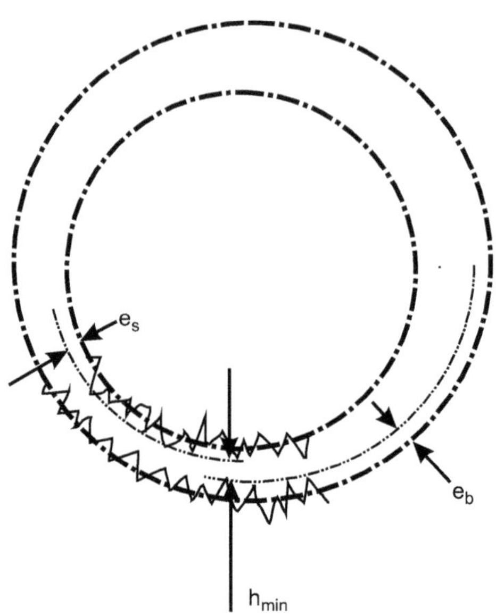

Figure 12.8 Typical surface asperities on a bearing and its mating shaft (e_s: maximum asperity height of the shaft; e_b: maximum asperity height of the bearing; h_{min}: minimum oil film thickness)

Some typical ratios for various lubrication conditions are:

- Boundary lubrication $\lambda < 1$ ($h < Ra$)
- Mixed (Partial) lubrication, $1 < \lambda < 3$ ($h \sim Ra$)
- Hydrodynamic lubrication $\lambda > 5$ ($h \gg Ra$)
- Elastohydrodynamic lubrication, $1 < \lambda < 10$

12.6.1 Boundary Lubrication

Both the bearing and the shaft come into closer contact at their micro-asperities; the heat developed by the localized pressure peaks provokes a condition which is called 'stick-slip' and some asperities break off. At the elevated temperature and pressure conditions, chemically reactive constituents of the lubricant react with the contact surface forming a highly resistant tenacious film (boundary film) on the moving solid surfaces, which is able to support the load, thus avoiding major wear or breakdown (Kopeliovich, 2010). Boundary lubrication is also defined as the lubrication regime in which the load is carried by the surface micro-asperities rather than by the lubricant. In the boundary lubrication mode of operation, the friction and wear characteristics of the lubricated contact are determined by the properties of the surface layers, often of molecular proportions and the underlying structure. The viscosity of the bulk lubricant has only an insignificant effect upon the performance of the boundary lubricated contacts and the frictional behaviour broadly follows the well-known laws for nonlubricated surfaces.

The oil film thickness is thin, and hence the asperity interaction of bearing and shaft takes place. The ratio, λ, drops below one. As the asperities carry an increasing portion of the load rather than the oil, the rate of heat generation increases. The rate is predominantly dependent on the bearing and shaft material properties, surface roughness, loading and geometry.

This regime is the most undesirable, since it is characterized by high friction coefficient (energy loss), increased wear, the possibility of seizure between the bearing/shaft materials and uneven distribution of the bearing load. This mode of lubrication is encountered in small end bushings and rocker lever bushings in normal operating conditions. Very severe engine bearing failures are caused by this boundary lubrication. Conditions for boundary lubrication are mainly at low speed friction (engine start and shut down) and high loads.

12.6.1.1 Material Properties

The bearing and shaft pair has to absorb and conduct away the heat generated without excessive temperature rise. Copper, lead and bronze with steel backing and a varying extent of lead and tin are typically used for the small end of the connecting rod and rocker lever or similar applications. The shafts are nitrided or induction hardened at the mating zone of the bearing surfaces and the hardness is typically 56 HRC. The hardness of the bearing material is typically in the range of 25 HV for white metal bearings, 45–100 HV for lead-based half bearings, 70–160 HV for connecting rod small end bushings, while aluminium-based materials are typically in the range 30–60 HV. The conventional lead-based overlays have hardness in the range 8–12 HV and sputtered overlays in the range 60–90 HV.

The compatibility of the bearing material with the shaft material is essential to ensure that this tribological pair functions successfully. If a softer material is selected for the bearing, heavy running-in wear of the bearing surface and ploughing of the shaft surface roughness asperities on the softer bearing surfaces would be expected. This is true even for connecting rod large end and main journal bearings subjected to frequent start stop conditions (steady loads and low speeds). To avoid such wear in nearly static conditions, especially in large engines, oil is pumped under pressure in the clearance space between shaft and the bearing and the two surfaces are separated by a static oil film (Figure 12.9). This hydrostatic condition of the bearing avoids the surface asperity contacts during start and stop.

12.6.1.2 Role of Oil and its Additives in Boundary Lubrication

In the boundary lubrication regime, the thickness of the protective layers formed by physical and chemical reactions between the solids and the surrounding bulk lubricant or the atmosphere is usually small

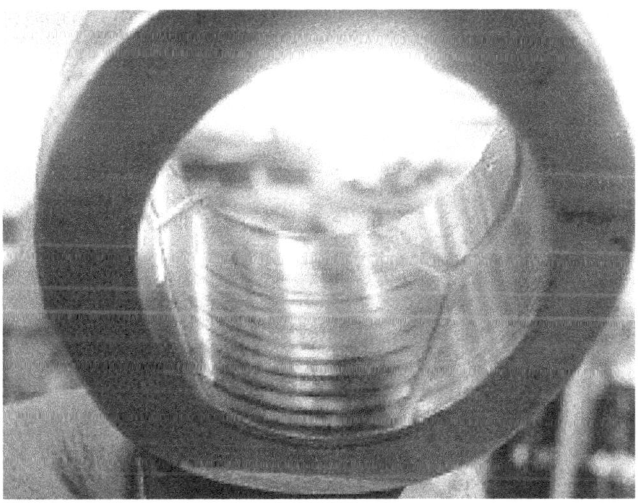

Figure 12.9 Small end bush showing wear on the surface

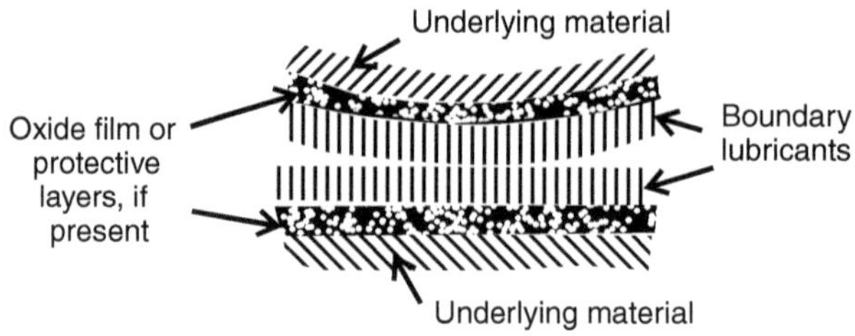

Figure 12.10 Boundary lubrication (Reprinted from *The Tribology Handbook*, 2nd edn., M.J. Neale, Copyright 1996, with permission from Elsevier.)

compared with the roughness of the solid surfaces. The extreme pressure additives present in the lubricants formulated for any specific application critically control the wear rates of parts in boundary lubrication conditions. The length of fatty acid molecules in boundary lubricants and the thickness of protective oxide films are often as small as two nanometres (Figure 12.10) (Neale, 1996). In addition, the oil carries away or flushes off the wear elements from the adjacent surfaces and dumps them in the oil sump.

12.6.1.3 Surface Roughness

It is essential to generate the surface roughness of the shaft carefully, as the bearing material is soft and replicates the finish of the shaft it mates with. Therefore, the surface roughness of the connecting rod small end pins are ground (especially because they face mixed/boundar lubrication) extra precision to surface roughness values in the region of 0.4 microns R_a or less.

12.6.1.4 Geometry Conditions

Manufacturing errors like taper on the shaft and housing bores, excessive ovality, deviations in the concentricity of the main bores or lack of parallelism of surfaces lead to early entry into boundary lubrication regime. The distortions of shaft and bore due to heavy loading, barrel shaped shafts and lobes in the shafts can give rise to local rubbing contact and an edge loading situation, which leads to defects, overheating and rapid wear, seizure and scuffing that can damage the bearings (Figure 12.11). Therefore, it is necessary to keep these deviations under control and also to have bearing materials that have the ability to withstand and sustain such conditions to certain designed levels. Various materials have been developed to accommodate these imperfections. The subsurface of the material plays an important role in this situation, for example, in the case of aluminium-based materials an aluminium foil separates the aluminium tin lining and the steel backing. The properties of the foil-like chemistry, hardness and thickness are so designed to accommodate edge loading behaviour and sustain it. In addition, profiling

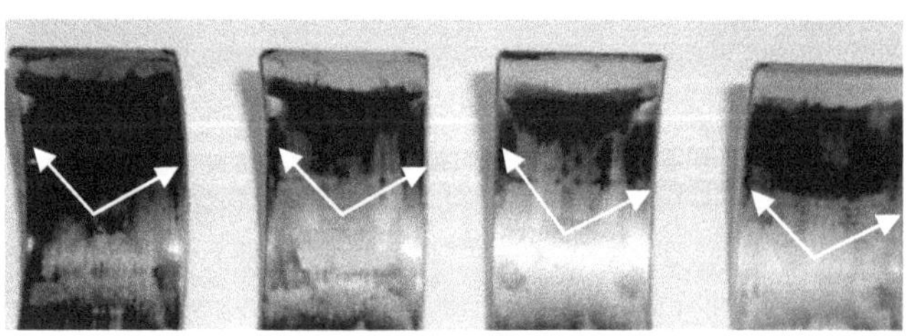

Figure 12.11 Bearings showing edge loading due to the deformation of the connecting rod and deflection of the shaft; edges show exposure and wear of copper lead lining below the overlay

(thinning) is done on the wall thickness of the bearings along the edges to accommodate the distortions in the housing and the shaft to accommodate the edge loading wear.

12.6.2 Mixed Film Lubrication

An intermittent contact between the rubbing surfaces at a few high surface points (micro-asperities) occurs at mixed lubrication. Mixed lubrication is the intermediate regime between boundary and hydrodynamic lubrication. Both the fluid film and the boundary lubrication are experienced in this regime of operation. The extent and duration of each is dependent on the parameters of load, speed and viscosity of the oil, along with material properties, surface roughness and geometry conditions in the tribology pair. The local hydrodynamic action accompanied by the local asperity contacts can result in lubrication of the asperities as well. Therefore, the properties of the material, the bulk lubricant as well as the additives and chemical interactions between them in a given environment decide the final mixed film lubrication condition. Such a condition is experienced by the connecting rod big end and main journal bearings at heavy loading conditions (maximum torque point or over loading conditions) and also by the bushings in the small end of the connecting rod or rocker lever. The flushing off of the smeared asperities of the bearing materials takes place in this regime and bearings are at lower temperatures than in the boundary lubrication regime. The initial oil change period is, therefore, heavily dependent on the running-in when the mating parts have seen each other for the first time in the engine. Bearing materials and surface topography along with oil properties and loading conditions play an important part in this condition.

Lead-based overlays have been popularly used on plain bearings with copper–lead lining and aluminium–tin lining for short durations of operation in this regime. Compatibility of the bearing materials, as in the case of the boundary lubrication, is as applicable here. Many materials have been developed to address this condition of operation, for example a molybdenum disulfide based lining.

12.6.2.1 Surface Topography

Bearing manufacturers applied various techniques to produce bearing topographies that allow a mixed film lubrication condition rather than a boundary lubrication condition. Historically, bearings have been either finish broached or bored on the internal diameter (Figure 12.12). Bearing surfaces with space for oil retention and flow at the same time, produced by special boring techniques, help achieve a mixed fluid film condition and avoid a boundary lubrication condition in severe operating conditions. The size of these grooves in the micro grooved bearings is optimally designed (Figure 12.13) (Watanabe *et al.*, 2000). There is a possibility of mixed lubrication whenever $\lambda < 5$ (Figure 12.14).

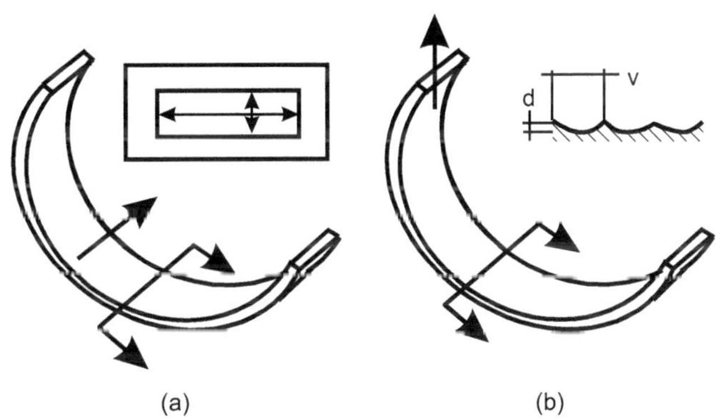

Figure 12.12 Surface generation processes. (a) Broached bearing; (b) bored bearing

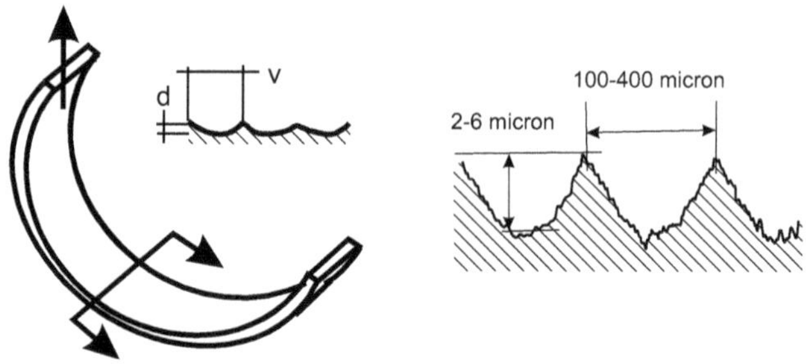

Figure 12.13 Special boring to obtain oil retention grooves in the micro-grooved bearings

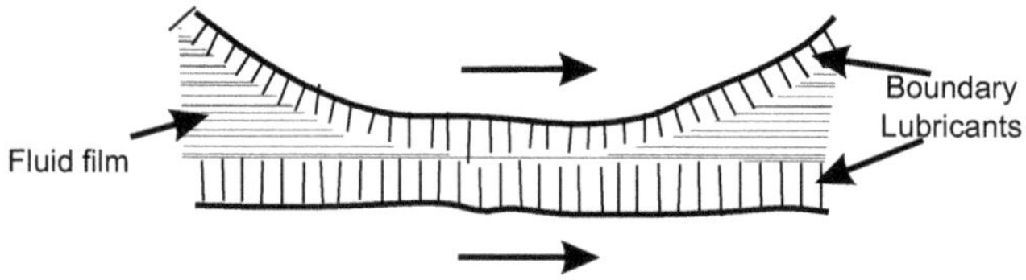

Figure 12.14 Mixed film lubrication (Reprinted from *The Tribology Handbook*, 2nd edn., M.J. Neale, Copyright 1996, with permission from Elsevier.)

12.6.3 Fluid Film Lubrication

High rotation speed at relatively low bearing loads results in hydrodynamic friction, which is characterized by the formation of a stable oil film between the rubbing surfaces. In this lubrication regime, no contact between the rubbing surfaces occurs. The pressurized film keeps the surfaces of the bearing and the shaft separate, as a result of the hydrodynamic forces acting in conjunction due to the relative movement of the surfaces and the viscous resistance of the fluid. Bearings working under the conditions of hydrodynamic lubrication are called hydrodynamic journal bearings. Elastohydrodynamic lubrication, when two surfaces (elastically deformed under stress) are separated by a very thin fluid film, is related.

The oil film is thicker than the surface roughness of the shaft and the bearing (Figure 12.15a). The frictional resistance arises solely from the viscous shearing of the fluid. Friction, and hence wear, in this regime is dependent on the peak oil film pressure generated in the fluid, the location of this pressure, the minimum oil film thickness generated in the bearing and its distribution and location for various operating conditions, the possible cavitation region generated in the fluid, change of load with respect to the shaft speed resulting in the ruptures of the oil film, dirt and wear material in the oil, elastic distortions of the housing, bearing and shaft and corrosion.

Although the lubricant flow is assumed to be laminar, in practice mixed flow regions can exist. Also, cavitation may be observed if the pressure drops below the vapour pressure of the oil. Since a requirement for successful fluid lubrication is the absence of asperity interaction, the film thickness is usually at least two to five times greater than the equivalent sum of the surface roughness values.

A special form of fluid film lubrication, in which the development of effective films is encouraged by local elastic deformation of the bearing surfaces, is known as elastohydrodynamic lubrication (Figure 12.15b). It is now recognized that this is the principal mode of lubrication in engine connecting rod and main journal bearings. Bearings in elastohydrodynamic contacts operate with film thicknesses which are small and there is often an associated mixed lubrication action. If local elastic distortion of the

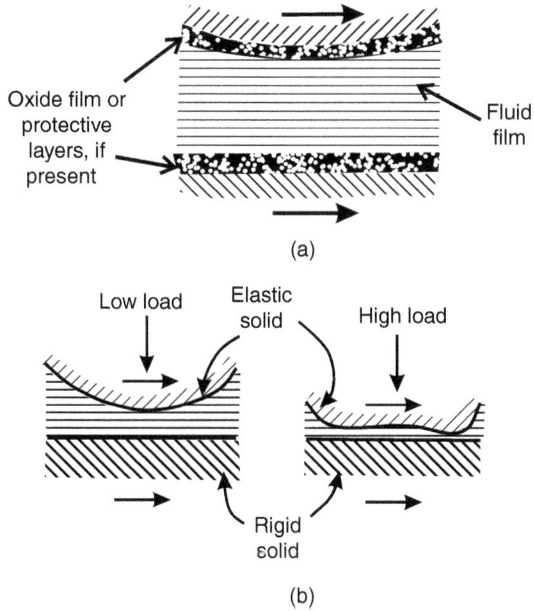

Figure 12.15 Regimes of hydrodynamic lubrication. (a) Fluid film lubrication; (b) elastohydrodynamic lubrication (Reprinted from *The Tribology Handbook*, 2nd edn., M.J. Neale, Copyright 1996, with permission from Elsevier.)

bearing surfaces occurs in the load carrying region, the range of effective fluid film lubrication is often extended well into the region normally associated with mixed or even boundary lubrication. A typical range of film thickness ratios for elastohydrodynamic conditions is given by:

$$1 < \lambda < 10.$$

12.7 Theoretical Friction and Wear in Bearings

When two rubbing surfaces slide in dry conditions, the real contact area, **A**, involves only sufficient properties of the softer material so that their yield pressure, **P**, will balance the total normal load, **W**, according to Equation 12.1, as illustrated in Figure 12.16:

$$A = W/P \tag{12.1}$$

With a bronze alloy bearing offering an asperity yield pressure of 500 MPa, for instance, a typical nominal bearing load of 3 MPa would produce asperity contact on less than 1% of the gross projected area. Yield pressure for many metal asperities is about three times their tensile yield strength.

12.7.1 Friction

During dry sliding, the force, **F**, needed to move an asperity of the softer material with shear strength, **s**, in Figure 12.16 becomes:

$$F = A\,s \tag{12.2}$$

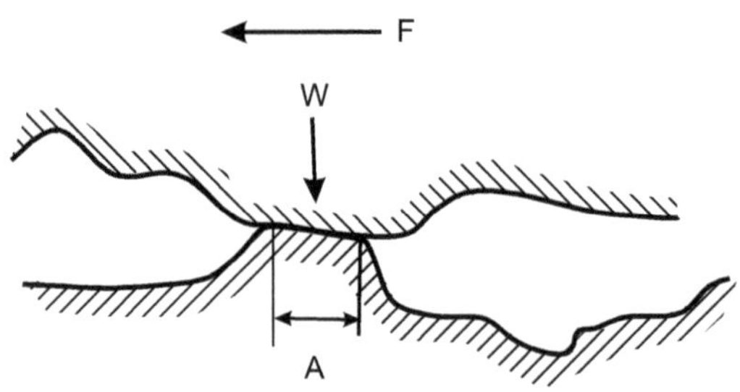

Figure 12.16 Asperity contact between two rubbing surfaces

Although this shearing of asperity contacts may represent 90% or more of the total friction force, other factors also contribute. Lifting force may be needed to raise asperities over the roughness of the mating surface. Scratching by dirt and worn particles can introduce ploughing. Internal material damping and surface films can also play a significant role. The coefficient of friction, **f**, is defined by the ratio of friction force, **F**, to the applied normal load, **W**. Combining Equations 12.1 and 12.2 yields the following, where f is the coefficient of friction:

$$f = F/W = s/P \tag{12.3}$$

Since shear strength, **s**, and compressive strength, **P**, are usually both related to the same material properties, such as lattice structure and bond strength, **f** often falls within the rather narrow range of about 0.20–0.35 for a wide variety of materials (carbon–graphite: 0.19, lead Babbitt: 0.24, bronze: 0.30, aluminium alloy: 0.33 and polyethylene: 0.33) for bearing materials varying several hundredfold in yield strength. Lower values down to about 0.05 can be obtained with low shear strength surface layers, such as those obtained with PTFE, solid lubricant surface treatments or additive layers provided by some lubricants (molybdenum disulfide). On the contrary, friction coefficients can rise to 10 or higher with welding of rubbing materials, such as two very clean copper surfaces in contact. After sliding is initiated, the friction force usually falls with increasing speed as the contact time diminishes for adhesion and related interactions between the surfaces.

The common coefficient of friction encountered with bearing materials can be dramatically diminished by minimizing sliding contact or avoiding it. In hydrodynamic bearings, a value of 0.001 is typical when operating with a full oil film.

12.8 Wear

This occurs to some degree whenever surfaces are in sliding contact. After the initial running operation of a bearing, involving usually higher than normal wear related to surface finish, misalignment and pollutant particles, sliding wear tends to reach a steady rate proportional to the total normal load, W, as approximated by the Archard equation:

$$V/d = kW/H \tag{12.4}$$

where **V** is the volume of worn material, **d** is the sliding distance and **H** is the hardness of the softer surface. Typical values of the dimensionless wear coefficient **k** in Figure 12.17 show the greatest wear for identical materials sliding without lubrication, and wear is then decreased by adding a lubricant and with different material pairs. Equation 12.4 indicates trends to be expected with changes in each of its factors, but its utility is limited by a typical uncertainty in values of wear coefficients. For evaluating wear of a shaft three or more times harder than its bearing, Rabinowicz (1984) suggested that surface wear volume will be less

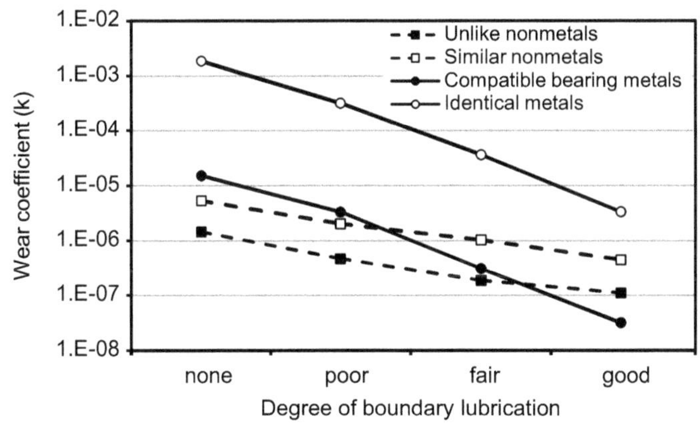

Figure 12.17 Representative wear coefficients for Equation 12.4

by three times the hardness ratio. For a shaft less than three times as hard as the softer bearing, the harder material wear volume would be less by a factor of the hardness difference squared.

Since clear definition of surface hardness is often difficult, Equation 12.4 is usually modified by elimination of the uncertain hardness term **H**, by including it with **k**, as a dimensional specific wear rate (mm^3/Nm) for use with various materials: metals, ceramics and self-lubricating materials. While Equation 12.4 is based mainly on adhesive wear contact of surface asperities, wear of most other types similarly increases in direct proportion to load and sliding distance.

12.9 Mechanisms of Wear

Wear is the progressive loss of substance resulting from mechanical interaction between two contacting surfaces and occurs because of the local mechanical failure of highly stressed interfacial zones. The failure mode will often be influenced by environmental factors (Pike and Conway-Jones, 1992). These surfaces are under load in relative motion, either sliding or rolling. Mechanisms of wear associated with failure signatures in engine bearings can be broadly divided as adhesive wear, abrasive wear, fatigue wear and wear by chemical reaction (Figure 12.18).

Adhesive wear: By this mechanism, a series of micro-seizures occurs resulting finally into scuffing or seizure. The wear of bearing and shaft surfaces is characterized by particle transfer or micro-welding of bearing surfaces to shaft surfaces leading into scuffing. Adhesive wear can be particularly problematic with two materials that have great affinity for each other or mutual solubility.

Abrasive wear: This is the ploughing of the bearing surface by a hard particle resulting in scoring.

Fatigue wear: The bearing surface plucked off with characteristic cracks networked together on the bearing surface resulting in accelerated dislodging of material is termed fatigue wear.

In all the three cases mentioned above, stress transfer is principally via a solid interface, but fluids can also impose or transfer high stresses when their impact velocity is high. Fluid erosion and cavitation are typical examples of fluid wear mechanisms

Wear by chemical reaction: A reaction does not itself constitute a wear mechanism; it must always be accompanied by some mechanical action to remove the chemical products that have been formed (Figure 12.19). After a period of severe wear, it is possible that the increase in oxidation strengthens the surface and the wear rate is restored to a slower initial level. However, chemical effects rarely act in such a simple manner; usually, they interact with an influence on the wear process, sometimes beneficially and sometimes adversely.

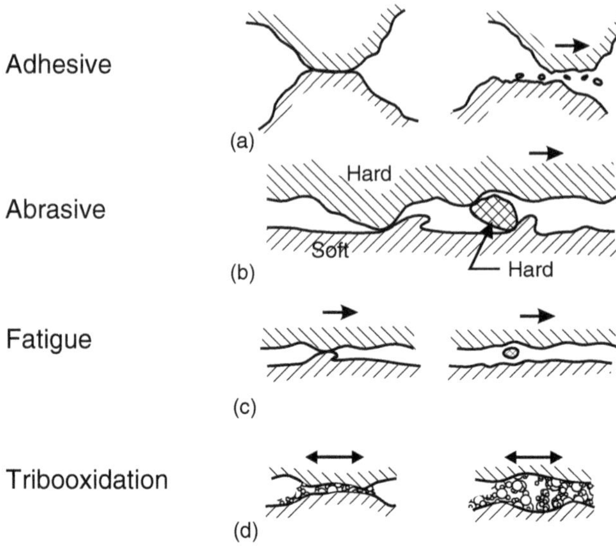

Figure 12.18 Wear mechanisms

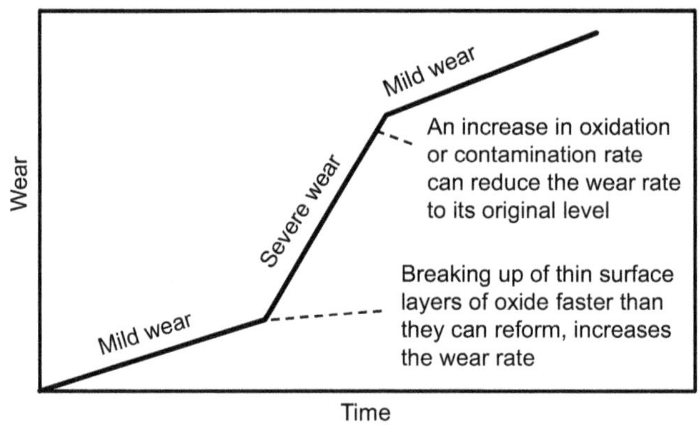

Figure 12.19 Wear rate (Reprinted from *The Tribology Handbook*, 2nd edn., M.J. Neale, Copyright 1996, with permission from Elsevier.)

12.9.1 Adhesive Wear

As in mixed film lubrication or a boundary lubrication regime, the bearing and shaft surfaces approach each other in intimate contact. Various wear mechanisms between the two surfaces potentially exist at different time instances depending on the operating conditions and the environment. As discussed earlier, the bearing surface hardness is much lower than the surface hardness of the shaft. Though the materials are selected to have good compatibility and seizure resistance, the mechanism of the adhesive wear remains complicated. Various theories have been put forth. Plasticity theory predicts that when a tangential traction is applied to a system already in a state of plastic contact, the junction area will grow as the two surfaces slide against each other.

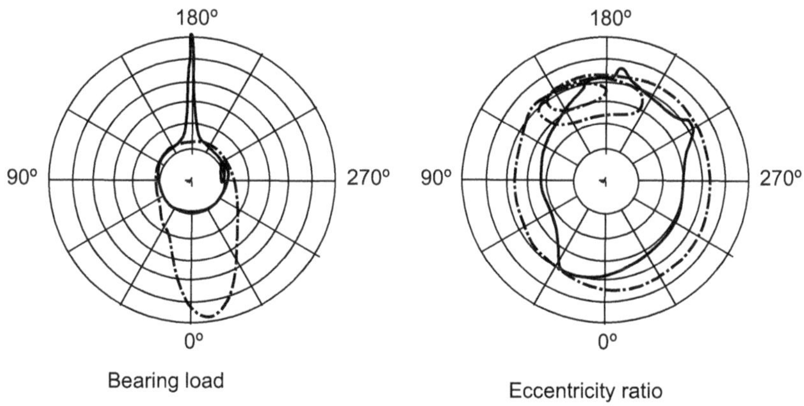

Figure 12.20 Polar plots for the bearing loads and eccentricity ratios for connecting rod bearings based on solution of Reynolds equation

In the analysis of the bearing design wherein the Reynolds equation is solved, a polar plot of the eccentricity ratio (Figure 12.20) is indicative of the extent of the wear region in the bearing circumference and the location of the wear. This is indicative of the minimum oil film thickness and its location on the bearing circumference. The asperity contact pressures along the circumference of the bearings are also computed to understand the probable mechanism of wear and its location in specific operating conditions (Figure 12.21).

The continuous growth of a localized weld is controlled by the influence of the interfacial layers, which reduce the shear strength of the interface to sufficiently contain the weld growth (Figure 12.22a). The weld formation is the first stage of the wear mechanism (Figure 12.22b), when the oil film or the oxidation layer ruptures. This welded region may be strengthened by the work hardening mechanism under the asperity contact pressures and shear may occur within the softer bearing component, thus allowing a fragment of bearing material to be transferred from one surface to another. If the result of a weld fracture is material transfer, then no wear occurs until some secondary mechanism causes this particle to break away. Often the material transferred from the bearing onto the shaft may adhere to the shaft and rotate with it and be transferred back to the original bearing surface. In the last stage of the wear mechanism, a group of particles is formed. The surface energy of these particles causes them to be retained on the surface. When the work done by the forces acting exceeds the energy limit, the particles break away from the surface as one unit.

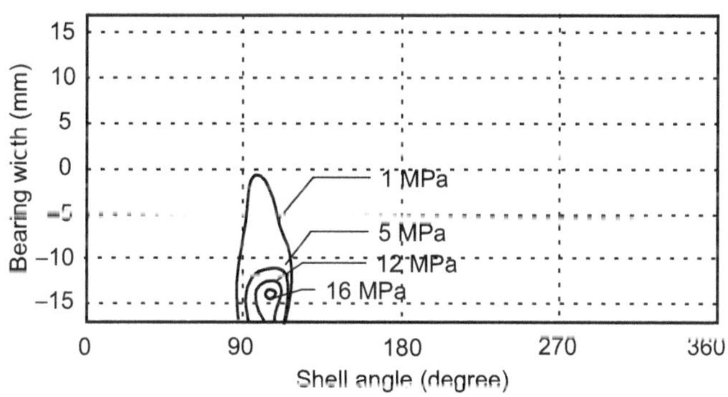

Figure 12.21 Estimate of the asperity contact pressures and location along the bearing width and circumference in elastohydrodynamic analysis of a multi-cylinder diesel engine

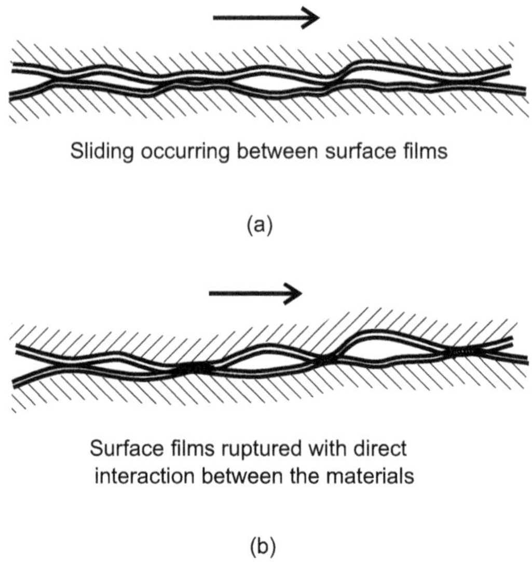

Figure 12.22 Simplified picture of adhesive wear (Reprinted from *The Tribology Handbook*, 2nd edn., M.J. Neale, Copyright 1996, with permission from Elsevier.)

The ecosystem in the tribological system of bearing, shaft, oil and the housing can be considered to be in dynamic equilibrium when the rupturing phenomenon is balanced by the healing one. The equilibrium is challenged every time when the operating parameters change until either a new equilibrium position is achieved or the situation deteriorates to higher rate of wear that may lead to seizure (Figure 12.23). This depends on the operating parameters, application and the material pairs employed. Increase in load, for instance, gives rise to additional asperity contact pressures and, hence, a higher rate of wear. The rate of change of load and the time available for the healing mechanisms to achieve a new equilibrium, due to the transition between loads, are important considerations. Also, in the case of a change from lower to higher speed, the same would be expected, and a rise in surface temperatures accelerates desorption of the weakly bound particles. However, cyclic hardening and quenching of the dislodged particles in oil near the loaded zone leads to abrasive wear due to hardened particles.

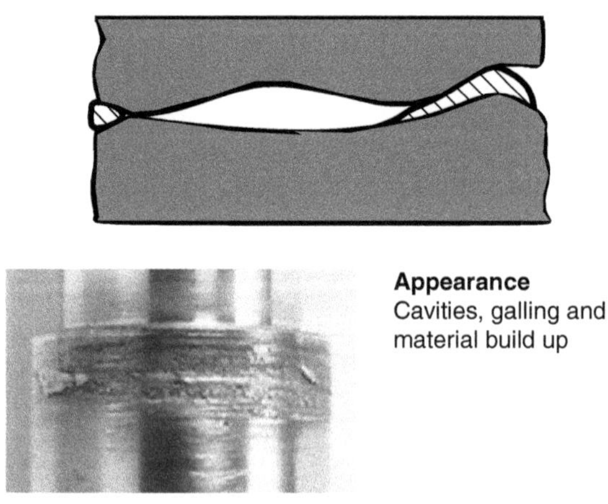

Figure 12.23 Adhesive wear

The initiation of scuffing can be begun by the plastic flow of one asperity. Plastic flow of asperities occurs by a combination of normal and tractive stresses, both of which can be transmitted through interposing films whether viscous or (soft) solid. By repeated plastic flow of some asperities, the surfaces scuff. After a few tens or hundreds of cycles of sufficient plastic flow, some deformed asperities break away by plastic fatigue. The debris of the particles agglomerate to form larger particles, which are repeatedly deformed and work hardened as they pass through the contact region. Eventually, a larger particle carries most of the load of the contact region, which produces a contact stress too large to sustain fluid film lubrication. Temperatures rise precipitously, which chemically alters any remaining protective films in the interfaces, and welding or adhesion occurs. The latter produces further severe plastic flow and heating, which propagates the damage at an ever-increasing rate, leading to scuffing (Ludema, 1984).

Wear on the outer diameter of the bearing surface and the housing which retains it can experience adhesive wear. Typically the housing and the bearing flex differently based on the stiffness, material pair and operating conditions. The housing materials are typically cast iron, forged steel or aluminium-based alloys and the bearing back steels are generally low carbon steel, for example SAE1010. The relative motion between the two surfaces which transfer the loads causes small slip excursion and can result in micro-welding and fracture of this weld leading to the formation of debris. This debris is in a highly oxidized state. However, the wear tends to slow down because of the debris buffers between the two surfaces. Subsequent wear is either by fatigue or abrasion based on the localized asperity pressures, nature of materials and operating conditions. Now it has been confirmed that fretting is related to the magnitude of the relative amplitude of the slip between the connecting rod big end bore and the bearing outer surface due to distortion of the connecting rod, and that finite element analysis effectively predicts the fretting. The difference in the magnitude of circumferential displacement between these two surfaces is defined as relative slip amplitude (Mizuno *et al.*, 1993). The elastic stress concentration factor which indicates the average load required on the part for the onset of plastic deformation or yielding also indicates the loads on parts that will cause fatigue fracture. When debris is generated the region of material loss at the location on the housing bearing interface can induce a crack, as it will act as a stress raising location. It leads to high local stresses and can lead to local plastic deformation surrounded by a region of elastic deformation.

12.9.2 Abrasive Wear

Wear caused by hard particles is termed abrasive wear. The bearing surfaces are much softer than the shaft surfaces (Figure 12.24). Typically, the overlay plating on the bearing surface has hardness in the range 8–16 HV, sputtered coatings of aluminium material have a hardness in the range 60–90 HV, aluminium bearings can have surface hardness in the range 35–65 HB and the shaft surface hardness is in the range 55–60 HRC.

When abrasive particles (generally from the oil circuit or the external environment) enter in between the soft bearing and harder shaft surface, the metal undergoes extensive work hardening. For this reason, initial hardness is not a particularly important factor in determining the resistance to abrasive wear, so long as the hardness of the abrasive is (as usual) always substantially greater than that of the metal surface (Figure 12.25). In this special condition, there is a relatively simple relationship between wear resistance and hardness. However, the coarse abrasive particle gouges out the bearing surface structure, and extensive fracture is formed along the track in the direction of rotation.

When the abrasive particle gets excessively work hardened with every cycle and approaches the hardness range of the shaft material, or a much harder particle enters the clearance space, then blunting of the

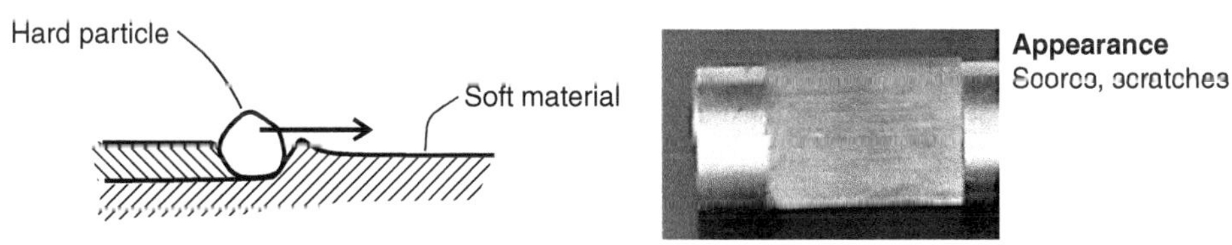

Figure 12.24 Some aspects of abrasive wear

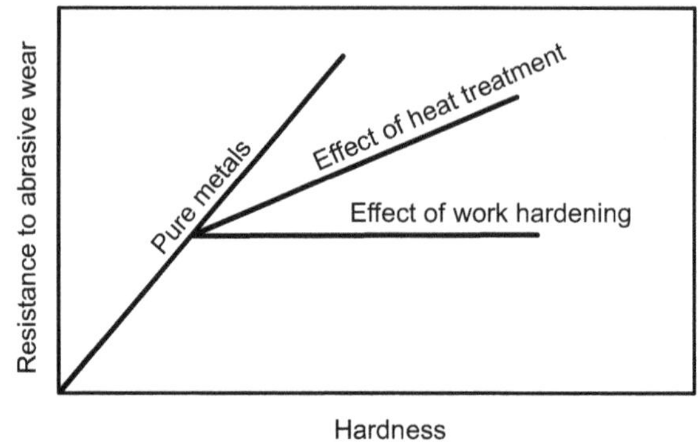

Figure 12.25 Resistance to abrasive wear (Reprinted from *The Tribology Handbook*, 2nd edn., M.J. Neale, Copyright 1996, with permission from Elsevier.)

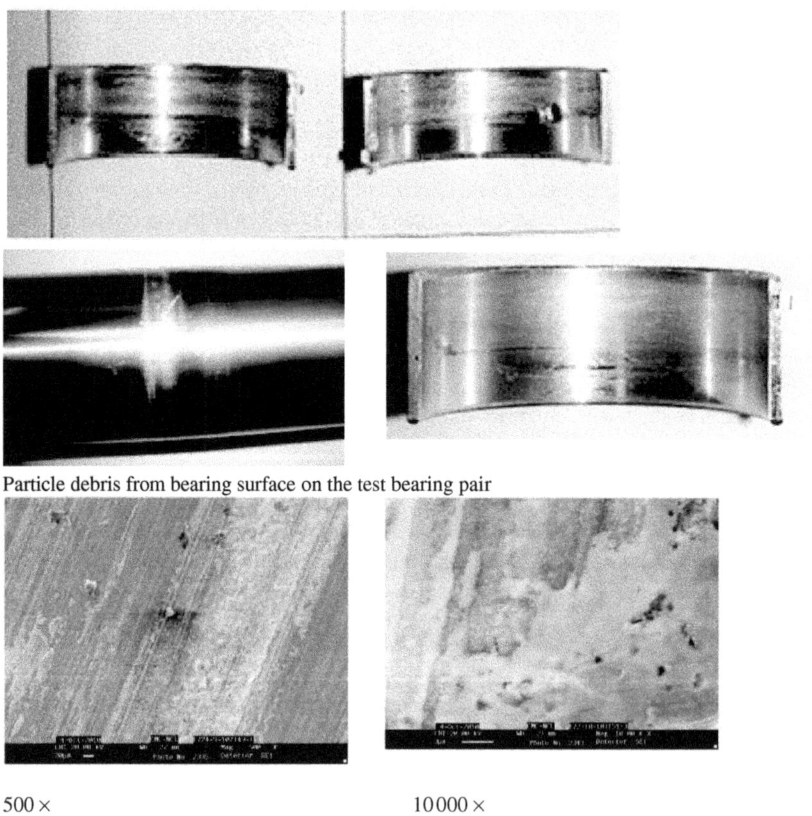

Figure 12.26 Scanning Electron Microscope images of the bearing surface showing the particle debris post plastic flow of asperities leading to scuffing and seizure

abrasive occurs. The particle can also create tracks of wear on the shaft and bearing surface (Figure 12.26). The form of the relationship between wear resistance and the relative hardness of the metal and abrasive is of considerable technical importance here. As an abrasive grain begins to blunt, the mode of wear changes from one of chip formation, perhaps aided by a plastic fatigue mechanism, to one which must be largely an adhesive fatigue process. The change is quite rapid and is usually fully accomplished over a range of $H_{metal}/H_{abrasive}$ of 0.8–1.3, where H is the actual surface hardness (Neale, 1996).

The mechanism of wear seems to be from the plastic fatigue. When the particle enters in the region of minimum oil film thickness or maximum oil film pressure, the abrasive particles gouge to greater depths, leaving behind a track and distressed zones, as can be seen in the scanning electron micrographs (Figure 12.26). The same particle will be quenched after moving out of the loaded zone. When such particles approach the areas near the relief section of the bearing or near the nick of the bearing, they tend to get embedded in the soft overlay. The hard embedment affects the zones at the interface of the particle and near it, resulting in removal of bearing surface and formation of debris. It may also result in contact loads resulting in adhesive fatigue in these regions.

12.9.2.1 Progression of Abrasive Wear Towards Seizure

The photographs in Figure 12.27a–e show five stages of progression of abrasive wear of an electroplated copper lead lined bearing in a diesel engine. Figure 12.27a indicates scoring of the bearing overlay with hard particles dislodged from the crankcase casting; Figure 12.27b shows progression of the abrasive wear along wider region of the bearing width accompanied by embedment of hard particles in the overlay; Figure 12.27c indicates abrasive wear progressing along the total bearing width with additional ingress of hard particles and associated fatigue wear in adjacent regions; Figure 12.27d shows heavy metal-to-metal contact and a large degree of scoring resulting in a small amount of exposure of the underlying copper–lead lining. Figure 12.27e highlights extreme wear of the overlay and copper–lead lining, revealing seizure of the bearing. Figure 12.27f is the scanning electron microscope image of the abrasive surface of the bearing, indicating scoring and scratching of foreign particles. Figure 12.27g shows the dislodged material embedded near the scoring zone of the bearing surface and Figure 12.27h is the Energy Dispersive X-ray (EDAX) analysis showing the presence of foreign particles from the crankcase on the bearing, namely iron and silicon (sand from the improperly filtered air breathed in or the casting), resulting in this type of scoring.

12.9.2.2 Fatigue

Fatigue mechanisms on the bearing running surface are associated with the maximum oil film pressure generated in the loaded section (Figure 12.28). In sliding wear conditions, during effective elastohydrodynamic lubrication where metal-to-metal contact, and hence adhesive interaction, is reduced or absent altogether, it is the peak oil film pressure, the rate of pressure rise (the pressure profile) with time and the number of cycles such loading dictates, for a particular material pair, that initiate fatigue. Ball and roller bearings, as well as gears and cams, are examples where a fatigue mechanism of wear is commonly observed and gives rise to pitting or spalling of the surfaces. In the present day scenario where there is ever increasing pressures on downsizing engines and increasing the peak cylinder pressures to develop more power from lesser engine space, the bearing system which sustains these loads has to have increased fatigue limits and build in reliability to the total system. The fatigue strength is a function of the bearing material construction. For selecting a bearing material for a particular application, it becomes critical to understand the contribution of individual layers in the bearing to the limiting fatigue value.

The crack nucleation often happens at the surface under extreme stresses due to the peak oil film pressure. The environment around it decides the rate of crack propagation, for example the oil entering into the crack and getting trapped inducing high fluid pressures and temperatures in the wedge of the opening and closing crack. The cracks propagate inside the surface reaching the high strength interface of

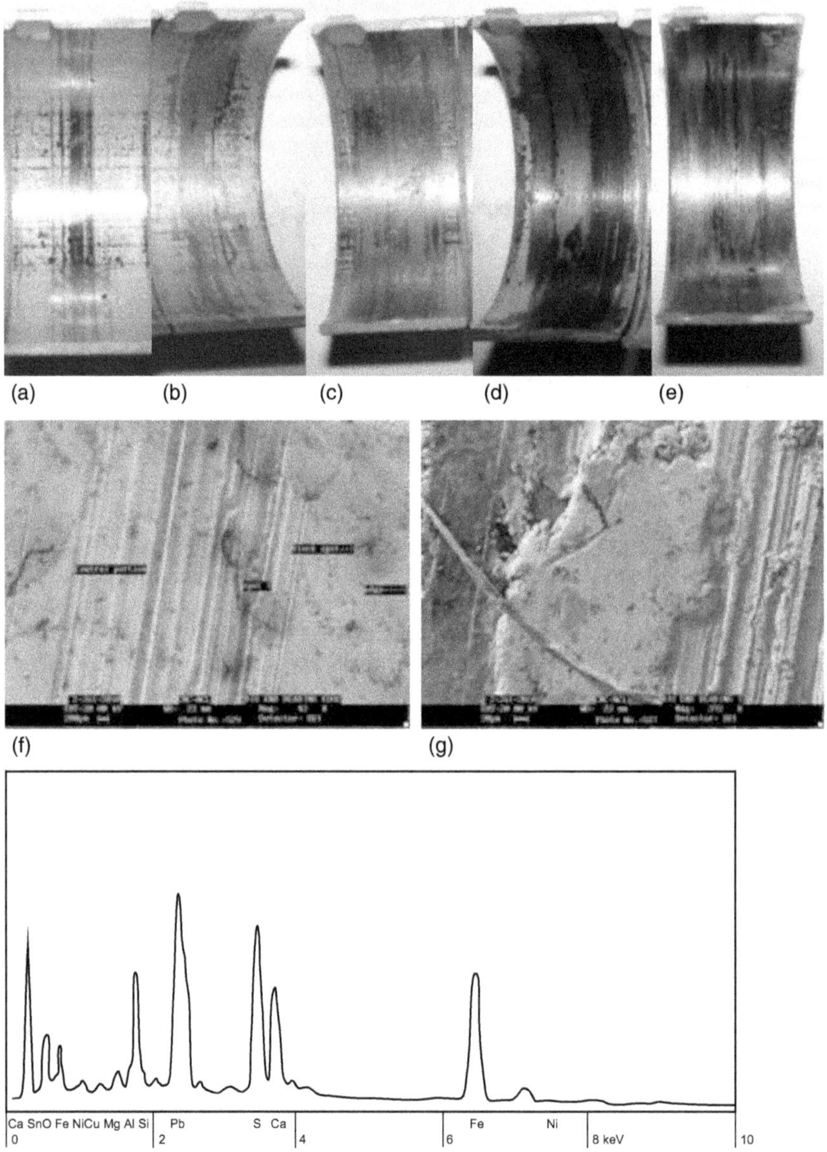

Figure 12.27 Stages of progression of abrasive wear of an electroplated copper–lead lined bearing in a diesel engine

the bearing lining and steel in the case of bi-metal bearings or between in the interface of the overlay and the lining in the case of the tri-metallic bearings. Continued use causes the cracks to turn parallel to the steel surface or the lining surface, until they join allowing the lining to flake.

There are many factors which contribute to the engine bearing fatigue: the relative strength of the lining to the strength of the overlay, the thickness of the lining and the overlay, the nature of the loading conditions, speeds of operation, adhesion between the overlay and lining and lining to steel backing in the

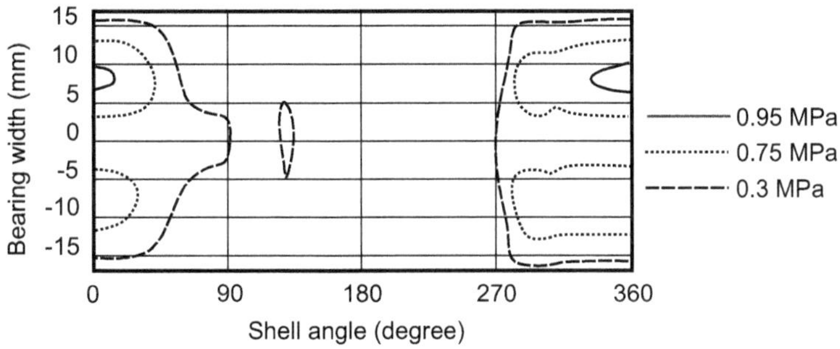

Figure 12.28 Hydrodynamic oil film pressure estimates (magnitude and location in the bearing width and circumference) from elastohydrodynamic analysis of a bearing simulation of a diesel engine

bearing construction, design aspects such as crankpin to journal diameter ratio, relative angular velocity of the load vector with respect to the journal speed, oil viscosity and rigidity of the housing and shaft.

The backing materials, that is the steel followed by the lining, need to have sufficient strength to support the overlay for fatigue resistance. If a high strength overlay is employed on a lower strength lining, then the fatigue crack initiation can happen in the lining itself. The fatigue limit for this bearing structure is limited by the weakest link, which in this case is the lining material, for example, a sputtered overlay deposited on a lower fatigue rated copper–lead lining material for high load applications. The tendency would be to use an overlay directly on the steel backing to support such high loads. However, it is also essential to take note of the cushioning effect that the lining material provides, to take care of edge loading scenarios typically encountered due to higher deflections in heavy loaded applications.

It is also well known in the bearings industry that the thinner the lining or overlay thickness, the higher the fatigue strength (generally, overlay thicknesses range from 8–20 microns, lining thicknesses can range between 0.15 and 0.6 mm). The explanation for this behaviour can be from the fact that the thin lining or overlay is prevented from straining to as great an extent as the surface of thicker lining by virtue of its closeness to the relatively harder lining and steel backing.

At higher engine speeds the load carrying capacity of the oil film increases, but the bearing should be able to deal with the increased inertia loads and sustain the deflections of shaft and increased oil film temperature which result in reduced fatigue strength of the bearing material.

Another factor associated with bearing manufacturing is the adhesion strength or bond strength between the overlay and the lining, as well as between the lining and the steel. A weaker bond can cause incipient fatigue damage to the bearing structure. Sufficient quality of surface preparation and cleaning are essential. The quantification of the bond strength is generally done by destructive methods, an important aspect to be considered for fatigue strength requirements.

Throughout the operational cycle, the maximum bending moments occur in the centre planes of the crankpins and their effects are reduced as the crankpin diameters are increased, hence offset connecting rods reduce significant load on the big end bearings, provided the offset is to the side which will permit the connecting rod to deflect conjugally with the deflection of the crankpin for a symmetrical load on the bearing

The relative angular velocity of the load vector with respect to journal speed is important to consider; loads rotating at half the journal speed, even for short durations, for example 50 degrees of journal rotation, are sufficient to reduce the thickness of the oil film to such an extent that the high rate of shear causes surface temperatures which result in lowering of fatigue strength. Thus, the relative mass centred at the big end compared with the mass of the piston and small end can affect the load carrying capacity of the big end bearings.

The fatigue loads imparted on the bearing surface are a function of the hydrodynamic oil film pressures generated in the oil film between the bearing and shaft. Elastohydrodynamic analysis using modern tools

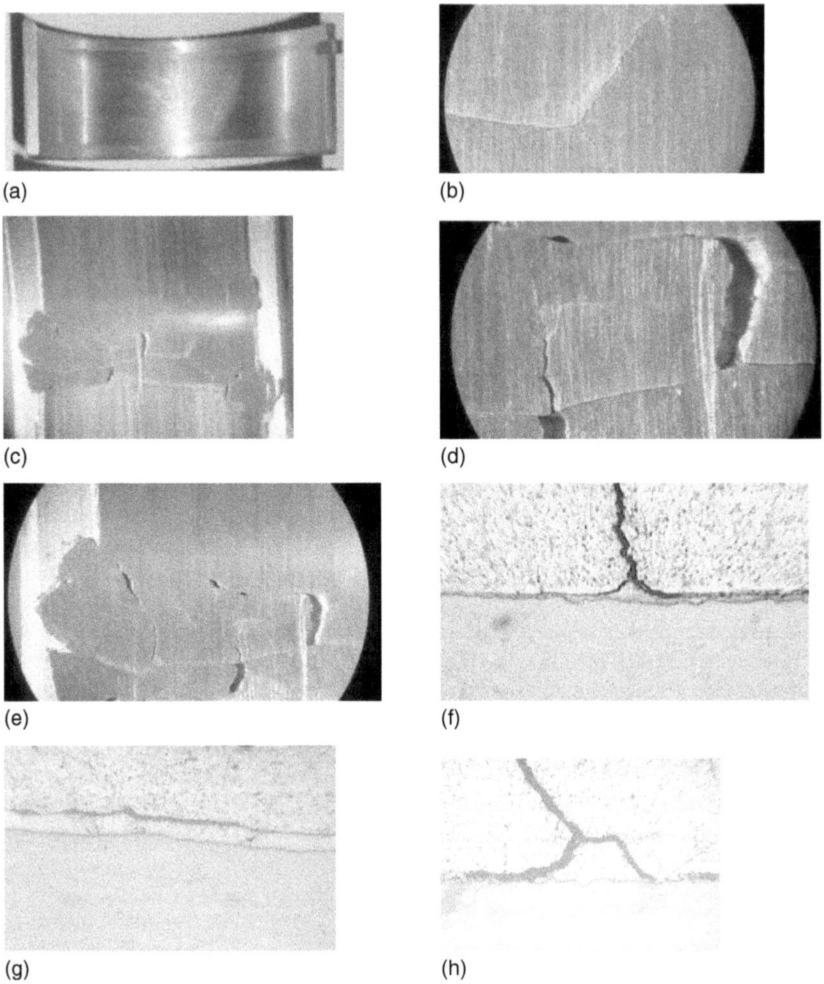

Figure 12.29 Fatigued surface and micrographs of a network of fatigue cracks

gives a good estimate of these pressures in the bearing circumference. The influence of the oil-hole path and the oil groove locations can be easily visualized from these analysis results. Various bearing materials are now characterized for their peak maximum oil film rating, which in turn characterizes the fatigue limit for the bearing materials. Figure 12.29 shows classical fatigue crack propagation on the bearing surface of an aluminium–tin–silicon bearing after reaching the material fatigue limit in 4×10^6 cycles of dynamic loading condition in a bearing fatigue test rig.

Figure 12.29a shows the bearing having crack generation at the point of maximum peak oil film pressure, that is the load zone. Figure 12.29b offers the zoomed view of the right edge of the bearing where the crack propagation is evident. Figure 12.29c presents extreme fatigue cracks resulting in plucking of the bearing material resulting in wear flow of the bearing material outside the bearing lining width. Figure 12.29d and Figure 12.29e offer zoomed view of the cracks.

Figure 12.29f shows a micro-cross-section of the fatigue bearing with the crack generation on the surface of the aluminium lining travelling down in the cross-section and turning at 90 degrees at the aluminium separation foil. Figure 12.29g shows the crack propagation parallel to the aluminium foil and Figure 12.29h shows heavy cracks propagated from the surface, bifurcating intermediately, passing through the foil and then turning at the steel interface of the bearing.

Similarly Figure 12.30 shows the fatigue mode of failure in a copper–lead lined bearing. Figure 12.30a presents the cross-section of a bearing having a sintered structure in which lead particles are distributed

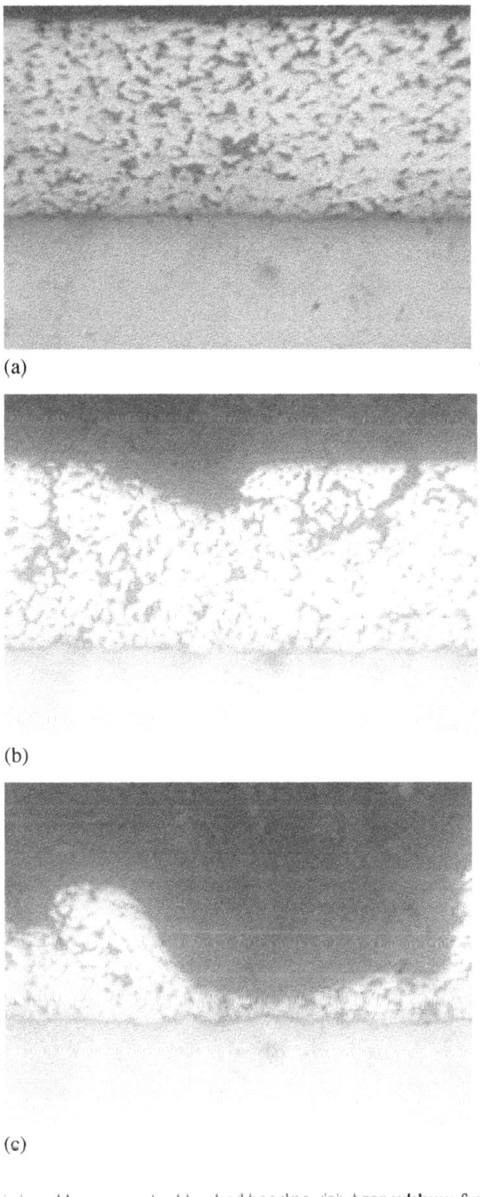

(a)

(b)

(c)

Figure 12.30 Copper–lead sintered lining on steel backed bearing. (a) Area without fatigue; (b) crack initiation at the surface and propagation towards the steel layer; (c) removal of lining material flakes after two different cracks initiated at the top meet at the steel interface uniformly in the copper–tin alloy. Figure 12.30b shows the cross-section of the bearing in the zone where the peak hydrodynamic pressures exceeded the fatigue limits of the bearing material for a substantial duration of time, resulting in crack initiation on the inner surface of the bearing at this location and then propagating inside the lining structure. Adjacent to the crack a chunk of removed material, due to fatigue cracks meeting each other midway, is also evident. Figure 12.30c shows large flakes of lining material removed from the bearing after cracks propagated up to the steel interface and joined each other.

The hydrodynamic pressures can accelerate the process of crack propagation and removal of the material as a result of the oil entering the cracks and creating additional sites for crack propagation.

12.9.3 Erosive Wear

12.9.3.1 Fluid and Cavitation Erosion

Both these wear mechanisms arise from essentially the same cause, namely the impact of fluids at high velocities. In the case of fluid erosion, the damage is caused by small drops of liquid, while in the case of cavitation the impact arises from the collapse of vapour or gas bubbles formed in contact with a rapidly moving or vibrating surface (Chena and Mongis, 2005).

In fluid erosion the duration of impact is extremely small, very sharp and intense compression pulses are transferred to the surface material by momentum transfer. Shear deformation is also observed in the peripheral areas where the oil flows off further. Repeated deformation of this nature gives rise to a fatigue type of damage and pitting or roughening of the surfaces soon becomes apparent.

In cavitation erosion, damage is caused by fluid cavities collapsing in regions of high pressure. The cavities may be vaporous or gaseous. The pressure differential across the bubble interface, which is influenced by the surface tension and fluid vapour pressure, dictates the physical stability of the bubble. Oil viscosity and surface energy of the bubble influence the extent of the damage (Figure 12.31). Nevertheless, the cavitation resistance as a material property was characterized as the materials response to this situation rather than an intrinsic property of the material. In addition to the nature of materials (e.g. chemical composition, micro-structure, heat treatment, geometry, surface roughness, residual stress etc.), cavitation resistance depends largely on liquid property, flow speed, vibration characteristics, temperature, hydrostatic pressure and so on. For certain materials, cavitation resistance is related to hardness, but for most metallic materials it is more closely related to fatigue strength of materials (Chena and Mongis, 2005).

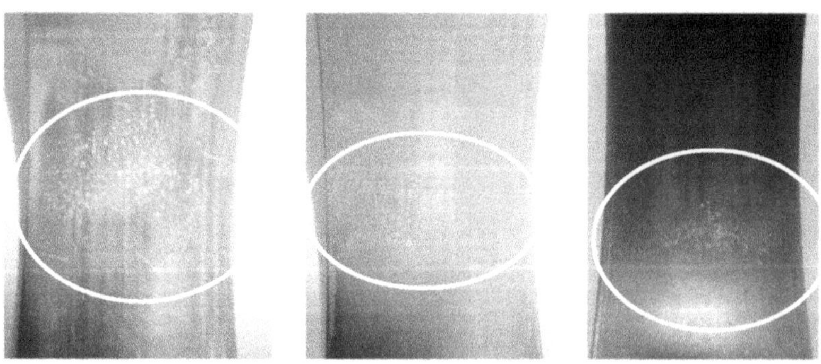

Bearing no. 1 Bearing no. 2 Bearing no. 3

Figure 12.31 Fluid cavitation erosion signatures on the bearing surface. Pitting marks observed on the bearing surface indicating fluid erosion from small droplets of oil on the surface, resulting in fatigue signatures on the overlay

So, it is very difficult to establish a universal rule for choice of materials to minimize cavitation erosion wear. The material's ultimate resilience characteristics have been correlated to the damage withstanding capability, which is related to the amount of energy that can be dissipated by a material before any appreciable cracking can occur. The physical damage to metals is pitting in nature and shows signatures of fatigue.

The photographs in Figure 12.32 show the phenomenon of cavitation in the connecting rod bearings in a heavy duty engine where the connecting rod is inclined cut and the load zone in the bearing is near the joint face, and hence just near the relief zone. Figure 12.32a shows a view of the bearing which indicates some amount of scoring wear and embedments to the naked eye. Figure 12.32b shows the stereomicroscope photograph under higher magnification at the nick region. It also indicates the extent of score and embedment in this region. Figure 12.32c is a view of the cavitation region in the crown of the bearing under the microscope, indicating the existence of the cavitation regions. Figures 12.32d and 12.32e show magnified views of the cavitation erosion of the bearing surface. The erosion of the bearing surface after the imploding bubble is clearly evident. Figures 12.32f and 12.32g offer the embedment of the eroded material just next to the points of removal.

Figure 12.32h shows the region near the bearing joint face area in the relief section of the bearing where the wall thickness is lower than that at the crown. The end eroded bearing material has been carried away by the oil flow and is embedded in the relief zone. It is also evident in the magnified view of the same region in Figures 12.32i and 12.32j, regions of cavitation and embedment. Figures 12.32k and 12.32l show microscopic images of cavitation erosion of the overlay of the bearing propagated to the inner copper lead lining surface. Figure 12.32k presents the initial stage of the failure and Figure 12.32l shows propagation of the failure over a wider distance.

Figure 12.33 shows the oil hole in a main journal of a slow speed engine. The cavitation erosion on the bearing surface has taken place due to interrupted oil flow from this oil hole by the edge of the oil outlet followed by solid particle erosion and transfer of the eroded material onto the shaft. Cavitation damage in oil lubricated plain bearings is currently observed both in medium and slow speed diesel engines for marine or power station applications and in high speed automotive engines.

Considering variations with respect to the fluctuations of forces, variation of oil pressures and instability of the flow and different operational parameters like torque, load and speed, it is difficult to analyse the real cause of cavitation damage. However, four main mechanisms of cavitation have been identified: suction, discharge, erosion and solid particle erosion or abrasion.

Suction cavitation is observed in the clearance space inside the bearing, where the pressures are extremely low, much below subatmospheric. The continuous movement of the shaft in the clearance space, accompanied by firing loads from the engine cycle, produces cyclic movements of the crankshaft resulting in aspiration of the oil and formation of bubbles in low pressure region. Therefore, these are normally found in the upper half of the main journal bearings. These bubbles then collapse to impinge on the bearing surface. Generally, high speed engines suffer from this phenomenon due to higher frequency of shaft motion.

High speed oil jets emerging from the loaded region in the direction of rotation lead to a drop in oil pressure and cavities are produced. This is termed discharge cavitation. Subsequent collapse of the cavities causes damage to the adjacent surface. Typically, a change in the geometry of the bearing surface (e.g. oil grooves, relief sections of bearings on the inclined cut connecting rods, edges of oil holes, eccentricity in the bore merging with the relief section) can often lead to flow instability. An example is a bearing grooved in the upper main bearing shell and without the oil groove in the lower (half) main bearing. The upper part experiences relatively low loads, and hence can afford reduction in bearing area caused by the provision of the groove. The oil groove helps in running the bearing cooler by allowing extra flow of oil. However, the lower bearing shell is heavily loaded and seldom can afford a reduction in area, and hence to increase bearing area the oil groove is not provided. Oil flows from upper shell (grooved) to the lower shell (not grooved) creating an area of low pressure in this region, and hence causing vapour bubble formation. Subsequent collapse of the bubbles leads to cavitation

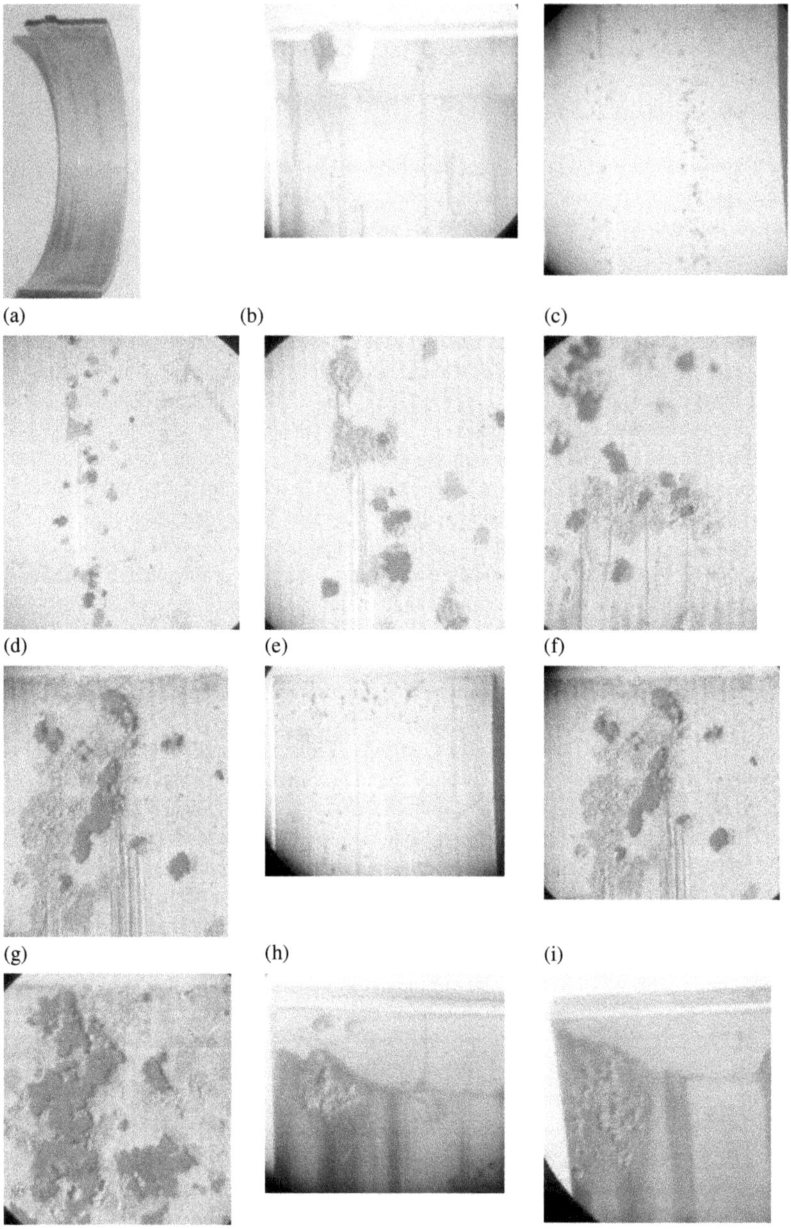

(a) (b) (c) (d) (e) (f) (g) (h) (i) (j) (k) (l)

Figure 12.32 Discharge cavitation in a connecting rod bearing

damage on the lower half shell of the bearings. This phenomenon is generally observed in medium and low speed engines.

Cavitation erosion is a result of impinging oil flow on the bearing surface at high pressures and can result in pitting damage or fatigue damage. For example, when oil flow through the oil hole in the shaft

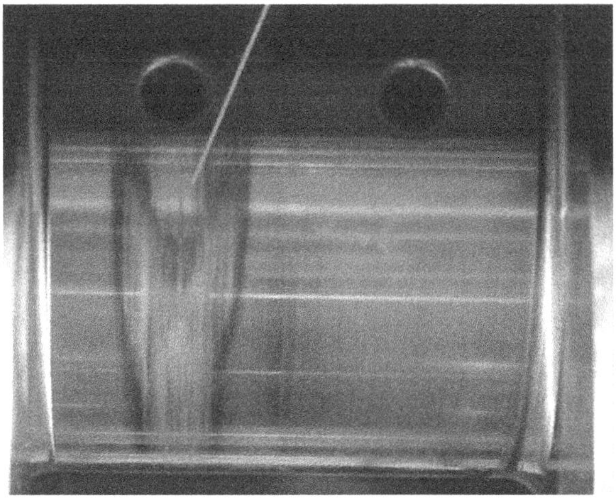

Figure 12.33 Cavitation erosion near the oil hole due to interrupted flow of oil

enters into the main bearing shell, extremely high pressures can be created in the oil jet due to the rotational inertia of the oil pockets in the oil holes of the crankshaft. In addition, a pressure wave is created in the oil hole of the crankshaft because of alternating oil flow direction. The oil flow from this hole, when interrupted by the edge of the oil outlet, can exert extreme pressures on the bearing surface and produce cavitation due to the difference in pressures.

12.9.3.2 Solid Particle Erosion or Abrasion

The abrasive particles may also be conveyed by an oil stream and the impact of the abrasive laden oil will give rise to erosive wear of any bearing surface. The magnitude and type of wear experienced depends very much upon the impingement angle (Figure 12.34) of the particles and the level of ductility, brittleness

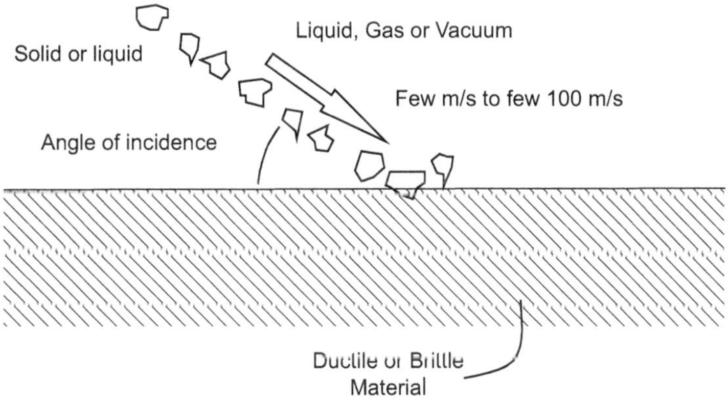

Figure 12.34 Erosion mechanism: wear by fluids containing abrasive particles (Reproduced by permisison of SG Roberts, University of Oxford.)

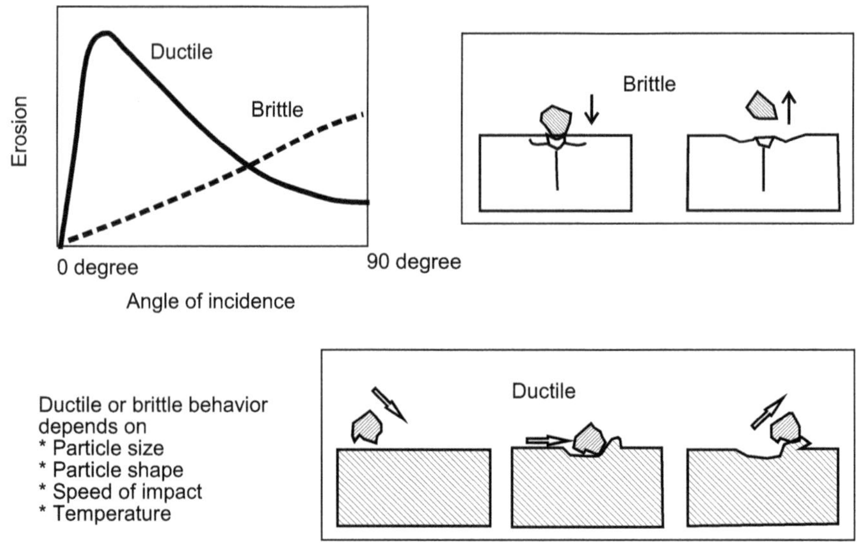

Figure 12.35 Erosion of ductile and brittle materials (Reproduced by permisison of SG Roberts, University of Oxford.)

or elasticity of the surface (Figure 12.35) (Watanabe *et al.*, 2000; Pike and Conway-Jones, 1992). The wear mechanisms are modified by the ability of the particles to rebound and the fact that the energy available is limited to that of the kinetic energy given to them by the fluid stream.

The solid particles dislodged by cavitation erosion, when accelerated by the oil velocity, bombard the bearing surface and can cause solid particle erosion or abrasion. This type of loading is similar to solid-to-solid contact with friction. When the stresses generated by this impingement, along with the oil film pressure, exceed the elastic limit of the material, it leads to plastic deformation at or near the surface; it shows a similar wear mechanism such as abrasion. If these stresses are less than the elastic limit they can lead to the cyclic fatigue mechanism. The velocity dependency of this mechanism decides the deformation rate, and hence is dependent on strain rate. The impact energy of the impinging particle can be absorbed either by elastic and plastic deformation, or by fracture and its propagation.

12.10 Requirements of Engine Bearing Materials

Considering the engine system as an ecosystem allows a better understanding of the interdependencies within the system and the expectations posed by various components of the system to deliver the desired performance. The inputs to this system are fourfold: the system design aspects, the properties of the system, the environment of the system and the contribution to the system from the bearing materials.

Considering the engine tribosystem (Figure 12.36), various aspects of the bearings under the purview of the designer for a particular application are the dimensions, geometry and features (concentric and eccentric bores, groove, location of oil hole, relief design), texture and mechanical motion. The essential properties of the system are the mechanical and thermal properties of the shafts, housings, heat treatment, surface treatment of the shafts, housings, temperature dependence of oil viscosity and so on. The environment surrounding the bearings is in relation to the nature of

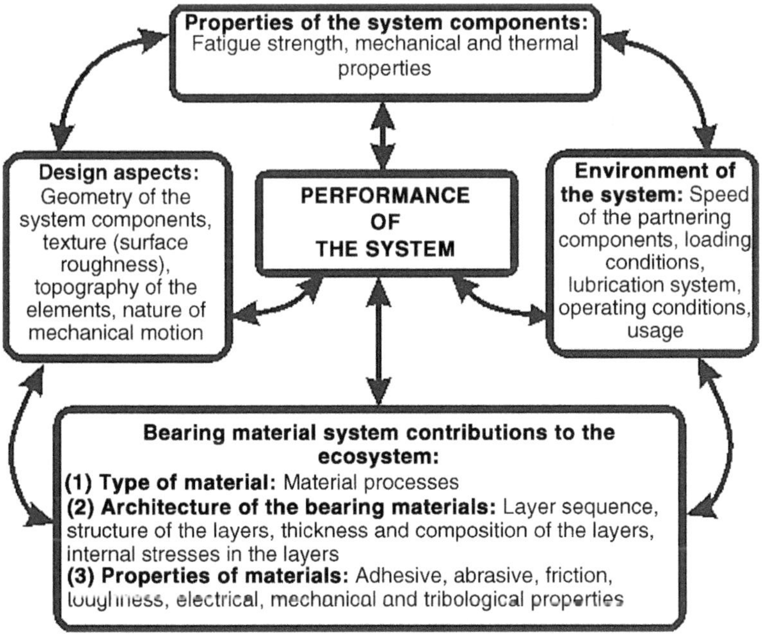

Figure 12.36 The engine tribological ecosystem

lubrication system, contamination levels, physical and chemical properties, dirt contamination, speeds of operation, specific application, the nature of usage are imposing their specific requirement on the nature of bearings to be used.

The selection of the bearing material is a very important aspect of the design process. The characterization of the bearing materials for basic mechanical properties in terms of hardness, tensile strength, fatigue strength and thermal properties is necessary, as is the case with many other parts of the engine. In addition, the environment, the speed, metallurgy and topography of mating parts, lubrication regime, loading conditions and the failure mechanisms dictate characterization.

The expectations on the bearing performance due to the tribosystem are summarized in Figure 12.37: embeddability, compatibility, conformability, corrosion resistance, fatigue resistance, seizure resistance, wear resistance, cavitation and erosion resistance. The requirements are important drivers for the design and selection of the bearing materials. To begin, fatigue and seizure resistance are generally covered under the unit load carrying capacity ratings for the bearing materials.

However, it is essential to have adequate experience and knowledge of the tribological ecosystem of the engine bearings before venturing into the design and selection of bearings for a particular application. The suitability of a particular bearing material and system needs to be ascertained based on various available inputs. Considering the complexity of characterizing the bearings in engines for a particular application, it is found suitable to have controlled test rigs to obtain individual tribological properties of bearing materials and then correlate these characteristics with actual engine test experience. With such data, bearing designs and materials are selected for a particular application. Figure 12.38 shows an inner circle of requirements which are primary for the design and then an outer circle of requirements for the constraints for the final design and selection of material.

However, the test history and engine experience are also indicative of failures of bearings for particular applications and the analysis of these failures form an important basis for the bearing design

The challenge is to achieve all of this simultaneously

Embeddability
The lining material surface should be able to embed micron-sized foreign particles and restrict their recirculation with the lubricating oil to avoid accelerated wear of lining material or shaft by scoring.

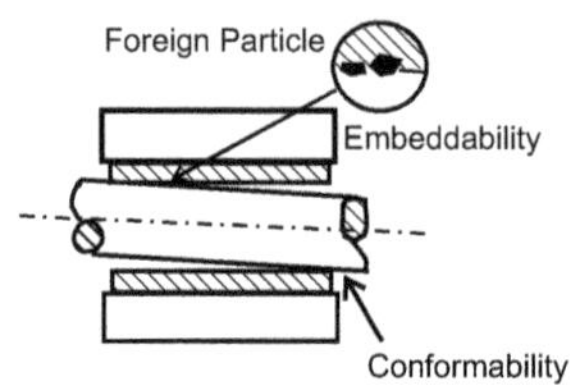

Conformability
The lining material should flex by itself in order to accommodate irregularities of manufacturing in either shaft or the housing to avoid disturbances in the oil film leading to either overloading or accelerated wear.

Load carrying capacity
The lining bears the loads in between the shaft and the housing. It should have sufficient strength to avoid plastic deformation and to avoid fatigue damage due to cyclic loading.

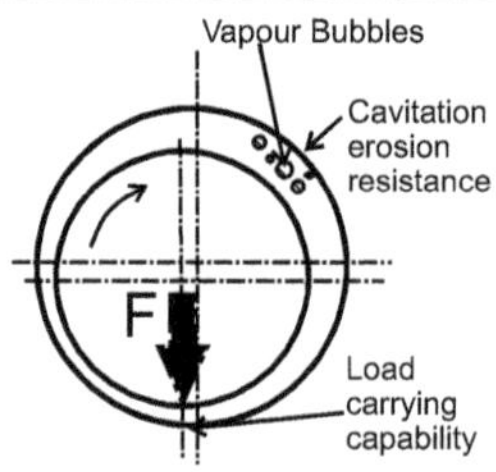

Cavitation erosion resistance
The lining should have sufficient resistance to cavitation erosion that imparts concentrated impact loading due to collapse of vapour bubbles of lubricating oil, formed due to sudden fluctuation in lubricating oil pressure.

Compatibility
The lining material should resist physically, possibilities of mutual welding, seizure or fusion to maintain relative motion between the shaft and the lining material against the conditions of rubbing and friction

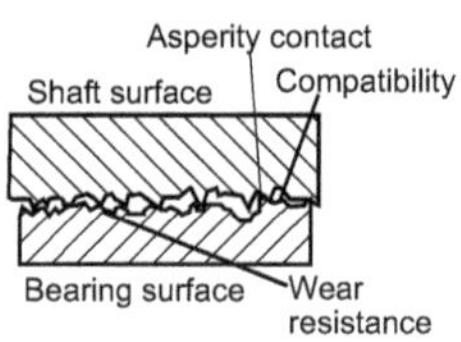

Wear resistance
Though a thin film of lubricating oil is maintained, the asperities of the lining material and shaft may contact intermittently causing wear. The lining should be able to sustain this to avoid accelerated wear in order to maintain the clearances for expected life.

Corrosion resistance
Additionally, the lining material should have sufficient resistance to the corrosion that arises due to deterioration of the lubricating oil.

Figure 12.37 Requirements of engine bearing materials

and material selection. The design criteria are then updated to accommodate this experience in the future. The mathematical models built to predict the bearing performance strive to accommodate the various possible predictions of failures. However, there are still some shortfalls in these simulation tools for predicting accurately. The tools indicate pressures, flow rates, minimum distances and shaft centre loci with respect to bearing geometry, and interpretation of these results into probable failure modes and correlation with experience are key to overcoming these shortfalls (Table 12.5).

Figure 12.39, as well as Table 12.6, shows fundamental requirements of the bearing coatings or overlays. It is clear that the overlays need to be hard and soft at the same time, which means it is necessary to select hardness and to balance opposing requirements of high fatigue and wear resistance against good embeddability and seizure resistance. Thus, selecting an overlay and bearing material is making the right compromises so that the tribological pair is successful throughout their intended life.

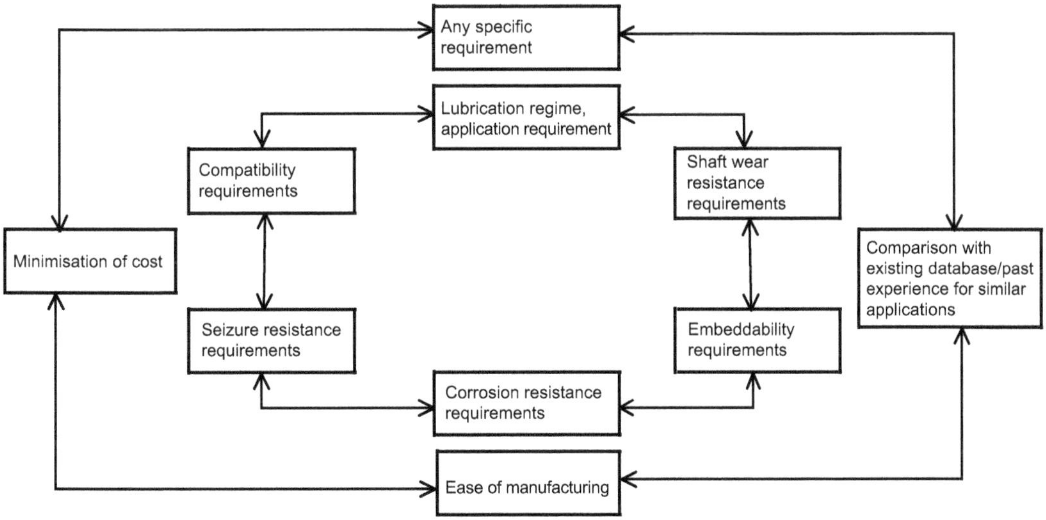

Figure 12.38 Requirements for the design and selection of materials

Table 12.5 Possibilities and shortfalls of mathematical models for explaining different failure modes of bearings

Failure mode	Possibilities/Shortcomings of mathematical models
Fatigue rupture of support shell	Arises only in connection with severe design errors. Stresses can be calculated
Fatigue rupture of inner layer	Calculation of maximum film pressure permits correlation
	Stresses rudimentary, local temperature inexact. Maximum specific load unusable
Cavitations	No quantitative models, correlation to certain operating characteristics in its beginning (oil flow and vibrations)
Abrasion/Erosion	Calculation of minimum film gap and correlation to R_z possible in connection with local temperature rise
Corrosion/Erosion	Influence of temperature barely quantifiable; effect of oil condition not at all
Fatigue/Erosion	Correlation from practical experience to certain h0min/pmax range possible
	No quantification of stress/temperature level
Creep of support shell, fretting	In all wear phenomena local energy distribution and residue time gives some correlation to operational experience
	Well defined stress and temperature limits for creep
	Limits of radial pressure for fretting from practical experience
	Large grey area, dynamic influences not sufficiently formulated
Disturbance of oil film gap geometry	No quantification as yet
	Minimum film gap, surface velocity and limits for particle size give some correlation from practical experience

Table 12.6 Role of the surface structure and coating

Basic demands on bearings	Role of coating
Provide favourable surface texture	Protect against adhesive wear
Reservoir of lubrication	Conserve surface structure against abrasive and adhesive wear
Removal of particles in grooves	Reduce friction
Wetting behaviour of the substrate on the coating	Modify surface energy

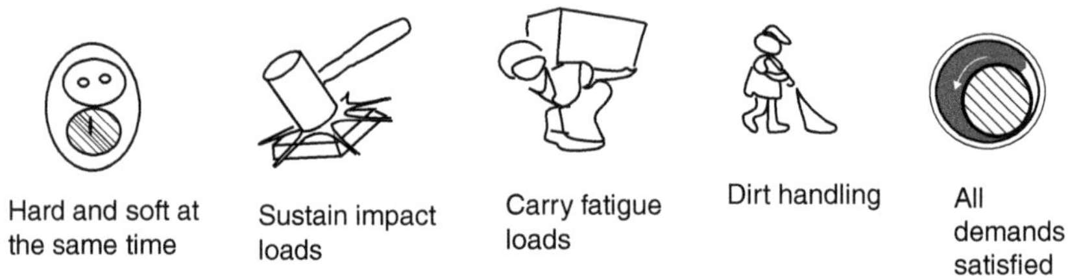

Figure 12.39 Fundamental requirements of bearing coatings

12.11 Characterization Tests for Wear Behaviour of Engine Bearings

To isolate various modes of failure and provide a control environment and operational conditions, and to allocate and analyse the contribution of bearing materials to the tribological performance, various test rigs have been designed to generate a particular failure mode and are in use:

(i) Ultra-complex rigs where loads and operating conditions are similar to those in the engines but having accurate control on design, dimension and geometry.
(ii) Test rigs where dynamic sinusoidal and static loads on the bearings are generated with other tribological conditions similar to the engine in order to produce fatigue, seizure and wear modes of failure on the bearings.
(iii) Test rigs using ultrasonic horns to generate cavitation in the fluid and characterize bearing materials for cavitation resistance.
(iv) For generating alternating bending fatigue loads on bearings to estimate the fatigue limits and fretting limits in the bearing form.
(v) Test rigs to generate accelerated chemical corrosion of the bearings and estimating their rating for these characteristics.
(vi) Test rigs to evaluate the alternate bending fatigue characteristics of the bearing strips in the strip form.
(vii) For mechanical and metallurgical properties characterization, for example tensile testing, hardness testing, microstructure and chemical composition testing.
(viii) For wear characterization using basic pin on disc or modified pin on disc to test bearing materials in lubricated and dry conditions; similarly modified test rigs to test thrust loads on thrust washers.
(ix) Special equipment to test micro- and nano-hardness of very thin overlays and coatings.
(x) Scratch test to quantify the adhesion or bond strength between overlay and lining.

Table 12.7 Fatigue strength of typical oil film bearing alloys (source: DeHart, 1984; Kingsbury, 1997).

Material	Load capacity (MPa)
Thick Babbitt	5–10
Thin Babbitt	
0.25–0.50 mm	14
0.10 mm	17
Copper–lead	38
Aluminium alloys	55
Silver	80
Bronze	100

Bearings characterized from each of these test rigs give insight into the behaviour of each material under specific test conditions and, therefore, indicate its potential behaviour in the engine under various operating conditions. These test rigs, however, do not eliminate the need to test the bearings in the engines, but increase the likelihood of making the bearings successful in the engine in the first iteration. Years of such parallel testing and data generation on the test rigs and engine tests as well as mathematical models enable the bearing performance of bearings in the engines to be predicted.

12.11.1 Fatigue Strength

Fatigue strength is one of the most important properties of bearing materials. Fatigue strength is particularly important for reciprocating loads encountered in the connecting rod and main bearings of reciprocating internal combustion engines. The need to achieve high fatigue strength often brings into use aluminium, copper–lead, bronze or silver bearings for automotive or diesel engines. Fatigue failures result from dynamic loads of the crank mechanism, especially dangerous at elevated temperatures. Application of the third layer, as is practiced in tri-metal bearings, can improve the fatigue strength. A thin Babbitt overlay is commonly employed for improved compatibility. Table 12.7 gives a rating of relative fatigue strength provided by various bearing materials.

A study carried out by Krzymién and Krzymién (2002) presented a probability of bi-metal bearing failure caused by scuffing or fatigue of the overlay. The example of a typical tri-metal bearing and the Rillenlager failure probability is presented in Figure 12.40. A definitely longer life for the Rillenlager with a worn sliding surface in comparison with the tri-metal bearing can be seen in this figure (Krzymién and Krzymién, 2002).

A number of new materials for individual layers spread over the steel support have been introduced in recent years. The efforts concentrate on an improvement of overlay fatigue strength, because fatigue failures are a serious problem for the operation of an engine crank mechanism. Anyway, tribological simulations are important to be carried out with new materials and overlays in order to know their friction and wear behaviour. Several important test procedures are described in detail.

12.11.2 Pin-on-disk Test

The pin-on-disk test is a basic test where the bearing materials are characterized for basic wear rating and base data for individual materials under different operating conditions are arrived at. A bearing coin can be prepared in a special way either from a strip or from the bearings. On the pin on disk machine (Figure 12.41), the coin is attached to the pin of the machine where the temperature, coefficient of

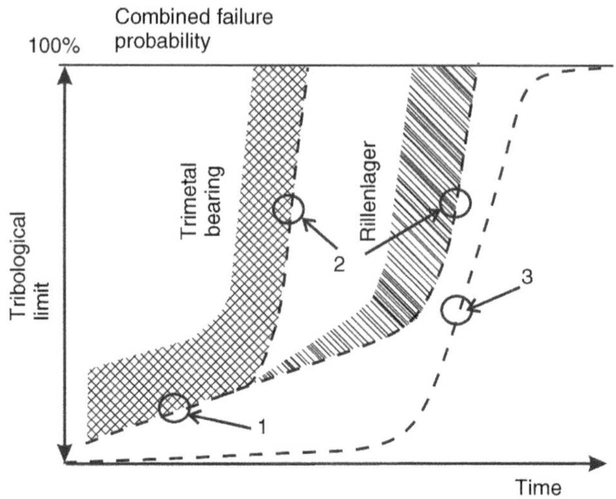

Figure 12.40 Failure probability of a tri-metal bearing. 1: Tribological limit for bearing with overlay; 2: tribological limit for bearing with worn overlay; 3: lining fatigue strength corresponding to the load level

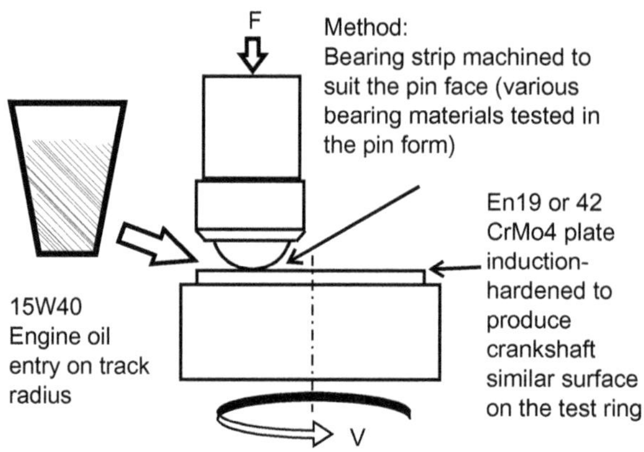

Figure 12.41 Pin-on-disk wear test (Reproduced by permission of CSM Instruments SA.)

friction and normal load are measured by the data acquisition system. The test ring (disk) is made in the crankshaft material and is given a similar treatment. The lubricant is metered at the radius of contact to generate various lubrication regimes from the curve. This simple test gives insight into the wear rates, emergency running properties and operation in mixed film conditions. However, it is difficult to generate heavy loads on this machine.

Table 12.8 Materials used for pin-on-disk wear test

Pin Material	SAE783, SAE788, SAE792, SAE793
Disk Material	42CrMo4
Lubricant	15W40
Temperature	Atmospheric

Table 12.9 Experimental parameters and observations: pin-on-disk wear test

Experimental parameter		Observations plotted in Figures 12.42 and 12.43
rpm of disk	1000	Initial mass of pin (g)
Lubrication rate	1 drop/2 seconds	Final mass of pin (g)
Track radius (mm)	60	Mass loss (g)
Velocity (m/s)	6.28	
Duration of test (min)	360	

Aspects that can be studied are:

- Running-in wear
- Wear rate at constant loads
- Wear rate at varying loads
- Different lubrication regimes
- Coefficient of friction analysis
- Edge loading behaviour
- Effect of roughness
- Various tribological material pairs and their compatibility.

An excerpt from tests conducted as per the test parameters mentioned in Tables 12.8 and 12.9 to compare wear characteristics of SAE783 against SAE788 can be seen in Figure 12.42 and SAE792 against SAE793 in Figure 12.43. These tests clearly show the friction behaviour in terms of:

- running-in process for short or long duration;
- study the coefficient of friction after running-in;
- study advent of seizure by observing the increase in load, temperature and friction;
- healing capability of the bearing material and the influence of the earlier stages by observing the decrease in load.

12.11.3 Scratch Test for Bond Strength

The technique involves generating a controlled scratch with a diamond tip on the bearing sample under test (Figure 12.44). The tip of either a Rockwell C diamond or a sharp metal is drawn across the coated (overlay) surface under either a constant or progressively increasing load. At a certain critical load, the coating starts to fail. The critical loads are detected precisely by means of an acoustic sensor attached to

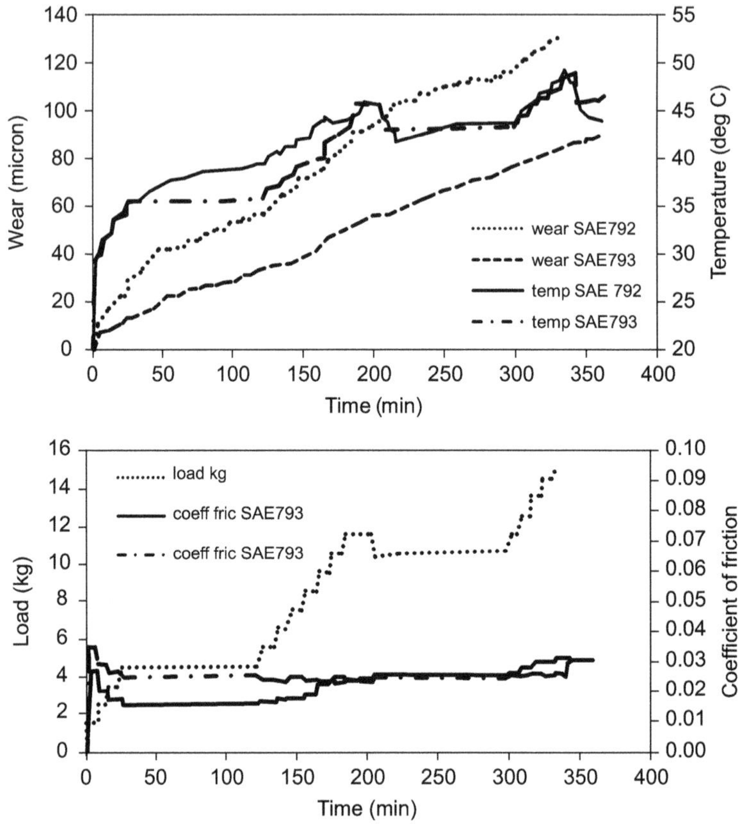

Figure 12.42 Comparison of SAE783 and SAE788 materials with short running-in process, stable running and load increase and decrease cycle (Reproduced by permission of CSM Instruments SA.)

the load arm together with observations from a built-in optical microscope. The critical load data are used to quantify the adhesive properties of different film substrate combinations. In addition to acoustic emission, the load normal to the coating, the tangential or friction force and the penetration depth are measured. These parameters as well as the acoustic emission constitute a unique signature of the coating system under test.

12.11.3.1 Sample Experiment

The following data are for an overlay coated bearing surface characterized by a progressive scratch test carried out in cooperation with CSM Instruments of Switzerland and using the instrument shown in Figure 12.44. Here, a load ramp is applied to the indenter during defined displacement of the sample beneath it. The standard operating parameters are given in Table 12.10. To have statistical analysis generally five scratches are done on the same sample at different locations. The first scratches are used to define the highest critical load (HLC) and subsequent scratches can then be limited to HLC + 10 N to prevent unnecessary wear of the indenter tip.

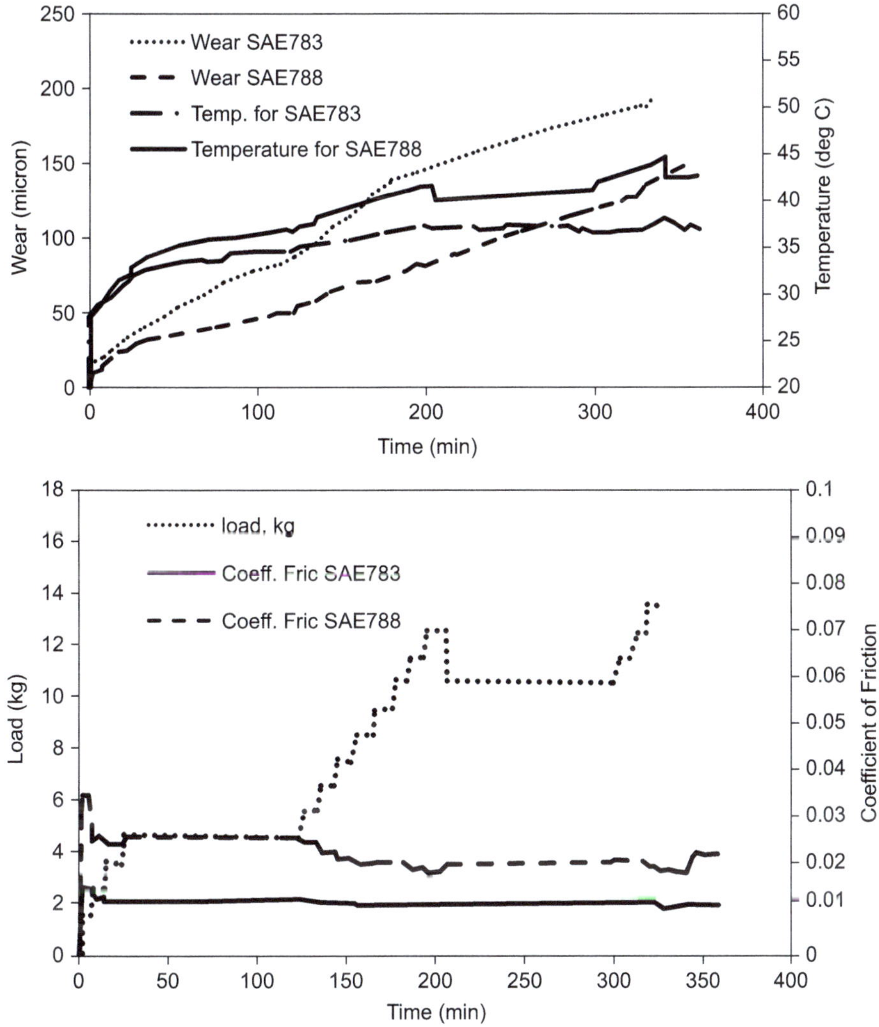

Figure 12.43 Comparison of SAE793 and SAE792 materials with short running-in process, stable running and load increase and decrease cycle (Reproduced by permission of CSM Instruments SA.)

Two bearing test samples, one with good bond strength and the other with weak bond strength were tested using the scratch tester. The observations and the results are summarized in Table 12.11. Figures 12.45 and 12.46 show the characteristic failure modes encountered with progressive load scratch testing of bearing coatings. Of particular interest is the manner in which the indenter causes small ruptures

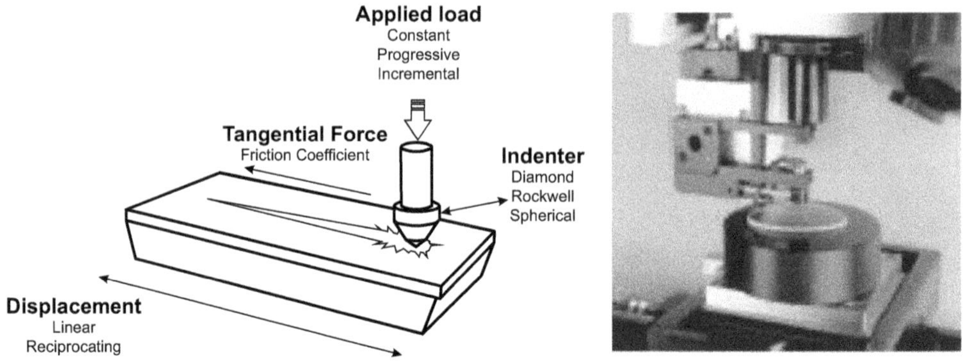

Figure 12.44 Scratch test for bond strength (Reproduced by permission of CSM Instruments SA.)

Table 12.10 Sample experiment operating parameters

Beginning load (N)	0.01	
Fn contact (N)	0.01	
End load (N)	20	
Fn speed (N/s)	2	
Loading rate (N/min)	40	
Fn remove speed (N/s)	2	
Diamond indenter radius (micron)	100	50

on the track path of the scratch (Figures 12.45c and 12.46c), after which the substrate, that is the backing lining, is reached. Pile up of material in front of the advancing indenter then allows the coating to support the applied load for a small distance until the substrate is again reached. The critical failure is defined by the first point at which the substrate is reached (Figures 12.45a and 12.46a). The corresponding

Table 12.11 Observations and results of typical experiments

	Load	N	Observation	View under a microscope	Traces of force and friction coefficient
Good coating	Critical, Lc1	2.72	First crack	Figure 12.45c	Figure 12.45a
	End	20	No substrate appearing end of crack within overlay		
Poor coating	Critical, Lc1	1.73	First crack	Figure 12.46c	Figure 12.46a
	Critical, Lc2	4.54	Substrate (lining) appears		
	End	20	Bond signatures from the scratch test		

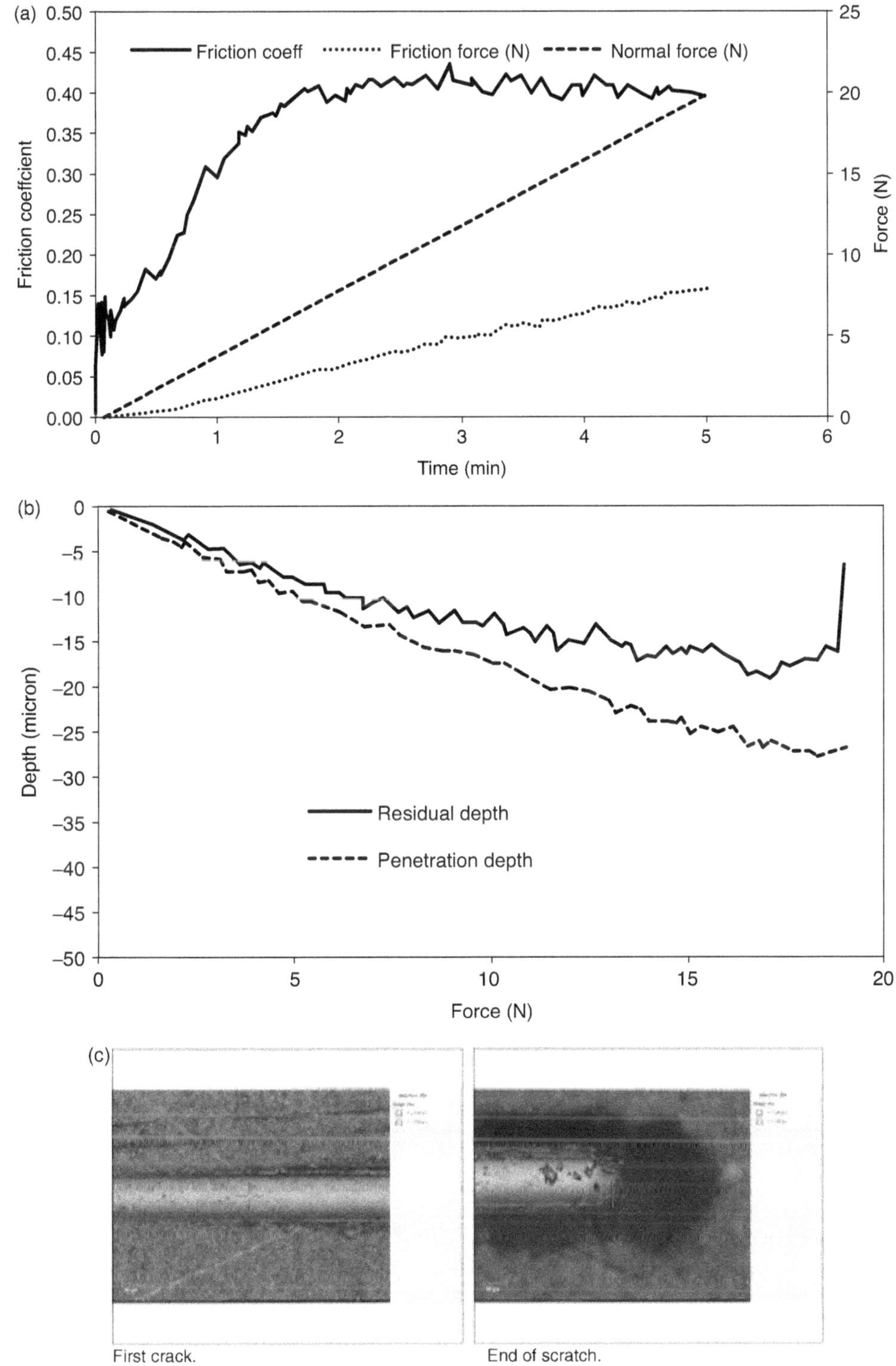

Figure 12.45 Good bond coating signatures (Reproduced by permission of CSM Instruments SA.)

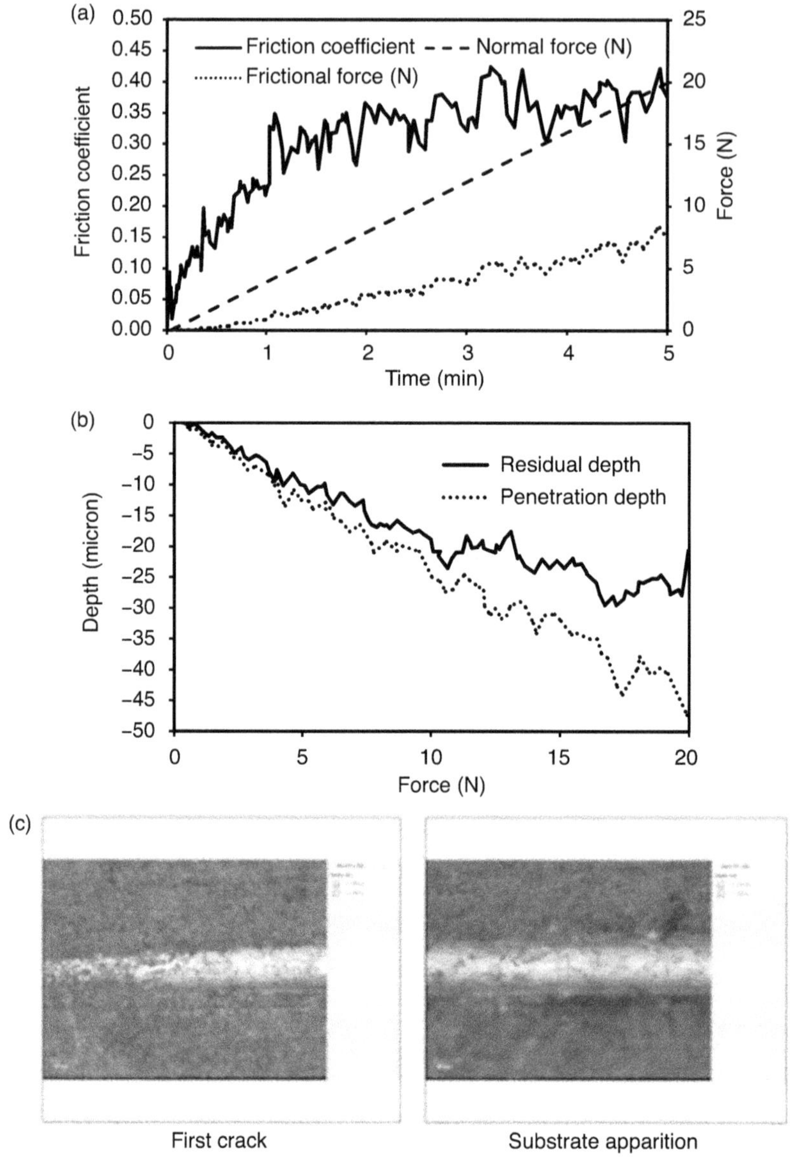

Figure 12.46 Poor bond coating signatures (Reproduced by permission of CSM Instruments SA.)

penetration depth and residual depth profiles are shown in Figures 12.45b and 12.46b. The critical point (4.54 N) at which the substrate is first reached can clearly be seen on the second profile, as can the increasing amplitude as a function of applied load, whereas in the first case the indenter did not reach the substrate.

Figure 12.47 Bush-on-bearing

12.11.3.2 Some Other Practical Examples of Tribological Simulations

Two types of tribological simulations were performed at Tekniker within the frame of a European project (Igartua *et al.*, 2009) to simulate the behaviour of journal bearings in engine applications 'simulated tribological tests in a SRV tribometer' and 'thrust washer' tests in a Falex tribometer.

Journal Bearing Simulation Tests in a SRV Tribometer

An Optimol Model SRV Tribometer was used to simulate reciprocating sliding motion, evaluating the coefficient of friction and the ability of different engine oils to protect against wear under similar test conditions. To get more realistic results, a new test in which the test specimens were cut directly from the original bearing parts was developed. Different bearing coatings were tested against a 42CrMo4 material for bush with one engine mineral oil. The contact represented was a line, following an oscillatory movement (Figure 12.47).

Typical steel sheets of bearing coated with AlSn20, G-CuPb22Sn and TiCN/DLC were tested as the lower specimen. The tests conditions were: time, 30 minutes; load, 50/250 N; frequency, 10 Hz; stroke, 1 mm; and temperature, 90 °C. The results obtained in bush-on-bearing tests for Journal Bearing Simulation are shown in Table 12.12.

Among the three coatings tested, the TiCN/DLC coating showed the best friction and wear behaviour with the engine mineral oil tested. In relation to the two bearing standard coatings, the CuPb22Sn overlay presented better friction and wear behaviour than the AlSn6 overlay (Figure 12.48). Theoretically, better tribological properties were expected with the leaded-copper layer than with the aluminium–tin coating.

With the TiCN/DLC coating, the friction coefficient was a bit unstable during the first minutes of the test, but after a few minutes it became stable and continues stable and lower than with the other two coatings.

Table 12.12 Results of bush-on-bearing tests for Journal Bearing Simulation

Bearing coating	Oil	Average CoF	Bearing mass loss (mg)
TiCN/DLC	Reference mineral oil	0.23	0.44
AlSn6		0.29	3.36
CuPb22Sn		0.26	1.65

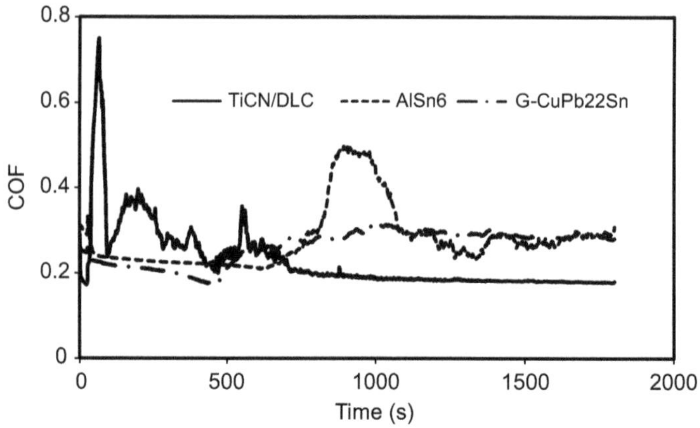

Figure 12.48 Journal Bearing Simulation tests; oscillatory movement; alternative materials with mineral oil

Thrust Washer Configuration Tests in the Falex Tribometer

The thrust washer test configuration from FALEX has been successfully used by Tekniker to simulate the friction and wear behaviour of bearing materials and lubricants (Figure 12.49). The tests conditions were: pressure 0.5 MPa (324 N); initial temperature, room (15–25 °C): speed, 1750 rpm (4.25 m/s).

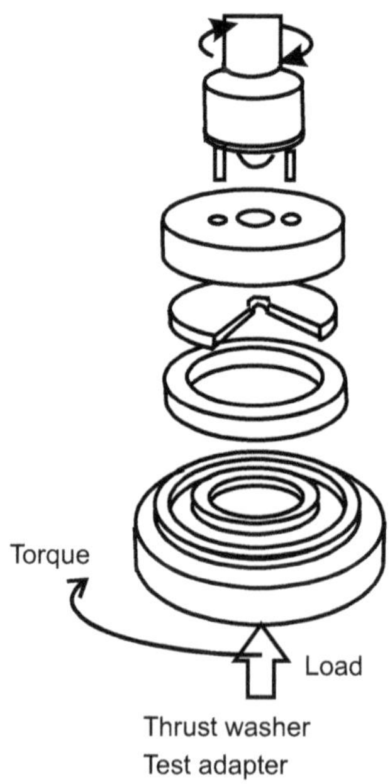

Figure 12.49 Thrust washer test for Journal Bearing Simulation

Figure 12.50 Refrigeration module developed by Tekniker

To perform the test a refrigeration module was developed by Tekniker (Figure 12.50). The refrigeration module reduces the working temperature, avoiding overheating during the tests, as shown in Figure 12.51.

The ester based bio-lubricant for passenger cars called EPC 49 developed by Fuchs in the frame of EREBIO Project (Igartua *et al.*, 2009) was compared with reference oil formulated with mineral oil. AlSn20 coated was selected as the bearing material against the 50CrV4 steel.

The new bio-lubricant has better friction behaviour than the reference motor oil 5W-30. This is clearly shown in Figures 12.52a and 12.52b by the lower temperature and lower friction torque achieved with the new lubricant. Also, Figure 12.52c shows marked reduction in the mass lost due to wear when the new lubricant is used. These results have been validated by IAV GmbH in Germany using a real size plain bearing model test bench and engine tests in a special torque loss test bench. The new EPC49 lubricant has been tested under boundary conditions with regard to its functional

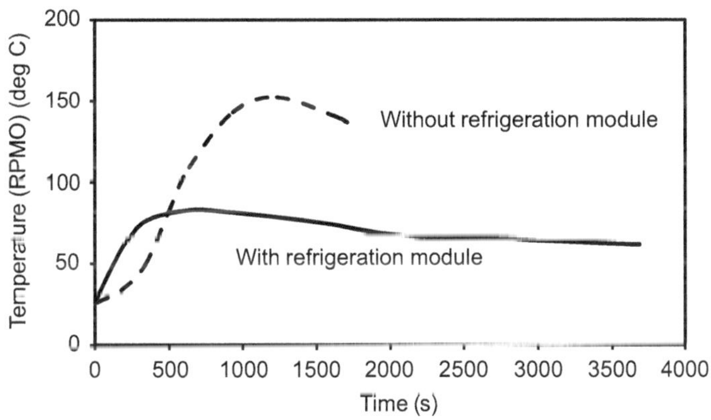

Figure 12.51 Temperature rise with and without refrigeration module

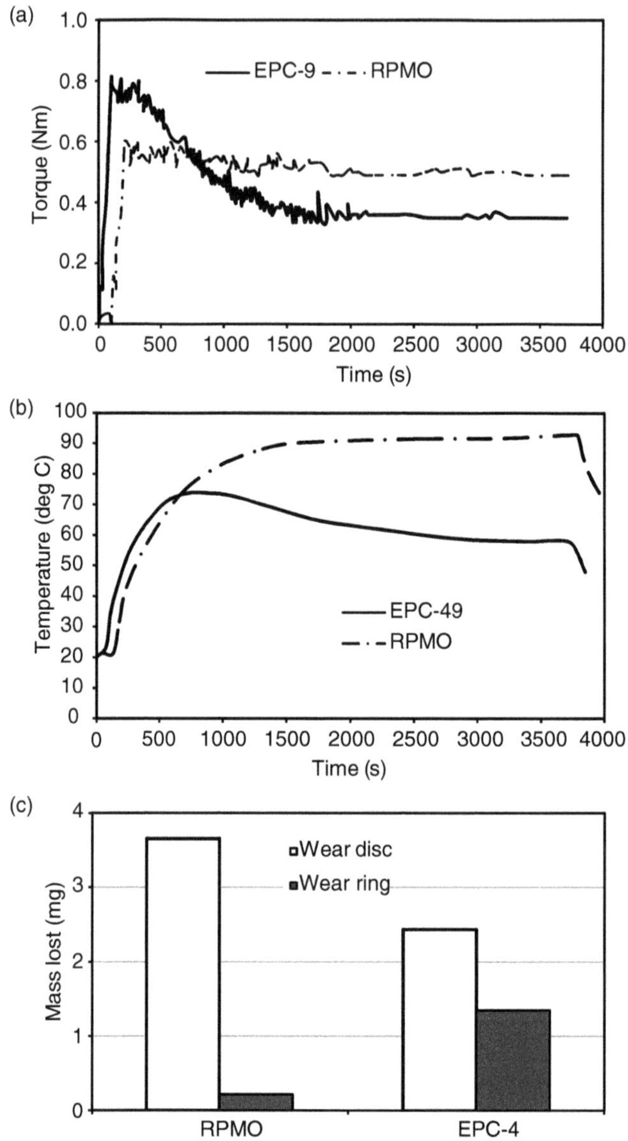

Figure 12.52 Journal Bearing Simulation – Thrust Washer test configuration: AlSn20 disc against 50CrV4 ring (Reference: 50CrV4//Typical sheet steel of bearing coated with AlSn20)

reliability to make statements on its friction behaviour (Figures 12.53 and 12.54). It was possible to reduce the friction of the examined engine with the developed oil, to a distinct degree especially at the operating points:

- Low temperatures and high engine speed
- High temperatures and low engine speed

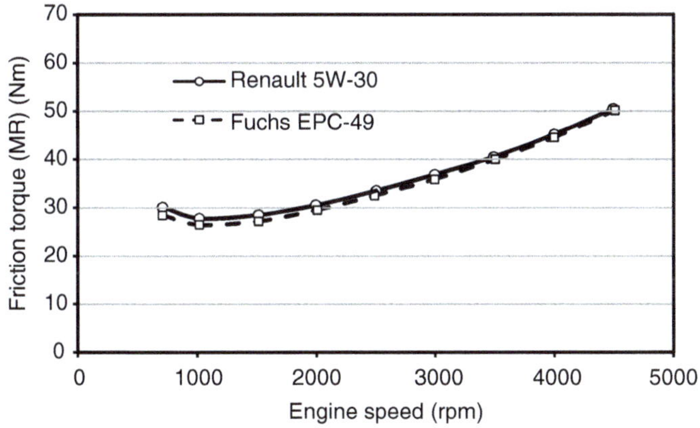

Figure 12.53 Friction moment at assembly level 1 (complete engine 1.9L TD, series) at 135 °C

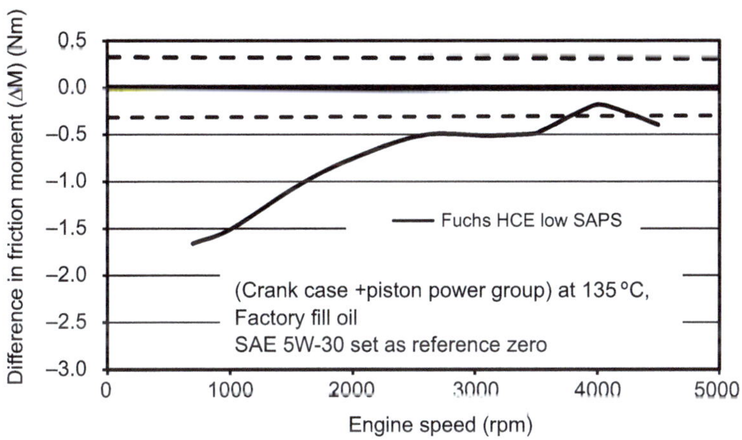

Figure 12.54 Reduction in friction with the low SAP oil at different engine speeds

The measured calculated reduction in friction coefficient corresponded to a reduction of fuel consumption of 2%. During the investigations, the test engine ran for all together more than 100 hours. On examining the engine after the tests, no serious signs of wear were found in the engine components.

12.12 Summary

A number of new materials for individual layers spread over the steel support have been introduced in recent years. The efforts concentrate on an improvement of overlay fatigue strength, because fatigue failures are the serious problem for operation of an engine crank mechanism. Another way to expand the bearing life is an application of inhomogeneous bearing material (Rillenlager) or sliding surface sputtered with AlSn, which is much more expensive and applied only to the bearings of highly loaded engines.

Bearing researchers have made attempts at characterizing each element of the tribology ecosystem to understand its characteristics and to study the interactions. Bearings are experimented on for various failure modes and properties. Tribological tests are a very promising and effective way to predict, at laboratory level, the bearing wear mechanisms, finding good correlations with real size plain bearing model and engine tests. On the basis of the data and good judgment, the bearing performance in the engine can be estimated using suitable models and a suitable bearing material for the given application is obtained.

References

Bhushan, B. and Gupta, B. K. (1991) *Handbook of Tribology: Materials, coatings and surface treatments.* McGraw-Hill, New York.

Budynas, R. and Nisbett, K. (2006) *Shigley's Mechanical Engineering Design*, 8th edn. McGraw-Hill, New York.

Chena, Y. M. and Mongis, J. (2005) Cavitation wear in plain bearing: Case study. *Mécanique & Industries*, **6**, 195–201.

DeHart, A.O. (1984) Bearing Materials, in *Handbook of Lubrication* (ed E. R. Booser), Vol. 2. CRC Press, Boca Raton, FL, pp. 463–476.

Igartua, A., Fernandez, X., Areitioaurtena, O. et al. (2009) Biodegradable and Tribo-reactive materials for automotive applications. *Tribology International*, **42**, 561–568.

CTP, Engine Bearings: Aluminium vs Copper/Lead. http://www.costex.com/Library/articles/IT01.pdf (accessed February 2011).

Khonsari, M.M. and Booser, E.R. (2001) *Bearings Design and Lubrication*, 2nd edn. John Wiley & Sons, Ltd, Chichester.

Kingsbury, G.R. (1997) Oil Film Bearing Materials, in *Tribology Data Handbook* (ed. E.R. Booser). CRC Press, Boca Raton, FL, pp. 503–525.

Kopeliovich, D. (2010) Bearings in internal combustion engines. SubsTech (Substances & Technologies) online. http://www.substech.com/dokuwiki/doku.php?id=bearings_in_internal_combustion_engines (accessed 3 June 2011).

Krzymién, A. and Krzymién, P. (2002), Three layer slide bearings for an IC engine crank mechanism. *Journal of KONES Internal Combustion Engines*, **1–2**, 134–135.

Ludema, K.C. (1984) A review of scuffing and running in of lubricated surfaces with asperities and oxides in perspective. *Wear*, **100**, 315–331.

Ludema, K.C. (2002) Failures of sliding bearings, in *Failure, Analysis and Prevention.* ASM Handbook Volume 11, ASM International, Materials Park, OH, pp. Introduction.

Mizuno, Y., Sugita, M., Okamoto, Y. *et al.* (1993) *Fretting Phenomenon on Outer Surface of Connecting Rod Bearings for Automotive Engines.* Technical Paper 931022, SAE, Troy, MI.

Neale, M.J. (1996), *The Tribology Handbook*, 2nd edn. Butterworth-Heinemann.

Pike, R. and Conway-Jones, J.M. (1992) Friction and Wear of Sliding Bearings, in *Friction, Lubrication, and Wear Technology* (ed. P.J. Blau). ASM Handbook Volume 18, ASM International, Materials Park, OH, pp. 741–757.

Rabinowicz, E. (1984) Wear Coefficients, in *Handbook of Lubrication* (ed E. R. Booser), Vol. 2. CRC Press, Boca Raton, FL, pp. 201–208.

Roberts, S., *Surface Engineering – Wear Materials.* University of Oxford, Oxford, UK. http://www-sgrgroup.materials.ox.ac.uk/lectures/surface_handout_4.pdf (accessed January 2011).

Watanabe, K., Hashizume, K., Kumara, Y. *et al.* (2000) *A Study of Microgroove Bearing Performance by using Numerical Analysis.* Technical Paper 2000-01-1338, SAE, Troy, MI.

Part V

Lubricating Oils for Modern Engines

13

Heavy Duty Diesel Engine Oils, Emission Strategies and their Effect on Engine Oils

Lawrence G. Ludwig, Jr
Schaeffer Manufacturing, Saint Louis, MO, USA

The main driving force since 1990s for the development of heavy duty diesel engine oils has been the concern over the environmental impact of diesel engine emissions, particularly of Nitrogen Oxides (NO_x) and Particulate Matter (PM) emissions. In 1997 the United States Environmental Protection Agency (EPA) adopted more stringent emissions standards for both NO_x and PM emissions reducing emissions to 0.2 grammes per brake horsepower hour (g/bhp-hour) for NO_x and 0.01 g/bhp-hour for particulate emissions by the 2010 model year. In addition to the emission regulations, the EPA set new limits on diesel fuel sulfur levels beginning in 2006 for on-road diesel engines and beginning in 2007 for off-road diesel engines in order to further control emissions. Globally these same trends in reduction in emissions and fuel sulfur, consumer demand for longer lasting oils and the concern over increased engine and oil sump temperatures due to current and future engine designs to meet the EPA emissions standards have further driven the development of new engine oil service categories.

This chapter reviews the drivers for the changes in heavy duty diesel engine oil formulations, the functions and performance requirements of diesel engine oils, how engine oil specifications are developed and set, as well as the impact emission design changes have on heavy duty diesel engine oil formulations.

13.1 Introduction

Heavy duty diesel engine oil formulations since the mid-1990s have been in a constant state of change in response to changing industry standard specifications, original equipment manufacturers (OEMs) requirements and consumers' needs. The industry standard specifications often are driven due to

Critical Component Wear in Heavy Duty Engines, First Edition. P.A. Lakshminarayanan and Nagaraj S. Nayak.

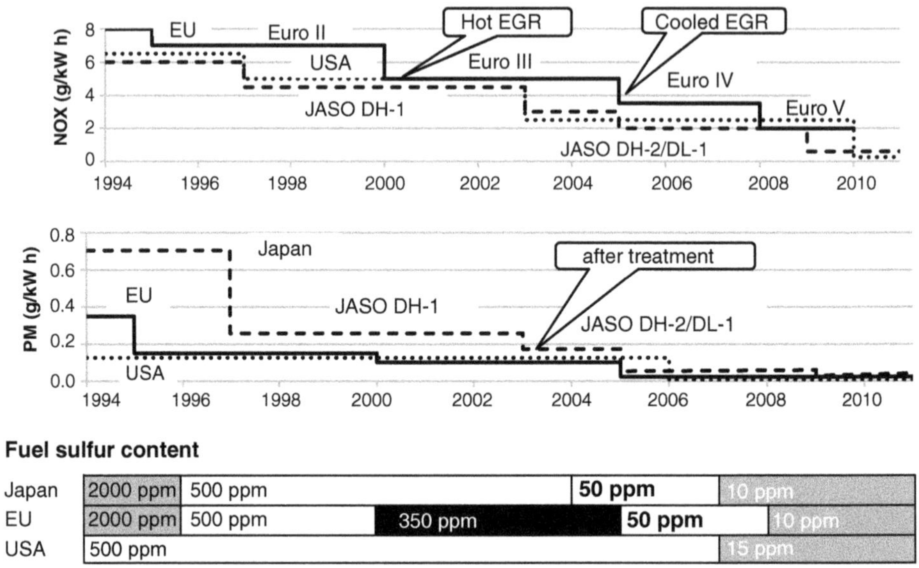

Figure 13.1 Emission regulations

governmental legislation (Figure 13.1), which usually results in new engine technologies and designs being developed. In turn, these new engine technologies and designs result in challenges being placed on the engine oil formulation in the areas of increased oxidative stability, as the result of increased engine and oil sump temperatures due to engine designs to meet emissions standards, handling of contaminants (e.g. soot), control or prevention of the formation of engine deposits at high operating temperatures, emissions systems compatibility, fuel economy and prevention of wear in newer and older engines, while still satisfying the consumer's needs and desires for increased engine oil drain intervals and engine durability and performance

13.2 What Drives the Changes in Diesel Engine Oil Specifications?

The main drivers for recent changes in engine oil formulation and specifications, particularly in the United States have come from three sources:

- The Government
- OEMs
- The Consumer

13.2.1 Role of the Government

Though the government does not have a direct role in the development of engine oil formulation and standards, it does have an indirect role that influences the development of engine oil formulations and standards through emissions and fuel efficiency regulations. To meet the stricter emissions and fuel efficiency requirements, OEMs develop new engine technologies and exhaust treatment or after-treatment designs that often necessitate the development of new engine oil formulations and specifications.

In the United States, the Environmental Protection Agency (EPA) monitors how engine oil standards and their outcome are developed by the lubricant industry in response to their mandates on fuel economy and emissions reductions. Additionally, for the EPA to be successful in achieving its goals, it influences consumer awareness for demanding that the latest engine oil is specified for a particular application and sways the market so that the oil is widely available from distribution channels. The EPA also requires the OEMs to educate the consumer about the use of the latest engine oil specification throughout the owner's manual. The OEM must also use the engine oil that is widely available for certifying new engines and factory fill these engines with it before it leaves the factory (Galligan, 2005). During the development of any engine oil specification the EPA's regulations pertaining to fuel efficiency and emissions has been the major driver.

13.2.1.1 Emissions Regulations

The Clean Air Act of 1990 mandated a vast reduction in emissions from both diesel and gasoline powered vehicles. In the United States the EPA sets and implements the requirements for both diesel and gasoline powered vehicles. To implement and meet these emission requirements OEMs have developed more complex emission control systems. For on-road and in some off-road diesel vehicles, OEMs have used retarded engine timing, improved combustion control through pulsed fuel injection, inlet air swirl designs and high pressure common rail injection, decreased piston crevice volume and the recently introduced exhaust gas recirculation (EGR) engines. All of the advancements in engine design have introduced more soot, higher engine operating temperatures and acids, thus necessitating the engine oil to be formulated to handle and deal with these issues. As diesel emissions standards have changed diesel, OEMs have or will introduce new exhaust after-treatment systems on their equipment, such as diesel particulate filters (DPF), selective catalytic reduction (SCR), oxidation catalysts (DOC) and NO_x absorbers or traps for reducing emissions. These devices will require engine oils that contain lower phosphorus, sulfur and sulfated ash to prevent damage to the system.

13.2.2 OEMs' Role

When OEMs make engine and vehicle design changes, new engine oil performance standards are often required to match them. These new designs can have a significant impact on the performance of the engine oil.

For on-road highway diesel engines, diesel OEMs have used retarded engine timing, improved combustion control through pulsed fuel injection, inlet air swirl designs and high pressure injection, decreased piston crevice volume and the recently introduced exhaust gas recirculation (EGR) engines to meet stricter diesel emission limits. These design change have resulted in:

Increased soot levels: Soot can cause abrasive wear as it accumulates in the engine oil. It can also increase the viscosity of the engine oil.

Increased acid levels: Higher acid levels in the engine oil accelerate the depletion of the engine oil's alkalinity reserve (also known as TBN) and increase the risk of corrosion.

Increased engine operating and oil temperatures: An increase in temperature can accelerate the rate of oxidation and deposits. Oxidation degrades the engine oil. It can also generate acidic by-products and cause the viscosity of the oil to increase.

Some diesel OEMs such as Caterpillar, Cummins, Mack, Volvo, Detroit Diesel and Mercedes Benz also create and recommend their own engine oil performance standards when needed or desired performance levels are not met by current American Petroleum Institute (API) engine oil standards. Often these OEM performance standards are developed in response to problems that may be occurring in the field or the need to protect older engine designs.

13.2.3 The Consumer's Role

The consumer of diesel engines often demands that the engine oil being used provides increased oil drain intervals and increased engine durability and life. Therefore, the engine oil must provide these benefits. It must have the ability to be used for longer drain intervals and still provide the protection against wear, deposit formation, resistance to oxidation and fuel economy benefits over the entire life of the vehicle.

13.3 Engine Oil Requirements

13.3.1 Overview and What an Engine Oil Must Do

Basically, engine oil reduces friction and prevents wear between the moving parts of the engine. Today's engines require the use of sophisticated engine oils. The development of more powerful engines is constantly changing the technology of engine oils formulated for use in today's modern engines. The efficient operation of any engine depends on the engine oil performing the following functions:

- Permitting easy starting and pumping
- Lubricating engine parts and preventing wear
- Reducing friction
- Protecting against rust and corrosion
- Keeping engine parts clean
- Minimizing combustion chamber deposits
- Cooling engine parts
- Sealing combustion pressures
- Being nonfoaming
- Aiding in fuel economy improvements.

13.3.1.1 Permit Easy Starting and Pumping

The ease of starting an engine depends not only on the condition of the battery, starter and ignition system, proper fuel volatility and air/fuel ratio, but also on the flow properties of the engine oil. An engine oil must reach a critical crankcase speed in order to start promptly and keep running. If the engine oil is too viscous, it will impose excessive drag on the moving parts, so that the critical cranking speed cannot be achieved and the engine will not start.

Viscosity is a measure of the engine oil's resistance to flow. The property that determines the ease of starting is its viscosity at the cranking temperature and critical cranking speed. Since it is this resistance to flow that is responsible for most of the drag on the starter during cranking, it is important to be able to measure a 'cranking viscosity'. The American Standard For Testing and Materials (ASTM) has developed a test method called the Cold Cranking Simulator Viscosity Test Method ASTM D-5293. The Cold Cranking Simulator Viscosity Test Method is used to measure an engine oil's ability to permit a satisfactory cranking speed to be developed in a cold engine at start up. The viscosity of the engine oil in engine journal bearings during cold temperature start up is a key to determining the lowest temperature at which an engine will start. The Cold Cranking Simulator Viscosity is determined under conditions similar to those experienced in engine bearings during start up. The Cold Cranking Simulator Test determines the apparent viscosity of the engine oil at temperatures from $-10\,^{\circ}\mathrm{C}$ to $-35\,^{\circ}\mathrm{C}$ and under high shear rate conditions. Viscosities measured by this test method have been found to correlate with engine speeds developed during low temperature cranking. The lower the engine oil's cold cranking viscosity, the easier for an engine to turn over during low ambient temperature conditions. The Society of Automotive

Engineers (SAE) has established maximum limits for each SAE W-grade engine oil. For example, SAE 10W-XX viscosity grade engine oils are tested at −25 °C and must have a viscosity below 7000 cP to pass the SAE J-300 viscosity specifications.

Since cold temperatures thicken all engine oils, an engine oil designed for winter use must have a sufficiently low cranking viscosity in order to permit adequate cranking speeds at the lowest anticipated starting temperatures. However, having the proper cranking viscosity does not ensure that an engine oil is fluid enough to flow quickly and continuously to the bearings and other moving parts immediately after the engine starts. In winter, parked, unattended vehicles are often subjected to extended periods of below freezing weather that can play havoc with an oil's ability to flow on start up. All oils thicken when the temperature drops, but under these conditions there is also a danger that wax and additive components of the engine oil, which are normally in solution, may come out of solution and convert the engine oil to a gel or even a solid. This structural modification results in increased viscosity and poor flow characteristics.

Two types of engine pumping failures result from this. One type of failure occurs when the engine oil that surrounds the pump inlet screen is sucked into the pump, but is immediately replaced by oil in the sump. Consider a conventional drinking straw partially immersed in a very thick milk shake. If the bottom of the straw is near the surface of the milk shake and it is sucked up the straw too quickly, a vortex forms around the straw and air becomes mixed with the milk shake in the straw. In the case of gelled engine oil, the oil pump inlet screen can become starved of fluid when the vortex forms and does not collapse. Oil pressure becomes erratic as air is entrained and oil flow through the oil suction tube is insufficient to maintain lubrication. This is called **air binding failure**.

Another type of failure mode is possible. Consider the straw and milk shake example again. This time, replace the conventional straw with one having a very small diameter. Now it becomes difficult to pull much fluid through the straw. Sometimes pumping failures occur because the engine oil becomes so thick in viscosity that it simply cannot be pumped through the inlet tube and through the narrow passages that convey the oil to the moving parts of the engine. When this happens, the failure is called a **flow limited failure**.

To measure the engine oil's pumpability characteristics at low ambient temperatures the ASTM has developed the MRC TP-1 Engine Oil Pumpability Test ASTM D-4684 This test method can be used to tell whether either type of failure is likely and at what temperature the failure may occur. The MRV TP-1 Viscosity is determined under low temperature cooling conditions similar to those experienced by an engine during actual field conditions. This test method covers the measurement of the yield stress and viscosity of engine oils after cooling at controlled rates over a period exceeding 45 hours to a final test temperature between −10 and −40 °C. The viscosity measurements are made at a shear stress of 525 Pa over a shear rate of 0.4–15 s^{-1}. The engine oil is held at 80 °C and then cooled at a programmed cooling rate to a final test temperature. A low torque is applied to the rotor shaft to measure the yield stress. A higher torque is then applied to determine the apparent viscosity of the sample. The test fixture consists of a mini-rotary viscometer, an apparatus that consists of one or more viscometric cells in a temperature-controlled aluminium block. Each cell contains a calibrated rotor-stator set. Rotation of the rotor is achieved by an applied load acting through a string wound around the rotor shaft. A temperature control system operates the heater in the mini-rotary viscometer block and regulates the coolant flow to the block. The temperature controller is the most critical part of this procedure. Thermometers measure the temperature of the block. A refrigeration device that is capable of maintaining a coolant to at least 10 °C below the lowest test temperature is used. A circulating system that will circulate the liquid coolant to the block as needed is also required. A chart recorder verifies that the correct cooling curve is being followed and also monitors the block temperature. The SAE has established maximum limits for each SAE W-grade engine oil. For example, SAE 10W-XX viscosity engine oils are tested at −30 °C and must have a viscosity below 60 000 cP to pass the SAE J-300 viscosity specifications.

13.3.1.2 Lubricate and Prevent Wear

Once an engine is started and pumping of the engine oil begins, the engine oil quickly circulates to all moving parts of the engine in order to prevent metal-to-metal contact that would result in wear, scoring or seizure of engine parts. Oil films in bearings and on cylinder walls are sensitive to movement, pressure and oil supply. These films must be continually replenished by adequate oil flow and proper oil distribution. Various regimes of lubrication exists in different parts of the engine under different conditions, namely full film lubrication, mixed lubrication and boundary lubrication, as shown in Figure 13.2 and described below.

As mentioned earlier, the viscosity of an engine oil must be low enough at starting temperatures to permit rapid cranking and starting, and high enough at peak operating temperatures to ensure adequate engine protection. Once the engine oil reaches the moving parts, the engine oil's function is to lubricate and prevent wear of the moving surfaces. In many parts of the engine, the oil is expected to establish – and consistently replenish – a complete, unbroken film between the surfaces of the moving parts. This is called full film or hydrodynamic lubrication.

Full film lubrication occurs when a film of oil continuously separates the moving surfaces. The determining factor in keeping these parts separated is the viscosity of the engine oil at operating temperature. The viscosity of the engine oil must remain high enough to prevent metal-to-metal contact. Since the metal surfaces do not make contact in full-film lubrication, wear is negligible unless the separated parts are scratched by particles thicker than the oil film itself. Crankshaft bearings, as well as connecting rods, camshafts and piston rings, normally operate with full film lubrication.

Under some conditions, it is impossible to maintain a continuous oil film between moving parts, and there is intermittent metal-to-metal contact between the high spots on sliding surfaces. This is called boundary lubrication. Under these conditions, the load is only partially supported by the oil film. The oil film is ruptured, resulting in significant metal-to-metal contact. When this occurs, the friction generated between the contacting metal surfaces can produce a sufficient amount of heat to cause one or both of the metal surfaces in contact to melt and weld together. Unless counteracted by the engine oil's additive

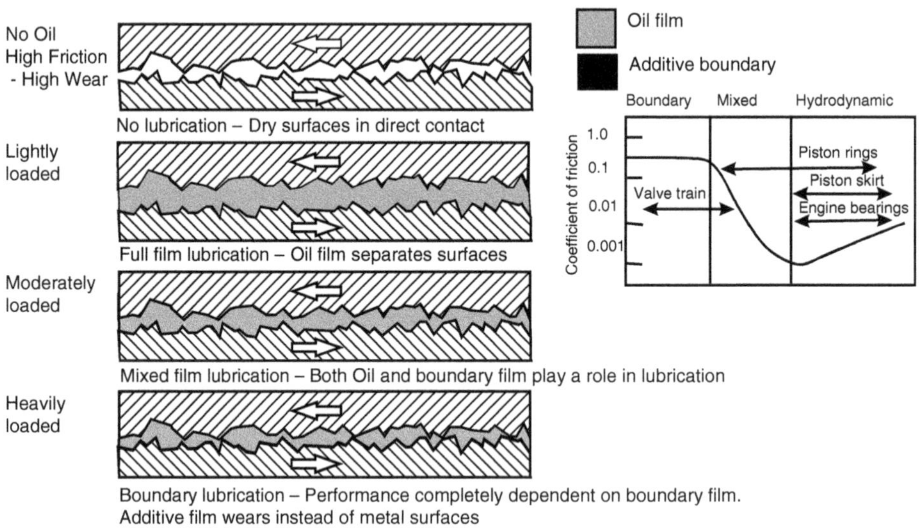

Figure 13.2 Engine lubrication regimes

system, the result is either immediate seizure or the tearing apart and roughening of the surfaces. Today's engine oils contain anti-wear and in some cases frictional modifier additive systems that minimize wear under boundary lubrication conditions.

Boundary lubrication conditions always exist during engine start up and often during the operation of a new or rebuilt engine. Boundary lubrication is also found around the top piston ring where oil supply is limited, temperatures are high and a reversal of piston motion occurs.

Extreme pressure conditions can develop between heavily loaded parts from lack of lubrication, inadequate clearance, extreme heat, and sometime as a result of using the wrong type or grade of engine oil. In today's engines the valve train, with its cams, valve lifters, push rods, valve stem tips and rocker arms, operate under conditions of extreme pressure because they carry heavy loads on a very small contact area. Unit loading may be as high as 1 378 960 kPa or more, which is many times greater than the loads found on the connecting rod bearings or on the piston rings.

13.3.1.3 Reduce Friction

Under full film lubrication conditions, a thick film of oil prevents metal-to-metal contact between the moving engine parts. Relative movement of these lubricated parts requires enough force to overcome the fluid friction of the lubricant. The viscosity of the engine oil should be high enough to maintain an unbroken film, but should not be higher than necessary, since this increases the amount of force required to overcome this fluid friction. If the engine oil's viscosity is too high, the engine must do additional work to overcome increased viscous drag. To do this work, additional fuel must be consumed, so engine fuel economy is reduced. This can also result in higher oil operating temperatures that put an increased strain on the engine oil's ability to impeded oxidation.

Various original equipment manufacturers of passenger cars and heavy duty diesel engines specify proper oil viscosity ranges by recommending the SAE grades to be used according to expected atmospheric temperature operating conditions. This is to ensure that the engine oil being used will provide adequate but not excessive viscosity at normal operating conditions. When an engine oil becomes contaminated, its viscosity will change. Viscosity will increase as the engine oil becomes contaminated with soot, dirt and sludge or is oxidized and decreases if it is contaminated with fuel. For this reason contaminant levels must be kept low and monitored through the use of a used oil analysis program.

The amount and type of chemical additives used in the formulation of an engine oil becomes more important to the effectiveness of an engine oil in reducing friction under the extreme pressure conditions of boundary lubrication. The use of the proper balance and the type of additive package used in the formulation of an engine oil is critical if all the lubrications of an engine are to be satisfied. When formulating an engine oil this careful balance is achieved through much research, with emphasis on proof testing in actual production engines, both in the laboratory and in field service.

One of the tests used to measure an engine oil's ability to provide an effective adequate viscosity in high shear components, such as the journal bearings and between the piston rings and cylinders under severe operating conditions, is the High Temperature High Shear Viscosity Test ASTM D-4683. The values obtained in this test provide an indication of the temporary shear stability of the viscosity index (VI) improver used in the formulation of multi-grade engine oils. If the engine oil is not able to maintain an adequate viscosity when high engine operating temperatures and high shear rates are encountered, wear to critical parts will occur. The values obtained in this test provide an indication of the temporary shear stability of the viscosity index improver used in the formulation of multi grade engine oils. The test fixture consists of a motor that drives a tapered rotor that is closely fitted inside a matched stator. The rotor exhibits a reactive torque response when it encounters a viscous resistance from an oil that fills the gap between the rotor and stator. Two oils, a calibration oil and a non-Newtonian reference oil, are used to determine the gap distance between the rotor and stator so that a shear rate of a $1 \times 10^6\,s^{-1}$ is maintained. Additional calibration oils are used to establish the viscosity/torque relationship, which is required for the determination of the apparent viscosity of test oils at 150 °C. The apparent viscosity to the nearest 0.01 cP (mPa s) at 150 °C and $1 \times 10^6\,s^{-1}$ shear rate is reported. Many OEMs, the SAE 'J' viscosity systems and the American Petroleum Institute (API) and Association des Constructeurs Européens dAutomobile (ACEA) engine oil specifications specify minimum High Temperature High Shear (HTHS) viscosity limits to ensure that the engine oil provides the proper viscosity in order maintain engine durability in high load, severe service applications.

13.3.1.4 Protect Against Rust and Corrosion

Under ideal conditions, fuel burns to form carbon dioxide and water. For a variety of reasons, an engine does not burn all of the fuel completely. Unburned and partially burned fuel, whether it is gasoline or diesel fuel, undergoes complex chemical changes during combustion and, under some conditions, forms soot or carbon. Some of this soot, water and partially burned fuel escapes through the exhaust in the form of black smoke, particularly when an engine is misfiring or running rich. However, some of these combustion by-products may escape past the piston rings into the crankcase. This can cause numerous problems, which an oil must control.

Water is one problem that the oil must deal with. For each gallon of fuel burned in an engine, more than one gallon of water is formed. Although most of this water escapes as a vapour out of the exhaust, some of it condenses on the cylinder walls or escapes past the piston rings and is trapped, at least temporarily, in the crankcase. This occurs most frequently in cold weather before the engine has warmed up.

In addition to water and the by-products from incomplete combustion of the fuel, other corrosive combustion gases, such as nitrogen oxides and sulfur oxides, also get past the rings and are condensed or dissolved into the engine oil. Add this to the acids formed by the normal oxidation of oil and the potential for rust and corrosion is significant.

The life of the engine parts depends in part on the ability of the engine oil to neutralize the effects of these corrosive substances. To combat these harmful effects an engine oil must contain a highly effective additive package that will neutralize the harmful effects of these corrosive substances in order to protect the engine from rust and corrosion.

13.3.1.5 Keep Engine Parts Clean

In the formulation of engine oils, a basic objective is to not only protect the engine from rust and corrosion, but to keep the engine clean and prevent the build up of sludge and varnish deposits from interfering with proper engine operation.

The formation of sludge in an engine is generally a problem of low engine operating temperatures. Combinations of water from condensation, dirt, the products of oil deterioration and the by-products of incomplete combustion of the fuel being burned form engine sludge deposits. Often, sludge-forming materials are materials so small initially that the engine oil filters cannot remove them. They can be much smaller than the thickness of the oil film on engine parts and, therefore, do not cause any wear or damage as long as they remain small and well dispersed. However, as their levels increase in the engine oil during normal engine use, they tend to join together to form larger masses, and oil flow can become restricted.

Sludge formation can be aggravated by water vapour that condenses in the crankcase during cold engine operation. The rate at which sludge-forming materials accumulate in the engine oil may be increased by several factors of engine operation, such as rich air–fuel mixtures which can occur during starting or from poor engine running, operation with dirty air cleaners or ignition misfiring.

Today's engine oils contain detergent/dispersant additive systems to keep engine parts clean and prevent the formation of sludge and deposits. Detergents and dispersants are also very effective in preventing varnish deposits within an engine. Varnish-forming materials react chemically or combine with oxygen in the crankcase to form complex chemical compounds. These compounds continue to react with each other and with oxygen and are baked by the engine heat into a hard coating on the hotter parts of the engine. The piston rings and pistons are particularly sensitive to varnish deposits. If varnish-forming materials are allowed to accumulate in these areas, engine operation is impaired or the engine may seize.

Engines cannot tolerate excessive amounts of sludge and varnish on sensitive parts. Sludge deposits collect on oil pump screens, limiting the flow of oil to critical engine parts, and result in rapid and destructive wear. Piston rings that are stuck or sluggish because of varnish accumulation prevent the engine from developing full power. Sludged or plugged oil control rings prevent the removal of excess lubricant from the cylinder walls and results in excessive oil consumption.

13.3.1.6 Minimize Combustion Chamber Deposits

In performing its lubrication function, some oil must reach the area of the top piston ring in order to lubricate the rings and cylinder walls. This oil is then exposed to the heat and flame of the burning fuel, and part of the oil actually burns off.

Modern refining techniques have produced oils that burn cleanly under these conditions, leaving little or no carbon residue. The detergent/dispersant additives in engine oils keep the piston rings free in their grooves, thereby maintaining compression pressures and minimizing the amount of oil reaching the combustion chamber. This not only reduces oil consumption but also, more importantly, keeps combustion chamber deposits at a minimum.

Excessive combustion chamber deposits adversely affect engine performance. Deposit build-up causes engine knocking or other combustion irregularities that reduce engine efficiency and economy. Because these deposits act as heat barriers, pistons, rings and valves are not properly cooled. This can result in damage or failure of the engine parts, necessitating premature engine overhaul. In preventing excessive combustion chamber deposits, it is important that engine oil accomplishes two things:

- The engine oil must keep the rings free so that they can minimize the amount of oil reaching the combustion chamber, and
- The portion of the oil reaching the combustion chamber should burn as cleanly as possible.

To further control combustion chamber deposits it is essential that the engine oil be formulated from base oils that exhibit low volatility characteristics. Volatility of an engine oil becomes important whenever the engine oil comes in contact with one of the high temperature zones in the engine, such as the turbocharger, cylinder walls, valves and the under-crown, and ring belt area of the pistons. An engine oil's volatility characteristics are directly related to the type of base oil used in the formulation of the engine oil. An engine oil that exhibits a high volatility not only increases the volume of oil consumed by the engine, but the loss can change the engine oil's effectiveness as a lubricant through the formation of deposits on critical engine parts, oxidation and thermal cracking of the engine oil and increased wear. Further, and of increasing importance, such volatile material can lead to an increase in particulate emissions and poison the after-treatment devices used to control particulate emissions.

To measure an engine oil's volatility characteristics the ASTM D-5800 Noack Volatility Test Method is used. In this test a measured quantity of sample is placed in an evaporation crucible or reaction flask that is then heated to 250 °C with a constant flow of air drawn through it for 60 minutes. The loss in mass of the oil is determined. The test method covers three procedures for determining evaporation loss. Procedures A and B use an electrically heated block to heat a steel crucible where Procedure C heats the glass reaction flask with an electric heating element. A vacuum pump is used to maintain a constant flow of air across the surface of the sample. Many OEMs and the API and ACEA engine oil specifications specify a maximum allowable evaporation loss limit.

13.3.1.7 Cool Engine Parts

Many people assume that engine cooling is accomplished only through the action of the water–antifreeze mixture used in the cooling system. In fact, this mixture does only about 60% of the cooling job. It cools several engine parts – the cylinder heads, cylinder walls and the valves. The crankshaft, the main and connecting rod bearings, the camshaft and its bearings, the timing gears, the pistons, the turbocharger and many other components of the engine are directly dependent on the engine oil for necessary cooling. All of these parts have finite temperature limitations that must not be exceeded. Some can tolerate fairly high temperatures, while others, such as the bearings, must run relatively cool to avoid failure. These parts must get an ample supply of cool oil by heat transfer to the surrounding air.

Some idea of the temperatures an oil experiences may be helpful in understanding the role of the engine oil as a coolant. Combustion temperatures are 1093–1649 °C (1999–3000 °F). Certain parts of the valves may reach temperatures of 538–1093 °C (1000–1999 °F). Piston temperatures can reach 538 °C (1000 °F) and this heat travels down the connecting rods to the rod bearings. Tin and lead, which are the more sensitive of the metals commonly used in bearings, become soft around 177 °C (351 °F). Tin melts at 232 °C (450 °F), while lead will melt at 327 °C (621 °F). After warm-up, engine oil temperatures reach 93–135 °C (199–275 °F) and the oil is supplied to the bearings at this temperature. The engine oil picks up the heat at the bearings and leaves them at a temperature around 121–149 °C (250–300 °F), well below the critical temperatures of bearing failure. The continual cooling by the oil by recirculation to the crankcase is essential to maximize bearing life. To keep this cooling process working, large volumes of oil must constantly be circulated to the bearings and other engine parts. If the oil supply is interrupted, either by deposit formation, oil pump problems or low crankcase oil levels, these parts can heat up rapidly from increased friction and combustion temperatures or decreased cooling by the oil. A bearing failure is often referred to a 'burned-out bearing' because temperatures were raised high enough to actually melt the bearing metal.

While only a small quantity of oil is required at any one time and any one place to provide lubrication, the oil pump must circulate many gallons of oil per minute in order to achieve lubrication and properly cool these parts. Chemical additives and the physical properties have little effect on its ability to provide adequate cooling. What is critical is the continuous circulation of large volumes of oil throughout the engine and over hot engine parts. This is made possible through the use of an oil of the right viscosity for that engine, high volume oil pumps and oil passages that adequately handle the required volume of oil. These oil passages cannot do the job properly if they are allowed to become partially clogged with deposits. When this happens, the oil cannot circulate or cool properly, and early engine failure may result. This is yet another reason for using the proper performance oil and for changing the oil before the contamination levels become too high. Proper cooling also requires that the oil level in the crankcase never be permitted to remain below the 'add oil' line on the dipstick.

13.3.1.8 Seal Combustion Pressures

The surfaces of the piston rings, ring grooves and cylinder walls are not completely smooth. If examined under a microscope, these surfaces would show minute hills (asperities) and valleys. For this reason, the rings alone can never completely prevent high combustion and compression pressures from escaping into the low pressure area of the crankcase, with the consequent reduction in engine power and efficiency. Engine oil fills these hills and valleys on the ring surfaces and cylinder walls and helps to seal the combustion chamber. Because the oil film at these points is rather thin – generally less than 0.025 millimetres thick – it cannot compensate for existing wear of rings, ring grooves or cylinder walls. Where such conditions already exist, oil consumption may be high. Oil consumption may also be high in a new or a rebuilt engine until the hills and valleys on these surfaces have smoothed out enough to allow the oil to form a tight seal.

13.3.1.9 Nonfoaming

Because of the many rapidly moving parts in an engine, air in the crankcase is constantly being whipped into the engine oil. This may produce foam, which is simply a lot of air bubbles that may or may not readily collapse. These air bubbles normally rise to the surface and break, but water and certain other contaminants slow down the rate at which this occurs, and the result is foam.

Some diesel engines use Hydraulically Actuated Electronically Controlled Unit Injector (HUEI) pumps. HUEI pumps use the engine oil from the main gallery and pressurize it in a plunger pump. The engine oil is used to operate the unit injectors which, when used with intensifiers, increases the fuel injection pressures up to 241 500 kPa independent of engine speed. The electronic controls permit varied injection timing and duration to provide optimum fuel economy and emissions reductions. This system can circulate all of the engine oil in the sump every 8.8 seconds, causing foaming of the engine oil.

If foaming occurs it can cause a multitude of problems. Foam is not a good conductor of heat, so if the amount of foaming is excessive engine cooling will be impaired because the heat will not be dispersed. Foam also does not have the ability to carry a load and prevent wear of many engine parts, such as bearings and lifters. This is because the engine oil contains air, and that air is easily compressed. On the other hand, oil that is free of air is virtually incompressible. Foam in oil may also cause the oil pump to be less efficient, which can lead to excessive wear and inadequate cooling. To help ensure proper engine protection, all engine oils contain anti-foam additives.

13.3.1.10 Aid Fuel Economy

Engine oils are available that have been formulated to provide fuel economy benefits. There are two ways engine oils may help to improve fuel economy. For both single grade and multi-grade engine oils, friction reduction may be achieved by the use of additives that lower the metal-to-metal friction that occurs during boundary lubrication. Since friction reducers are surface active materials, engine oil formulators take great care to ensure that these additives do not interfere with anti-wear/extreme pressure agents that are needed to prevent wear between heavily loaded parts.

Another way in which engine oils may help to improve fuel economy is to reduce friction in areas of the engine that incur full film hydrodynamic lubrication. This can be accomplished by changing to an engine oil that has a lower viscosity, such as an SAE 50 to an SAE 30 grade engine oil. The SAE 30 grade oil, being less viscous, causes less friction, resulting in improvements in fuel economy. Also, multi-grade oils that contain polymeric additives known as viscosity index improvers have the ability to provide a variable viscosity when lubricating high shear zones in an engine. This results in reduced resistance to flow, lower operating temperatures and better fuel economy. By careful selection of the type of polymer chemistry used, a lubricant formulator can develop multi-grade engine oils that provide better fuel economy than their single grade counterparts.

13.4 Components of Engine Oil Performance

There are three principal components of engine oil performance:

- Viscosity, or resistance to flow
- Protection against wear, deposits, and oil deterioration
- Fuel Economy.

13.4.1 Viscosity

The viscosity of an oil is the resistance to flow and is a basic criterion for predicting engine oil performance. At cold temperatures, an engine oil must be thin enough to permit easy starting and rapid lubricant flow to critical engine parts. At high temperatures, an engine oil must be thick enough to provide a tough protective film between sliding metal surfaces. To ensure that the proper viscosity grade is used the Society of Automotive Engineers (SAE) has developed a viscosity grade classification for engine oils.

13.4.1.1 SAE Viscosity Classification System

The standard system for defining engine oil viscosity is the SAE viscosity classification system for engine oils SAEJ300. This system assigns an SAE grade number called a viscosity grade to a specific viscosity range, as measured at a specified temperature – the higher the viscosity the higher the grade number. All engine oils are classified according to this system worldwide (Table 13.1).

Single grade oils in common use today are labelled SAE 5W, 10W, 15W, 20W, 20, 30, 40, 50 and 60. Thick flowing oils have high numbers while thin, free flowing oils have high numbers. The W denotes engine oils that are suitable for use at low ambient temperatures. To ensure that the W oils do, in fact, have proper flow characteristics, their viscosities are defined at appropriate low temperatures. For example, SAE 10W engine oils are tested in the Cold Cranking Simulator at −25 °C, since they are designed for use at lower ambient temperatures. Even lower temperatures are used to establish that engine oils have good pumpability characteristics. The Mini-Rotary Viscosity test for a W grade is always 5 °C below the designated Cold Cranking Simulator test temperature. Thus, SAE 10W oils are tested in the Mini-Rotary Viscosity test at −30 °C and SAE 5W oils are tested at −35 °C (−31 °F). SAE designated oils without W designations are measured for viscosity at 100 °C to ensure adequate viscosity at normal engine operating temperatures.

The High Temperature, High Shear viscosity relates to the viscosity under heavy loads and high temperatures. The test is conducted at 150 °C under shear stress conditions that are similar to the very thin film lubrication area of the engine, such as those found at the piston ring-to-cylinder wall interface. As part of the SAE J300 Engine Oil Viscosity Classification minimum High Temperature/High Shear Viscosity Limits have been set to each of the non W Grades (20, 30, 40, and 50) in order to ensure that multi-grade engine oils do not create a situation in which they will not provide an adequate viscosity during use. For example, the SAE has set a minimum High Temperature High Shear Viscosity for SAE 15W-40 engine oils of 3.7 cP minimum. If the HTHS viscosity falls below 3.7cP this indicates that the engine oil will not perform as SAE 40 grade engine oil at engine operating conditions.

13.4.1.2 Multi-grade Engine Oils

When consulting the owner's manual, recommendations may be found that include engine oils labelled SAE 5W-40, 10W-30, SAE 15W-40, or other combinations. However, single grade oils are also recommended for some engines. The first type of oils are known as multi-grade oils. They can be used over a wider temperature range than single grade oils. Their temperature/viscosity characteristics provide for easy starting and pumping at low temperatures; yet they are viscous enough at high temperatures to lubricate like thick, single grade oils. Multi-grade engine oils meet the viscosity requirements and do the work of two and usually three or more SAE viscosity grades. For example, an SAE 10W-30 engine oil is formulated to meet the Cold Cranking Viscosity and Mini-Rotary Viscosity low temperature viscosity requirements of an SAE 10W and the high temperature viscosity requirements of an SAE 30 grade engine oil.

Often multi-grade oils are also called by names that imply all-season usage, since they perform satisfactorily in both winter and summer. This capability is made possible by the use of polymeric additives called viscosity index (VI) improvers, which enhance lubricant flow characteristics over a wide temperature range. An engine oil with a higher viscosity index is one that shows less change in viscosity over a wide temperature range.

Multi-grade engine oils can offer significant advantages over single grade engine oils. Multi-grade engine oils reduce the need for seasonal oil changes. They can provide easy starting and fast lubricant flow to engine parts in cold weather and a tough lubricating film at high operating temperatures.

Multi-grade engine oils also offer a fuel economy advantage over single grades. The benefits of multi-grade engine oils have long been recognized in petrol powered passenger cars and light trucks and over the last 30 years in heavy duty diesel engines.

Table 13.1 SAE Viscosity Grade Classification System SAE J-300 (November 2007)[a]

SAE viscosity grade	Viscosity (mPa s) at low temperature (°C)[c,d,e]		Kinematic viscosity (cSt) at 100 °C (API, 2011a, p. 40)[b,c,f]		High shear rate viscosity (HTHS) (mPa s) at 150 °C (API, 2011a, p. 40)[g]
	Max. (*API Bulletin*, 1509, 2005, p. 37)	Max. with no yield stress (*API Bulletin*, 1509, 2005, pp 37, 39, 40)	Min.	Max.	Min.
0W	6200 at −35°	60 000 at −40°	3.8	N.R.	N.R.
5W	6600 at −30°	60 000 at −35°	3.8	N.R.	N.R.
10W	7000 at −25°	60 000 at −30°	4.1	N.R.	N.R.
15W	7000 at −20°	60 000 at −25°	5.6	N.R.	N.R
20W	9500 at −15°	60 000 at −20°	5.6	N.R.	N.R
25W	13 000 at −10°	60 000 at −15°	9.3	N.R.	N.R
20	N.R.	N.R.	5.6	<9.3	2.6
30	N.R.	N.R.	9.3	<12.5	2.9
40	N.R.	N.R.	12.5	<6.3	3.5 (0W-40, 5W-40 and 10W-40)
40	N.R.	N.R.	12.5	<16.3	3.7 (15W-40, 20W-40, 25W-40 and 40 grades)
50	N.R.	N.R.	16.3	<21.9	N.R
60	N.R.	N.R.	21.9	<26.1	N.R.

[a] N.R = No requirement.
[b] 1 cP = 1mPa s^2; 1 cSt = 1 mm^2/s.
[c] All values are critical specifications as defined by ASTM D3244.
[d] ASTM D-5293: standard test method for apparent viscosity of engine oils and base stocks between −5 and −35 °C using cold-cranking simulator.
[e] ASTM D-4684: standard test method for determination of yield stress and apparent viscosity of engine oils at low temperature. Note that the presence of any yield stress detectable by this method constitutes a failure regardless of viscosity.
[f] ASTM D-445: standard test method for kinematic viscosity of transparent and opaque liquids (and calculation of dynamic viscosity).
[g] ASTM D4683 or CEC-L-36-A-90 (ASTM D-4741 or ASTM D-5481): standard test method for measuring viscosity of new and used engine oils at high shear rate and high temperature by tapered bearing simulator viscometer at 150 °C.

Multi-Grades in Heavy Duty Engines

Multi-grade engine oils also provide important benefits in heavy duty diesel engine applications. Most of the manufactures of heavy duty diesel engines approve or specify the use of multi-grade engine oils. In fact, some manufacturers expressly recommend the use of multi-grade engine oils over single grades. This is because multi-grade engine oils, such as SAE 5W-40, SAE 10W-30 and SAE 15W-40 viscosity grades, offer significant advantages in improved oil consumption and fuel economy, faster cold starting and reduced bearing wear.

Improved Oil Consumption

Multi-grade engine oils can provide low oil consumption than single grade engine oil in heavy duty diesel engines. This is because the higher viscosity of a multi-grade engine oil at the high ring zone temperatures encountered in a diesel engine reduces the tendency of the engine oil to leak past the piston rings and into the combustion chamber, where it would be burned.

Fuel Economy

Multi-grade engine oil offers fuel economy benefits due to its superior low temperature fluidity, which reduces energy robbing fluid friction. For example, while an SAE 10W-30 multi-grade and a SAE 30 straight grade both act like a SAE 30 grade at high temperatures, the multi-grade also exhibits the fluidity of an SAE 10W at low temperatures. Additional fuel economy benefits derive from the multi-grade engine oil's ability to reduce shear friction losses over a broad temperature range due the presence of VI improvers.

Easier Cold Starting

Multi-grade engine oil can provide easier cold starting than a single grade engine oil because a multi-grade is thinner at cold temperatures and flows more quickly to the engine parts. For example, it has been demonstrated in cold room tests that an SAE 15W-40 engine oil can lower the minimum starting temperature of an engine by as much as 3–4 °C compared to an SAE 30 grade engine oil.

Reduced Bearing Wear

In many cases multi-grade engine oils may reduce bearing wear due to their flow characteristics. These flow characteristics ensure that an adequate flow of lubricant is supplied to the all critical areas of the engine in order to reduce friction and wear.

13.4.2 Protection against Wear, Deposits and Oil Deterioration

The second component of engine oil performance is protection against wear, deposits and oil deterioration. Excessive engine wear increases fuel and oil consumption, reduces power and shortens engine life. Engine deposits such as sludge and varnish cause valve and ring sticking, clogged oil passages and piston and cylinder wear. Oil deterioration – the result of high temperature oxidation and reaction with combustion by-products – can increase the engine oil's viscosity and impede oil flow to critical engine parts.

These serious threats to engine life and performance are minimized by incorporating additives to improve the engine oil's properties. For example, anti-wear additives help to reduce wear between heavily loaded engine parts; detergents and dispersants help to prevent the build up of contaminants, sludge, soot and varnish; and oxidation inhibitors help prevent lubricant breakdown at high operating temperatures.

13.5 How Engine Oil Performance Standards are Developed

In the United States the American Petroleum Institute (API) and International Lubricant Standardization and Approval Committee (ILSAC) establish engine oil performance standards in cooperation with

OEMs, oil marketers, additive companies and testing laboratories. Each of these groups plays a different role and provides significant input during the various stages of the development, the definition and adoption of the standards and test methods used for a particular engine oil performance standard. A number of standard guidelines have been developed to define the overall capability for gasoline and diesel engine oils to prevent wear, deposit formation and oil deterioration. The principal guidelines are those established by:

- The American Petroleum Institute (API) in conjunction with the Society of Automotive Engineers (SAE), the American Standard for Testing Materials (ASTM) and ILSAC and the Diesel Engine Oil Advisory Panel (DEOAP).
- ILSAC (International Lubricant Standardization and Approval Committee) composed of North American and many Japanese automobile manufacturers.
- Individual OEMs

The development of any new diesel API Service Classification is done in three phases:

- Category Request and Evaluation
- Category Development
- Category Implementation.

13.5.1 Phase 1: Category Request and Evaluation (API, 2011a, pp. 36, 37)

An OEM, individual, company or association which is referred to as the sponsor of the new category can request new performance standards when there are changes in customer's needs or engine and design changes to meet government regulations. Any requests for a new standard are submitted to the Chairpersons of the Joint API/Engine Manufacturers Association (EMA) Diesel Engine Oil Advisory Panel (DEOAP). The DEOAP consists of a committee of representatives from the API and EMA member companies who deal with lubricant manufacturers that affect these two trade associations. The DEOAP guides and facilitates the introduction of the new proposed category and works with liaison representatives from allied organizations, such as the American Chemistry Council (ACC), The Society of Automotive Engineers (SAE), The American Standard for Testing Materials (ASTM), the Independent Lubricant Manufacturers Association (ILMA) and the United States Army. Once the Chairpersons of the DEOAP acknowledge the receipt of the new category request it works with the sponsor to furnish the DEOAP with the information that is needed to make a decision. The DEOAP has six months from the date that all the required information has been presented by the sponsor to make a decision to either accept or reject the request. If no decision is made within the six month timeframe, the request is forwarded to the API's Lubricant's Committee for its member's information and disposition (API, 2011a, pp. 36, 37).

The sponsor of the oil category must provide to the DEOAP data and justification for the proposed category. This request must demonstrate a need for significant oil performance changes to meet the requirements not met by any of the existing and current categories. The justification has to include but is not limited to one or more of the following:

- Likely or impending government regulations
- Consumer driven needs
- New hardware design or service requirements.

The DEOAP will ask the API, EMA and the ACC to appoint representatives to serve on an *ad hoc* review team called a New Category Evaluation Team (NCET) to formally evaluate the request. The NCET is limited to the minimum number needed to accomplish the work, while remaining consistent with full technical representation. The number serving on this *ad hoc* committee can vary. The API,

EMA and ACC can have up to three representatives on the NCET. The NCET develops the working rules, elects a chairperson, decides who to invite as liaison representatives and requests a meeting with the sponsor. During the evaluation process the NCET works to reach a consensus position on the following questions:

- What is the proposed change and why is it needed?
- Do data presented support the request?
- When is it needed in the marketplace?
- What are the potential impacts on engines?
- What are the potential impacts on consumers?
- What are the potential impacts on the environment?
- How could the change affect existing API categories?
- Are performance tests available that properly evaluate the performance needs requested?
- Do the perceived benefits outweigh the projected costs?
- How much will it cost to develop test procedures and determine precision and define, if necessary, Base Oil Interchange (BOI) and Viscosity Grade Read Across (VGRA) Guidelines for the proposed category?
- What is the estimated total cost to carry out projected work for the new category if the need is approved (API, 2011a, pp. 37, 39, 40)?

The DEOAP is responsible for calculating the estimated total costs for developing the proposed category and ensuring an agreement is reached on category development funding before submitting a request to the API Lubricant Committee. In order to do this the DEOAP establishes an *ad hoc* committee that consists of representatives from the principle stakeholders in the process: API, EMA, ACC, independent test laboratories and other parties, such as lubricant additive manufacturers and lubricant manufacturers (API, 2011a, p. 40).

The NCET can also solicit additional industry input and data at any time during the review process to assist it in reaching a decision. Any industry group, such as the SAE, API Detroit Advisory Panel (DAP), EMA and ILMA, can be asked to provide supplemental information. The NCET's specific charge during the decision process is to make one of the following decisions:

- Either support the request for the new category and recommend to the DEOAP that the request be forwarded to the API Lubricant's Committee for consideration to proceed with the development of the category. This recommendation has to identify the need for the category, recognize its feasibility, provide a timetable for category development, suggest draft language for the category and identify the proposed method of funding for development of the new category. The API Co-Chairman of the DEOAP then presents the recommendation, along with the appropriate documentation to the API Lubricants Committee for consideration.
- Deny the request, explaining to the sponsor in writing the reasons for the denial. The sponsor has the option of resubmitting the request with additional information.
- Not reaching a consensus: If the NCET cannot reach a consensus on the request, the API Co-Chairman has to provide the API Lubricants Committee with the vote outcome and a summary of the reasons for this action.

Once the need for a new engine oil performance standard has been established and passed on, the API Lubricants Committee must approve or deny the recommendation from the DEOAP by a formal vote. If the API Lubricants Committee approves the NCET's recommendation for the new category, the API DEOAP Co-Chairpersons moves the process forward to further develop the new category. Whether the API Lubricants Committee approves or denies the request, the *ad hoc* NCET disbands at this point (API, 2011a, p. 40).

13.5.2 Phase 2: Category Development (API, 2011a, pp. 41, 42)

Once the API Lubricants Committee approves the request, the API DEOAP Co-Chairpersons convene an *ad hoc* New Category Development Team (NCDT). The NCDT functions under the same guidelines at the NCET. The NCDT decides on working rules, selects a chairperson or facilitator and invites liaison representatives from the ASTM, SAE, ILMA, independent test laboratories or others as required. Other national, regional or international bodies, such as the Japan Automobile Manufacturers Association (JAMA) or the Coordinating European Council (CEC), may also be asked for input during the category implementation. Once the NCDT is formed, the DEOAP Co-Chairpersons are responsible for ensuring that funding sources are established to cover specific costs for development of new engine and bench tests and precision matrix testing. The DEOAP Co-Chairpersons establish a Task Force to arrive at an agreement on funding and the dispersal of the funds. The NCDT manages and coordinates the new process toward final approval within the timetable and budget set up by the DEOAP. The NCDT manages all phases of category development through four functional work groups chaired by NCDT members: an API member, an EMA member, an ACC member and an ASTM and SAE member.

13.5.2.1 API Function (API, 2011a, pp. 42, 43)

The API member functions in the following capacities:

- Ensures that no conflicts develop between existing categories and the one proposed.
- Coordinates API Base Oil Interchange/Viscosity Read Across Guidelines Task Force on its development of base oil interchange and viscosity grade read across guidelines based on data (including ASTM matrix testing), engineering judgment and field experience.
- Drafts a timetable to enable licensing at the earliest practicable date. The timeline will indicate the dates at which the first allowable licensing can occur. Normally, the first allowable licensing date for a new category is one year after ASTM Subcommittee B formally approves the new performance standard used to define the category. The delay allows all oil marketers equal opportunity to meet category requirements.
- Develops draft Consumer User Language. The final version of that language is approved by the API and EMA Lubricants Committees.
- Ensures the emergent marketing or consumer issues that arise during category development are brought to the attention of responsible groups for resolution.

13.5.2.2 The Engine Manufacturers Association (EMA) Function (API, 2011a, p. 43)

The Engine Manufacturers Association member's function is to:

- Guide the selection process for appropriate reference oils as well as low and high discrimination oils. At least one reference oil must be identified that meets all of the bench and engine tests contained in the new category. The reference oil is used in the test development and reformulated as necessary to ensure the best measure of performance. Before any new minimum performance category can be established by ASTM, at least one reference oil must be able to meet all category requirements. This reference oil must be tested in accordance with the ACC Code of Practice. The new category sponsors or their designee has the primary responsibility for recommending oil selection. The DEOAP provides feedback and formally approves the selections, and the selections are reviewed with ASTM.
- Recommend and/or provide relevant engine tests and hardware, with or without a test procedure.
- Stay abreast of changes that may occur (government-, industry- or consumer-generated) and, when necessary, suggest modifications to the new category to ensure it will meet the predetermined target. Coordinate any necessary modifications in language and tests with the NCDT.

13.5.2.3 American Chemistry Council (ACC) Function (API, 2011a, p. 43)

The ACC member's function in this process is to:

- Assess the new test against the criteria of the ACC Code of Practice Template with the objective of optimizing cost effective engine testing quality. Test precision and discrimination are examples of qualities to be assessed. Provide analysis of these assessments to the DEOAP and NCDT.
- Incorporate the new engine tests that meet the Template into the ACC Code together with accompanying test scheduling and registration procedures.

13.5.2.4 ASTM and SAE Function (API, 2011a, pp. 43, 44)

The American Standard Materials and SAE functions in this process are to:

- Work through the ASTM's D02.B0.02 Heavy Duty Engine Oil Classification Panel (HDEOP) to select or develop methods that evaluate the needs defined by the NCDT. The membership of the HDEOP is made up of representatives from the EMA, ACC, oil additive manufacturers and lubricant manufacturers.
- Ensure the bench and/or engine tests selected for the new category will satisfy the requirements of the draft consumer language approved by the API Lubricants Committee. The NCDT and ASTM develop a timetable that contains planned dates for reference oil selection, bench and engine test selection, and test method completion. The dates for the timeline must agree with the timeline that was approved by the API Lubricants Committee. The tests developed also need to correlate with field experience.
- Provide input, as requested, to the new category sponsors in the selection of appropriate discrimination reference oils for the individual test in the new category.
- Coordinate with other appropriate technical societies to develop and approve written test procedures and limits for tests not within the ASTM system that will be published as standards and specifications.
- Once a test shows satisfactory discrimination of oil performance, matrix testing is conducted to determine test precision and assess base oil and viscosity grade effects. For example, if an engine test is developed by ASTM, it is ASTM's responsibility to have a functioning task force or surveillance panel in place to coordinate activities and analyse test data. For bench tests the ASTM has to provide a method for referencing and/or calibrating each bench test that does not have an assigned surveillance panel.
- Implement and coordinate through the appropriate ASTM groups the funding mechanism recommendation by the NCDT and approved by the API Lubricants Committee for the development of tests, precision and base oil interchange. Also establish the high reference 'passing' category oil for the ACC's Test Monitoring Center.
- Establish pass/fail limits for each test and the entire category in accordance to the ACC's Multiple Acceptance Testing Criteria (MTAC).
- Update the appropriate SAE 'J' documents.

The tests used to certify the engine oil against a performance level standard consist of a set of engine sequence, laboratory bench tests, physical and chemical tests.

The engine sequence tests consist of a series of laboratory engine tests using production engines. These engine tests assess one or more performance parameters, such as piston deposits, valve train wear, fuel efficiency or oil consumption.

The laboratory bench tests determine the engine oil's relative performance parameters of corrosion, elastomer compatibility, oxidation stability, rust protection, foaming tendency and evaporation loss.

The physical tests, which consist of a number of laboratory bench tests, determine the physical properties of the engine oil. These physical tests assess engine oil properties such as viscosity, pour point, foaming tendency and evaporation loss due to volatility.

The chemical tests consist of laboratory bench tests, which are used to determine the chemical properties of the engine oil such as additive elements, total base number (TBN) and sulfated ash. Engine oils may also be tested for phosphorus, sulfur and sulfated ash content to determine if they are compatible with emission treatment systems.

13.5.2.5 Category Completion (API, 2011a, p. 44)

At the end of the development of the new category the NCDT undertakes a number of actions to bring the process to a conclusion. These actions are to review the output of four function groups that advise the NCDT as necessary to ensure the completion as well as harmony among the discrete parts. These specific actions are:

- For the ASTM functional group review the appropriateness of the test data developed for discrimination and precision. Agree on the final description for each new performance test and that the optimum test methods and performance limits have been chosen. Choose at least one 'demonstration' reference oil capable of meeting all minimum performance criteria required for use.
- For the ACC functional group, ensure that the ACC Codes include each of the new engine performance tests.
- Obtain from the SAE and other cooperating agencies any standards, codes and publications that are necessary parts of the new category.

Once the NCDT is in agreement that all of the goals and objectives have been met, it forwards all procedures, facts, data and information that are pertinent to the new category to the DEOAP. The DEOAP then convenes and together with the NCDT ensures that (1) the tests developed under NCDT guidance satisfy the need expressed by the sponsor, (2) the performance targets contained in the proposed consumer language are met by the tests proposed for the category, (3) the timetable is acceptable, and (4) the test methods chosen to define the new standard represent the most cost effective means of establishing the new performance level. All input, including API Base Oil Interchange and Viscosity Grade Read Across Guidelines. The complete package is then presented by the DEOAP Co-Chairpersons, with a recommendation for formal approval, to the API Lubricant Committee. The API must approve the complete package, including the final consumer language.

13.5.3 Phase 3: Category Implementation (API, 2011a, p. 45)

The implementation of the new category can take two different paths. If there are any unanticipated problems or situations that have arisen during the development process that cannot be overcome and that delay the category development or prevent it from being developed within the timelines that had been set, the EMA can choose to develop minimum performance requirements or a new category for consideration by the API through a process of their own choosing or outside of the current API developmental process. However, before this or any new minimum API performance category can be adopted it has to be approved by the API Lubricants Committee (Figure 13.3a).

If, upon agreement between the NCDT and DEOAP that all parameters of the new category that were approved by the API Lubricant Committee during the evaluation phase have been met, the final approval procedure begins to be implemented. However, if for some reason, full complete approvals have not been obtained, the DEOAP can carry out negotiations among the interested parties to resolve any differences. When all the differences have been resolved, the final specification includes its API Category Designation, a description of performance parameters, pass/fail limits, Base Oil Interchange and Viscosity Grade Read Across Guidelines, ACC Code of Practice requirements and any consumer language. Timelines for licensing are also designated by the API.

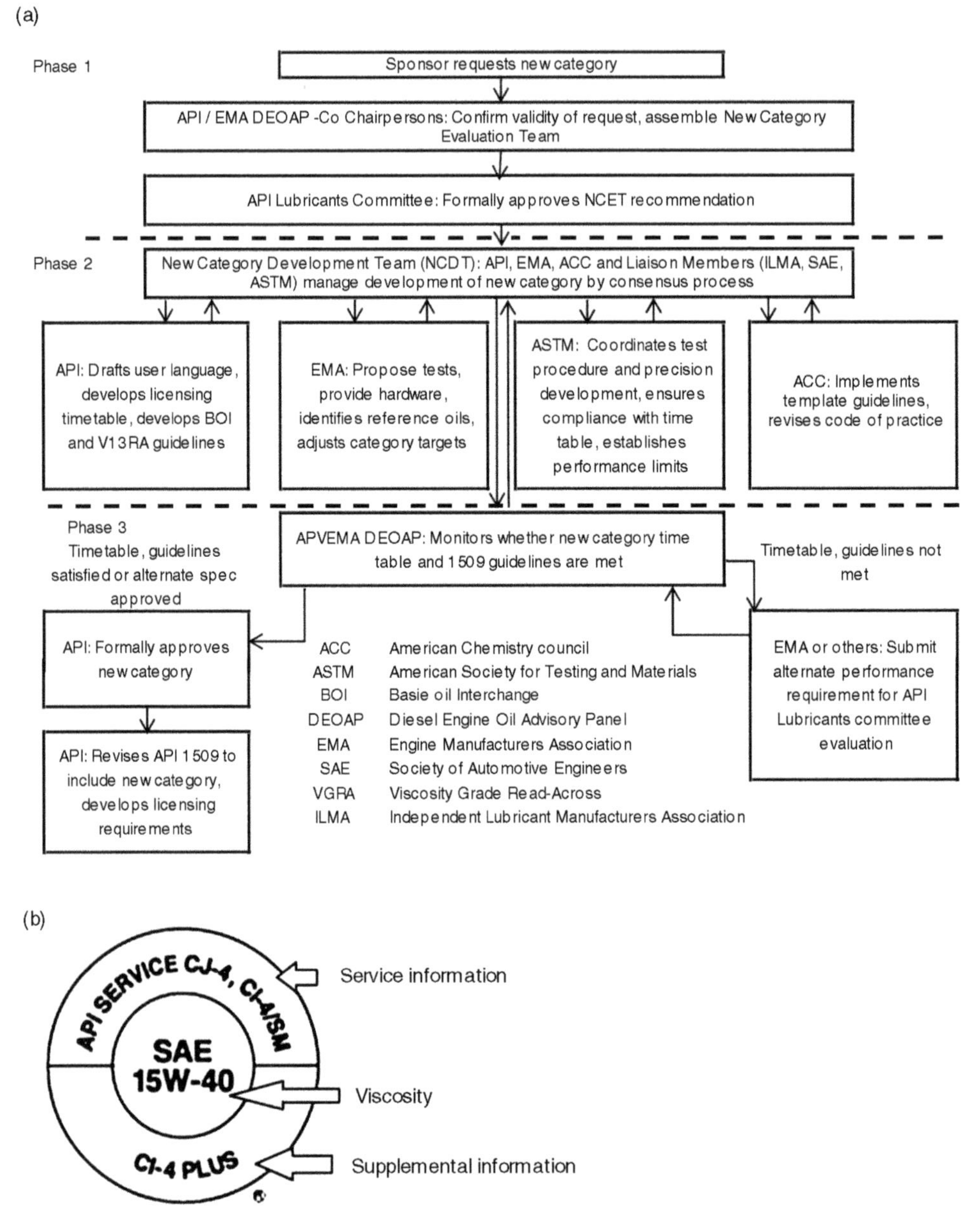

Figure 13.3 (a) Heavy Duty Category Processes (*API Bulletin*, 1509, 2011a, p. 38) (b) API Service Symbol

After the final approval has been obtained the API establishes the licensing protocol for the new category, publishes any revisions to the API 1509 Engine Oil and Licensing publication and advises oil marketers and other affected parties of the new diesel engine oil service classification. The API then administers its voluntary Engine Oil Licensing and Certification System (EOLCS) Program. The EOCLS system includes a formal licensing agreement executed by oil marketers with the API. Within this system, candidate oils must be supported by engine testing in accordance with the ACC's Code of Practice for Multiple Testing Acceptance Criteria (MTAC) and base oil read across guidelines. The API also administers the Aftermarket Audit Program to prevent mislabelling on oil containers and to ensure that licensed engine oils adhere to and meet the technical specifications of the engine oil standard.

13.5.4 API Licensing Process

The API administers the licensing and certification of engine oil performance standards through the Engine Oil Licensing and Certification system. This system's purpose is to define, certify and monitor engine oil performance (Lubrizol K2005a).

To qualify for a license or certification, engine oil marketers must submit an application in which they identify each product's brand name, viscosity grade and API category being licensed They must attach data sheets reporting the chemical and physical properties of each viscosity grade for each brand name being submitted. The candidate oil must be supported by engine testing using the ACC's Code of practice and must comply with the API's base oil interchange/viscosity grade read across guidelines. The marketer must sign an affidavit that test data are available to support the performance claims.

If the candidate oil qualifies, the oil marketer must enter into a formal licensing agreement to display the API Certification Mark (Starburst) and/or API Service Symbol (Donut) on their oil containers. They must also pay licensing and annual royalty fees associated with engine oil licensing and certification. Licensed and certified engine oils are subject to review by the API's Aftermarket Audit Program.

13.5.4.1 API Symbols

Once an engine oil is licensed the marketer can then display the API's engine oil symbols on its containers. The API labelling system consists of two symbols: the API Service Symbol and the API Certification Mark (for petrol engine oil service classifications).

API Service Symbol

The API Service symbol (Figure 13.3b), which is commonly called the Donut, is used for both diesel and petrol engine oils. The Donut is designed to provide specific information to the consumer regarding the engine oil's viscosity grade, the service classification and specific information, such as if the engine oil meets energy or resource conserving capabilities, or supplemental specifications, such as API CJ-4 Plus. The Donut can be placed anywhere on the container.

The API Donut is divided into three parts:

(i) The top half describes the engine oil's service classification and performance level.
(ii) The centre identifies the engine oil's viscosity.
(iii) The bottom half tells whether the oil has demonstrated energy conserving properties in a standard test in comparison to a reference oil or meets a supplemental specification.

Service Classifications and Performance Levels

The top of the Donut shows the engine oil's performance level for gasoline and/or diesel engines. The letter 'S' followed by another letter (for example SM) refers to an engine oil suitable for gasoline engines. The letter 'C' followed by another letter/and or number (for example CJ-4) refers to engine oils suitable for diesel engines. If an engine oil meets both a 'S' and a 'C' classification the 'C' classification is always displayed first.

SAE Viscosity Grade

The centre of the Donut shows the engine oil's SAE viscosity grade. Viscosity is a measure of an engine oil's flow characteristics, or thickness at certain temperatures. The low temperature viscosity (the first number 15W in a SAE 15W-40 and 5W in a SAE 5W-40 engine oil) indicates how quickly an engine will crank in winter and how well the engine oil will flow to lubricate critical engine parts art low temperatures. The lower the number the more easily the engine will start in cold weather. The high temperature viscosity (the second number, 40 for a SAE 5W-40 and SAE 15W-40 engine oil) provides thickness, or body, for good lubrication at operating temperatures.

A multi-grade engine oil provides good flow capabilities for cold weather but still retains thickness for high temperature lubrication. A single grade engine oil (as single number in the centre of the donut for example SAE 40) is recommended for use under a much narrower set of temperature conditions than a multi-grade engine oil.

Energy or Resource Conserving or Supplemental Categories
The bottom of the donut tells whether the engine oil has energy conserving properties when compared against a reference engine oil in the Sequence VIB test or Sequence VID tests. Oils labelled as 'Energy Conserving or Resource Conserving' have passed these tests. Widespread use of engine oils with this designation should result in an overall savings of fuel in the vehicle fleet as a whole, but a particular vehicle operator may not experience a fuel savings as a result of using these oils due to many different variables, such as the mechanical condition and maintenance of the engine, operating conditions and driving habits.

The 'Energy Conserving or Resource Conserving' designations apply only to engine oils intended for use in passenger cars and other light duty vehicles, and primarily to SAE 0W-20, 0W-30, 5W-20, 5W-30 and 10W-30 viscosity grade engine oils.

The bottom of the donut can also be used to display if the engine oil meets a supplemental specification to an existing API Service Classification. Currently on the API Supplemental Service Classification CI-4 Plus can appear at the bottom of the donut.

13.6 API Service Classifications

The API Service Categories are named with an alphanumeric system that consists of two letters that sometimes are followed by a number. The first letter is always either 'S' for gasoline engine service or 'C' for commercial diesel engine service. The second letter increases sequentially with each new category as engine oil evolve to match new performance requirements. The number 2 or 4 may also follow the two letters in the 'C' classification to identify if the engine oil is formulated for two-stroke or four-stroke diesel engines. The API 'C' service classifications for diesel engines are shown in Table 13.2.

Of these API Service categories only four are now considered to be active categories. Most 'C' categories usually include the performance properties of all the earlier performance categories and are 'backward' compatible. The exception to this rule is the API CF-2 service category for severe duty, two-stroke cycle diesel engine service. Engine oils that meet API CF-2 may be used in place of engine oils from the earlier API CD-II, but may not meet the requirements of earlier service categories such as API CE and CF-4, which were designed for four-stroke engine applications.

13.7 ACEA Specifications

Unlike the United States, European diesel engine oils are not driven by one industry standard. Each oil can be approved against a number of different specifications that are issued by industry bodies such as the ACEA and the API or individual vehicle manufacturers.

The Association des Constructeurs Européens d'Automobile (ACEA) is the industry body that creates the classification system used to define engine oil quality for European service fill applications (Table 13.3). The ACEA Engine Oil Sequence System is based upon a schedule of physical, chemical and engine tests similar to those used in the United States but using both the ASTM and the Coordinating European Council (CEC) test methods. All the performance claims against a particular ACEA requirement must be supported by data generated under the European Engine Lubricant Quality Management

Table 13.2 The API 'C' service classifications for diesel engines (Source: API (2010))

Category	Status	Service
CJ-4	Current	Introduced in 2006. For high speed, four-stroke engines designed to meet 2007 model year on-road exhaust emission standards. CJ-4 oils are compounded for use in all applications with diesel fuels ranging in sulfur content up to 500 ppm (0.05 wt-%). However, use of these oils with greater than 15 ppm (0.0015 wt-%) sulfur fuel may impact exhaust after-treatment system durability and/or oil drain interval. CJ-4 oils are effective at sustaining emission control system durability where particulate filters and other advanced after treatment systems are used. Optimum protection is provided for control of catalyst poisoning, particulate filter blocking, engine wear, piston deposits, low and high temperature stability, soot handling properties, oxidative thickening, foaming and viscosity loss due to shear. API CJ-4 oils exceed the performance criteria of API CI-4 with CI-4 Plus, CI-4, CH-4, CG-4 and CF-4 and can effectively lubricate engines calling for those API service Categories. When using CJ-4 oil with higher than 15 ppm sulfur fuel, consult the engine manufacturer for service interval.
CI-4	Current	Introduced in 2002. For high speed, four-stroke engines designed to meet 2004 exhaust emission norms implemented in 2002. CI-4 oils are formulated to sustain engine durability where exhaust gas recirculation (EGR) is used and are intended for use with diesel fuels ranging in sulfur content up to 0.5 wt-%. It can be used in place of CD, CE, CF-4, CG-4 and CH-4 oils. Some CI-4 oils may also qualify for the CI-4 Plus designation.
CH-4	Current	Introduced in 1998. For high speed, four-stroke engines designed to meet 1998 exhaust emission standards. CH-4 oils are specifically compounded for use with diesel fuels ranging in sulfur content up to 0.5 wt-%. Can be used in place of CD, CE, CF-4 and CG-4 oils.
CG-4	Obsolete	Introduced in 1995. For severe duty, high speed, four-stroke engines using fuel with less than 0.5 wt-% sulfur, CG-4 oils are required for engines meeting 1994 emission standards. Can be used in place of CD, CE and CF-4 oils.
CF-4	Obsolete	Introduced in 1990. For high speed, four-stroke, naturally aspirated and turbocharged engines. Can be used in place of CD and CE oils
CF-2	Obsolete	Introduced in 1994. For severe duty two-stroke engines. Can be used in place of CD-II oils.
CF	Obsolete	Introduced in 1994. For off-road, indirect injected and other diesel engines including those using with over 0.5 wt-% sulfur. Can be used in place of CD oils.
CE	Obsolete	Introduced in 1985. For high speed, four-stroke, naturally aspirated and turbocharged engines. Can be used in place of CC and CD oils.
CD-II	Obsolete	Introduced in 1985. For two-stroke engines.
CD	Obsolete	Introduced in 1955. For certain naturally aspirated and turbocharged engines.
CC	Obsolete	Caution: Not suitable for use in diesel powered engines built after 1990.
CB	Obsolete	Caution: Not suitable for diesel powered engines built after 1961.
CA	Obsolete	Caution: not suitable for use in diesel powered engines built after 1959.

Table 13.3 Current ACEA Specifications (Source: ILMA (2010))

Properties in ACEA 08 Heavy Duty					
Sequences		E4-08	E6-08	E7-08	E9-08
HTHS	mPa s	≥ 3.5	≥ 3.5	≥ 3.5	≥ 3.5
Sulfated ash	% m/m	≤ 2.0	≤ 1	≤ 2	≤ 1
Phosphorus	% m/m		≤ 0.08		≤ 0.12
Sulfur	% m/m		≤ 0.3		≤ 0.4
TBN		≥ 12	≥ 7	≥ 9	≥ 7
Wear	OM646LA	≤ 140	≤ 140	≤ 155	≤ 155
Piston cleanliness	OM501LA	≥ 26	≥ 26	≥ 17	≥ 17
Soot in oil		Mack T-8E	Mack T-8E	Mack T-8E	Mack T11
Soot induced wear				Cummins ISM	Cummins ISM
Liner–ring–bearing wear			Mack T12	Mack T12	Mack T12

System (EELQMS). The EELQMS system was developed jointly by engine manufacturers, oil marketers, additive companies and test developers and defines the process for developing, testing and reporting the necessary performance data (Lubrizol K2M, 2005b).

The ACEA specifications define a minimum quality level of a product. However, performance parameters other than those covered by the tests shown in the different ACEA classifications can be further supplemented by more stringent limits or even different engine sequence tests as specified by a European OEM. This results in a tiered engine oil system in which European OEMs use as a springboard to issue their own proprietary specifications. Since their introduction in 1996, the ACEA Oil Sequences have been updated approximately every two years.

For heavy duty diesel engine oils the ACEA specifications are referred to as 'E' Sequences. They define the requirements for service fill usage in heavy duty diesel engines. Each of the E Sequences has limits for:

- Sulfated ash content
- % Sulfur
- % Phosphorus
- % Noack Volatility
- High Temperature High Shear Viscosity
- TBN.

13.7.1 Current E Sequences

E4, E6, E7 and E9 sequences are described below. They provide excellent wear control, soot handling and lubricant stability. They are recommended for highly rated diesel engines meeting Euro I, Euro II, Euro III, Euro IV and Euro V emission requirements and running under severe conditions, for example extended oil drain intervals according to the manufacturer's recommendations. However, recommendations may differ between engine manufacturers, so Driver Manuals and/or Dealers shall be consulted if in doubt (API, 2011b).

E4: Stable, stay-in-grade oil providing excellent control of piston cleanliness, wear, soot handling and lubricant stability. It is suitable for engines without particulate filters, and for some EGR engines and some engines fitted with SCR NO_x reduction systems.

E6: Stable, stay-in-grade oil providing excellent control of piston cleanliness, wear, soot handling and lubricant stability. It is suitable for EGR engines, with or without particulate filters, and for engines fitted with SCR NO_x reduction systems. E6 quality is strongly recommended for engines fitted with particulate filters and is designed for use in combination with low sulfur diesel fuel.

E7: Stable, stay-in-grade oil providing effective control with respect to piston cleanliness and bore polishing. It is suitable for engines without particulate filters, and for most EGR engines and most engines fitted with SCR NO_x reduction systems.

E9: Stable, stay-in-grade oil providing effective control with respect to piston cleanliness and bore polishing: It is suitable for engines with or without particulate filters, and for most EGR engines and for most engines fitted with SCR NO_x reduction systems. E9 is strongly recommended for engines fitted with particulate filters and is designed for use in combination with low sulfur diesel fuel (ACEA, 2010).

13.8 OEM Specifications

United States automotive manufacturers and foreign automotive manufacturers that export their engines express requirements for their engines in terms of API or ILSAC Engine Service Categories and by viscosity grade. Manufacturers of heavy duty diesel engines either publish their own specifications or employ API Categories to define the desired performance levels for engine oils used in their engines.

Some OEMs create and recommend their own engine oil performance standards when a need or a desired performance level is not met by a current API Service Classification. API standards represent a voluntary benchmark for minimum engine oil performance. While the API tries to encompass all the OEM's engine oil performance needs, this is not always possible, due to constantly changing equipment designs and how these engine oils are actually performing in the field. OEM specifications represent a higher level of performance than the API standards and in many areas exceed those required for a particular service category. The standard or specification sets tougher engine sequence limits than those found in a particular API Service Classification or additional engine sequence testing in order to protect the OEM's engine during service and provides performance benefits beyond those of standard API categories.

OEM performance standards are primarily found in heavy duty diesel applications. Current heavy duty diesel engine oil standards are:

- Caterpillar ECF-1-a (equivalent to API CH-4)
- Caterpillar ECF-2 (equivalent to API CI-4 or CI-4 Plus)
- Caterpillar ECF-3 (equivalent to API CJ-4)
- Cummins CES 20078 (equivalent to API CI-4 or CI-4 Plus)
- Cummins CES 20081 (API CJ-4 requirements with more restrictive limits)
- Detroit Diesel PowerGuard 93K214 (equivalent to API CI-4 or CI-4 Plus)
- Detroit Diesel PowerGuard 93K218 (API CJ-4 requirements with additional test for bore polishing)
- Mack EO-N Premium Plus-03 (API CI-4 or CI-4 Plus requirements with more restrictive limits)
- Mack EO-N Premium Plus-07 (API CJ-4 requirements with more restrictive limits)
- Volvo VDS-3 (API CI-4 requirements with more restrictive limits)
- Volvo VDS-4 (API CJ-4 requirements with more restrictive limits).

13.9 Why Some API Service Classifications Become Obsolete

New engine oil service categories have to be created whenever engine performance requirements change. Older service categories become obsolete if a new category is developed and the new category includes performance properties required to replace them. For the petrol service classifications the latest category is designated as an improvement on or a replacement for all prior categories.

For the diesel engine oil service categories, new categories cannot always include the performance properties of all of the prior categories. API currently has three active categories to match different types of equipment and operating conditions. For example, diesel engine oils can be formulated for two- or four-stroke applications, for moderate or for severe service, or for high or low sulfur fuels. Also, some of the older service classifications are kept active if there is a market demand. This is because many older diesel engines are still in service in some markets, and engine oils that meet older service categories may be preferred for these older engines because they cost less than engine oils that meet the newer engine oil service classification.

13.10 Engine Oil Composition

Heavy duty diesel engine oils comprise approximately 75–85% base oil with the remainder made up of additive systems. In the formulation of a heavy duty diesel engine oil, different types of base oils and additive system can be used.

13.10.1 Base Oils

The base oils used in the engine oil's formulation make up a significant portion of the finished lubricant and contribute significantly to the performance characteristics of the finished oil in areas such as thermal stability, viscosity, volatility, the ability to dissolve additives and contaminants (oil degradation materials, combustion by-products etc.), low temperature properties, air release/foam resistance and oxidation stability.

A base oil is defined as a refined petroleum fraction or a synthetic material that is produced to a given set of specifications by a single manufacturer. Base oils can be manufactured using a variety of different processes, including but not limited to distillation, solvent refining, hydrogen processing, oligomerization, esterification and re-refining (API, 2011b). The types of base oils used in the formulation of heavy duty diesel engine oils are derived from either mineral base oils or synthetic base oils. Each of these base oils, which are also known in the lubricant industry as base stocks, are available in different viscosities regardless of if they are made from the same refining or chemical processes, belong to the same base stock grouping and come from the same manufacturer. A lubricant's quality can depend upon the type of base oil used, the refining and/or the method used to produce the base oil.

13.10.1.1 Mineral Base Oils

Mineral base oils are derived from petroleum base oils. Petroleum base oils are derived from crude oil, which is also used to produce petrol, diesel fuel, kerosene, asphalt, gases and petrochemical feedstocks. Approximately 95% of all lubricating base stocks come from petroleum. There are three different types of crude that can be used to produce mineral base stocks. They are: paraffinic, naphthenic and aromatic (Figure 13.4).

When crude oil is taken from the ground, it is composed of many different types of chemical materials, mostly based on carbon. The best way to think of the different hydrocarbon compositions present in crude oil is to compare them to different types of pasta. Crude oil is analogous to a mixture of every kind of pasta, such as shells, spaghetti, ziti, lasagne, rotini, linguini, noodle and so on, placed in the same pot. It is all flour but in many different forms. Of these types of crudes the most favoured types used for producing petroleum base oils are the paraffinic type crudes, which are analogous to long straight pasta, such as spaghetti, because of their high viscosity index. The two basic types of refining processes used for obtaining petroleum lubricant base stocks are those for separation and conversion. Base stocks produced by these methods are sometimes referred to as *conventional base oils* and *unconventional base oils*, respectively. Conventional refining technology involves the separation of the crude oil into selected desirable components by distillation, solvent extraction and solvent or catalytic dewaxing. Additional steps such as hydrofinishing, hydrotreating and hydrocracking can be added to the process and still considered to be conventional refining. Unconventional base stock refining uses severe hydrocracking refining techniques, such as two-stage hydrotreating, hydroisomerization or isodewaxing/hydrocracking processes.

During the refining process hydrocarbons are separated into different types through the process of distillation (the analogy being separating the very big and little pieces of pasta), cracking (like breaking

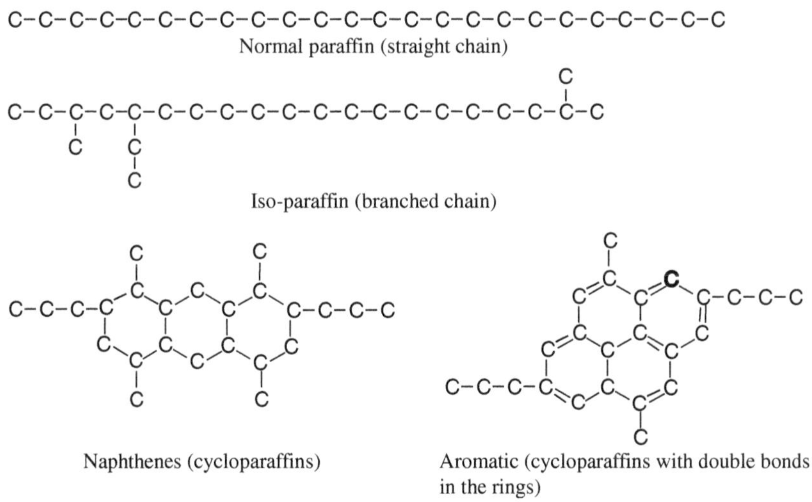

Figure 13.4 Hydrocarbon types used to produce mineral base stocks

the very long pieces of pasta like linguini and spaghetti into shorter pieces), hydrogenation (converting some pieces into others) and dewaxing (removing certain large pieces like lasagne) and so forth.

These processes produce approximately two-thirds of the world's paraffinic base stocks. As was stated previously, paraffinic base oils are the most widely used base oils in the formulation of lubricants for use in industrial and automotive lubricant applications. They contain varying levels of paraffinic molecules, have a relatively high viscosity index, good to very good oxidation stability and high amounts of wax.

Crude oils also contain heteroatom compounds of sulfur, oxygen and nitrogen. These compounds need to be removed during the refining process because they can affect the base oil's stability and viscosity–temperature characteristics.

13.10.2 Refining Processes Used to Produce Lubricating Oil Base Stocks

As was previously mentioned, the two basic processes used for obtaining petroleum lubricant base stocks consist of separation and conversion processes. These separation and conversion processes used are:

1. Separation type procedures
   - Distillation
     - Atmospheric
     - Vacuum
   - Propane deasphalting
   - Solvent extraction
   - Solvent dewaxing
2. Chemical conversion procedures
   - Hydrofinishing (Hydrotreating)
   - Hydrocracking methods
     - Two-stage hydrotreating
     - Isodewaxing/hydrotreating
     - Raffinate hydroconversion
   - Hydroisomerization
   - Catalytic dewaxing

13.10.2.1 Separation Type Procedures

Distillation Processes

The distillation process is a two-step process that consists of atmospheric and vacuum distillation. In the atmospheric distillation process the crude oil is heated in a distillation tower known as a pipe still at atmospheric pressure of 101.35 kPa (14.7 psi) to boil off the lube oil fractions from the crude.

Vacuum distillation is used to separate lube fractions having different boiling point temperature ranges in the presence of a vacuum from the heavy residual oils and asphalt containing compounds. This process controls the base oil's initial and final boiling point range. Narrower boiling point ranges lead to lower base oil volatility characteristics. This process also separates the lube oil fractions into its different viscosity fractions.

The base oil lube cuts produced from this process are classified as neutral distillates, while the heavy residual oils are called vacuum residuals.

Propane Deasphalting

Vacuum residuals contain a considerable quantity of high viscosity components useful in the manufacturing of lubricating oils in combination with undesirable constituents, such as asphaltenes and resins. These undesirable constituents can darken the colour of the oil and cause the formation of carbonaceous deposits when the oil is exposed to high temperatures. To improve the stability of vacuum residuals the propane deasphalting process is used. In this process the vacuum residuals are pumped to the propane deasphalting unit. The propane deasphalting unit consists of an extraction column which is used to mix the vacuum residuals with propane (usually about 6–8 volumes per volume of residual oil). The propane is charged to the bottom of the column. The residual oils, which are denser than the propane, flow down the column with the propane rising up in a counter flow. The mixing of the propane and vacuum residuals is achieved through the use of either perforated steel plate baffles or a rotating disc contractor at temperatures of 50–80 °C (120–180 °F) and pressures of 3447 kPa (500 psi). The rising propane dissolves the soluble components of the vacuum residuals, which are carried out of the top of the column with the propane. The insoluble, asphaltic and resin containing materials settle to the bottom of the column. The deasphalted vacuum residuals and the residue are separated from the propane using evaporators called flash towers and strippers. The propane is recovered and recycled after cooling and compression. The deasphalted vacuum residuals are sent for further processing.

Solvent Extraction

The solvent extraction process is used to further process the neutral distillates and vacuum residuals to remove undesirable components such as aromatics and unsaturated polar compounds, and sulfur and nitrogen-containing species, through the use of various solvents. Though aromatics have good solvency, their presence in the lube oil base stock can affect the oxidative stability and viscosity–temperature characteristics of the base oils. Aromatics have relatively low viscosity indexes and can start oxidative chain reactions, which can significantly shorten the useful life of the base oil.

Aromatics in the base oils present in the neutral distillates and vacuum residual are removed by feeding the raw feedstocks into a solvent extractor unit, where it is contacted in a counter-current flow with a solvent at a controlled temperature and atmospheric pressure. The solvents most commonly used in this process are furfural, n-methyl pyrrolidone (NMP) and DUO-SOL™. The solvents extract the aromatics and other undesirable materials from the feedstock and settle them to the bottom of the extraction tower. The resulting product, which is known as a raffinate, and the residue are separated from the propane using evaporators called flash towers and strippers.

Solvent extraction typically removes 50–80% of the undesirable impurities from the neutral distillates and vacuum residuals and raises the viscosity index and the thermal and oxidative stability of the base oils.

Dewaxing

All lubricating oil base stocks contain a certain percentage of wax. The deasphalting and solvent extraction processes concentrates the wax in the lube feedstocks. Removal of wax from these feedstocks is necessary in order to improve the low temperature properties of lube oil base stocks. If it is not removed from the base oils used in the blending and compounding of heavy duty diesel engine oils, the engine oil will not be able to properly flow at low temperatures.

Wax is removed from the raffinates by either solvent dewaxing or catalytic dewaxing methods. Of these processes, solvent waxing is the most widely used method.

In the solvent dewaxing process the raffinates are mixed with either methyl ethyl ketone/toulene or a methyl isobutyl ketone/toluene solvent mix and then chilled to −10 to −20 °C (14 to −4 °F). The wax crystals that form are kept in suspension by stirring during cooling and are removed by the use of drum type rotary vacuum pressure filters. The solvent is recovered from both the wax and the oil and is recycled for reuse.

Catalytic Dewaxing

In catalytic dewaxing, a shape selective catalyst is used to crack the paraffins in the wax structure into smaller more volatile segments that can be fractionated by distillation. The catalytic dewaxing process transforms the wax into highly branched isoparaffins, which can be used as feedstocks for the production of Group II and Group III base oils. Catalytic dewaxing is a continuous process and is more efficient in removing wax and other branched paraffins than solvent dewaxing. The resulting base oil from this process is slightly lower in viscosity index than those produced by solvent dewaxing. However, there are several processes that have been developed that isomerize the wax into highly branched isomers that will not crystallize as readily as linear paraffinic molecules.

Once the dewaxing process is completed, the lube oil base stocks can either be sold to lubricant blenders and compounders or finished off by chemical conversion methods, such as hydroprocessing methods, to further remove any small quantities of sulfur, nitrogen, organic acids and aromatic compounds that can cause the base oil to exhibit an increased tendenciy to form sludge and other insolubles. These methods are also used to improve the lubricant base stock's oxidative and thermal stability and viscosity index.

13.10.2.2 Chemical Conversion Procedures

Hydroprocessing Methods

To further improve the performance of the lubricant base stock performance there are several hydroprocessing techniques that can be used. These hydroprocessing methods use hydrogen in the presence of a catalyst to further remove impurities and improve performance. Each hydroprocessing process varies in severity and the way it changes the chemical structure of the base oil. These hydroprocessing methods include:

- Hydrofinishing (Hydrotreating)
- Hydrocracking
- Hydroisomerization

(i) Hydrofinishing

Hydrofinishing, which is also known as hydrotreating, is a process where hydrogen is added to the base oils in a reactor vessel maintained at temperatures above 316 °C (600 °F) and pressures above 3447.4 kPa (500 psi) in the presence of a bed of catalysts, which can consist of molybdenum oxides on alumina, nickel on precious metals, zeolite or palladium or platinum. Depending upon the severity of the hydrotreating process, the majority of undesirable components are removed from the base stocks.

The sulfur is converted to hydrogen sulfide, which can be separated from the oil upon exit from the reactor. The hydrogen sulfide can be processed to produce elemental sulfur for industrial or agricultural use. Aromatic and any unsaturated compounds react with the hydrogen and are converted into saturated hydrocarbons that can be highly paraffinic in nature. This results in base oils that are significantly lower in sulfur and have improved thermal and oxidative stability and viscosity indexes over 100.

Hydrotreating is sometimes used prior to solvent refining to increase the yield produced from the extraction process. This is accomplished by saturating the aromatic compounds that would otherwise be extracted by the solvent extraction refining process into more saturated hydrocarbons that are more useable as a lubricant base stock feed.

(ii) Hydrocracking

Hydrocracking is used to produce Group II and Group III base stocks which were introduced to meet the increasing demands for lubricants with improved oxidation stability and higher viscosity indexes. Hydrocracking can be used to replace solvent extraction.

In this process hydrogen is added to the base oil feed, which flows over a high activity catalyst bed at temperatures above 343 °C (650 °F) and pressures above 6894.8 kPa (1000 psi). The base oil feedstock molecules are reshaped and some side chain large molecules are split into smaller ones. This results in the formation of straight chain and highly branched paraffinic hydrocarbon structures, and aromatic compounds and naphthenic rings being converted into saturated paraffinic compounds. Because the hydrocracking process due to the cracking of side chains of molecules produces smaller molecules that have relatively low boiling points, which are of no use in the manufacturing of lubricants, these smaller molecules are removed from heavier components by the use of strippers and fed into the petrol–kerosene and diesel manufacturing process.

The significant reforming of the base oil's molecular structure by the hydrocracking process produces base oils that contain 97–100% saturate levels, contain very little or no sulfur and have viscosity indexes that can range from 110 to 130, improved thermal and oxidative stability and low pour points. This process removes most of the sulfur and completely removes polar impurities. However, the process does not affect or remove the waxy compounds that still may be present in the base oil feedstock. These waxy components must be removed in a subsequent process, such as catalytic or solvent dewaxing, to further improve the base oil's pour points.

After hydrocracking the base oil feed stocks can be further hydrotreated at temperatures as high as 400 °C (750° F) and pressures as high as 20 684.4 kPa (3000 psi). The use of the hydrocracking process also allows for the production of high quality lube base stocks from nonlube quality crudes such as wax and waxy feedstocks. The different types of hydrocracking processes that can be used by different refiners are:

- Two-stage hydrotreating
- Isodewaxing/hydrotreating
- Raffinate hydroconversion

Two-stage Hydrotreating

In the two-stage hydrotreating process the distillate feed stock from atmospheric and vacuum distillation is fed to the hydrocracking unit and heated to 371–538 °C (700–1000 °F). Hydrogen is added to the reactor vessel under high pressures ranging from 13 789.6–20 684.4 kPa (2000–3000 psi) in the presence of a bed of catalysts. At these temperatures and pressures, the base oil's large molecules are split into smaller molecules and completely rearranged. Once rearranged the base oil's molecules are absorbed onto the catalyst's surface and converted into a highly reactive state.

In this highly reactive state, aromatic and naphthenic rings are opened up saturated with hydrogen and converted into highly branched paraffinic molecules. This process can produce base oils that are 97–100% saturated. After leaving the hydrocracker the base oil feedstock undergoes atmospheric and vacuum distillation to produce the desired viscosity grades. These various viscosity grades then undergo solvent dewaxing and severe hydrotreating to remove any remaining sulfur, nitrogen and polar compounds.

Isodewaxing and Hydrotreating Process

The first step in this process is atmospheric and vacuum distillation. The second step is severe hydrocracking to remove the majority of sulfur, nitrogen and polar compounds and saturate most of the aromatic and naphthenic compounds that may be present. The next step is isodewaxing.

The isodewaxing process, which is patented by Chevron, takes *n*-paraffins and other molecules with waxy side chains and converts them into isomerized branched paraffinic molecules (isoparaffins). The isodewaxing process can also use wax produced by solvent dewaxing as one of its feedstocks. The fourth step is severe hydrotreating, which is used to remove any remaining sulfur, nitrogen, polar compounds and unsaturated molecules.

The final step is atmospheric and vacuum distillation to produce the desired viscosity grades. This process produces base oils that have saturate levels of 98–100%, virtually no sulfur and viscosity indexes approaching 130.

Raffinate Hydroconversion

This process was developed by ExxonMobil. In this process the first step involves atmospheric and vacuum distillation. The second step is severe solvent extraction to produce base oil feedstocks that have a VI of 90–95. In the third step of the process the waxy raffinates are placed into a proprietary hydroconversion reactor that saturates the aromatic and naphthenic compounds into highly paraffinic molecules. Next the lube oil factions are stabilized and fractionated in a finishing reactor to further improve the VI and volatility characteristics of the base oils. In the final step, the lube oil fractions undergo solvent dewaxing. The final product has a VI of 105–120 and is classified as an API Group II base stock. The final products are marketed as EHC™, where EHC refers to 'Exxon Hydro-converted'.

(iii) Hydroisomerization

Hydroisomerization is the most severe of the hydrotreating processes. In this process, wax produced by catalytic or solvent dewaxing and waxy petroleum feedstocks can be used. First the feedstock undergoes atmospheric and vacuum distillation. Next the isodewaxing process takes *n*-paraffins and other molecules with waxy side chains found in the wax and waxy petroleum feedstocks and converts them into isomerised branched paraffinic molecules (isoparaffins). The third step is severe hydrotreating, which is used to remove any remaining sulfur, nitrogen, polar compounds and unsaturated molecules. The final step is atmospheric and vacuum distillation to produce the desired viscosity grades. This process produces Unconventional Base Oils (UCBO) or Very High Viscosity Index (VHVI) Group III base oils that have saturate levels of 98–100%, virtually no sulfur, viscosity indexes over 130 and very low pour points.

These base stocks have synthetic like properties and are classified in the United States as synthetics because of the severity of the chemical alteration, which results in properties that are similar to polyalphaolefin (PAO) synthetic base stocks.

13.10.3 Synthetic Base Oils

Synthetic base oils are base stocks in which a chemical conversion of one complex mixture of molecules into another complex mixture has taken place. The most common synthetic base oils used in the formulation of heavy duty diesel engine oils are polyalphaolefin (PAO), diesters and polyol esters. Synthetic base stocks have higher viscosity indexes, flashpoints, lower coefficients of traction, (the coefficient of traction is the lubricants' molecular resistance to motion of one layer of fluid sliding over or along another layer of fluid) and lower pour points than mineral base stocks. These characteristics make them valuable in the blending of lubricants used for extreme service at both ends of the temperature extreme and for providing energy efficiency.

13.10.4 Synthetic Blends

Synthetic blend base stocks are base stocks made from a mixture of synthetic and petroleum base stocks. Synthetic blend base stocks use varying degrees of synthetic base oils ranging from 10 to 40% in the formulation of the base stock mix. A synthetic blend combines many of the performance features of synthetic base stocks with the cost economics of petroleum base stocks. The synthetic portion provides better oxidation and thermal stability, low temperature properties and improved viscosity–temperature characteristics.

13.10.5 API Base Oil Categories

Base oils used in the formulation of engine oils have different physical and performance properties depending upon their crude source, refining methods and the chemical conversion methods used to produce them. The API has defined five categories of lubricant base stocks in order to try separate conventional, unconventional, synthetic and other classifications of base stocks that can be used in the formulation of engine oils (Table 13.4). These base oil categories were developed for the purpose of creating interchanging guidelines for base oil read across performance parameters used for the licensing of engine oils by engine oil blenders. The API recognized that not all base oils have similar physical or chemical properties or provide equivalent engine performance in engine testing. These categories identify base stocks by their physical and chemical characteristics. Although these categories are independent of crude source or processing, they can generally indicate how the base oil was refined or produced and its properties.

Groups I, II, III are mineral oil base stocks and are classified by the amounts of saturates and sulfur content and by the viscosity index of each. Group IV is reserved for polyalphaolefins (PAO). Group V is reserved for other base stocks not included under Groups I to IV, such as diesters, polyol esters, polyalkylene glycols and naphthenic and aromatic base stocks.

The API Base Oil Categories do not mean or imply one Group is better than another. For example, if a base stock falls in the Group I category, it does not mean that it is better or worse than a base stock that falls in the Group II category. Although the Group II base stock contains lower levels of sulfur and aromatics, an increased potential for improved oxidation stability and a higher viscosity index, it may provide poorer solubility of additives and contaminants than a Group I base stock. The real measurement is the performance of the finished lubricant in the field. In recent years the Group I, II and III categories have been informally subdivided into Group I+, Group II+ and Group III+ subcategories.

What are Saturates?

Saturates are the types of molecules commonly found in a base oil. The most common types of saturate molecules are paraffinic and naphthenic hydrocarbon molecules. These molecules do not contain any double bonds. This is important because the presence of a double bond in a molecule can be a site for chemical reactions. Substances that contain double bonds are more vulnerable to oxidation and other chemical reactions.

Table 13.4 API base oil categories

Group	Sulfur (wt-%)		Saturates	V.I.
I	>0.03	and/or	≤ 90	80–120
II	≤ 0.03	and	≥ 90	80–120
III	≤ 0.03	and	≥ 90	≥ 120
IV	All PAO			
V	All base oils not included in Groups I–IV (naphthenic, esters and polyglycols)			
In recent years, these categories have been informally subdivided into Group I+, Group II+ and Group III+				

Base oils that contain high amounts of saturates are less reactive and more resistant to oxidation. Some saturates are present naturally in base oils, but their presence is usually achieved during the refining process. Refining techniques such as hydrocracking and hydroprocessing methods to convert any olefinic and aromatic compounds present in the crude oil into saturates. These refining methods can also convert naphthenic hydrocarbons into highly branched isoparaffinic hydrocarbons to further improve the base stock's oxidative and thermally stability and viscosity–temperature characteristics.

Sulfur

Sulfur is naturally present in crude oils in the form of organosulfur compounds. The sulfur content of the crude oil can range from 0.1 to >2 wt-% depending upon the source of the crude oil. After refining these organosulfur compounds can typically range for zero to as high as 1 wt-%. How these organosulfur compounds impact the performance of the lubricant base stock is very complex and depends to a large extent on the nature of the additive systems used in the lubricant formulation and the application the lubricant is going to be used in. The presence of sulfur can have both a positive and negative effect on the base oil.

Certain organosulfur compounds that may be present in the lubricating base stock can act as natural antioxidants at high temperatures, thereby improving the base stocks oxidative stability, while others can enhance the rate of oxidation that can lead to degradation of the lubricant.

The presence of sulfur in the lubricant base stock can also have detrimental effects on exhaust after-treatment devices used to reduce emissions. During normal operation, a small percentage of the engine oil consumed by heavy duty diesel engines travels past the rings and burns in the combustion chamber. Once in the exhaust stream sulfur can inhibit the effectiveness of the particulate filters by poisoning the catalysts. This poisoning can increase the conversion of sulfur oxides to sulfates, which increases particulate emissions and accumulation of particulate material. Accumulation of particulate material can lead to reduced engine performance, due to increased back pressure and, ultimately, failure of diesel particulate filters and other devices that are used to control emissions.

Viscosity Index

The degree to which a base stock resists viscosity changes due to temperature is expressed as the viscosity index. The higher the viscosity index of a base oil, the less is its rate of change with respect to temperature. All lubricating oils increase in viscosity with decreasing temperature and decrease in viscosity with increasing temperature. Lubricants that have a high viscosity indexes are generally preferred over those with low viscosity indexes because they retain their viscosity characteristics over a wider temperature range. High viscosity index engine oil, for example, will be less likely to thicken in viscosity on a cold morning and will provide better cold starting and pumpability. At high operating temperatures it will better retain its viscosity and continue to provide lubrication to critical engine parts. The type of refining methodology used to produce the base stock can enhance its viscosity index.

13.10.5.1 Group I Base Oils

Group I base oils are mineral base derived by less severe processing than Group II base oils. Group I base oils have sulfur contents that range typically from 0.1–1%, saturate levels that are typically 65–80% and viscosity indexes of 90–100. Most base oil refineries use solvent extraction to remove aromatics and other unsaturated compounds. The use of solvent extraction improves the base oil's viscosity index and oxidation stability. Group I base oils can also undergo solvent or catalytic dewaxing to remove wax and lower its pour point. In addition, these base stocks can also undergo hydrofinishing (hydrotreating) to improve their physical and performance characteristics.

Some Group I base stocks that have undergone severe hydrofinishing (hydrotreating) can have saturate levels greater than 90% but their sulfur level can be greater than the minimum limit of 0.03% sulfur for Group I base oils. This category is the only base oil category that contain an and/or clause because of this aspect (Table 13.4).

API Group I base oils are used in automotive and industrial applications. However, with the introduction of demanding automotive performance specifications, especially in the areas of volatility characteristics and oxidation stability, and restrictions on the amount of sulfur that can be present in the final engine oil formulation their use has declined in these applications

13.10.5.2 Group II Base Oils

Group II base oils are mineral oils that have a sulfur content of less than 0.03%, saturate levels of greater than 90% (typically 97–99%) and viscosity indexes that range from 95 to as high as 115. All Group II base oils typically undergo hydrocracking and catalytic dewaxing. The processes convert almost all of the aromatics in the oil into saturates and remove sulfur and other impurities. These high saturate levels provide the Group II base oils with very good oxidation and thermal stability

Because of the severity of the refining process used in the production of Group II base oils they are only available in a limited viscosity range. They also have lower solvency than Group I base oils, which requires formulators of heavy duty diesel engine oils to have to select a careful balance of additive systems for use in the formulation of the oil.

13.10.5.3 Group III Base Oils

Group III base oils are mineral base oils produced through the use of severe hydrocracking, advanced catalytic dewaxing and/or hydroizomerization. Group III base oils contain a sulfur content less than 0.03% (typically less than 10 ppm), saturate levels that are typically 98–100% and viscosity indexes that range from 120 to 135. Some Group III base oils, depending upon the hydrocracking process used, can have viscosity indexes greater than 140.

Because of the severity of the refining process used in the production of Group III base oils they are only available in a limited viscosity range. They also have lower solvency than Group I base oils, which requires formulators of heavy duty diesel engine oils to have to select a careful balance of additive systems for use in the formulation of the oil.

These base stocks have synthetic like properties and are classified in the United States as synthetics because of the severity of the chemical alteration, which results in properties that are similar to polyalphaolefin (PAO) synthetic base stocks.

13.10.5.4 Group IV Base Oils

The Group IV base oil category is the category that is reserved for polyalphaolefin (PAO) synthetic base fluids. In the United States and Europe PAOs are the most widely used synthetic base stocks in the formulation of lubricants.

PAOs are synthetic saturated hydrocarbons that are synthesized by a two-step reaction process from linear alphaolefins (LAO), which are derived from ethylene. The source of the ethylene comes from the steam cracking of hydrocarbons that are derived from either crude oil or natural gas sources.

In the first step of the synthesis, the ethylene undergoes the process of oligomerization to develop 1-decene (C_{10}) and 1-dodecene (C_{12}). The 1-decene or 1-dodecene in the presence of a catalyst (either a barium trifluoride (BaF_3) and protic co-catalyst such as water, alcohol or a weak carboxylic acid or a Ziegler Natta aluminium chloride base catalyst) undergoes oligomerization to form mixtures of dimers, trimers, tetramers and higher oligomers. The crude reaction products are washed with water or caustic, allowed to settle and washed again with water to remove any traces of the catalysts

The second step of the synthesis process entails hydrogenation of the unsaturated oligomers and distillation of the saturated oligomers to remove any unreacted monomers and separate the different viscosity grades of PAOs available (2–3200 cSt). The hydrogenation process is typically performed over a supported metal catalyst such as nickel/kieselguhr or palladium/alumina. Hydrogenation of the unsaturated oligomers is necessary to completely saturate the oligomers in order provide the final product with increased chemical inertness and additional oxidative stability. The hydrogenation process can take place either before or after the distillation process

PAOs are completely 100% saturated hydrocarbons that do not contain any sulfur and have viscosity indexes that can range from 135 to as high as 300 (for the heavier viscosity grades). Because they are wax free, PAOs typically have pour points that can range as low as –60° C (−76 °F) to as high as −18 °C (0 °F) depending upon the viscosity of the PAO. PAOs possess the following performance characteristics when compared to mineral oils:

- Greater oxidative stability
- Greater thermal stability
- Excellent low temperature viscosity characteristics
- Excellent hydrolytic stability
- Very good oxidation and thermal stability
- Very good to excellent low volatility characteristics
- Very low coefficients of traction
- Compatibility with mineral oils and ester-based synthetics
- Good to very good corrosion resistance
- Excellent lubricity and viscoelastic behaviour
- Availability in a wide range of viscosity grades

13.10.5.5 Group V Base Oils

The Group V base oil category is used for those base oils that are not included in any of the other categories. This category includes base oils with a wide range of performance properties. The Group IV base oil category can include the following types of base oils:

Naphthenic Mineral Oils
These oils are refined from crudes that have high levels of naphthenes and aromatics, but low wax contents. Naphthenic base oils typically have good solvency and low pour points, but possess low viscosity indexes of 0–75, saturate levels of 35–50%, sulfur contents up to 1% and poor oxidative and thermal stability.

Esters
Esters are synthetic base fluids. There are two main types of esters: diabasic acid esters and polyol esters. Esters are synthesized by reacting an acid an alcohol. The term polyol ester refers to a molecule with two alcohol functions in its structure, examples being trimethylpropane (TMP), neopentylglycol (NPG) and pentaerythritol (PE). They are primarily used in the formulation of high temperature industrial applications because of their thermal stability. Esters like PAOs do not contain any sulfur or wax. They possess very good to excellent thermal stability, typically have low pour points that range from −30° to −73 °C (−22 to −99 °F) (depending upon the type of ester chemistry used), viscosity indexes typically of 120–165 (some diabasic acid esters have viscosity indexes as low as 75), excellent solvency, poor to fair hydrolytic stability and greater tendencies to swell certain types of elastomeric materials. Some esters can compete with additive systems for binding sites on the metal surfaces because of their highly polar nature.

Polyalkylene Glycols
Polyalkylene glycols (PAGs) are polymers of alkylene glycols. They are produced by the polymerization of one or more alkylene oxides, such as ethylene oxide, propylene oxide and butylene oxide. They can vary widely in performance depending upon the type of monomers used in their synthesis, their molecular

weight and the terminal or end group. PAGs have very high viscosity indexes, typically ranging from 175 to 200 (some can be >250), and very good high temperature stability. They exhibit low deposit formation, excellent solvency and have very low coefficients of traction. Some PAGs are water soluble and they have poor compatibility with mineral oils, PAOs and certain types of ester chemistries.

13.11 Specific Engine Oil Additive Chemistry

Unadditivated base oil, regardless of how well it is refined or if it is a synthetic base stock will not provide satisfactory lubrication for a heavy duty diesel engine. Because of this aspect, heavy duty diesel engine oils contain additives to provide the extra performance required for the engines.

Additives can be classified as materials that impart new properties to or enhance existing properties of the lubricant into which they are incorporated. In an engine oil formulation the additive system can make up anywhere from 10 to 30% of the finished product. Each additive or combination of additives is incorporated into an engine oil for a specific purpose based upon the type of service expected from that engine oil. Additives enhance or impart properties in three ways:

- Protect engine surfaces
- Modify oil properties
- Protect the type of base stock used.

An engine oil may contain some or all of the types of additives that are discussed in this section. Each of these additives can include a number of different types of chemical compounds. These various chemicals are chose depending upon their specific performance and compatibility with other chemicals present in the engine oil. It takes a great deal of time, money and research to develop these additives and to determine the correct additive balance needed. Engine sequence tests have been devised to measure the performance of oils in the kinds of service to which they will be exposed.

These tests duplicate in the laboratory the wide range and variety of operating conditions to which the engine will be subjected to. Care is taken to ensure that the tests accurately predict engine oil performance. These laboratory tests are further supported by field tests operated under a variety of conditions and vehicle types. These laboratory and field tests, which have been conducted in all types of weather conditions, demonstrate the necessity of using chemical additives in engine oil. They also show the significant importance in performance achieved by using fortified oils.

The type of additives found in the formulation of engine oils fall into the following categories:

- Detergent–Dispersant Additives
- Anti-wear Additives
- Friction Modifiers
- Rust and Corrosion Inhibiting Additives
- Oxidation Inhibitor Additives (Anti-oxidants)
- Viscosity Index (VI) Improvers
- Pour Point Depressants
- Foam Inhibitors.

13.11.1 Detergent–Dispersant Additives

Detergent and dispersant additives are used to keep the engine's metal surfaces clean, prevent the formation of deposits and to neutralize the harmful effects of corrosive acids that are formed by the combustion of diesel and petrol fuels.

All engine oils, as they deteriorate either due to oxidation or by contamination, will form insoluble sludge and varnishes and resins that can become deposited on the surfaces of the engine. Once deposited, the sludge, varnishes and resins can block oil lines and oil passages causing the engine oil from reaching the parts to be lubricated. This in turn can result in increased wear, heat build up and eventual engine failure.

Further, engine oils can be exposed to fuel soot due to incomplete combustion of the fuel or carbon being introduced into the engine by various emission controls, acids formed by the combustion of the fuel and the ingestion of moisture and dirt from the engine's air intake system. If these contaminants are allowed to build up in the engine oil they can result in:

- Increased engine wear, especially in the valve train area.
- Increased deposits, especially in the piston rings and crown land of the piston.
- Increased cylinder bore polishing.
- Thickening of the engine oil's viscosity.
- The formation of oxidation precursors in the engine oil.
- Poor oil pumpability, especially during cold weather conditions.
- Plugging of oil filters.
- Rapid depletion of the engine oil's additive system.
- Decreased engine durability and life.

To prevent the formation of these deposits, sludge and resins, it is important that the engine oil contain an effective detergent/dispersant additive system.

Detergents (Figure 13.5) are oil soluble bases that are derived from the organic soaps or salts of calcium, magnesium or sodium or barium carboxylates, sulfonates, or phenates. Primarily, they are calcium or magnesium based. These materials are often referred to as organo-metallic compounds and they are polar in nature, which allows them to cling to the surfaces of particles. Detergents serve two principal functions.

Firstly, they lift any deposits from the surfaces of the engine which they adhere to and then chemically combine to form a barrier film, which keeps the deposits from coming out of suspension and coagulating. Detergents form two kinds of barrier films. On small particles (generally less than 0.02 micron in size), detergents form an absorbed film which slows down coagulation of the particles. On much larger particles (ranging from 0.5–1.5 microns in size), detergents cause the particle surfaces to acquire an electrical charge of the same sign so they can repel each other.

The polar metallic heads of detergents have a great affinity for each other. These molecules attract each other like magnets and form clusters called 'micelles'.

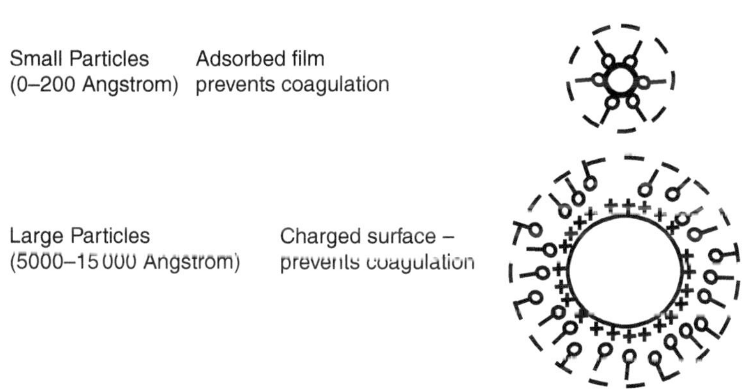

Figure 13.5 Action of organo-metallic detergents

The deposit precursors, being oil insoluble, have a greater affinity for the detergent molecule than the oil molecules. They are attracted to the detergent micelles (much like iron fillings are drawn to a magnet) and trapped within them. Thus, they are kept in solution in the engine oil and cannot settle out to form deposits in the engine.

The number of particle that can be contained in a micelle is limited. When the number of particles exceeds the capacity of the type of detergent chemistry being used deposits can form. Therefore it is necessary that the engine oil be drained before this happens if engine cleanliness is to be maintained.

Secondly, detergents neutralize any acids formed by the combustion of the fuel by chemically reacting with the acids to form harmless neutralized chemicals.

Dispersants are polar additives that are used to disperse sludge and soot particles for the purpose of preventing agglomeration, settling and deposits. Dispersants envelop particles and keep them finely divided. Dispersants are polymeric and ash-less compounds. These compounds are based on long chain hydrocarbons, which are acidified and then neutralized with a compound containing basic nitrogen. The most common types of dispersant used consist of succinimides, succinate esters and alkylphenol amines.

The hydrocarbon portion provides oil solubility, while the nitrogen portion provides an active site that attracts and holds potential deposit forming materials to keep them suspended in the engine oil. This dispersant molecule combines a compact, electrically polar head and a long, oil soluble tail (Figure 13.6).

Of the different types of chemistry that can be used, it is the succinimide type dispersants that is the most popular type; the piece containing nitrogen (N) is the polar head; the piece both containing nitrogen (N) and oxygen (O) is the connecting link; 'R' is the long, oil soluble tail. The polar heads attach themselves to any deposits or acids that may be formed by the combustion of the fuel to form micelles, which are taken into solution in the oil by the hydrocarbon tails. These micelles can trap deposit precursors up to 0.05 micron in size by proving a thick absorbed barrier film or they can also hold larger particles up to 0.1 micron in size by electrical charge repulsion. In this state, the acids and deposits cannot see the engine's metallic surfaces.

Dispersants can hinder agglomeration of soot, sludge and dirt particles via two different mechanisms, steric and electrostatic stabilization. In the first case, once the polar head group has adsorbed onto the surface of the dirt particle, the tail provides a physical barrier to attraction (Figure 13.7). This separates the small particles and prevents them from coming into contact with others; they increase in size, as shown in Figure 13.8.

Dispersants can also induce a charge on the soot, sludge and dirt particles. This results in contaminants such as sludge that have acidic groups and dispersants that contain basic amine sites interacting, forming salts. Agglomeration is then inhibited by electrostatic repulsion of the negatively charged dirt particles (Figure 13.9).

As previously stated in this discussion on detergents and dispersants, one of their functions is to help neutralize any acids that are formed by the combustion of the fuel. Each of these additives contributes to the neutralization of these acids by providing the engine oil with a Total Base Number (TBN), which measures an engine oil's ability to neutralize acids that are formed by combustion. As long as an engine oil's TBN stays above a certain limit during use (generally $^1/_2$ of its original TBN number), the engine oil is

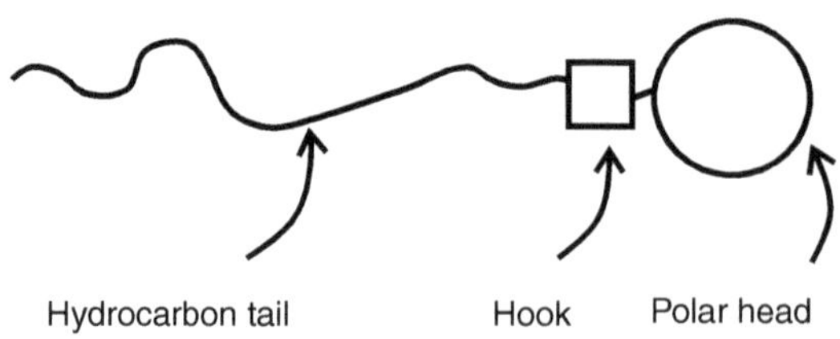

Figure 13.6 Dispersant molecule

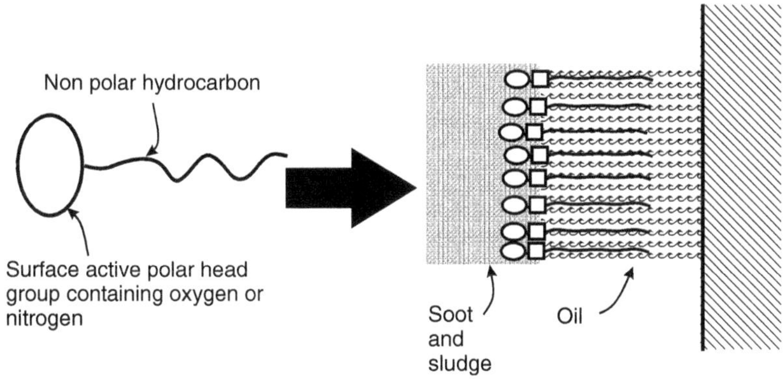

Figure 13.7 Polar head adsorbed on to the dirt particle providing a physical barrier to attraction

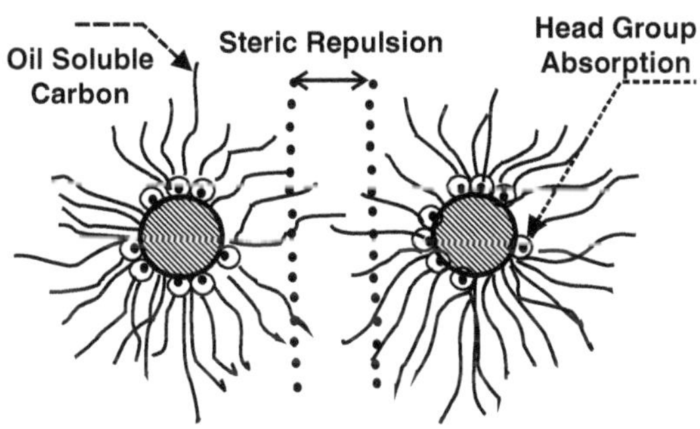

Figure 13.8 Separation of small particles and prevention from coming into contact with others and increasing in size

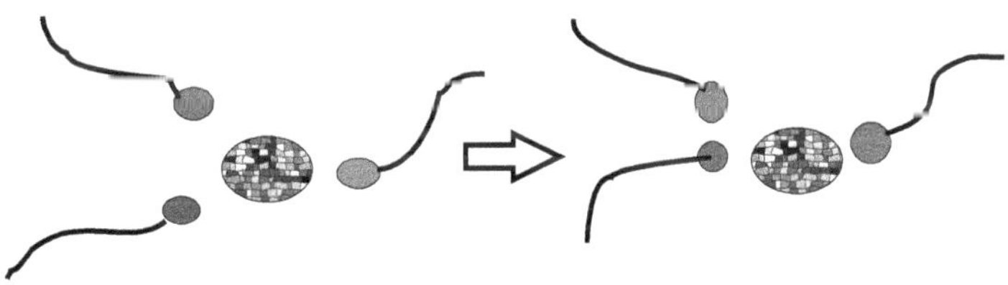

Figure 13.9 Electrostatic repulsion of charged particles

still fit for service. In fact, a new oil's TBN is less important than the TBN during service, as this protects the engine. An engine oil must have the ability to retain its TBN reserve (alkalinity reserve), which is contributed to by both the detergents and dispersants during its entire drain interval.

Of these two additives, it is detergents that offer the best alkalinity reserve. Though dispersants are a necessary additive for the formulation of engine oils, an engine oil's total TBN that is derived through the use of high dispersant chemistry does not offer the adequate protection an engine needs against the corrosive attack of acidic combustion by-products. Dispersants are more rapidly depleted than detergents because of the way they react chemically with acids that are formed and by the way they react with other particulate contaminants.

Detergents have the ability to retain their TBN reserve over longer periods, thus providing a more protective form of TBN over the entire drain period of the engine oil.

Therefore it is important that the engine oil contain a careful balance of detergent and dispersant additives.

13.11.2 Anti-Wear Additives

One of the primary functions of engine oil is to reduce wear. Normally metal-to-metal contact is prevented by the engine oil's lubricant film being thick enough to keep the contact metal surfaces of the engine separated (Figure 13.10a). However, during periods of high shock loading, high pressures, high speeds, heavy loading or at cold start up, the lubricant film between the two metal surfaces is either squeezed out or ruptured. This not only causes the two metal surfaces to come into contact with each other, but also causes the entire load to be carried by the contacting metal surfaces of the two mating parts. Once this occurs, severe wear, galling and eventual failure of the metal surfaces can take place, unless some means is found to prevent metal-to-metal contact.

The means employed to combat metal-to-metal contact is the incorporation of anti-wear additives into the formulation of the engine oil. Anti-wear additives prevent metal-to-metal contact by adding film forming compounds that protect the metal surfaces, either by a physical absorption or a chemical reaction with the metal surface; just as temperatures begin to rise due to the initial metal-to-metal contact they form a low shear film at the point of contact. These films are weaker than the underlying metal, so they can easily slide over each other without welding or causing other damage. Further, by reducing wear the films formed by anti-wear additives also prevent excessive friction energy losses.

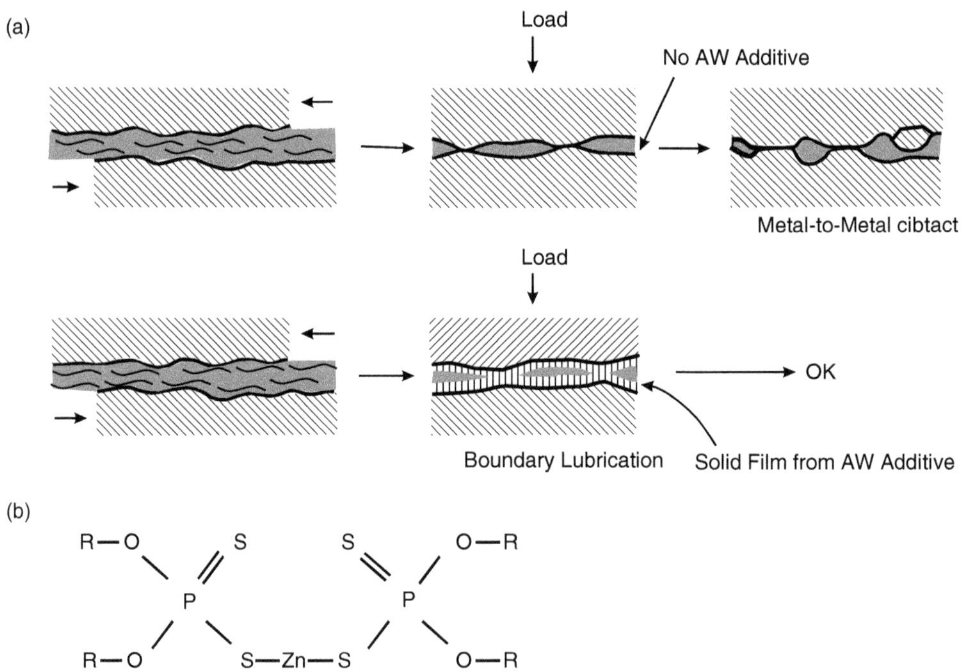

Figure 13.10 (a) Anti-wear additives: mode of action of rust inhibiting additive; (b) chemical structure of ZDP

There are many types of anti-wear additives, but the most widely used and the predominant type of anti-wear additive that has been used in the formulation of engine oils for the past 50 years is zinc diorgano dithiophosphate (ZDP). ZDP is a jack of all trade type of additive, since it not only functions as an anti-wear additive but also as an oxidation and rust and corrosion inhibiting additive. Its chemical structure is given in Figure 13.10b.

The R's can be either alkyl (straight or branched hydrocarbon chains) or aryl (aromatic hydrocarbon rings) or a combination of both. Their main purpose is to make oil soluble all of the inorganic compounds in the molecule (such as the zinc) so that they can be carried by the engine oil to where it is needed.

ZDP comes apart at high temperatures in engines to form protective films of zinc sulfides and zinc phosphates. These films bond to the metal surfaces and prevent the metal surfaces from contacting with each other. As fresh metal is exposed by rubbing, ZDP forms new films and so on until the anti-wear additive is used up.

13.11.3 Friction Modifiers

Friction modifiers were originally a minor class of additives used in the formulation of engine oils. However, friction modifiers have recently come into more prominence, owing to the desire for improved fuel economy and the discovery that these additives can make significant contributions to fuel economy benefits.

Friction modifiers are additives such as organo-molybdenum compounds, long chain polar compounds and so on that reduce the coefficient of friction between contacting metal surfaces when boundary lubrication conditions occur. Boundary lubrication occurs in an engine between the piston rings and the cylinder walls when the movement of the piston ceases for an instant and begins to reverse its direction of motion. At this point, pressures behind the piston rings are the greatest and the temperature of the cylinder wall is at its highest. Boundary lubrication also occurs between the severely stressed parts of the valve train at the junction between the cams, lifters, push rods and rocker arms.

When boundary lubrication conditions occur, the lubricant film is no longer sufficiently thick enough to keep the moving parts separated. Once this occurs, friction between these contacting surfaces, unless it is reduced, increases wear to the metal surfaces and increased engine drag.

Friction modifiers work by plating and absorbing to the metal surfaces of the engine, forming a long lasting solid lubricant film (Figure 13.11) which reduces contact between the contacting asperities, thus

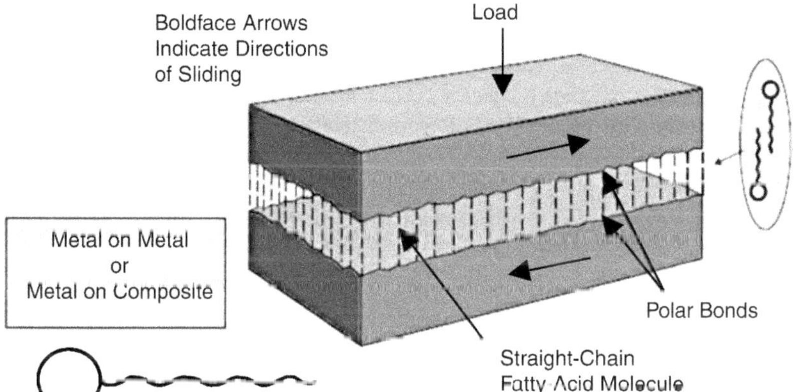

Figure 13.11 Friction modifiers plate and are adsorbed to the metal surfaces to form a long lasting solid lubricant film

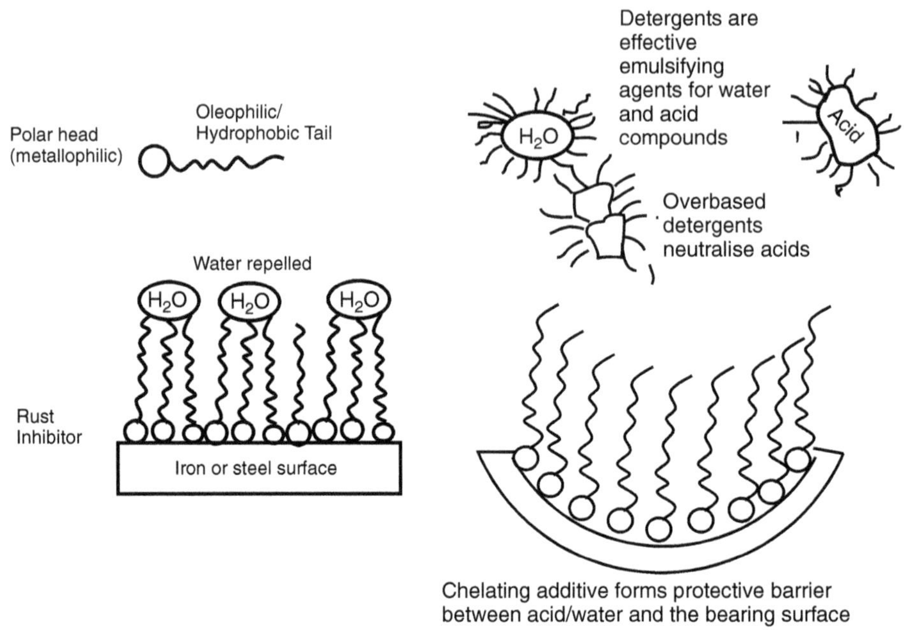

Figure 13.12 Action of a rust inhibitor

allowing them to slide over each other without contacting each other. This results in less heat being generated and less frictional drag, resulting in increased fuel economy and better engine efficiency.

13.11.4 Rust and Corrosion Inhibitors

Corrosion and rusting of metals, particularly ferrous metal surfaces, occurs when oxygen and moisture act together on the metal surfaces. This condition can be further aggravated by the presence of acids, especially those acids formed by the oxidation of the engine oil or the combustion of the fuel. If allowed to occur, rusting and corrosion can promote oxidation of the lubricant, metal-to-metal contact, wear and eventual engine failure.To prevent rust and corrosion of the metal surface by moisture and acidic components, the engine oil must contain rust and corrosion inhibitors.

Rust inhibitors are chemical compounds that absorb on the ferrous metal surface of the engine to form a polar adherent film that not only repels water but also neutralizes any acidic components that may come into contact with the engine oil (Figure 13.12). Corrosion inhibitors are chemical compounds that function by neutralizing and inhibiting the formation of acidic compounds by forming a protective film that seals the metallic surfaces from contact with water and corrosive acids.

The different chemical compounds that are used as rust and corrosion inhibitors are multifunctional and can be used as detergents, dispersants, oxidation inhibitors and anti-wear agents.

13.11.5 Oxidation Inhibitors (Antioxidants)

Oxidation of the engine oil occurs when it comes into contact with air, usually in the presence of catalytic metals such as copper and copper alloys during high temperature operation. As temperatures increase, oxidation of the engine oil is accelerated. At high engine operating temperatures and in the presence of catalytic metals, the oxygen in the air reacts with the oil to form unstable compounds known as free radicals and peroxides. Once they are formed these free radicals and peroxides start a chemical chain reaction that causes the engine oil to deteriorate. Once deterioration occurs, the formation of harmful sludge, varnish and resin deposits, acidic components and vast increases in viscosity begin to occur. If allowed to go unchecked, oxidation of an engine oil can lead to corrosive bearing wear, abrasive wear to various parts of the engine, plugged filters and oil passages, lack of lubrication and eventual engine failure. Because of these aspects, it is important that the engine oil contain additives known as oxidation inhibitors or antioxidants. Antioxidants prevent the formation of acids, gums, resins, varnish and sludge that normally results from oxidation.

The use of antioxidants can extend the life of an engine oil 10–100 times. Antioxidants are a class of various chemicals that terminate the oxidation of the engine by either destroying free radicals (known as chain breaking) as they are formed or by reacting with peroxides as they are formed during the oxidation process. Among the most widely used antioxidants in the formulation of engine oils are phenolic and aromatic amine types and zinc dithiophosphate. Another class of antioxidants that has begun to be used in the formulation of engine oils is the soluble molybdenum compounds that are used as frictional modifiers.

Zinc dithiophosphate and organo-molybdenum compounds are peroxide decomposers, while the phenolic and aromatic amine types are chain breakers and free radical traps. All of these types of antioxidants are highly effective in preventing oxidation at high and low temperatures. Further, some types of antioxidants, such as the zinc dithiophosphate and soluble molybdenum compounds, also form protective films on the metal surfaces of the engine, making them impervious to attack. The mode of action of antioxidants is shown in Figure 13.13.

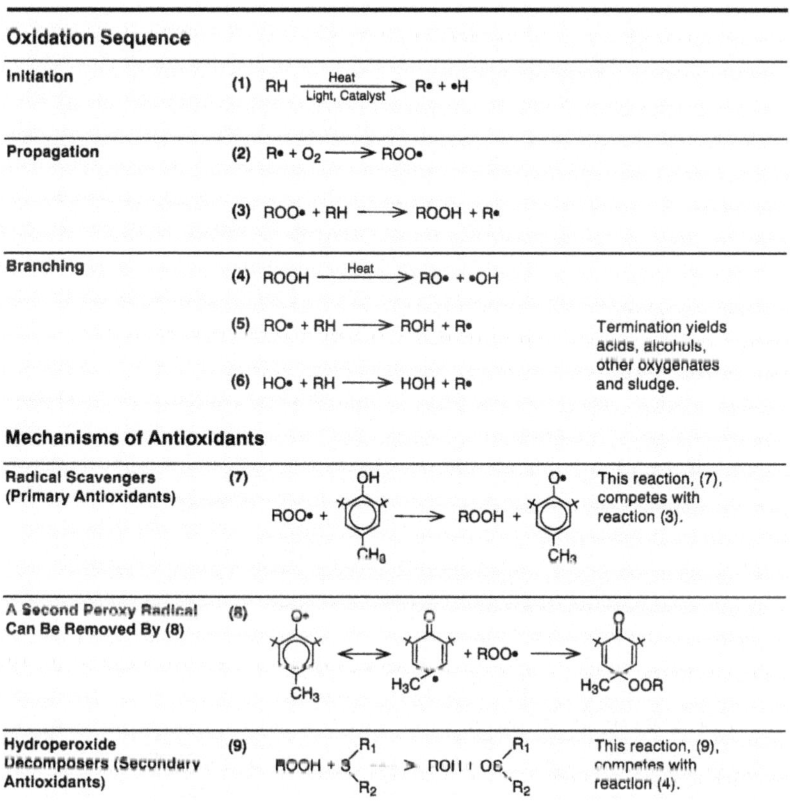

Figure 13.13 Mode of action of antioxidants

Finally, though the oxidation of engine oils is inhibited by the use of antioxidants, the extent of the rate of oxidation of engine oil is also dependent upon the following factors:

- The quality and type of base stock used. The use of base stocks that are highly paraffinic in nature or synthetic can greatly enhance the oxidation stability of the engine oil, because their molecular structures are more closed and impervious to attack.
- The amount of time the engine oil is in service.
- The operating temperature to which the engine oil is exposed.
- The type and amount of contaminants the engine oil is exposed to.
- The type of service and operating conditions to which the engine oil is exposed.

13.11.6 Viscosity Index Improvers

Having the appropriate engine oil viscosity is essential for satisfactory performance. The ideal engine oil would have a constant viscosity that is adequate to prevent contact of the metal surfaces of the engines at high temperatures, while reducing viscous drag at low temperatures. Heavy duty diesel engine oils can be exposed to ambient and operating temperatures that can extend well below 0 °C and well above 100 °C. When used at a wide range of ambient and operating temperatures the engine oil's change in viscosity with respect to temperature variations should be as small as possible. To minimize the adverse effects of variations of viscosity with respect to temperature, polymeric materials known as viscosity index improvers or viscosity modifiers are often used in the formulation of the heavy duty diesel engine oil. The presence of these polymeric materials has the ability to decrease the rate at which oils change viscosity with respect to temperature and allows the engine oil to take on multi-grade properties. Viscosity index improvers are polymers that consist of ethylene-propylene-diene copolymers, ethylene-propylene copolymers, styrene-butadiene copolymers, polymethacrylates and hydrogenated polyisoprenes that have molecular weights of 10 000–20 000.

Without viscosity index improvers, engine oils normally low enough in viscosity to meet the low temperature requirements of SAE 5W, 10W or 15W engine oils would be unable to meet the high temperature requirements of an SAE 30 or 40 weight oil. The reason for this is that the normal rate of viscosity change over the required range of ambient starting to engine operating temperatures is simply too large. With the use of viscosity index improvers, these light oils can have the correct viscosity over a wider temperature range. In addition, without the use of viscosity index improvers start up at low temperatures can cause the engine oil not to be distributed to critical engine parts, and once the engine has reached operating temperature can cause the engine oil to be not viscous enough, resulting in increased engine wear and impaired fuel economy.

Viscosity index improvers cause an increase in the viscosity of lubricating oil at all temperatures, except that at low temperatures the thickening in viscosity is less than that at higher temperatures. This is because at low temperatures the viscosity index improver is less solvated, occupies a smaller volume and tends to contract and be attracted to itself, rather than to the base oil's molecules, by folding upon itself to form small coils or clumps of the polymer that interfere less with the flow of oil, thus offering less resistance to oil flow and pumpability at low temperatures (Figure 13.14). At high temperatures the situation is reversed because the polymer chains extend or expand into the base oil as temperatures increase. The polymers are said to be more solvated, which means they are surrounded by the engine oil's base oil molecules and extend into the oil interfering with its flow and thereby increasing the engine oil's viscosity considerably over what it would otherwise be at elevated temperatures.

Variable thickening of the engine oil's viscosity by viscosity index improvers at low and high temperatures allows the formulation of multi-grade engine oils that are designed to have the proper viscosity–temperature characteristics to provide adequate viscosity at high temperatures for engine protection and low viscosity at low temperatures for easy start up and pumpability.

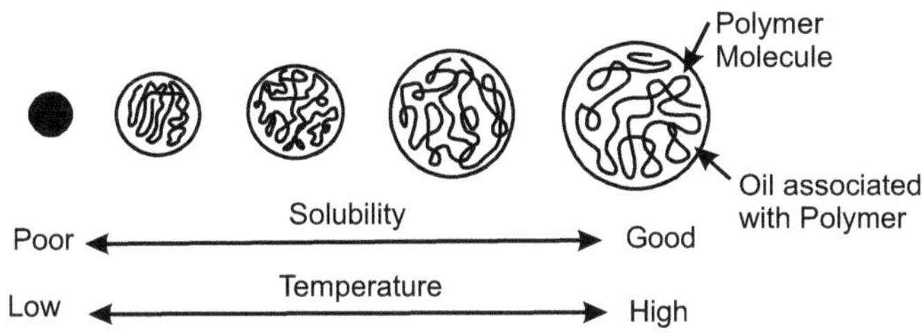

Figure 13.14 Mode of action of viscosity improvers

Though the use of viscosity improvers are critical in the formulation of multi-grade engine oils, their use can lead to problems if they do not exhibit shear stability (resistance to mechanical shear or tearing apart of the polymer structure) and chemical and thermal stability.

Unlike straight grade oils, multi-grade oils exhibit non-Newtonian viscoelastic rheological characteristics. That is, their viscosity depends upon the degree of mechanical shear stresses they are exposed to. In a diesel engine different parts the engine subject the multi-grade engine oil to different mechanical shear rates. These different shear rates cause the engine oil's viscosity to vary with location, even when they are at the same operating temperature. Where shear rates are high, such as in the rod bearings and behind the piston rings, these high shear rates cause the engine oil's polymeric viscosity index improvers either to be squeezed to a more compact form or to align themselves in the direction of motion, reducing the resistance within the oil film. This temporarily brings the viscosity of the engine oil to that of the base oil, thereby reducing fluid friction for better fuel economy or power output. When the engine oil leaves the high shear zone, the polymers revert to their former arrangement and the engine oil recovers its higher viscosity.

High engine shear rates can also cause a permanent loss in viscosity to take place (Figure 13.15). Viscosity loss is considered to be permanent if, after the shear forces are removed, the viscosity does not revert to its prior value. Permanent viscosity loss occurs when the polymers breakdown and rupture to lower molecular weight fragments and into smaller molecules. This causes the engine oil to lose its 'viscous grip' and no longer stay in its SAE viscosity grade. In addition to permanent loss of viscosity due to mechanical shear stresses, polymers can also undergo thermal and oxidative degradation. Thermal degradation can occur when the polymers break down by chain scission to form lower molecular weight

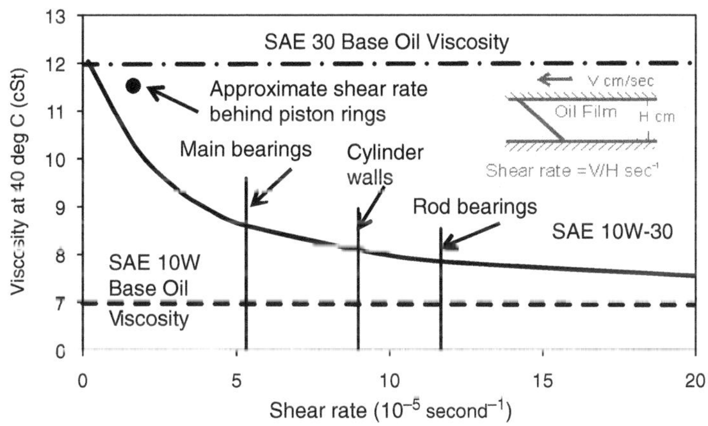

Figure 13.15 Loss in viscosity due to shear rate

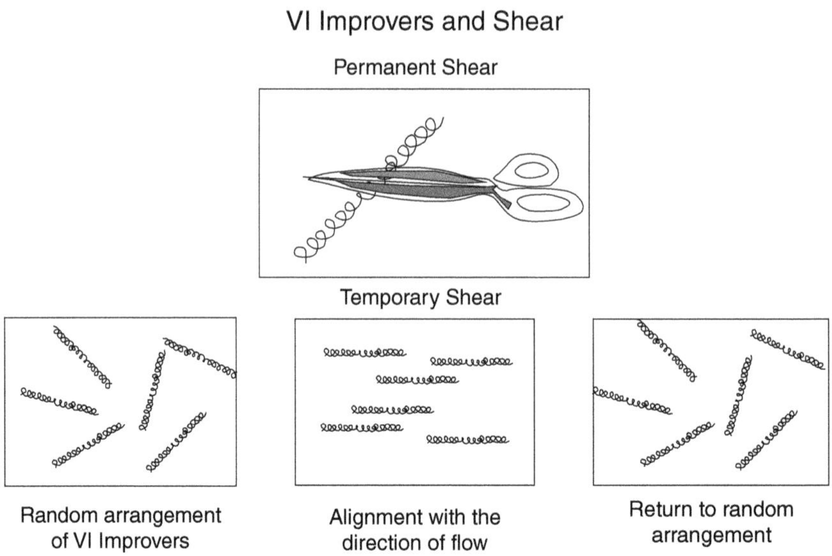

Figure 13.16 Viscosity index improvers: permanent and temporary shear

fragments resulting in a loss of viscosity (Figure 13.16). Oxidative degradation occurs when weak carbon-to-hydrogen bonds in the polymers react with oxygen to form hydroperoxides and peroxide radicals. Once these chemical species are formed they can cause the polymers to form lower molecular weight compounds that result in a loss of viscosity (Rizvi, 2003, pp. 225, 227, 228).

Once a loss in viscous grip occurs, whether it is due to mechanical shearing or degradation, this can result in:

- Increased oil consumption
- Increased oil leakage
- Increased engine noise
- Improper fuel injection, precision and timing (Hydraulic Electronic Controlled Unit Injectors found in some diesel engines)
- Formation of deposits on critical engine parts
- Loss of fuel economy
- Increased wear
- Increased emissions

There are all different types of polymer chemistries that can be used as viscosity index improvers in the formulation of engines oils. Some types are more shear stable and resistant to thermal and oxidative degradation than others. Therefore, it is important that right type of chemistry is chosen in order to provide not only shear stability but also improved viscosity–temperature characteristics, particularly low temperature viscosity, and protection against the formation of high temperature deposits.

13.11.7 Pour Point Depressants

Pour point is the lowest temperature at which the lubricating oil will pour when cooled under defined test conditions, which are found in the prescribed ASTM test methods used to measure pour point. In general,

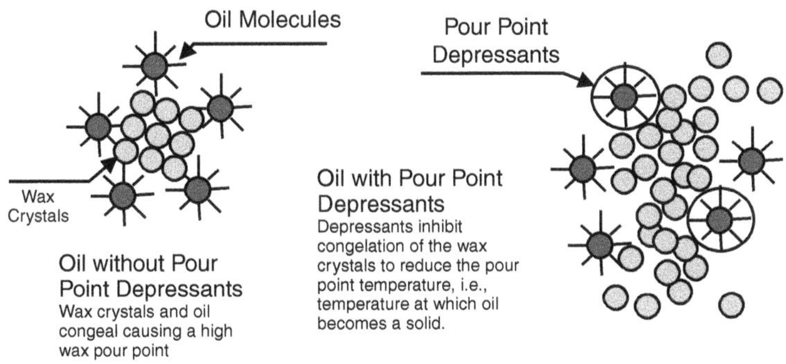

Figure 13.17 Mode of action of pour point depressants

pour point is indicative of the amount of wax (straight chain paraffins) present in the type of petroleum base stock used in the engine oil formulation. At low temperatures, these straight chain paraffins have a tendency to separate as crystals and form a lattice honeycomb-like structure (Figure 13.17) that can trap a substantial amount of oil via association, thus inhibiting proper oil flow and circulation to critical engine parts at start up (Rizvi, 2003, p. 230). Engine oils must begin to circulate as soon as the engine is started at low temperatures otherwise catastrophic wear can occur.

Base oil refiners and suppliers of petroleum base oils remove most of the wax through the use of either solvent dewaxing or catalytic dewaxing processes during the refining process. However, complete dewaxing of the petroleum base oils is not practical because of process limitations, economics and the desirable presence of wax due to its high viscosity index characteristics (Rizvi, 2003, p. 230). To control problems with the wax crystallization, chemical additives called pour point depressants are used in the formulation of engine oils. Most pour point depressants are high molecular weight organic polymers that are either alkylated naphthalenes, oligomized alkylphenols, polyalkylmerthacrylates, ethylene-vinyl acetate copolymers or phthalic acid esters.

Depending upon the type of base oil responsiveness to the pour point depressant they can lower the pour point of the lubricating oil by as much as 28–40 °C, although a lowering of the pour point by about 20–30 °C is more common. Pour point depressants have virtually no effect on the temperature at which the wax crystals begin to precipitate or the amount of wax that separates. They essentially act as wax crystal modifiers and function by altering crystal size and shape. Pour point depressants do this by either absorbing onto the surface of the wax crystals as they form or by co-crystallizing with the precipitating wax crystals to prevent wax crystal growth. Both types of mechanisms inhibit lateral crystal growth and lower the temperature at which the rigid lattice structure forms (Rizvi, 2003, p. 230).

13.11.8 Foam Inhibitors

Air can be whipped into the engine oil by the rapid movement of the engine. Air can also be trapped in the engine oil during high pressure release or when the oil pump sucks in air with the engine oil. This can result in a mass of oily froth called foam. The presence of small of amounts of water in the engine oil can further increase the problem. The basic cure for this problem is an engine design that prevents air from being whipped into the engine oil and excludes water entering the system. Even with the best engine designs, this problem cannot be totally eliminated.

A mixture of engine oil and air does not make for a good lubricant, cannot support bearing loads and does not provide adequate cooling to the various parts of the engine. Because it contains air, engine oil

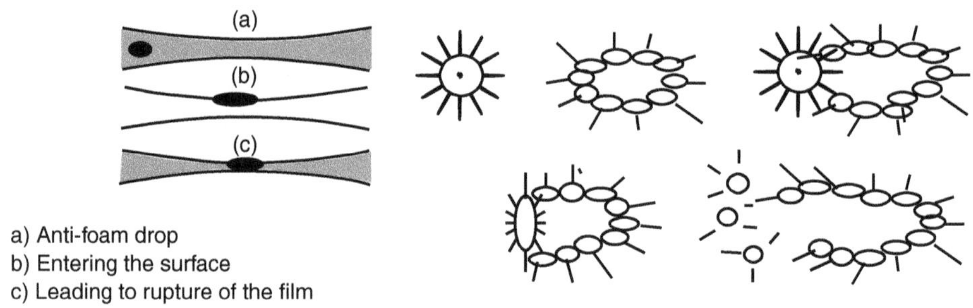

Figure 13.18 Mode of action of anti-foam additives

foam is compressible. This impairs its ability to prevent wear, and can result in the collapse of hydraulic valve lifters, improper functioning of hydraulically actuated electronically controlled unit injector pumps and subsequent poor engine operation.

To prevent foaming, engine oils contain chemical additive systems known as foam inhibitors (Figure 13.18). Foam inhibitors are surface active chemicals that attach themselves to the air bubbles that are formed by the whipping of the engine oil.

The chemical compounds used as foam inhibitors are polymeric materials composed principally of either polysiloxanes (silicones) and polyalkylacrylates. Silicones are the most commonly used anti-foam agents. The type of silicones used in engine oil formulations are in the form of polydimethylsiloxane.

When used at a few parts per million silicones control foam formation by altering the surface tension of the oil and by facilitating the separation of air bubbles from the oil phase. In general, polydimethylsiloxanes have limited solubility in oil, hence they are added as very fine dispersions. Being a surface active chemical and in a fine dispersion in the oil silicones attach themselves to the air bubbles and can either spread or form unstable bridges between the bubbles, which then coalesce into larger bubbles, which rise readily to the surface of the foam layer, where they collapse, thus releasing air.

The use of different additive systems makes possible the heavy duty diesel engine oils that are available in the marketplace today. Not all engine oils contain all of the individual additives that have been discussed in this section, since some additives can fulfil more than one requirement. The use of the proper and carefully balanced additives systems that can be used and the amount of each that is required can only be determined through the use of the various engine sequence tests that simulate the type of service for which the engine oil is intended. Only fully formulated and fortified engine oils can provide the protection required by modern, high precision heavy duty diesel engines.

The most important thing to understand about engine oil additives systems is that no matter what their chemical make-up or the amounts in which they are used in the formulation and the types of base oils, they are only effective for a certain length of time. They can be used up as they perform their various functions and can be only replenished by oil changes at specified intervals. Technology has developed these additives, which are effective until they are depleted, to protect the engine and prolong its life. All an equipment owner has to do to provide the engine with quality lubrication is to use the correct API Service Classification and SAE Viscosity Grade of engine oil that is specified by the OEM and to change the engine oil at either the recommended intervals in the owner's manual or the OEM's Lubricants Service Bulletins or through the use of a qualified engine oil analysis laboratory program.

13.12 Maintaining and Changing Engine Oils

The use of the right diesel engine oil is only the first step in sound oil use practices. Even the best heavy duty diesel engine oil cannot perform its job indefinitely without regular oil changes, nor can it perform with full effectiveness if the oil filter, air filter and other components are not maintained or if the proper oil level is not maintained.

13.12.1 Oil Change Intervals

Nothing lasts forever. Engine oils are no exception. A dominant factor in the control of wear and deposits is the interval at which the engine oil is drained and replaced. Engine oil drain intervals are governed by the 'severity' of use and by the amount and types of contaminants which the engine is exposed to.

In any engine oil the detergent/dispersant, anti-wear, oxidation inhibitors, rust and corrosion inhibitors eventually become depleted, and the engine oil will eventually lose its ability to prevent the formation of deposits on critical engine parts. Depletion of anti-wear additives will result in wear of close tolerance parts. Additionally, contaminants can alter the viscosity of the engine oil, preventing proper lubrication. Soot, dirt and glycol contamination and oxidation by-products can increase the engine oil's viscosity. The presence of fuel dilution or severe shearing of the engine oil's viscosity improvers can cause engine oil to lose viscosity.

Among the critical engine components affected by improperly maintained engine oil are the piston rings, which may become stuck and lose their sealing ability. As a result, engine oil gets past the rings and into the combustion chamber, where it is burned. This is a major cause of oil consumption, as well as a source of contaminants that can reduce the effectiveness of the engine's emission control devices. Also, stuck piston rings cause a loss of power and allow combustion products ('blow-by') to enter the crankcase, further contaminating the engine oil and accelerating deterioration. If such problems are allowed to progress until they become severe, they can cause engine failure and costly repairs. For this reason, it is essential to drain the engine oil before contaminant levels and other stresses on the oil exceed the engine oil's ability to handle them.

The best guide to proper oil drain intervals in a heavy duty diesel engine is the engine manufacturer's published recommendations. Operators of heavy duty diesel engines can exceed the recommended drain intervals and not void any product warranties as long as a rigorous maintenance programme that includes the use of an used oil analysis program, close adherence to maintenance schedules and consistent make-up oil additions is followed according to the engine manufacturers published recommendations. These published recommendations can be found in the operator's manual for the equipment, which should be kept on file, or in the OEM's published Fuel and Lubricant Service Bulletins, which can be obtained from either the OEM's equipment dealers or, in many cases, from the OEM's website.

13.12.2 Used Engine Oil Analysis

Oil analysis is a scientific way to ensure that the engine oil is performing as expected. A regular used engine oil analysis program lets the equipment owner or operator to build a historic database and watch for trends in a variety of areas such as:

- The condition of the engine oil and if it is still serviceable or ready for change.
- The presence of contaminants such as dirt, water, soot or process contamination.
- The presence of wear to identify the possibility of failure or to assist in making critical maintenance decisions.

The benefit of oil analysis is an early warning system that alerts the equipment owner or operator to problems before they can become costly headaches. Used oil samples can tell if water, coolant or fuel is mixing with the engine oil, indicating a head gasket leak or a leaking fuel injector. Spectrochemical analysis detects the presence of metals in the engine oil – unusually high amounts of metals can be a sign of abnormal wear – and the amount and condition of the detergents and anti-wear additives present in the oil. The analysis will also measure the engine oil's viscosity, total base number and the condition of the detergents and anti-wear additive system.

Because laboratories analyse thousands of samples, they can quickly spot 'red flags' and notify the equipment owner or operator immediately. Final reports can either be received by the customer of the laboratory through the use of a web-based electronic system or via the mail. These reports can include trend analyses, warning alerts comments regarding the condition of the engine oil.

A very good oil analysis program provides information on both engine condition and engine oil quality. The oil analysis report will provide information regarding the following:

- If the unit is able to go to the next maintenance or sampling interval without any issues.
- If there is a problem what is it?
- If any excessive wear or contamination taking place.
- If the lubricant suitable for continued use.
- If engine oil drain intervals be extended.
- Any wear trends that may be occurring.
- What corrective actions may or need to be taken.

At the minimum, the laboratory used for the analysis of the heavy duty diesel engine should perform the following tests:

- Viscosity at 100 °C ASTM D-445 to detect an increase in viscosity, which may indicate the presence of suspended solid materials, such as wear particles, contamination and soot, the additions of a higher viscosity oil or engine oil oxidation or a decrease in viscosity, which may indicate contamination from water or fuel, the addition of a lower viscosity oil or shearing of Viscosity Index improvers (multi-grade engine oils).
- Fourier Transform Infrared (FT-IR) spectroscopy to measure degradation by-products (oxidation, nitration, sulfate), percentage soot contamination, the presence of water or glycol contamination.
- Total Base Number ASTM D-4739 and Total Acid Number ASTM D-664 to measure the presence and build-up of acidic components in the engine oil.
- Inductively Coupled Plasma Spectroscopy (ICP) or Atomic Emission Spectroscopy to detect the presence of wear metals, contaminant metals, additive metals.

Used engine oil analysis is a very cheap and effective early warning system. It can detect problems such as trace amounts of coolant and fuel dilution contaminating the engine oil, and these problems once the oil analysis laboratory notifies the equipment owner or operator can be quickly fixed before major engine damage occurs.

Any engine oil analysis program has to be based on two fundamental procedures that often do not get the attention they deserve:

- Making sure the oil sample is taken properly.
- Making sure that the laboratory is provided with all of the necessary information that is needed.

13.12.2.1 Taking a Proper Oil Sample

If an engine oil sample is taken when the oil is being drained, the sample must be taken while the engine is still warm. Catch the oil sample midstream; by waiting that way, heavy metals and other deposits lying at the bottom of the oil pan are not caught with the sample. Samples taken from the bottom of the oil pan do not tell the complete condition of the engine or the engine oil.

Another method is through the use of either an inexpensive vacuum sample pump or a sample valve port mounted in the oil gallery. The oil sample should be taken while the engine is still warm to ensure a representative sample. When taking samples with a vacuum pump, precautions need to be taken so that an accurate sample is retrieved. Make sure the sample hose does not curl and touch the side or bottom of the sump. Try to take samples from the centre of the sump. Always flush the hose out by dumping at least two bottles of oil. It is a good idea to flush the sample bottle out regardless, since they are easily contaminated when left open for any length of time. If samples are taken using a sample valve port it should be placed on return lines before filters where possible.

Samples should be taken the same way every time no matter who takes them and the sample bottles provided by the laboratory should be used, since they are clean and hold the proper amount for testing to be performed. The use of other types of containers, such as rinsed out antifreeze containers, oil containers or even glass jars that held other types of products, can contaminate the engine oil sample. Snap-shot samples or sampling after hearing a noise provide only limited information. Oil analysis result trends have to be followed. If a sample is taken once in a while, there are too many variables between the samples to form much of an idea about the health of the system. Testing at regular intervals, the same way every time and so on lets the operator track the engine oil's progress and performance and allows the operator to see if any corrective action worked. No matter how the sample is taken or who takes it, the owner or operator needs to make sure it is taken the same way every time. If one person takes the sample from the drain valve, the next person from the filter, the next time someone cracks a fitting and the next time it is taken from a test port, the results are nearly meaningless and will not be consistent.

13.12.2.2 Providing the Necessary Information

Besides, taking the sample properly it is important that the label or form that accompanies the bottles into which the oil sample is placed is completely filled out. Without essential information that tells the laboratory precisely what they are looking at, all the high technology analysis in the world will not be worth the report it is in.

Mandatory information includes:

- The engine type
- The engine make and model
- The total miles or hours on the engine
- The unit number or vehicle number
- The date when the sample was taken
- Miles or hours on the engine oil when drained or when the sample was taken
- Amount of make-up oil added between drain intervals
- The brand of engine oil
- The SAE Viscosity Grade of the engine oil
- Comments about any engine problems that may have been experienced or any maintenance work that has been done (e.g. replaced main seal).

A very good oil analysis laboratory has a vast database on different types of engine oils and engines, but it is all meaningless without an exact fix on the 'life story' of the oil being analysed.

Finally, oil drain intervals and equipment condition cannot be determined by only one oil analysis sample. A trend must be developed. At least three oil samples must be taken to establish a trend and samples should be taken at the same oil drain intervals and by the same oil sampling method. Only a series of samples will show a trend. Many factors can influence the final analysis and interpretation of the test results, such as new oil handling methods, storage, sampling methods, oil drain intervals, the amount of time in hours or miles on the engine oil, since it determines the severity of wear and contamination, the amount of make-up oil added, since higher consumption means the sump is always being replenished with fresh oil resulting in a decrease of wear and contamination levels, equipment operation and maintenance conditions and the presence of any contaminants. The OEM's published oil analysis warning limits should be followed or used as guideline. Oil analysis is a maintenance tool and setting up trends on the equipment being analysed is the only way to properly use this tool.

13.13 Diesel Engine Oil Trends

Heavy duty diesel engine oils have historically evolved over time to meet the demands that have been placed upon them by OEM oil service requirements. Currently there are three active API service classifications for diesel engine oils. These API service classifications were developed to help consumers of heavy duty diesel engines select the most appropriate oil for use in their engine. To develop and establish each of these categories the API works with the OEMs, oil marketers, additive companies and testing laboratories to decide on the performance requirements and the engine and laboratory bench tests needed to qualify engine oils within each category. Each of the API service classifications was developed to meet a specific engine design or designs and for use with certain types of diesel fuels. For example, API CI-4 was designed specifically for EGR engines designed to meet the 2002 exhaust emissions standards and burn low sulfur diesel fuel (less than 500 ppm sulfur), while API CJ-4 was designed for engines designed to meet the 2007 exhaust emission standards and burn ultra low sulfur diesel fuel (less than 15 ppm sulfur). The main driving force since 1990 for the development of the entire 'C' numbered 4 service classifications (e.g. CF-4, CG-4 etc.) is the concern over the environmental impact of diesel engine emissions, particularly NO_x. In 1997 the EPA adopted more stringent emissions standards for both NO_x and particulate emissions to reduce NO_x and particulate emissions to 0.2 grammes per brake horsepower hour (g/bhp-h) and 0.01 g/bhp-h respectively by the 2010 model year. As well as the emission regulations, the EPA has set new limits on diesel fuel sulfur levels beginning in 2006 for on-road diesel engines and beginning in 2007 for off-road diesel engines in order to further control emissions. In addition, consumer demand for longer lasting oils and the concern over increased engine and oil sump temperatures due to current and future engine designs to meet the EPA emissions standards have further driven the development of new engine oil service categories. Globally these same trends in reduction in emissions and fuel sulfur, consumer demand for longer lasting oils and the concern over increased engine and oil sump temperatures due to current and future engine designs to meet the emissions standards have further driven the development of new engine oil service categories. Figures 13.19 and 13.20 illustrate these driving forces.

To achieve these emissions limits OEMs are using or will use a combination of cooled Exhaust Gas Return (EGR) at high return rates and exhaust after-treatment devices such as catalytic Diesel Particulate Filters (DPF), NO_x absorbers or traps (LNT), Lean NO_x Catalysts (LNC), Diesel Oxidation Catalysts (DOC) and Selective Catalytic Reduction (SCR). The use of these devices has resulted in a new generation of engine oils that will provide emission control system durability, prevent catalyst poisoning and particulate filter blocking while still offering optimum protection for control of piston deposits, oxidative thickening, oil consumption, high temperature stability, soot handling properties, foaming and viscosity loss due to shearing.

13.14 Engine Design Technologies and Strategies Used to Control Emissions

The need to reduce NO_x and particulate emissions has resulted in a careful balancing act being performed by heavy duty diesel OEMs. This balancing act results in trade-offs between these two controlled pollutants. Generally, any engine design used to improve the combustion process will reduce particulates. Unfortunately, this can lead to very high peak flame temperatures, which generate

(a)

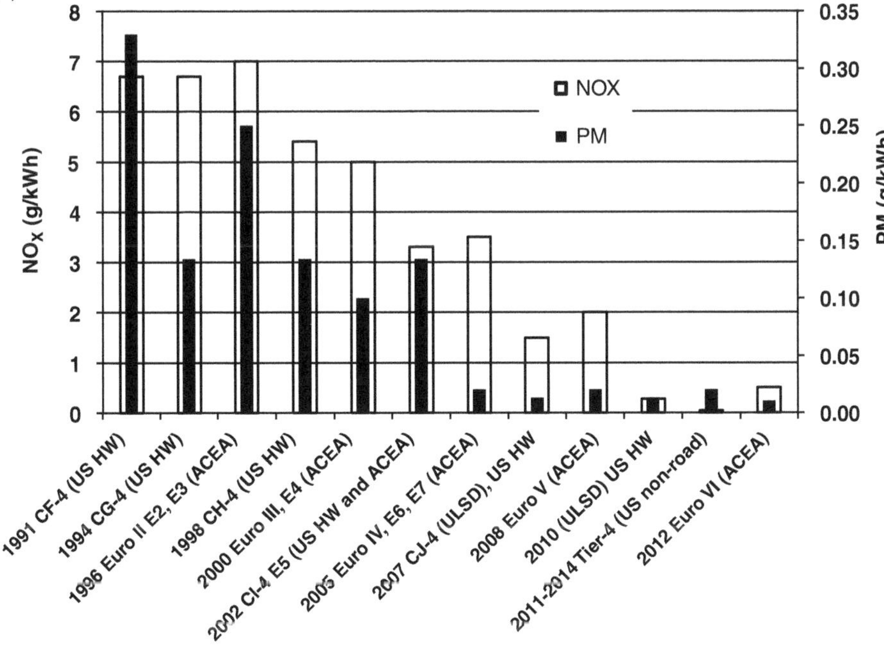

(b)

	1996	1997	1998	1999	2000	2001	2002	2003	2004	2005	2006	2007	2008	2009	2010	2011	2012
Bangladesh							5000										
Cambodia					2500					1500							
Hong Kong, China		500					50					10					
India (nationwide)	5000				2500							500			350		
India (metro cities)	5000				2500		500			350					50		
Indonesia	5000										2000				350		
Japan					500					50		10					
Malaysia	5000		3000								500						50
Pakistan	10000						7000										
Philippines	5000					2000							500				
PRC (nationwide)	5000						2000			500							
PRC (Beijing)	5000						2000		500	350			50				
Singapore	3000					500					50						
South Korea	500							430	100		30		15				
Sri Lanka	10000							5000			500						
Taipei, China	3000				500		350		100				50				
Thailand	2500					500			350		150				50		
Vietnam	10000											500					
European Union							500			50			10				
United States	500										15						

Figure 13.19 (a) Challenge of emission compliance; (b) diesel fuel sulfur levels: current (as on 27 March 2008) and proposed sulfur levels in diesel in Asia, the European Union and the USA

higher levels of NO_x. Conversely, anything done that affect the combustion process can bring down both the peak flame temperature and the generation of NO_x, but at the same time directly increases particulate emissions. The techniques that have been used and their effects on NO_x and particulate emissions are illustrated in Table 13.5.

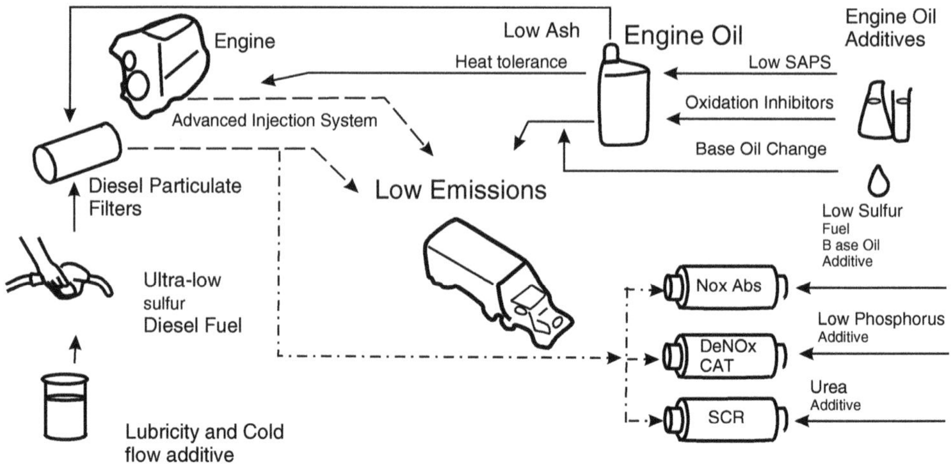

Figure 13.20 Changes in fuels, lubricants, additives and after-treatment to achieve heavy duty diesel emissions targets

Table 13.5 NO_x vs. particulate emissions

Reduce Particulates – Aggravate NO_x	Reduce NO_x – Aggravate Particulates
• Improved combustion – Inlet air swirl – High pressure injection – Advanced timing – Reduced crevice volume	• Less effective combustion – Reduced injection timing
• Higher combustion peak temperatures.	• Lower combustion peak temperatures.

With a current focus on the reduction of NO_x and particulates emissions for on-road and off-road diesel engines, the different heavy duty diesel OEMs have used or are planning to use the following engine design technologies:

- High Pressure Common Rail (HPCR) Fuel System
- Combustion optimization (retarded injection timing, swirl designs)
- Advanced Turbocharging
  - Variable Geometry Turbocharging
  - Series Turbocharging
- Exhaust Gas Recirculation (EGR)
- Advanced Combustion Emissions Reduction Technology (ACERT)
- Crankcase Ventilation
  - Closed Crankcase Filters (CCF) (also known as open system)
  - Closed Crankcase Ventilation (CCV)
- Exhaust after-treatment
  - Diesel Oxidation Catalysts (DOC) (up-stream of DPF)
  - Diesel Particulate Filters (DPF)

Table 13.6 Emission design strategies

	2002/2004	2005/2006/2007/2008	2010+
North America	EGR	EGR + DPF/DOC/CCV ACERT/CGI + DPF/CCV	EGR + DPF/DOC + SCR EGR + DPF/NO_x Absorbers/SCR (Cummins) MEGR + DPF/DOC/NOx-Absorbers (Navistar)
European Union	SCR	SCR SCR + DPF (2008)	EGR + DPF/DOC + SCR
Japan	EGR + DPF/DOC	EGR + DPF/DOC + SCR	EGR + DPF/DOC + SCR

  - NO_x Absorbers (Navistar, Cummins)
  - Selective Catalytic Reduction (SCR)
- On-Board Diagnostics (OBD)

The emission design strategies used in North America (United States and Canada), the European Union and Japan by heavy duty diesel engine OEMs since 2002 to meet emissions requirements for NO_x and PM emissions are summarized in Table 13.6.

13.14.1 High Pressure Common Rail (HPCR) Fuel System

The fuel injection system is the heart of the engine. It must consistently deliver precise quantities of fuel at high pressure to the combustion chamber. The flow rate is similar to that of a garden hose, but this flow passes through holes of size of a human hair for a period of the order of a millisecond. Diesel engines previously had, and many still have, unit injectors, which are individual piston pumps combined with a nozzle. These can be mechanically driven by a camshaft, either directly or through rocker arms and pushrods, or they can be hydraulically driven. Less expensive diesel engines tended to have pump-line nozzle injection systems, in which one pumping element for each cylinder is contained in a block and actuated by a camshaft gear driven from the engine's crankshaft, with individual lines from each pumping element to each injection nozzle. Unit pumps were also used in which the individual pumping elements were on the engine's camshaft, thereby avoiding a separate box with a camshaft, with fuel lines connecting each pumping element to an injection nozzle. The least expensive diesel engines used rotary injection pumps. This system shared the pumping elements, and a distributor head connected each injection line to the pumping elements at the appropriate time (Xinqun, Deu and Winsor, 2010).

Today's low emission on-road and off-road diesel engines employ the use of electronic injection systems that include electronically controlled rotary pumps and electronic unit injectors and unit pumps. The type of injection system that is becoming the most commonly used is the High Pressure Common Rail (HPCR) injection system (Figures 13.21, 13.22, 13.23 and 13.24). HPCR rail injection is an advanced high pressure fuel pump technology, driven from the engine's crankshaft in a controlled manner, to maintain a desired fuel pressure in a pressure vessel, which is called the 'rail'. This design allows the use of higher injection pressures. The use of a common rail from which the injectors are directly fed allows the pressures and injection timing to be precisely controlled by electronic means. By directly feeding the injectors from a single pump, the electronic system optimizes the amount of fuel pressurized prior to injection. HPCR can provide multiple injections and post injection. The extremely high pressures (up to 241 MPa) allow for finer atomization of the fuel droplets, which means a far more homogenous mixture with air in the cylinder. This results in more complete combustion and reduced exhaust emissions. These injectors can be either hydraulically electronic controlled unit injectors (HEUI) or mechanically electronic control unit injectors (MEUI).

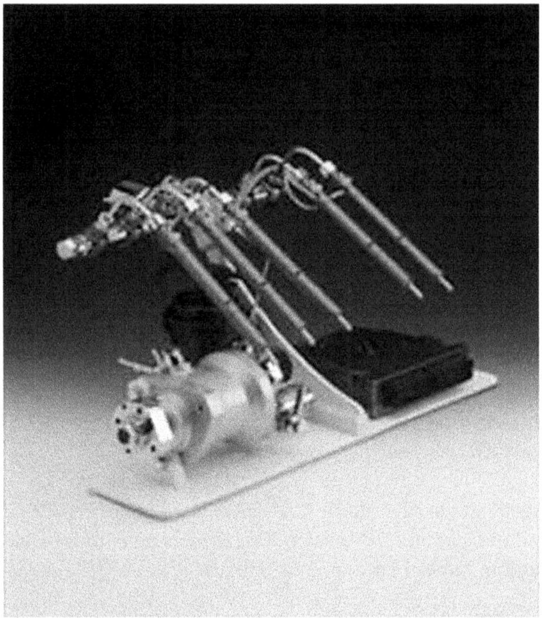

Figure 13.21 High pressure common rail system for cars

13.14.2 *Combustion Optimization*

The configuration of the combustion chamber can increase the efficiency and effectiveness of the combustion process. Many OEMs use a swirl geometry design and variable valve timing to ensure the air and vaporized fuel are thoroughly mixed. This lowers the combustion temperature and thereby prevents

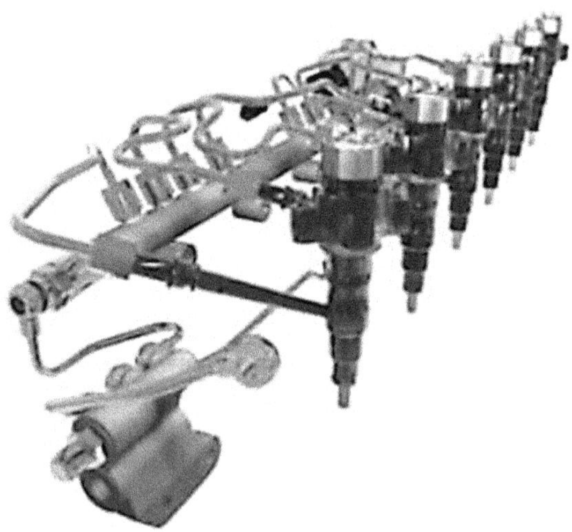

Figure 13.22 High pressure common rail system for heavy duty vehicles

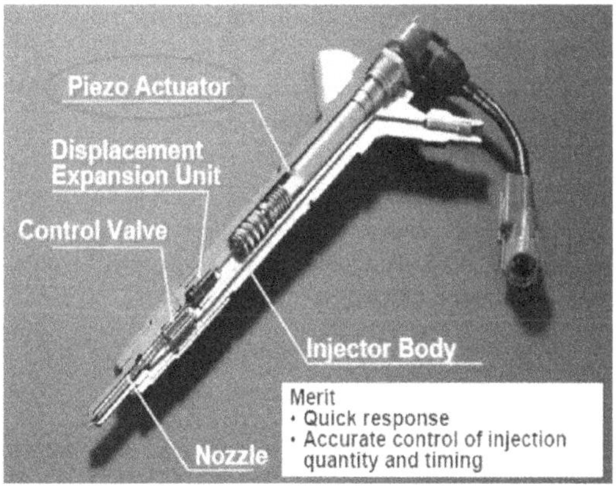

Figure 13.23 Photograph of a sectioned common rail injector

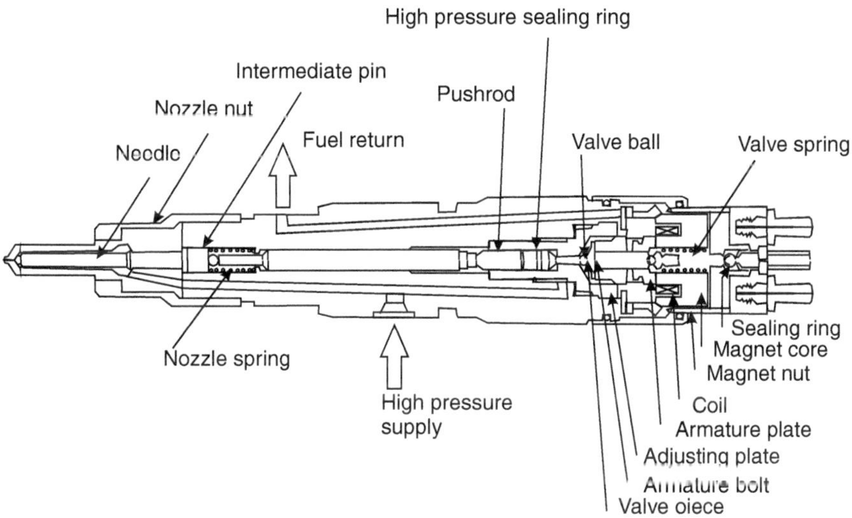

Figure 13.24 Section of a common rail injector showing details

the formation of NO_x. The shape of the combustion bowl at the top of the piston helps to increase turbulence and air fuel mixing, thereby burning all available fuel during combustion.

The addition by some OEMs of valve guide seals limits particulates by reducing oil consumption. Directed top liner cooling reduces oil consumption and enhances combustion efficiency by reducing wear in the area around the top ring and improving ring performance.

In addition to the above mentioned strategies, OEMs of heavy duty diesel employ the following schemes:

- Retarded fuel injection timing lowers peak flame combustion temperature, thus reducing NO_x formation by displacing the combustion event until later in the combustion stroke.
- Variable Injection Timing, which is an innovation in new fuel control systems, consists of an electronically controlled high pressure fuel injector that releases the precise amount of diesel fuel at the moment of maximum compression when the piston reaches the top of the cylinder.

- Rich engine operation is the use of an air/fuel mixture with more fuel than air versus what would occur in a natural (stoichiometric) burning condition.
- In traditional diesel combustion, the burning occurs in the rich regions of the spray, resulting in high temperatures and high NO_x emissions. Pre-mixed Compression Ignition uses multiple fuel injections of the diesel fuel to lower combustion temperatures and reduce NO_x emissions.
- The air-to-air charge air cooler is a device that is designed to reduce the temperature of the intake air compressed by the turbocharger. The air-to-air cooler reduces the inlet charge air temperature by using ambient air as the cooling medium. This allows higher specific power output, with better fuel economy, lower internal stresses and lower NO_x formation.

13.14.3 Advanced Turbocharging

Besides providing fuel to the combustion chamber it is important that the necessary amount of air for combustion is also provided. This air flow is provided by the use of a turbocharger, which consists of a turbine wheel in the exhaust gas stream that drives and turns a compressor wheel in the intake air stream. The compressor wheel pressurizes the air forcing more air into the cylinders, allowing the engine to induct more air, resulting in the engine burning the diesel fuel more efficiently and increasing the engine's power density. Today's low emission engines use turbocharging and cooling of the air compressed by the turbocharger. With turbocharging, a turbine waste gate (a valve that opens to allow exhaust to bypass the turbine wheel) can be used to limit the boost at high speeds. Presently, there are two different types of charging being used in modern low emissions engines. These types are Variable Geometry Turbocharging (VGT) and Series Turbocharging.

Variable geometry turbocharging (VGT) is used to control the boost across the engine's operating range. There are two common types of VGT: the swing vane and the sliding vane (Figures 13.25 and 13.26). The moving parts change the velocity and direction of the air flow entering the turbine wheel in order to change the work extracted from the exhaust. There are two passages in the turbine housing. Exhaust gas enters the turbine through the inner passage at low engine speed and load and through both inner and outer passages at high engine speed and load. The opening of the outer passage is controlled by a simple actuator that is similar to a waste gate actuator (Xinqun, Deu and Winsor, 2010, pp. 6, 7).

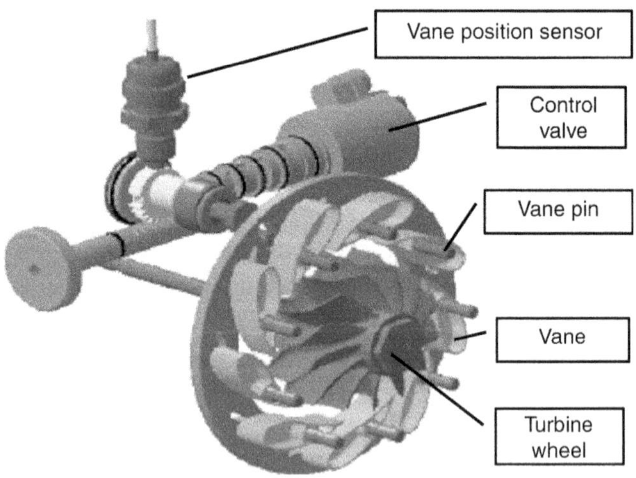

Figure 13.25 Swing vane type variable-geometry turbocharger

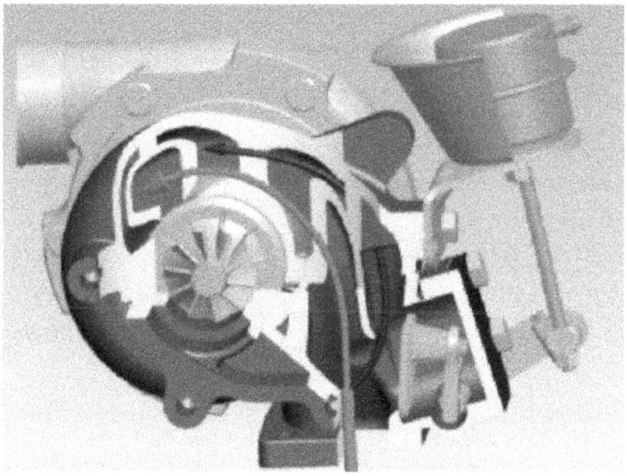

Figure 13.26 Sliding vane type variable-geometry turbocharger

Another form of turbocharging that can be used is Series Turbocharging. In Series Turbocharging both a low pressure and high pressure turbocharger are used to increase engine output and altitude capability. Filtered ambient air is drawn into the low pressure turbocharger. The pressurized air is then passed to the high pressure turbocharger compressor, cooled and introduced into the intake manifold. These two compressor stages enable much higher intake air pressure than that which is possible with a single turbocharger.

Turbocharging increases air flow and allows more fuel flow to increase engine power density. In addition, fuel efficiency is improved because the engine can be smaller and have less friction, and because the power needed to compress the cool inlet air is obtained by the expanding hot exhaust, which is an efficient process. The extra air provided by turbocharging reduces smoke emissions and allows the use of EGR to be added to the cylinder while maintaining adequate air to the cylinder (Xinqun, Deu and Winsor, 2010, p. 7).

13.14.4 Exhaust Gas Recirculation (EGR)

The use of Exhaust Gas Recirculation has been in use to reduce emissions in on-road diesel engines since 2002 and off-road engines since 2008 in the United States. Globally, its employment to meet emissions requirements is growing. EGR re-introduces a portion of the exhaust normally emitted by the vehicle and uses it as part of the fresh air intake air to provide inert mass in the cylinder to reduce peak flame combustion temperature, and thereby reduce NO_x emissions. There are many ways of adding EGR. The simplest is internal EGR, whereby some of the exhaust is retained in the cylinder or leaked into the fresh air charge while it is being inducted. However, internal EGR is not cooled and, therefore, the NO_x reduction is relatively small. External EGR allows the recirculated exhaust gas to be cooled as it flows through a pipe, or more aggressively by a heat exchanger using engine coolant or cooling air. Cooled EGR is much more effective at reducing NO_x, and larger amounts of EGR can be used without displacing as much air. Therefore, cooled EGR has become a popular method of NO_x control (Xinqun, Deu and Winsor, 2010, pp. 10, 11).

In cooled EGR the exhaust gases are cooled and then diverted through the intake manifold into the cylinders along with filtered air and air from the outside of the vehicle by using of a variable geometry turbocharger, so as to balance the fresh air/EGR ratio (Figure 13.27).

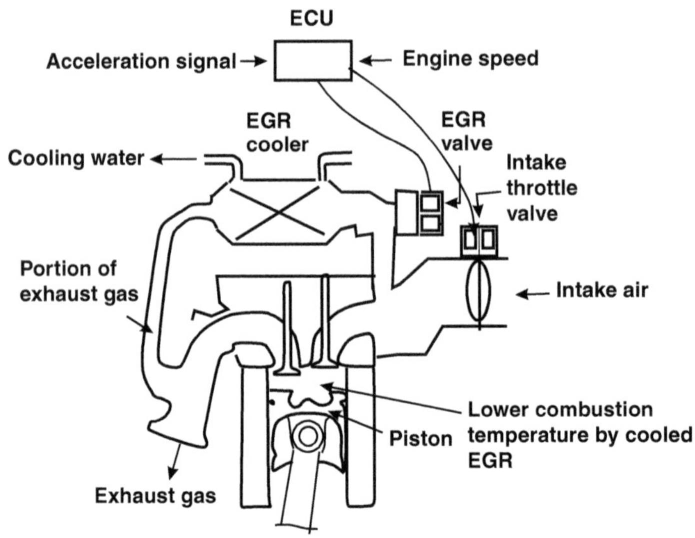

Figure 13.27 EGR technology

The purpose of cooled EGR is to lower the peak flame temperature inside the combustion chamber. The EGR device recycles as much as 40% of the engine exhaust back into the engine air intake and uses the engine's coolant to cool the recycled exhaust gases. This raises the temperature of the coolant and when it flows back into the engine it can increase the engine's operating temperature. The oxygen depleted exhaust gas is mixed into the fresh air that enters the combustion chamber. This dilutes the oxygen content of the intake air in the combustion chamber, resulting in a lower peak combustion temperature. A valve-controlled tap is placed in the high pressure exhaust area upstream of the turbocharger. Either a variable geometry turbocharger or a series turbocharger is used to balance the fresh air/EGR ratio. Once the valve is opened, filtered output air from the compressor side of the turbocharger is ducted through the after-cooler. The EGR valve is modulated to control flow of the exhaust gas by ducting it through the EGR cooler and into the pressurized inlet air as it leaves the inlet cooler. The cooled exhaust gas is mixed with the incoming inlet air as it leaves the inlet air cooler in the EGR mixer, then the mixture of fresh inlet air and exhaust gas is drawn into the engine on the intake stroke.

13.14.5 Advanced Combustion Emissions Reduction Technology

Advanced Combustion Emissions Reduction Technology (ACERT) is a technology that has been used by Caterpillar since 2007 in it on-road and off-road diesel engines to meet emissions regulations. It is a system that reduces emissions at the point of combustion. This technology includes a combination of four existing engine systems:

- Improved air management
- Advanced fuel injection technology
- Advanced electronic controls
- After-treatment.

The air management system uses VGT to force cool clean air into the combustion chamber by using electronic control mechanisms and variable valve timing and actuation to control the air volume required at various loads and speeds to provide complete combustion of the diesel fuel.

The advanced fuel management system, which is a combination of hydraulic and mechanical electronic injection systems, provides for a multiple injection process. The small, multiple shots of fuel into the combustion chamber at the appropriate time result in fuel economy and lower emissions. The engine's Electronic Control Module (ECM) determines the amount of diesel fuel injected and the timing required. The advanced electronics integrates the systems to achieve reduced emissions.

The after-treatment uses diesel oxidation catalysts to remove particulates, hydrocarbons and carbon monoxide.

In conjunction with ACERT, Caterpillar has also employed the use of Clean Gas Induction (CGI) to meet stringent particulate emission standards since 2007. CGI employs the use of remote EGR, a closed crankcase ventilation system and diesel particulate filter system with active regeneration. Further reductions in NO_x are achieved by the CGI system. CGI draws clean inert soot-free gas from downstream of the particulate filter and then puts the clean gas into the intake system. This clean gas does not induce the engine wear that cooled EGR produces and the low intake manifold gas temperature of the CGI contributes to lower NO_x emissions.

13.14.6 Crankcase Ventilation

During normal operation, a small percentage of the engine oil can travel past the rings and then burns in the combustion chamber. The remainder of the consumed oil is lost through evaporation by being emitted through a crankcase ventilation tube and is not combusted. In addition, vapours generated in the crankcase can be discharged into the engine's intake system, usually via the intake manifold, where they are burned as part of the combustion process rather than being discharged into the atmosphere.

If an engine oil that contains a sulfur level greater than 0.4% (once sulfur is in the exhaust stream) it can inhibit the effectiveness of the diesel particulate filters and diesel oxidation catalysts by poisoning the catalysts. This poisoning can increase the conversion of sulfur oxides to sulfates, which increases particulate emissions and accumulation of particulate material. Accumulation of particulate material can lead to reduced engine performance, due to increased back pressure, and ultimately to failure of the after-treatment devices. Since 2007 the use of Closed Crankcase Ventilation has been employed to recover the evaporated oil and carry it through the exhaust stream.

There are two types of crankcase ventilation that can be used:

- Closed Crankcase Filters (CCF) (aka as open system)
- Closed Crankcase Ventilation (CCV).

Closed Crankcase Filters (CCF) are used to reduce emissions from the crankcase breather tubes in turbocharged after-cooled diesel engines by using a multi-stage coalescing filter to collect, coalesce and return the emitted engine oil to the engine's sump (Figure 13.28). Closed Crankcase Ventilation (CCV) removes harmful vapours generated in the crankcase. These vapours are discharged into the engine's intake system (usually via the intake manifold), where they are burned as part of the combustion process rather than being discharged into the atmosphere.

13.14.7 Exhaust After-Treatment

Since 2007 exhaust after-treatment has been employed globally to further control NO_x and particulate emissions. The exhaust after-treatment devices being employed consist of:

- Diesel Oxidation Catalysts (DOC) (upstream of DPF)
- Diesel Particulate Filters (DPF)
- NO_x Absorbers (Navistar, Cummins)

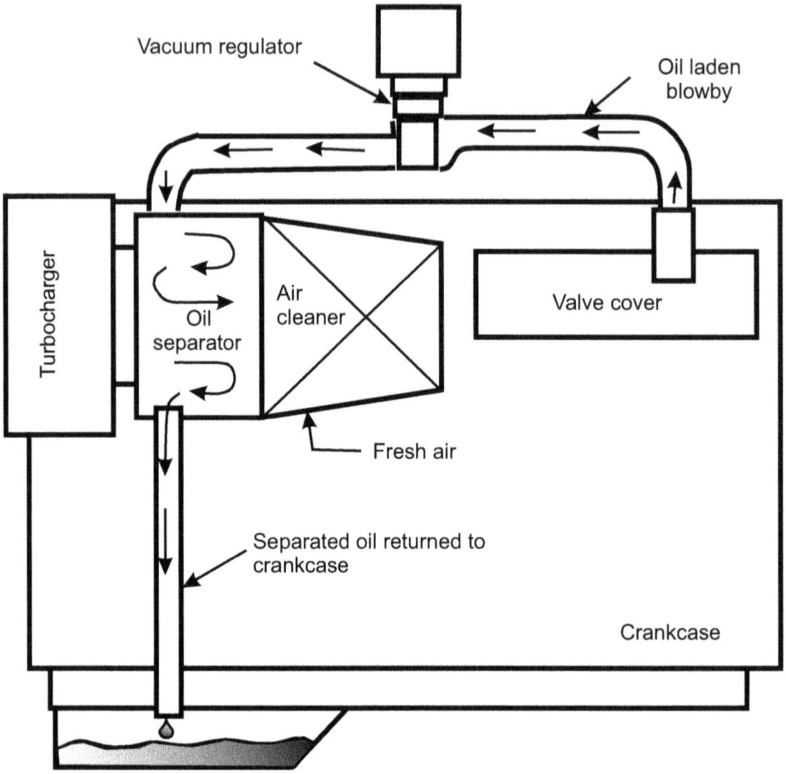

Figure 13.28 Separation of oil from the blow-by gases and recirculating the gases to the engine inlet

- Selective Catalytic Reduction (SCR)
- Diesel Oxidation Catalysts (DOC)

The primary function of the diesel oxidation catalysts (DOC) is to oxidize hydrocarbons delivered by the hydrocarbon injector (HCI) to raise the exhaust temperature during DPF regeneration (Figure 13.29). DOC consists of a catalytic coating of platinum or palladium on a honeycomb substrate. DOC are used to remove organic particulate matter (which come from unburned fuel and the engine oil), hydrocarbons and carbon monoxide from diesel exhaust. Hydrocarbons are injected upstream of the DOC and oxidized into carbon dioxide and water by the DOC. The fuel energy released heats the DOC and raises the exhaust gas temperature. The catalysts help to promote oxidation of organic particulates, hydrocarbons and carbon

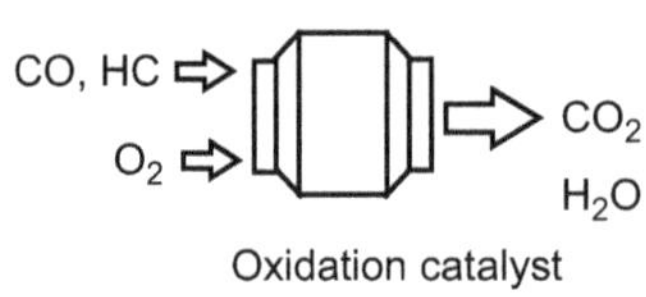

Figure 13.29 Diesel oxidation catalyst

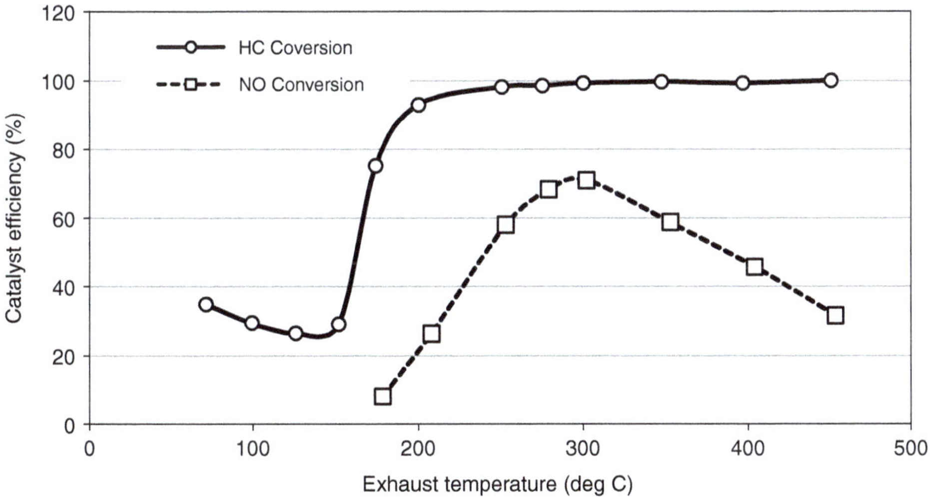

Figure 13.30 Typical DOC hydrocarbon conversion light-off curve and NO to NO_2 conversion efficiencies

monoxide, thereby converting them to carbon dioxide and water. The DOC requires a minimal exhaust temperature to light-off and be active, as shown in Figure 13.30. Above the DOC light-off temperature, the DOC stays hot and hydrocarbons are oxidized at greater than 90% efficiencies, which ensures very little hydrocarbon slippage. High efficiency and robustness with varying oxygen levels and flow rates enable precise DOC outlet temperature control.

Under normal operating conditions, the DOC cleans up carbon dioxide, hydrocarbon and soluble organic factions of PM emitted from the engine. The DOC also oxidizes NO to NO_2 to promote passive soot oxidation. The DOC operates in a passive-only mode without active regeneration, thus making them less efficient at particulate matter reduction than DPF. Regeneration with a DOC requires a controlled amount of diesel fuel to be injected upstream. To reduce heat loss, the DOC is integrated with the DPF (Xinqun, Deu and Winsor, 2010, pp. 10, 11).

13.14.7.1 Diesel Particulate Filters (DPF)

Diesel particulate filters (DPF) are after-treatment devices that work in conjunction with the Diesel Oxidation Catalyst to remove NO_x and particulate matter from burned diesel fuel. DPF, which resemble a large muffler, contain a porous honeycomb structure, usually made of a high temperature ceramic structure or a densely packed ceramic and metal fibres, that physically capture particulates in the diesel exhaust as the exhaust passes through it, so preventing their discharge from the exhaust pipe (Figures 13.31, 13.32 and 13.33).

Collected particulates, such as soot, as they build up over time are removed from the filter by oxidizing the soot (known as regeneration). Regeneration can be done either actively or passively, and involves either the use of electrical heaters, passive heat from the exhaust or the injection of a small amount of diesel fuel into the exhaust stream into the filters to burn off the collected particulate matter completely.

Active regeneration refers to the injection of extra fuel to raise temperatures to accomplish diesel particulate filter regeneration. Active regeneration is used when the engine's exhaust temperature is not hot enough during the work cycle to burn off the soot being collected. Active regeneration requires a

Figure 13.31 A sectional view of a diesel particular filter combined with exhaust silencer (Galligan, 2005, p. 22)

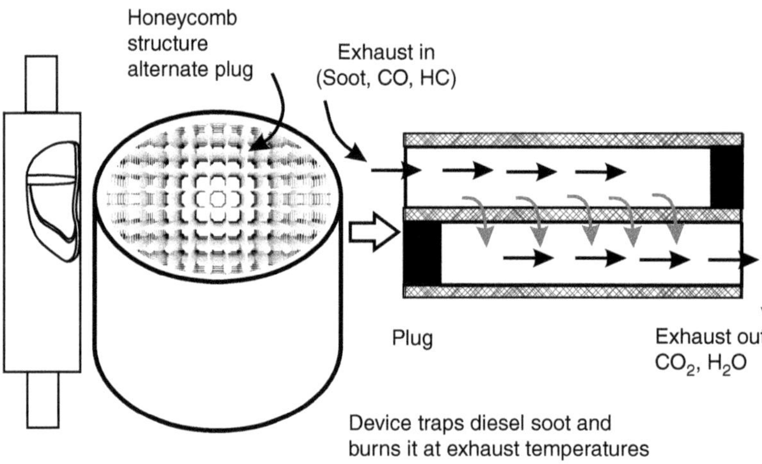

Figure 13.32 Functioning of a diesel particulate filter (Galligan, 2005, p. 28)

means of raising the exhaust temperatures above 550 °C (1022 °F). Two primary heating technologies for active regeneration are exhaust diesel fuel burners and DOC. The diesel burner can be integrated either with the engine or with the diesel particulate filter. A major advantage of a burner is its ability to regenerate the DPF at idle, even during cold ambient temperature conditions (Xinqun, Deu and Winsor, 2010, p.10)

Fuel required for DPF active regeneration can either be introduced by a late post injection during the power stroke or by an exhaust fuel doser. If fuel is introduced by an engine late post injection, then the hydrocarbon is fully evaporated and is well mixed with exhaust gas. The DOC/DPF can therefore be close-coupled with the turbocharger to further reduce heat loss. Since fuel is injected late in the combustion cycle, frequent post injections increase the oil dilution with fuel, which can lead to

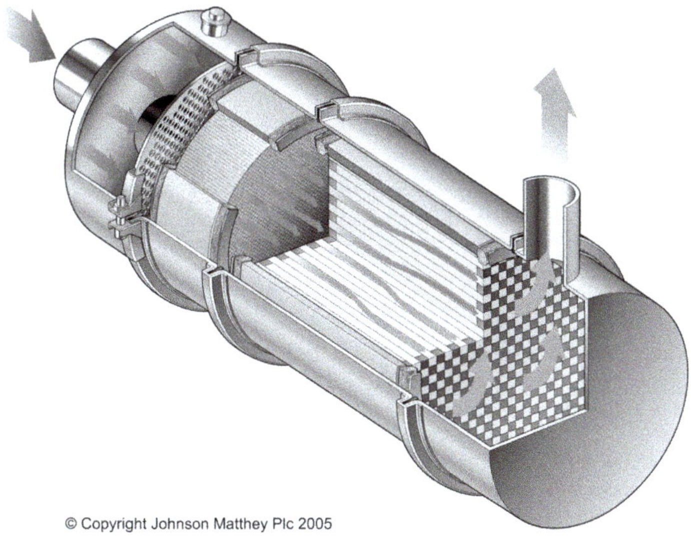

Figure 13.33 Diesel particulate filter with a diesel oxidation catalyst (Reproduced by permission of Johnson Matthey, USA.)

shorter oil change intervals. An example of a fuel dosing system in the exhaust stream is shown in Figure 13.34. Sufficient mixing distance is required to ensure fuel evaporation and uniform hydrocarbon distribution before the exhaust gas enters the DOC (Xinqun, Deu and Winsor et al, 2010, p. 12).

Passive regeneration occurs when the soot in the DPF chemically reacts to form carbon dioxide during the normal work cycle when the exhaust temperatures are sufficiently hot. Passive regeneration occurs at normal engine operation above 300 °C (572 °F) without any special engine control intervention. Passive regeneration is preferable to active because it does not consume extra fuel.

Outboard diagnostics are used to monitor the DPF's condition and manage regeneration. If regeneration is needed and the vehicle is at idle, on-board fuel reformers that convert diesel fuel to a more hydrogen rich, hotter burning fuel will be activated by on-board diagnostics that signal the operator to initiate regeneration.

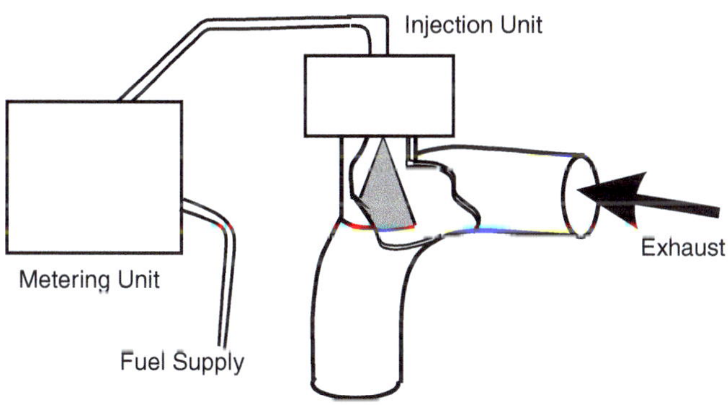

Figure 13.34 Fuel dosing system

Any remaining residue and ash left from the regeneration process is blown against the exhaust flow and into a container trap for disposal. These traps need to be cleaned out periodically to keep the diesel particulate filters from clogging. The United States EPA requires that DPFs have the capability go at least 241 395 km (150 000 miles) before they need cleaning. Most OEMs have designed their systems to go 321 860–643 720 km (200 000–400 000 miles) depending upon the engine's duty cycle and service application before the traps need to be cleaned. The main types of diesel particulate filters are:

- Catalysed DPFs (cDPF)
- Fuel additive-catalysed DPFs
- Continuously Regenerating Traps.

Catalysed DPFs (cDPF)

Catalysed DPFs have the surface of the inner walls of the honeycomb filter coated with a catalyst such as platinum, palladium or rhodium to promote oxidation of the particulates. These catalysts help reduce the temperature that is required to burn off the particulates and soot.

Fuel Additive-catalysed DPFs

In fuel additive-catalysed DPF systems, an additive is held in a separate tank from which it is dispersed into the diesel fuel. This additive is trapped with the particulates and lowers the temperature required for burning soot. The ash remains from the additive remain in the filter after the particulates have burnt off, thereby contributing to ash build-up in the filter.

Continuously Regenerating Traps (CRT)

CRT combine DOC upstream of the DPF. The DOC converts NO_x emission to nitrogen dioxide before the exhaust enters the DPF. The nitrogen dioxide lowers the temperature required to oxidize/burn off the particulates allowing the CRT to continuously clear itself of trapped particles.

13.14.7.2 NO_x Absorbers

Depending upon the engine's NO_x levels, several NO_x after-treatment technologies are being employed for use by some OEMs. NO_x absorbers are after-treatment technologies that use a base metal oxide and a precious metal compound as a catalyst to transform NO_x to nitrogen gas and water vapour. The technologies used are:

- Lean NO_x Traps (LNT),
- Lean NO_x Catalysts (LNC).

Lean NO_x Traps (LNT)

Lean NO_x Traps (LNT), which are also known as NO_x adsorbers or NO_x Storage Reduction Catalysts (NSRC), are one of the latest advances in removing NO_x. LNT remove lean exhaust NO_x by catalytically oxidizing it to nitrogen dioxide and storing it in an adjacent trapping site as a nitrate. LNT remove nitrogen oxides from emissions in two stages. During the first stage precious metal catalysts convert nitrous oxides to nitrogen dioxide and store the nitrogen dioxide in an alkaline earth oxide (usually barium oxide) as a nitrate salt. The stored nitrogen oxides are released by creating a temporary rich exhaust condition and then catalytically reduced to nitrogen gas using unburned fuel as the reducing agent. To operate effectively the NO_x adsorber must remain stable for extended periods, during which time the exhaust environment modulates between rich and lean conditions.

Lean NO_x Catalysts (LNC)

Lean NO_x Catalysts (LNC), which are also known as $DeNO_x$ catalysts, use noble metal catalysts of platinum and palladium, or a mixture, in the presence of hydrocarbons to reduce nitrogen oxides into harmless nitrogen gas. They can be passive or active. Passive LNC uses hydrocarbons from the exhaust stream. Active LNC requires enrichment of the exhaust stream by the addition of fuel injected into the exhaust stream. Lean NO_x Catalysts are designed to reduce nitrogen oxides from diesel or spark-ignited engine exhaust gases under net oxidizing conditions (e.g. in the presence of abundant amounts of oxygen). They employ a ceramic honeycomb monolithic substrate featuring many small parallel channels running in the axial direction.

13.14.7.3 Selective Catalytic Reduction (SCR)

United States EPA 2010 emissions regulations required a 90% reduction in NO_x emissions for on-road heavy duty diesel engines. To meet this further reduction in emissions, major North American and European OEMs are committed to the use of Selective Catalytic Reduction (SCR) with urea treatment (Figure 13.35), not only to meet emissions requirements in the United States but also those in the European Union. Navistar and MAN on some of their on-road diesel engines are the only exceptions. These two OEMs employ the use of Massive EGR to meet the 2010 emission requirements. SCR is a bolt on device in addition to DPF and DOC. SCR has to be tamper-proof in accordance with EPA rules.

SCR is an after-treatment technology for reducing NO_x by injecting a finely atomized Diesel Exhaust Fluid (DEF) from a storage tank into the exhaust stream before it enters a catalyst system (Figure 13.35). The system uses a DOC to convert NO_x to nitrogen dioxide and remove hydrocarbons at reasonably high temperature. The DEF solution is an aqueous solution of 32.5% urea and 67.5% deionized water that is injected into the exhaust stream and thermally decomposes into ammonia and isocyanic acid (HNCO). The isocyanic acid is further hydrolysed into ammonia and carbon dioxide through the use of a hydrolysis catalyst to generate ammonia. The SCR catalyst system, which contains titanium, vanadium or zeolite, uses the ammonia as a reducing agent to convert NO_x to nitrogen gas and water, both which are clean components of air. In the final stage an oxidation catalyst removes any unreacted ammonia. SCR systems require temperature window of 150–500 °C to operate properly.

Diesel Exhaust Fluid (DEF)

As previously mentioned, Diesel Exhaust Fluid (DEF) is an aqueous solution of 32.5% high purity urea in 67.5% deionized water. It was originally developed for use to control NO_x emissions in Europe in approximately 2005. In the European market it is sold under the trade name of AdBlue. Urea is the carrier

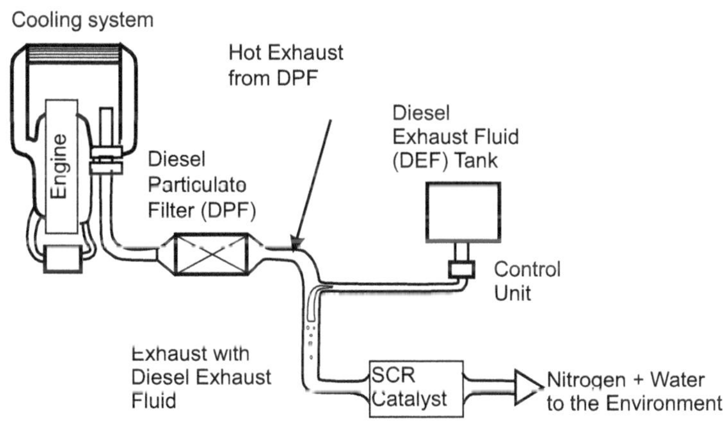

Figure 13.35 Schematic of SCR and set-up

for the ammonia needed to reduce nitrogen oxide (NO_x) emissions from vehicles into nitrogen and water through the following chemical reaction:

$$(NH_2)_2CO + H_2O \rightarrow CO_2 + 2NH_3$$

$$2NH_3 + NO + NO_2 \rightarrow 2N_2 + 3H_2O$$

DEF is nontoxic, biodegradable and is not classified as being hazardous by the United States Department of Transportation, the United States EPA and the European Union. DEF solution freezes at −11 °C. Therefore, the DEF tank must be designed with an active heating function. The DEF storage tank is typically thawed by engine coolant due to the high energy requirement, while DEF transfer lines and the DEF pump are commonly thawed by electrical power. Furthermore, the DEF solution decomposes above 65 °C, which creates additional vehicle integration, storage and transport challenges. The DEF dosing system includes a supply pump and an injector. Pressurized DEF is metered and injected into the exhaust as a fine spray. Injection at low exhaust temperatures is avoided, since DEF cannot be fully decomposed into ammonia and the catalyst is not active (Xinqun, Deu and Winsor, 2010, p. 16).

The purity of the DEF is essential to vehicle performance. DEF quality levels have been set by the:

- SAE AUS specifications
- ISO 22141-1 (Table 13.7).

In addition to these quality levels, the API has set up licensing programme for DEF. It is similar to engine oil licensing programme. The API will conduct random sampling of DEF to monitor quality. The license is valid for one year and renewable. Additional information on the API programme can be found at www.apidef.org

The DEF tank must be refilled periodically, as DEF is typically consumed at rate of two to four gallons for every 100 gallons of diesel fuel the vehicle consumes. As an example, an over-the-road truck that does

Table 13.7 Composition of DEF urea solution

Specification	Limits	Properties/Behaviour affected in engine
Urea (wt-%)	31.8–33.2	
Density at 77 °F (lb/gal.)	9.05–9.09	
Refractive Index at 77 °F	1.3814–1.3843	
Alkalinity (as ammonia) (wt-%)	max. 0.2	Shelf life
Insolubles (wt-%)	max. 0.002	Clog DEF Injection Nozzles
Calcium (wt-%)	max. 0.00005	
Magnesium (wt-%)	max. 0.00005	
Aldehydes (wt-%)	max. 0.0005	Lacquer
Sodium (wt-%)	max. 0.00005	Poison SCR Catalyst
Potassium (wt-%)	max. 0.00005	
Phosphates (wt-%)	max. 0.00005	
Biuret (wt-%)	max. 0.3	
Aluminium (wt-%)	max. 0.00005	
Iron (wt-%)	max. 0.00005	
Copper (wt-%)	max. 0.00002	
Zinc (wt-%)	max. 0.00002	
Chromium (wt-%)	max. 0.00002	
Nickel (wt-%)	max. 0.00002	

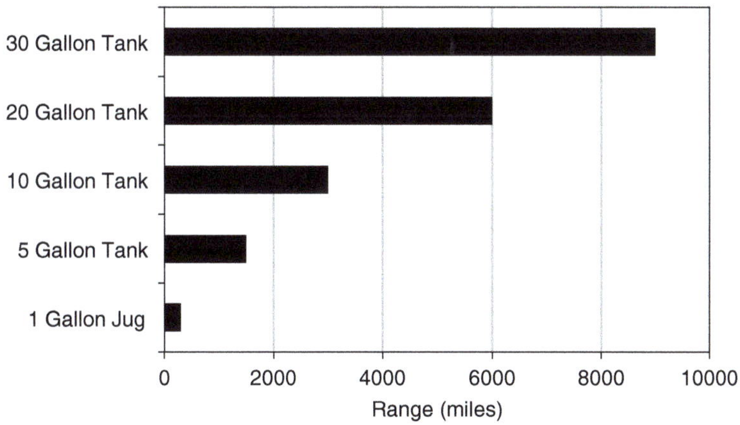

Figure 13.36 DEF mileage (Source: www.factsaboutscr.com/def/convenience.aspx)

120 000 miles/year and uses 20 000 gallons of fuel/year at six miles per gallon would use 400 gallons of DEF (2% rate). The DEF tank size is scaled to approximately 2.5 times the size of the vehicle fuel tank capacity. Assuming 3% dosing tank capacity equals $3 \times 2.5 = 7.5$ gallons. Tank sizes will generally range from 5 to 30 gallons for medium duty vehicles. A typical fleet sleeper tractor will need to fill the DEF tank once every ten days or top-off every two or three days (Source: www.factsaboutscr.com/def/convenience.aspx) and the range is graphically illustrated in Figures 13.36 and 13.37.

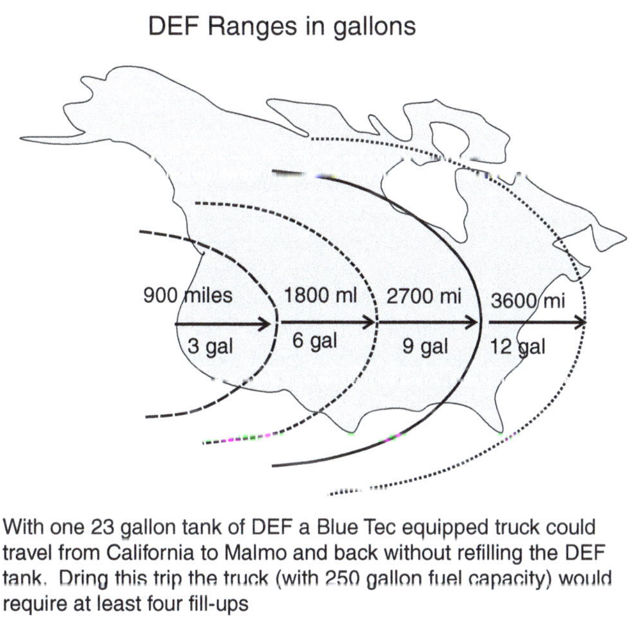

Figure 13.37 DEF mileage on a map of the USA (Source: Detroit Diesel, 2010)

Lamp	Lamp name	Description	Results	Lamp solid	Lamp flashing
	Amber Warning lamp	Indicates fault with the engine controls	Vehicle can be driven to end of shift. Call for service.	At the start of every ignition cycle (check bulb) or When an electronic system fault occurs	Diagnostic request switch is used to activate the AWL to flash inactive codes. Flashes last 90 seconds before idle shutdown if programmed for override. Flashes when idle shutdown or the optimised idle shutdown occurs.
! CAUTION STOP ENGINE	Stop lamp	Indicates a major engine fault that may result in engine damage. Engine derate and/or shutdown sequence will be initiated	Move the vehicle to the nearest safe location and shut down the engine. Call for service.	At the start of every ignition cycle (check bulb) or When a potential engine damaging fault occurs	Flashes when engine protection shutdown occurs. Diagnostic request switch is used to activate the RSL to flash active codes.
	DPF Regeneration lamp	Solid yellow indicates regeneration is required. Blinking yellow, derate and/or shutdown are possible as soot load continues to increase. Lamp will shut off during parked regeneration	Lamp solid- regeneration is required. Lamp flashing- regeneration is required immediately.	At the start of every ignition cycle (bulb check) or when regeneration is required.	When a regeneration is required immediately (if the lamp flashing is ignored), derate and/or shutdown could occur.
	High Exhaust system temperature lamp	Lamp is yellow. Indicates exhaust temperature is above a preset limit and unit is operating at low vehicle speed. When the engine speed is increased for a parked regeneration, lamp will flash once every 10 seconds	Vehicle can be driven. Lamp solid for extended period -call for service.	At the start of every ignition cycle (bulb check) or when vehicle speed is less than 5 mph and the DPF outlet temperature is greater than 525° C.	Flashes every 10 seconds when the DPF outlet reaches a temperature greater than 525°C.
	Malfunction Indicator lamp	Yellow lamp indicates a failure of an emission control device. May illuminate at the same time as the amber warning lamp.	Vehicle can be driven to end of the shift. Call for service.	At the start of every ignition cycle (bulb check) or for any emission related fault	Never flashes

Figure 13.38 OBD diagnostic warnings

13.14.8 On-Board Diagnostics (OBD)

On-Board Diagnostics (OBD) are engine computer monitoring system that are used to detect emission related malfunctions, alert operator of the malfunction, store diagnostic records and support off-board communications to diagnostic service tools. Their use has been mandated by the EPA for 2010 and later heavy duty diesel engines used in highway applications over 14 000 lb gross vehicle weight rating (GVWR). The EPA mandates that one engine family per OEM be certified from 2010–2013. By 2013 all on-road engines for all OEMs must contain OBD, and 2010 and later model year highway heavy duty applications under 14 000 GVWR must use OBD to detect catastrophic failure of the emissions system. An example of the readings from a diagnostic OBD system can be found in Figure 13.38.

13.15 Impact of Emission Strategies on Engine Oils

The use of different emissions strategies by heavy duty diesel OEMs has impacted the performance of the engine oil and has placed increased stresses on it (Figure 13.39). Low emission engines generate more soot and have higher peak cylinder temperature due to the high rates of EGR that are used to help control NO_x. This will cause the engine to run hotter and require engine oil with improved oxidation resistance. Further, to protect the after treatment devices against plugging and poisoning, the engine oils have to contain lower sulfated ash, sulfur and phosphorus contents, while still offering optimum

		Heat/ Temperature	Soot Levels	Acid Levels	Deposit formation	Oil Degradation	Corrosion	Ring/Liner Wear
ULSD	Ultra Low Sulfur Diesel		↓	↓			↓	↓
Higher EGR	Exhaust gas recirculation	↑↑	↑↑	↑↑	↑↑	↑↑	↑↑	↑↑
CCV	Closed crankcase Ventilation				↑	↑		

Figure 13.39 Impact of emission devices on the engine oil

protection for control of piston deposits, oxidative thickening, oil consumption, high temperature stability, soot handling properties, foaming and viscosity loss due to shearing.

Tighter emissions legislation have led to changes in diesel fuel quality, engine hardware and the need for advanced exhaust emission control systems. The use of ultra low sulfur diesel fuel also will be much more than use of low sulfur diesel fuel; also, the ultra low sulfur fuels have a lower heat (BTU) content, resulting in increased fuel consumption, and could cause premature injector failure due to its lack of lubricity. The roles played by combustion and oxidation by products are even greater, due to increasing power densities in modern engines and the higher levels (up to 40%) of Exhaust Gas Recirculation (EGR) used to control NO_x emissions

13.15.1 Impact of Cooled EGR on Engine Oil

EGR, although it has helped reduce exhaust emissions by diverting exhaust gases back into the engine, has resulted in a more hostile environment for the engine oil. EGR has an impact on the engine oil in three main ways (Figure 13.40):

- Increased soot levels
- Increase acid levels
- Increased heat levels.

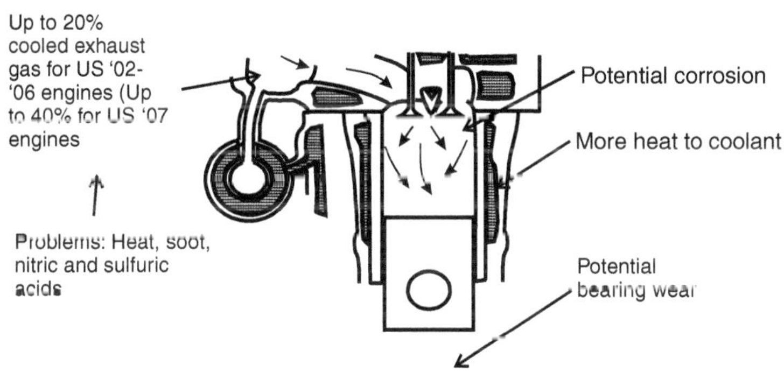

Figure 13.40 EGR control and challenges

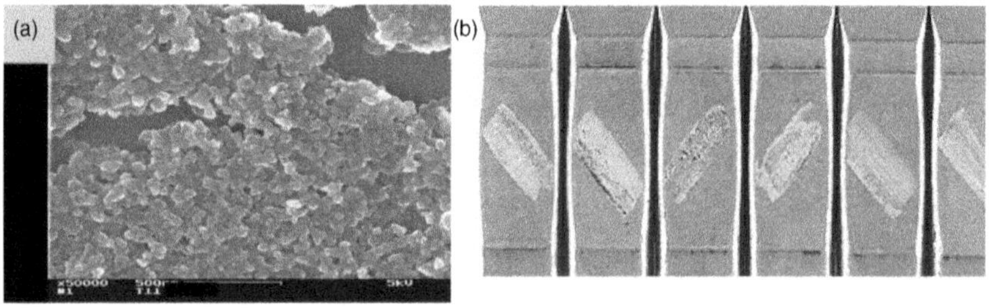

Figure 13.41 (a) Soot agglomeration; (b) soot induced wear

13.15.1.1 Increased Soot Levels

With the adoption of the Clean Air Act of 1990 in the United States and other global emissions regulations, such as the European Union's Euro IV and V and Japan's JPN '07 and '09 regulations, heavy duty diesel OEMs were mandated to significantly reduce oxides of nitrogen and particulate emissions. To control NO_x and PM emissions heavy duty OEMs have employed advanced emission reduction strategies, such as retarded engine timing, improved combustion control through pulsed fuel injection, inlet air swirl designs and high pressure injection common rail injection, decreased piston crevice volume, exhaust gas recirculation, ACERT technology (Caterpillar) and, most recently, closed crankcase ventilation (CCV), diesel oxidation catalysts (DOC) and diesel particulate filter (DPF) after-treatment combinations (Figure 13.41). Each of these emission control devices, with the exception of the DOC and DPF, has a tendency to increase the levels of soot that enter the engine oil. Soot levels as high as 10% have been reported in the field with the use of EGR. Of these devices it is the use of EGR that generates the most hydrocarbon soot, which can be trapped into the engine oil and agglomerate into sticky fused masses. As EGR rates increase to meet ever more demanding control of NO_x emissions, the risk of soot build up in the engine oil has increased significantly. The type of soot that is produced by EGR engines is different to that is produced by non-EGR low emissions engines. The type of soot that is produced is more prone to agglomeration and build-up in the engine oil. If soot is not properly managed and dispersed by the engine oil's additive system but is allowed to agglomerate together it can cause abrasive wear to important engine parts, such as the crosshead of the slider cam follower or cam roller followers, camshaft lobes, pistons, piston rings and cylinder walls. Excessive soot build up can in the long term also lead to the formation of deposits and sludge on the valve decks, in the oil pan and on the pistons and rings, which can lead to excessive bore polishing and loss of oil control.

Higher amounts of soot can also lead to an excessive increase and thickening of the engine oil's viscosity. This excessive increase in viscosity can make the engine oil too viscous to flow during high and low temperature operation. This means decreased low temperature pumpability, more start-up wear and more work for the starting system at low temperatures and less flow to critical engine parts at high temperatures, resulting in the engine working harder and increased fuel consumption

13.15.1.2 Increased Acid Levels

Cooled exhaust gases create acidic materials in the intake systems and combustion chambers. These acids are formed when nitrogen from the air and sulfur from the fuel are converted during the combustion process into nitrogen and sulfur oxides that come into contact with the water vapour present in the exhaust stream. The water reacts with the nitrogen and sulfur oxides to form a mist of nitric and sulfuric acids. When fed back into the power assembly of the engine, these acids can become concentrated, resulting in corrosive wear.

High acid levels can also accelerate the depletion of the engine oil's alkaline reserve, which is known as total base number (TBN), and reduce its usefulness

13.15.1.3 Increased Heat Levels

EGR engines run hotter since the engine's coolant is used to cool the recycled exhaust gases. EGR technology has as the engine ingest not only the intake air, which is normally only 37.8 °C (100 °F), but also up to 40% of the exhaust gases at temperatures well above 93 °C. As a result, the engine oil's temperature is increased. This increase in engine oil temperature can accelerate the rate of engine oil oxidation. Oxidation can degrade the engine oil causing the creation of acidic components that can further deplete the engine oil's alkaline reserve and lead to increased deposit formation, and a dramatic increase in the engine oil's viscosity.

13.15.2 Effects of Post-Injection on Engine Oils

Post-injection is the introduction of fuel late in the combustion cycle as part of an advanced control strategy to reduce emissions. Some OEMs use post-injection in their strategies to 'regenerate' or burn off soot accumulated in diesel particulate filters (DPF).

Post-injection of fuel into the cylinders is intended to vaporize in the cylinder but not combust, exiting then through the exhaust valves and travelling downstream, where the introduction of the unburned fuel to the catalyst creates an exothermic reaction incinerating the collected soot in the DPF. Inevitably the heavier fractions of fuel will not vaporize during post-injection and in liquid form can adhere to the cylinder walls. Through the slapping motion of the pistons and oil rings, the unburned fuel from post-injection can make its way through the tight, hot quarters between the piston, rings and cylinder walls. The fuel accumulates in the crankcase and dilutes the oil, which is a major concern regarding engine wear and longevity (Figure 13.42).

Post-injection fuel injection results in elevated levels of fuel dilution regardless of what type of diesel fuel is being used. However, the rates of fuel dilution and the effects on the engine oil are more pronounced when biodiesel fuel blends are burned. The rate of fuel dilution with the use of biodiesel is higher due to biodiesel's viscosity, density and surface tension characteristics. These characteristics increase the fuel's droplet size as it is being injected into the combustion chamber. The droplet characteristics and the lower volatility characteristics combined with the spray pattern and wall impingement patterns used in post-injection allow any noncombusted biodiesel to go past the rings, contact the cylinder liner and be scraped into the oil sump (Figure 13.43).

Because biodiesel is less volatile and has a narrower boiling range than petroleum base diesel fuel, it will have a tendency to become concentrated in the engine crankcase, with the levels of unburned fuel

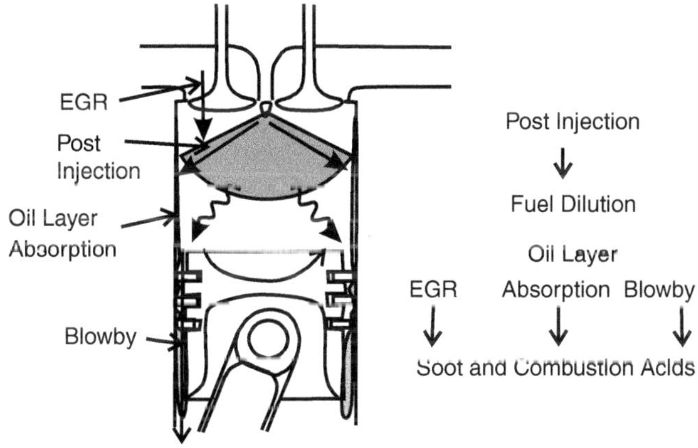

Figure 13.42 The combustion chamber is the source of most oil contaminants

	Diesel	RME, biodiesel
Surface tension	22.5	25.4
Specific gravity	0.852	0.874
Viscosity (cSt) at 100 °C	1.3	2.4
Drop size	Normal	Large
Crankcase dilution	Increases	Very high

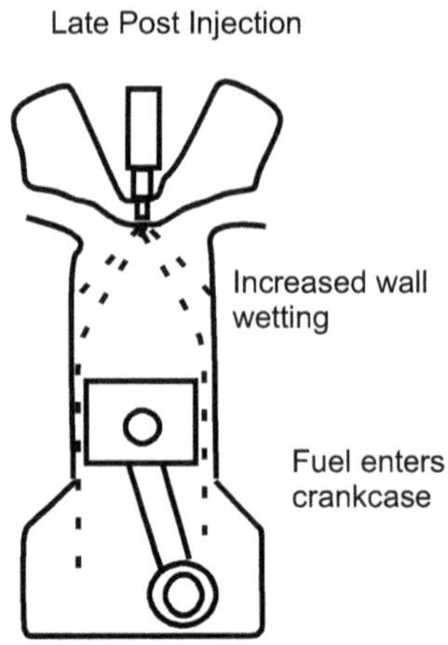

Figure 13.43 Fuel dilution by mineral diesel and biodiesel

building up over time. In the field some OEMs have seen anywhere from 15 to 30% fuel dilution. This build up can reduce the engine oil's viscosity, resulting in a higher risk of engine component wear. In addition, fuel injector deposits that may have been formed by the use of biodiesel can disrupt the fuel spray patterns, further exacerbating the rate of fuel dilution.

Other issues associated with the use of biodiesel fuel that can affect the performance of the engine oil are its impact on the ability of the engine oil to protect against corrosion, the formation of engine deposits and the effect of oxidation.

The increased risk of corrosion to engine bearings is a significant concern. The oxidation products and the presence of unsaturated free fatty acids in the biodiesel are known to be aggressive towards journal bearings that contain lead and copper, such as the engine's main and rod bearings

Biodiesel fuels that contain significant quantities of unsaturated and polyunsaturated fatty acid esters can exhibit poor oxidative stability. In the crankcase these unsaturated and polyunsaturated fatty acid esters can undergo oxidative polymerization, resulting in degradation of the engine oil that can lead to the formation of engine deposits and sludge and the thickening of the oil. The formation of deposits, especially on the pistons, can not only lead to increased wear but also to piston ring sticking. Stuck rings allow more blow-by gases and soot to enter the sump, thereby promoting further viscosity increases and degradation of the engine oil. This increase in viscosity also affects the engine oil's low temperature pumpability. The level of oxidation is dependent upon the rate of fuel dilution, the concentration of biodiesel used (e.g. B20 which is a blend of 20% biodiesel and 80% petroleum diesel fuel), the quality level of the diesel fuel, the source of the biodiesel's feedstock (e.g. soybean, canola, palm, animal fats), the engine operating conditions and the quality of the engine oil.

13.16 How Have Engine Oils Changed to Cope with the Demands of Low Emissions?

To combat the effects of low emissions engine technology, heavy duty diesel engine oils have been and are being formulated to handle the by-products of the combustion process and to maintain the durability and effectiveness of the engine. They have been improved in the following areas:

- Improved Soot Handling Capability – Increased use of dispersant and detergent chemistry to prevent the clumping together of soot particles. Some of the types of dispersant and detergent chemistries used react

directly with the surface of the soot particles by thoroughly coating them. This causes the soot particles to repel each other and prevent agglomeration and clumping, resulting in:
  - Significantly less soot induced wear to critical engine parts
  - Significantly less bore polishing
  - Significantly less soot induced oil thickening
  - Significant reduction in engine deposits.
- Improved acid neutralization
- Improved corrosion protection – Increased use of acid neutralizing and corrosion inhibiting additives to combat the effects of increased acid and to form a protective barrier film to protect the engine parts from the corrosive effects of acid.
- Improved temperature and oxidation stability – The engine oils contain higher amounts of antioxidant additives. These additives are carefully balanced to help control the oxidation process that results from the higher engine temperatures. The engine oils are also being formulated using Group II hydrocracked, Group II severely hydrocracked and polyalphaolefin (PAO) synthetic base oils in their formulations in order to provide lower volatility limits and to improve the engine oil's oxidation and thermal stability. These lower volatility limits and increased oxidation and thermal stability result in a minimum amount of engine oil being consumed, thus leading to a decrease in emissions, and the prevention of the formation of deposits on critical engine parts.
- Increased protection of after-treatment devices – To ensure protection of the after-treatment devices chemical limits for sulfated ash, phosphorus and sulfur content (SAPS) have been set for heavy duty diesel engine oils. In addition to these chemical limits, a volatility limit of 13% maximum has also been set. This has led to an increased use in the use of Group II hydrocracked, Group II severely hydrocracked and polyalphaolefin (PAO) synthetic base oils. Group II hydrocracked, Group II severely hydrocracked and polyalphaolefin (PAO) synthetic base oils contain low amount of sulfur and have low volatility characteristics.

13.17 Most Prevalent API Specifications Found In Use

Today the most widely used engine oil specifications used to define heavy duty on-road and off-road diesel engine oils are:

- API CH-4
- API CI-4
- API CI-4 PLUS
- API CJ-4.

Each of these specifications offers their own attributes to the performance of the heavy duty engine oil.

13.17.1 API CH-4

The API CH-4 service classification was introduced in 1998. The service category describes oils formulated for use in high speed, four-stroke cycle 1998–2001 low emission engines as well as previous model years.

Engine oils that meet this service category are specifically compounded for use with diesel fuels having a sulfur content up to 0.5% and provided improvement in the following areas:

- Wear control and high temperature stability
- Soot handling
- Nonferrous corrosion protection to two-piece pistons

- Oxidative stability
- Foaming
- Shear stability.

13.17.2 API CI-4

The API CI-4 service classification was first licensed on 5 September 2002. The service category describes engine oils that are formulated for high speed, four-stroke cycle diesel EGR containing engines designed to meet the emissions standards implemented in October 2002. API CI-4 engine oils are compounded for use in all applications with diesel fuels having a sulfur content up to 0.5 wt-%. API CI-4 quality engine oils are especially effective at sustaining engine durability where EGR is used, but can be used in older engines that call for API performance oils CD, CE, CF-4, CG-4 or CH-4.

API CI-4 quality engine oils offer performance benefits over previous API categories, including:

- Superior protection against corrosive wear
- Improved soot handling abilities and piston deposit control
- Greater low and high temperature stability
- Greater thermal and oxidative stability to protect against oxidative thickening
- Better protection against foaming
- Better shear stability
- Better TBN reserve.

13.17.2.1 Differences Between CH-4 and CI-4

Engine oils that meet the API CI-4 service classification are designed to function in a more hostile environment than those that meet the API CH-4 classification, since EGR containing engines as previously discussed generate more soot, acid and heat than older engines. In order to be licensed as meeting the API CI-4 service classification, an engine oil must pass a total of 15 laboratory and engine sequence tests. Three of these tests are engine sequence tests specifically designed for the CI-4 category:

- Caterpillar 1R: 504 hour test that evaluates the engine oil's ability to control piston deposits and oil consumption.
- Mack T-10 (EGR): 300 hour test used to measure multiple engine oil performance properties. It measures the engine oil's ability to prevent piston ring and liner wear, lead bearing corrosion and high temperature engine oil oxidation.
- Cummins M-11 EGR: 300 hour test that measures the engine oil's ability to protect the boundary lubrication areas in the engine, particularly the valve train, under high soot-laden conditions.

In addition to these three new engine sequence tests, the API CI-4 service classification has stricter passing limits for the 300 hour Mack T8-E test, which is used to measure the engine oil's ability to resist soot induced viscosity thickening, stricter volatility limits reduced from 17% evaporative loss for CH-4 to 15% evaporative loss as determined by the ASTM D-5800 NOACK Volatility Test, the gasoline engine oxidation test Sequence IIIF and two new laboratory bench tests:

- An elastomer seal compatibility test using four different seal materials: nitrile, polyacrylate, silicone and fluroelastomer. The test compares the seal compatibility performance of the candidate engine oil to the performance of a known reference fluid. To pass the test, the candidate engine oil must provide equivalent or better performance than the reference oil in the areas of volume change, hardness, tensile strength, elongation and reversion.

- A low temperature pumpability test for heavily sooted engine oils: The test uses the ASTM D-4684 MRV TP-1 and is performed on a used oil sample taken from the Mack T-10 (EGR) test after 75 hours of running time. To pass the oil must not exhibit a viscosity greater than 25 000 cP at −20 °C.

13.17.2.2 Benefits of Using CI-4 Engine Oils

The use of engine oils that meet the API CI-4 service classification will provide complete protection of both EGR containing and non-EGR containing engines that are used in either on-road or off-road applications in the areas of engine durability and oil life (drain intervals).

Engine Durability
- Improved viscosity control of soot laden engine oils
- Shorter oiling times to critical areas during start up at low temperatures
- Extra protection by controlling crosshead valve train wear
- Enhanced bearing protection
- Longer seal life.

Oil Life (Drain Intervals)
- Greater oxidation stability at higher engine oil temperatures
- Better oil consumption control
- Improved protection against premature exhaustion of oil filter.

The relative performance of CI-4 engine oils in relationship to other API 'C' service classifications and the number of laboratory bench and sequence tests used to qualify and engine oil as meeting these service classifications are illustrated in Figure 13.44.

13.17.3 API CI-4 Plus

The API CI-4 Plus is a supplemental category to the API CI-4 service classification. API CI-4 Plus was first licensed on 1 September 2004. The API CI-4 Plus category was developed in response to field problems being reported by some OEMs, particularly Mack, in the areas of soot dispersion and shear stability.

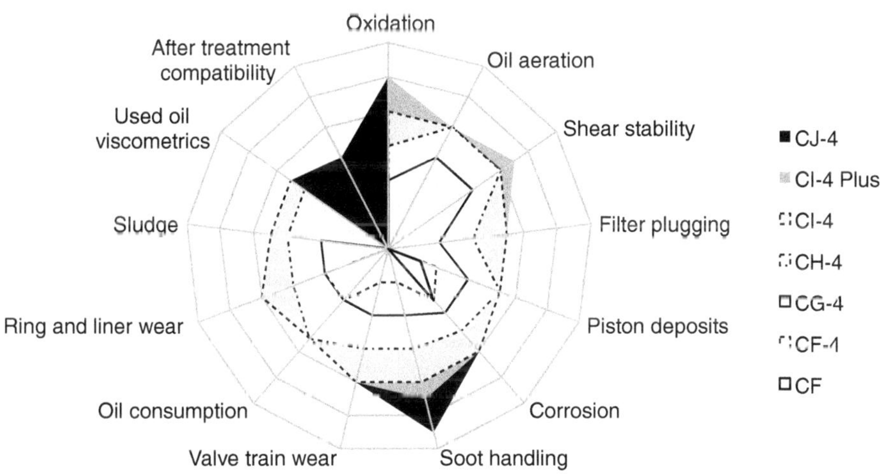

Figure 13.44 Performance of CI-4, CI-4 Plus, CJ-4 and outdated oils (Lubrizol K2M, 2005c)

OEMs were concerned that the API CI-4 service classification used early assumptions about hardware and operating conditions. The tests used and limits for these tests were set without field correlation. Because of the time frame implemented by the EPA in the consent decree with heavy duty diesel engine OEMs to meet the 2002 emission requirements, it was virtually impossible to test and work out all of the issues facing the various engine builders with the use of EGR.

The field problems Mack was experiencing consisted of high soot levels at low drain intervals, high viscosity increases at low soot levels, high levels of acid condensate in the intake system and shearing of the engine oil out of viscosity grade in its ASET vocational engines, which are used in severe vocational applications such as refuse and dump trucks. These engines employ the use of internal EGR, which allow a certain percentage of the exhaust gases to remain in the cylinder from the previous cycle. This results in the formation a higher amount and different kind of soot particles that can thicken the engine oil much more rapidly and is more abrasive.

In addition to these field problems there were concerns by Caterpillar, which uses ACERT Technology rather than EGR, that the high ash, high TBN formulations used to meet CI-4 would lead to higher piston deposit levels. These concerns lead to Caterpillar to introduce the CAT ECF-1 specification, which puts a 'cap' on the engine oil's sulfated ash content and total base number for engine oils that meet the API CI-4 service classification. Companies that employed the use of EGR responded with oil specifications designed to make sure that the engine oil's TBN did not drop below a certain level, which would rob the engine oil of its ability to provide the proper acid neutralization capabilities to protect the engine from the effect of acidic corrosion. Cummins Engine Company, for example, revised its CES 20078 specification to place a limit of 10 TBN minimum in order to assure the proper acid neutralization capacity (Lubrizol K2M, 2005c).

Rather than open the door to an endless proliferation of engine model and brand specific oils or the development of totally new service category of oils, the engine OEMS, lubricant manufacturers and marketers and additive companies asked the API for an enhancement to its CI-4 category. The result was CI-4 Plus. The CI-4 Plus supplemental category took just shy of 12 months to come into being compared to the typical 2–3 year timeframe taken to develop an API Service Classification, and is the result of a 'gentlemen's agreement' of sorts among all of the engine manufacturers to develop the supplement to address soot induced viscosity thickening problems and be the last API service classification until late 2006 early 2007, when the EPA's new emission regulations came into effect (Kilcarr, 2004).

13.17.3.1 Differences between CI-4 Plus and CI-4

To pass the CI-4-Plus Supplemental Category the engine oil must pass all of the laboratory bench and engine sequence tests for CI-4 and meet the following requirements:

- Pass the Mack T-11 test or be listed on Mack's EO-N Premium Plus-03 specification. The Mack T-11 tests is a 300 hour test that uses a Mack E-Tech V-Mac III engine with EGR that measures the engine oil's soot dispersancy and viscosity control at high soot levels (6% soot).
- Pass the Bosch Injector Shear Test ASTM D-7109, which is used to evaluate the percentage viscosity loss of multi-grade engine oils resulting from degradation of the engine oil's viscosity index improvers. The engine oil is passed through a diesel injector nozzle at a shear rate that causes the engine oil's viscosity index improvers to degrade. Typically in this test the engine oil is cycled through the diesel injector for 30 cycles in order to closely represent the severity that is seen during engine operation. However, in the field there has been a growing concern among OEM's, especially those that produce heavy duty diesel engines, that this 30 cycle limit is not severe enough to protect equipment in the field. It had been observed that certain products meeting the 30 cycle limit have been shearing out of viscosity grade during service and not providing the proper lubricant film needed to protect the engine from increased wear, especially under high soot level conditions. Therefore, the engine oil must not exhibit greater than a 12 cSt increase in viscosity at 100 °C over the fresh oil after 90 passes through the Bosch Injector Shear Test ASTM D-7109.

- Exhibit less than than a 13% evaporative loss as determined by the NOACK Volatility Test ASTM D-5800. This lower evaporative loss protects the engine from the formation of deposits on critical engine parts.

13.17.4 API CJ-4

API CJ-4 represents the latest in a series of engine oil upgrades for heavy duty diesel engine oils. API CJ-4 came into effect in May 2006 and first licensing occurred on 15 October 2006. This timing coincided with the October 2006 legislated introduction of ultra low sulfur diesel fuel for the on-road market in the United States, as well as the early production of EPA 2007 compliant engine hardware outfitted with particulate filters. API CJ-4 quality engine oils were compounded for use in all applications with diesel fuel sulfur content up to 500 ppm (0.05 wt-%).

To ensure protection of the after-treatment devices that are used on 2007 and 2010 US EPA and EU Euro IV compliant engines, for the first time ever chemical limits were set for heavy duty diesel engine oils by the ASTM's Heavy Duty Engine Oil Classification Panel. In previous API heavy duty diesel engine oil categories there were no chemical limits on the engine oil, because exhaust after-treatment devices were not used. In the previous category, API CI-4, which was introduced in 2002, the engine oils' sulfated ash was generally in the range 1.3–1.5%, phosphorus levels in the range 0.12–0.14% and sulfur range 0.45–0.8% (McGeehan *et al.*, 2006). The chemical limits for API CJ-4 target the engine oil's sulfated ash, phosphorus and sulfur content, which are commonly referred to as 'SAPS'. These chemical limits, which represent a 'chemical box' for API CJ-4, are as follows:

- 1.00% maximum sulfated ash (per ASTM D-874)
- 0.12% maximum phosphorus (per ASTM D-4951
- 0.40% maximum sulfur (per ASTM D-4951 or ASTM D-2622).

In addition to these chemical limits, a volatility limit of 13% maximum as determined by the NOACK Volatility Test Method ASTM D-5800 has been set for API CJ-4. Though there was no previous field data to support chemical limits, research data from dynamometer tests that indicated a direct relationship between sulfated ash levels and incombustible material in the DPF, along with limits imposed in the ACEA E6 and JASO DH-2 engine oil specifications, influenced the ASTM's Heavy Duty Engine Oil Classification Panel's task force that researched these limits to agree upon the chemical limits for API CJ-4 (Table 13.8) (McGeehan *et al.*, 2006, p. 4).

SAPS are found in or derived from components in engine oil formulations. These various components are used to help provide extended oil drain intervals, TBN retention and to protect against wear, oxidation, corrosion and piston deposits. Though they contribute significant performance benefits they can cause problems in the 2007 and 2010 United States EPA and EU Euro IV complaint engines.

The most concern to proper functioning of the 2007 and 2010 low emission complaint engine lies in the impact sulfated ash has on after-treatment devices such as diesel particulate filters (Figure 13.45).

Table 13.8 Comparison of API CJ-4, ACEA E-6 and JASO DH-2 Chemical Limits

Oil Category	API CJ-4	ACEA E6	JASO DH-2
Sulfated ash (%)	1.0	1.0	1.0
Phosphorus (%)	0.12	0.08	0.12
Sulfur (%)	0.4	0.3	0.5
Volatility (%)	13	13	16

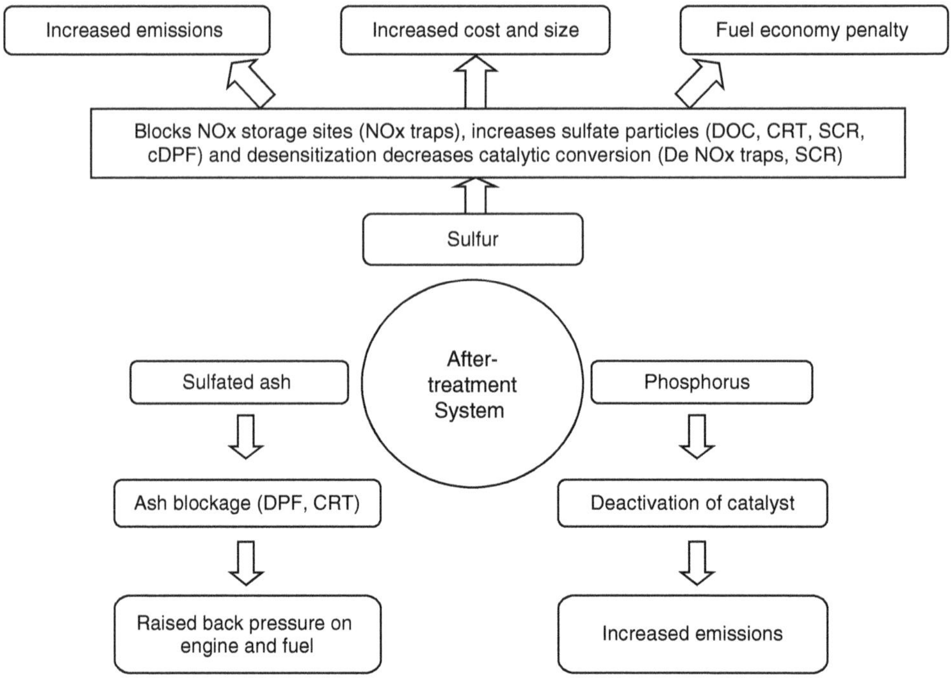

Figure 13.45 Impact of SAPS on after-treatment devices

13.17.4.1 What is Sulfated Ash and its Impact on Diesel Particulate Filters?

The term sulfated ash relates to the amount of metallic elements in the engine oil, which are mostly present in the engine oil's detergent and anti-wear additive chemistry. Modern engine oil additive packages contain multiple components based on metals such as calcium, magnesium, zinc and so on. Since a 100% seal between the piston rings can never be achieved a certain amount of engine oil will enter the combustion and be burned.

When the engine oil containing these elements enters the combustion chamber and is burned the residue that is left behind is an ash-like material. This ash-like material can contribute to deposits in the crown land above the piston ring as well as to deposits in the ring grooves. These deposits can lead to rubbing wear on the cylinder liner and piston rings that do not operate freely. Ultimately, as the cylinder liner-to-ring interface is compromised high oil consumption can occur (McFall, 2005). In addition to these deposits inorganic compounds from the additives of lubricating oil can become oxidized in the combustion and generate metal oxide particles. These particles can be carried downstream with the exhaust and collect on the diesel particulate filter. These metallic oxide ash particles are not removed by filter regeneration because they are not combustible. As the ash particles accumulate, they reduce the porosity of the filter resulting in filter blockage (Figures 13.46 and 13.47). This filter blockage increases back pressure to the engine, increases fuel consumption and decreases power. Furthermore, reactions with the ash in the filter and high temperatures, along with the increased pressure drop across the filter, can lead to permanent filter substrate cracking. Ash particle build up also results in the particulate filters having to be cleaned more often by mechanical means, such as with compressed air or water.

The engine oil's sulfated ash content is determined by the use of the ASTM D-874 Test Method. In this test up to an 80 gramme sample of oil is weighed into an either a porcelain or fused silica dish.

Figure 13.46 Plugged DPF

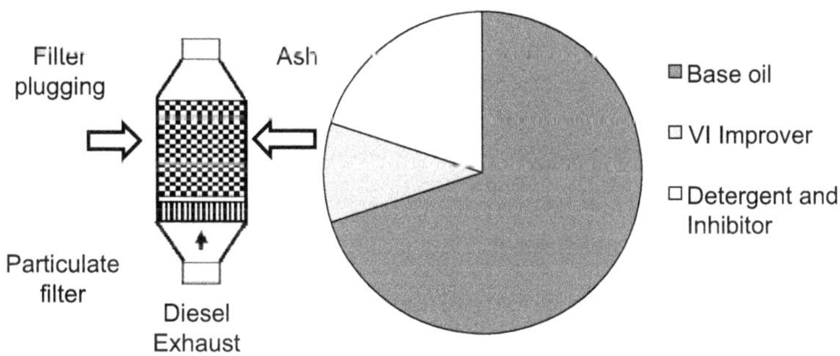

Figure 13.47 Contribution of VI improver, detergents and inhibitor in oil to ash formation and filter (DPF) plugging (McGeehan, 2006)

The dish is placed into an electric muffle furnace and heated until it is ignited and burned. After cooling, the residue is treated with sulfuric acid and distilled water and heated to 775 °C (1427 °F) until oxidation of the residue is complete. The ash is then cooled, retreated with three drops of water and ten drops of sulfuric acid and heated at 775 °C (1427 °F) for 30 minutes. At the end of 30 minutes the sample is cooled and weighed. The sulfated ash content is expressed as a percentage by mass and is determined by the use of the following equation:

$$\%\ weight\ ash\ content = \frac{Weight\ of\ residue\ left}{Weight\ of\ sample} \times 100$$

The engine oil's sulfated ash content also directly relates to the engine oil's acid neutralization capabilities (TBN), since most of an engine oil's TBN comes from the metal-containing detergent additives. Generally, the higher an engine oil's TBN, the higher its ash content and the greater its ability to prevent acidic corrosion in the engine.

13.17.4.2 Source of Sulfur and its Impact

Heavy duty diesel engine oils are comprised of approximately 75–85% base oil with the remainder made up of additive systems. The sulfur concentration in the base oil can range from essentially zero (synthetic base fluids such as PAOs) to as high as 0.5 wt-% (Group I base stocks). Sulfur content in a base oil can be reduced by the use of hydrotreating and hydrocracking methods to levels ranging from less than 0.1 wt-% to less than 0.3 wt-%. The additive systems used are also major sources of sulfur. The sulfur-containing additives used in the formulation of heavy duty diesel engine oils include the detergents, anti-wear agents primarily from zinc dithiophosphate, corrosion inhibitors, friction modifiers and antioxidants. The sulfur in these additives is in the form of sulfonates, phenol sulfide salts and thiophosphonates.

The sulfur content of API CH-4, CI-4 and CI-4 Plus engine oils can range from 2500 to as high 8500 ppm sulfur by weight. It has been estimated by the EPA that anywhere from one to seven ppm of sulfur can be contributed to the diesel engine's exhaust, when the engine oil enters the combustion chamber and is burned. The worst case estimate of seven ppm is based upon nominal heavy duty diesel vehicle fuel and oil consumption rates of six miles/gallon and one quart per 2000 miles respectively. During normal operation, only a small percentage of the engine oil consumed by open crankcase ventilation heavy duty diesel engines travels past the rings and burns in the combustion chamber. The remainder of the consumed oil is lost through evaporation by being emitted through the crankcase ventilation tube and is not combusted. If an engine oil that contains a sulfur level of greater than the 0.4% maximum limit for API CJ-4 were used in a 2007 and 2010 compliant engine, the closed crankcase ventilation system would recover the evaporated oil and carry it through the exhaust stream (Whitcare, 2000; EPA, 2000; DECSE, 2000, p. 3).

Once in the exhaust stream sulfur can inhibit the effectiveness of the particulate filters by poisoning the catalysts. This poisoning can increase the conversion of sulfur oxides to sulfates, which increases particulate emissions and accumulation of particulate material. Accumulation of particulate material can lead to reduced engine performance, due to increased backpressure and ultimately failure of the trap.

13.17.4.3 Phosphorus Source and Impact

The primary source of phosphorus in heavy duty diesel engine oils comes from the anti-wear agent zinc dithiophosphate. Phosphorus can also come from corrosion inhibitors, friction modifiers, and antioxidants. The phosphorus in theses additives is in the form of dialkyldithiophosphates, thiophosphonates, phosphoric acid esters and amine phosphates. Typically, heavy duty diesel engine oils contain 0.11–0.15 wt-% phosphorus.

Once in the exhaust stream, phosphorus can reduce the efficiency and deactivate the noble metal catalysts by coating and building up on the active catalyst sites, causing irreversible damage that accumulates over time. As a result, increased levels of harmful emissions such as NO_x, carbon monoxide and hydrocarbons pass through the catalytic converter unchanged, resulting in an increased level of NO_x, CO and hydrocarbon emissions.

13.18 Paradigm Shift in Engine Oil Technology

Though API CJ-4 is backward compatible with previous heavy duty diesel engine oil API Service Classifications such as CI-4 and CI-4 Plus, CH-4, CG-4 and CF-4, the restriction in SAPS for API CJ-4 has resulted in a paradigm shift in engine oil technology. The reduction in ash levels from the norm of 1.3–1.5% to the mandatory maximum of 1% in addition to the reduction in sulfur levels to 0.4% maximum required replacing some of the conventional metal-containing detergent chemistry with alternative detergent and dispersant chemistries that are low in metallic content, sulfur and, in some cases, ash less. The use of these alternative detergent and dispersant chemistries has resulted in a reduction of the engine oil's previous TBN levels of 10–14 to 7–10 TBN for API CJ-4 quality engine oils. Though the lowering of TBN can normally result in a reduction of oil drain intervals in off-road diesel engine applications that burn low sulfur diesel fuel (500 ppm maximum), it has been documented by the lubricants industry and suppliers of additive chemistries for heavy duty engine oil lubricants that this has not been the case.

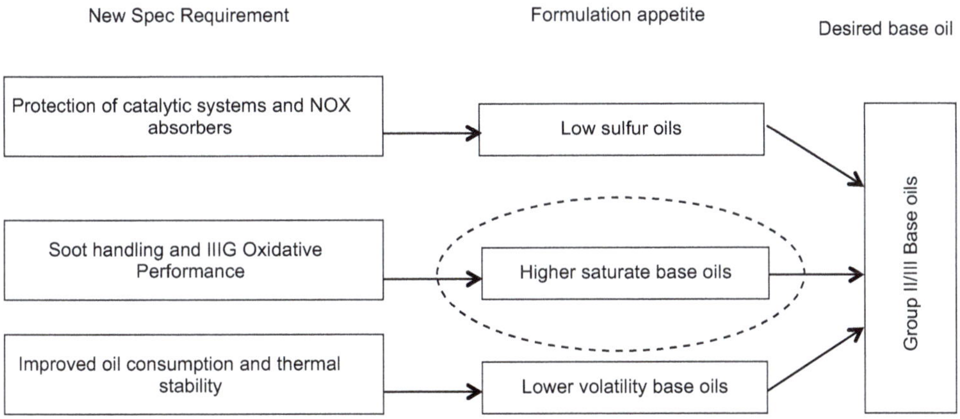

Figure 13.48 Implications of base oil effects

Field test evaluations have shown that lower TBN oils if properly formulated and if additive chemistry is balanced will provide very good to excellent engine performance and TBN retention, even with high sulfur fuels provided that oil drain intervals are properly adjusted to the fuel sulfur level by the use of used oil analysis. In addition, off-road fuels in North America, Europe and parts of Asia Pacific are being reduced to less than 500 ppm sulfur, and that of specification Ultra Low Sulfur Diesel fuel (<15 ppm by weight of sulfur) is being sold as high sulfur fuel in the United States, Canada and parts of the European Union.

To meet the API CJ-4 limit of 0.12% a reduction in the amount of zinc dialkyldithiophosphate used in the formulation of heavy duty diesel engine oils has occurred. This reduction in zinc dialkyldithiophosphate has resulted in the addition of ash-less anti-wear agents and organo-molybdenum compounds in order to protect the engine's valve train from wear.

The reduction in sulfur levels to 0.4% maximum along with the NOACK volatility limits of 13% maximum and the need for increased oxidation stability (due to the increased thermal stress placed on the engine oil from the use of heavy EGR rates and after-treatment), have resulted in the use of Group II and Group III hydrocracked base stocks as well as Group IV synthetic base stocks. Because of their sulfur content, lower oxidation stability and higher volatility characteristics, the use of Group I base stocks in the formulation of engine oils that meet API CJ-4 has been eliminated. The implications of these base oil effects are illustrated in Figure 13.48.

13.18.1 Backward Compatibility and Engine Tests

Whenever a new heavy duty diesel engine oil category is developed the API and ASTM have always mandated that backward compatibility with previous existing service categories occur. The process of upgrading engine oil categories is designed to keep the existing tests from the previous category, which have successfully eliminated oil-related failures, and address the EMA's concerns with engine oil performance. As engine oil designs change to meet emission and field performance requirements, these previous engine tests are juxtaposed with new engine tests that address the expected performance issues of newer engines. To ensure backward compatibility the performance tests for API CJ-4, a combination of existing engine and laboratory bench tests from API Service Classifications API CJ-4, CI-4 and CI-4 were included in API CJ-4. In addition to these tests, the Bosch Injector Shear Stability Test ASTM D-7109 for 90 cycles, five new diesel engine tests and one gasoline engine sequence test were developed and/or included into API CJ-4. The five new engine sequence tests were developed to address not only the EMA's concerns about valve train wear performance for old and new engines due to the reduction in phosphorus levels from 0.14% to the maximum level of 0.12% for API CJ-4, but also to address the OEMs and EMA's concerns regarding oil consumption control, valve train wear at high soot levels, wear of ring, liner and bearing, oil oxidation control and soot dispersion. The five new engine sequence tests and the one gasoline engine test used in API CJ-4 are:

- Caterpillar ACERT C-13 engine test for piston deposits and oil consumption
- Cummins ISB engine test for valve train cam and tappet wear
- Cummins ISM engine test for valve train wear, filter plugging and sludge formation at high soot levels
- Mack T-12 for ring/liner and bearing wear and oil consumption
- Mack T-11 for viscosity increase due to soot and the Mack T-11A for low temperature pumpability of sooted engine oils
- GM Sequence IIIG or IIIF for oil oxidation.

The existing engine sequence and laboratory bench tests for API CJ-4 are:

- Caterpillar 1N Diesel Engine Test for Deposits and Oil Consumption
- GM 6.5L engine test for roller follower wear
- Navistar EOAT for aeration
- Laboratory bench tests
  - ASTM D-892 laboratory bench test for foaming
  - High Temperature Corrosion Bench Test ASTM D-6594 for corrosion
  - Bosch Injector Test ASTM D-7109 for shear stability
  - Noack Volatility ASTM D-5800
  - High Temperature High Shear Viscosity ASTM D-4683
  - Seal Compatibility tests.

13.18.2 New Engine Sequence Tests

13.18.2.1 Caterpillar ACERT C-13

The Caterpillar C-13 test is a 500 hour test run on 15 ppm sulfur fuel that uses a dual-stage turbocharging and ACERT inline six-cylinder C-13 engine with one-piece steel piston technology and closed crankcase ventilation. The test is used to evaluate the engine oil's ability to protect against excessive oil consumption and the formation of piston deposits. Test parameters are: 1800 rpm, 1200 g/min fuel flow, 40 °C intake manifold temperature, 88 °C (190 °F) coolant out temperature, 40 °C (104 °F) fuel in temperature, 98 °C (208 °F) oil gallery temperature, 280 kPa (41 psi) intake manifold pressure for 500 hours At the end of the test the pistons are rated for top land carbon, top groove carbon and carbon on the top face of the second rectangular ring. The oil consumption experienced in this test is calculated by averaging the oil at the 100 and 150 hour points of the test and comparing this to the average oil consumption at 400–500 hours to determine the increase. The pass/fail limits are found in Table 13.9.

13.18.2.2 Cummins ISB Engine Test

The Cummins ISB engine sequence tests is a 350 hour test run on 15 ppm sulfur fuel that uses a Cummins 5.9L ISB medium duty diesel engine equipped with EGR and variable geometry turbocharger. The engine is certified at 2.0 g NO_x/bhp-hour and is rated at 300 bhp (215 kW) at 2600 rpm. This procedure is used to evaluate the engine oil's ability to reduce camshaft lobe and valve train wear. After 100 hours of steady state operation at 1600 rpm to accumulate 3.25% soot in the oil, the engine is operated for 250 hours on a multi-step 27 second cycle simulating cyclic operation from low speed idle, to rated load and speed, to

Table 13.9 Pass/fail limits for the Caterpillar ACERT C-13 test

	Pass/fail criteria			
	Maximum	Merit weight	Anchor	Minimum
Delta O/C	31	300	25	10
ATLC	35	300	30	15
ATGC	53	300	46	30
R2TCA	33	100	22	5
Merits	1000			

peak torque. At the end of the test the camshaft, mushroom slider tappet and crosshead are measured and rated for wear. To pass this engine sequence test the engine oil cannot exhibit more than 100 mg of tappet wear and 55 microns of cam wear.

13.18.2.3 Cummins ISM Engine Sequence Test

The Cummins ISM engine sequence test is a 200 hour test that uses a 2002 Cummins 11 litre ISM diesel engine that is equipped with electronic controlled unit injectors, cooled EGR and a variable geometry turbocharger. The engine is rated at 330 bhp (236 kW) at 1800 rpm. High load, heavy duty field conditions with high soot and EGR flow rates using a 2007 emission compliant engine are simulated in this test. The engine test uses 500 ppm sulfur fuel for backward compatibility. This evaluates an engine oil's ability to protect turbocharged, after-cooled four-stroke cycle diesel engines equipped with EGR against valve train wear, cylinder and liner wear, filter plugging and deposit formation under soot-laden conditions. The 200 hour test is done in four 50 hour stages. During stages one and three, the engine is operated with retarded fuel injection timing and is over fuelled to generate excess soot in the range of 5 to 6.5% soot loading. During stages two and four, the engine is operated at conditions to induce valve train wear (Figure 13.49).

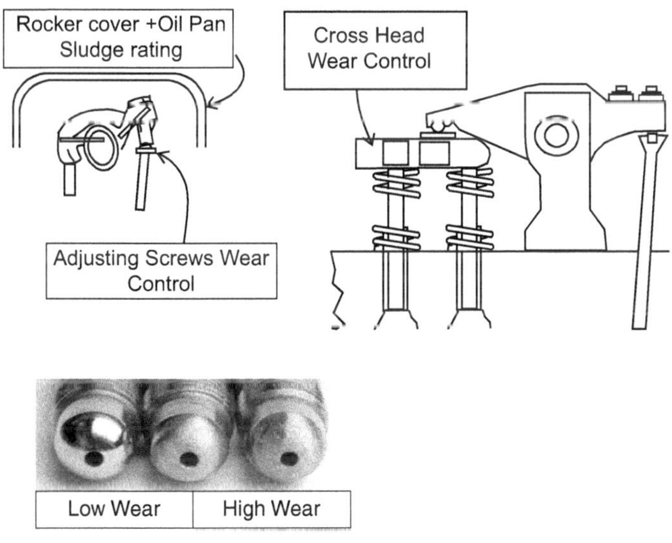

Figure 13.49 API CJ-4: Cummins ISM, valve train wear, filter delta P and sludge control (McGeehan *et al.*, 2006, p. 7)

Table 13.10 Pass/fail limits for the Cummins ISM sequence test

	Pass/fail criteria			
Parameter	Anchor	Merit weight	Maximum	Minimum
XHD	5.7	350	7.1	4.3
RWL	100	0	100	0
Ofdp	13	150	19	7
IAS	27	350	49	16
Sludge	9	150	9.3	8.7
Merits	1000			

Critical parts evaluated are injector adjusting screw, sludge, crosshead, top ring wear and oil filter plugging. Pass/fail limits for this test are listed in Table 13.10.

13.18.2.4 Mack T-12 Engine Sequence Test

This procedure evaluates an oil's ability to minimize cylinder liner, piston ring and bearing wear in engines with heavy exhaust gas recirculation (EGR). The test uses a modified Mack E7 E-Tech 460 12 litre engine rated at 460 bhp (329 kW) and 1800 rpm, with EGR, two turbochargers, one which is a variable geometry turbocharger, and 2002 low swirl with combustion system. Heavy duty on-road truck operations after 2007 are simulated in this test.

The test is a 300 hour engine procedure operated in two phases lasting 300 hours at a temperature of 127 °C (261 °F). In phase one, which is the first 100 hours, the EGR is operated at a rate of 35% and the engine is operated at rated speed and power to generate a soot level of 4.3%. In phase two, which is the last 200 hours of the test, the EGR level is operated at 15% and the engine is operated at over-fuelled peak torque at which the peak cylinder pressure is 240 bar (3500 psi) to produce ring and liner wear at 1200 rpm to maximize the wear rates on the rings and liner. In addition, the oil gallery temperatures are held at 116 °C (240 °F) to produce oxidation of the engine oil, which can result in lead corrosion of the bearings. The soot level generated at the end of the test is 6%.

Piston ring wear, cylinder liner wear, lead bearing corrosion, oil consumption and oxidation are evaluated. Ring and liner wear are measured only at the end of the test. Lead in used oil is monitored throughout the test, with limits on the lead increasing after 300 hours and between 250 and 300 hours. The lead increase is due to oil oxidation that causes corrosive wear of the copper–lead bearings.

13.18.2.5 Mack T-11 Engine Sequence Test

The Mack T-11 engine sequence test is a 252 hour test that is used to evaluate the engine oils to disperse soot and its soot handling performance, as measured by viscosity increase. The Mack T-11 engine sequence test uses a Mack E-Tech V-MAC III 12 litre in-line six-cylinder engine, electronically controlled fuel injection with six electronic unit pumps, using 2002 low swirl cylinder heads, cooled EGR with normal low swirl cylinder head twin turbochargers that burns 15 ppm sulfur fuel (Figure 13.50b). The engine is operated at 1800 rpm and 350 bhp (250 kW) for 252 hours, with soot control targets identified at three different points: 96 hour soot window of 2.5–3%, 192 hour soot window of 5.1–5.85% and 228 hour soot window of 6.09–6.97%. These limits were imposed to ensure that the rate of soot build up is controlled. At the end of each (test procedure window), oil filter plugging is examined. Oil samples are taken every 12 hours and analysed for soot content and viscosity. During the development of the test there were concerns both from Cummins and Mack about viscosity increase due to the build up of soot in the engine oil in both pre-2007 and 2007 engines. It was agreed to define three pass/fail limits for viscosity

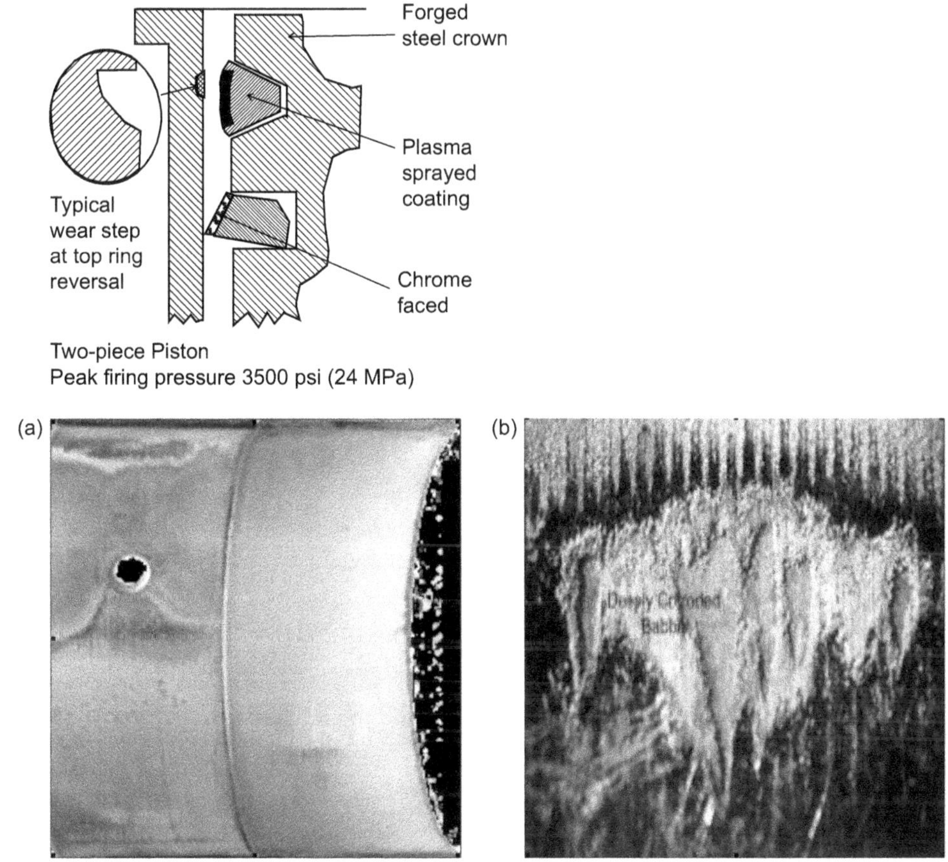

Figure 13.50 (a) Mack T-12 piston and ring configuration at top ring reversal. (b) Mack T-11 and Mack T-11A engine tests

increase after the 90 cycle shear down in the Bosch Shear Injector Test ASTM D-7109. These limits are (Figure 13.51):

- 3.5% soot viscosity increase at or below 4 cSt at 100 °C
- 6.0% soot viscosity increase at or below 12 cSt at 100 °C
- 6.7% soot viscosity increase at or below 15 cSt at 100 °C.

At 180 test hours a sample is taken for low temperature pumpability as defined by MRV TP-1 ASTM D-4684 at −20 °C (−4 °F). It has been found that at low ambient temperatures heavily sooted engine oils can impede the pumpability of the engine oil. In order to pass the Mack T-11 test the engine oil cannot exhibit a viscosity of greater than 25 000 cP at −20 °C (−4 °F).

During the development of API CJ-4 it was decided by the ASTM's HDEOCP that if an oil passed the T-11 viscosity limits but failed the MRV test, that the engine could run the Mack T-11A engine sequence test. The Mack T-11A test has the same operating conditions as the Mack T-11 test. An oil sample is taken at 180 hours where the soot level is nominally at a 5.2% level, within a soot window of 4.82–5.49%.

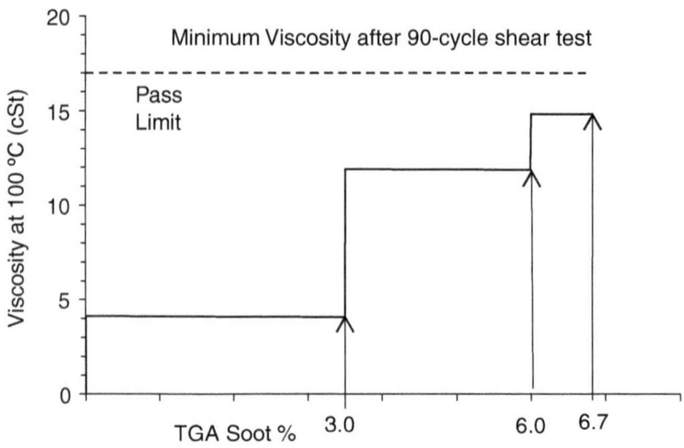

Figure 13.51 Mack T-11 passing limits

13.18.2.6 GM Sequence IIIF or Sequence IIIG Tests

The Sequence IIIF and IIIG tests, which are petrol engine sequence tests, are used to measure the engine oil's ability to resist oxidation, limit deposit tendencies and prevent valve train wear due to high temperature operation. These tests were chosen for inclusion into API CJ-4 to supplement the Mack T-12 engine sequence test's measurement for oxidation control, since the Sequence IIIF and IIIG tests operate at oil temperatures of 155 °C and 150 °C (311 and 302 °F), respectively. Either test can be used to qualify the engine oil as meeting the API CJ-4 specification. These sequence tests are designed to simulate 'a six-cylinder pick-up truck with a flat-tappet cam shaft pulling a horse trailer across Texas in the summertime' (Farania, 2005).

The Sequence IIIG test (used as part of the petrol engine oil specifications ILSAC GF-4 and ILSAC GF-5) replaced the Sequence IIIF test, which was used previously used as part of the (now obsolete) ILSAC GF-3. The Sequence IIIG test is roughly as difficult to pass as the Sequence IIIF test, when it is run at twice its prescribed interval (Farania, 2005). Both engine sequence tests use a 1996/1997 231 C.I.D (3800 cc) Series II GM V-6 fuel injected engine operated at 3600 rpm. Using unleaded petrol, the engine runs a ten minute initial oil levelling procedure followed by a 15 minute slow ramp up to speed and load conditions. The engine is then operated at the specified bhp, rpm, oil temperatures and test hours for each test procedure. The test procedure is interrupted at hour specific intervals for oil level checks and to take oil samples. The operating parameters for each test are shown in the Table 13.11.

Kinematic viscosity increases at 40 °C are compared to those of the used engine oil's viscosity at 40 °C. The unused engine oil's viscosity is used as a baseline to determine the percentage viscosity increase. To pass these tests the engine oil cannot exhibit more than a 150% increase in viscosity at 40 °C at the end of 80 or 100 test hours respectively.

Table 13.11 Operating parameters for Sequence III-F and III-G tests

Test	Duration (hours)	bhp	Operating temperature (°C)	Oil check interval (hours)
IIIG	100	125	150	20
IIIF	80	100	155	10

13.18.3 Previous Engine Oil Sequence Tests

To ensure backward compatibility the performance tests for API CJ-4 a combination of existing engine and laboratory bench tests from API Service Classifications API CJ-4, CI-4 and CI-4 were included into API CJ-4. The existing engine sequence tests used for API CJ-4 are:

- Caterpillar 1N diesel engine test for deposits and oil consumption
- GM 6.5L engine test for roller follower wear
- Navistar EOAT engine test for aeration.

13.18.3.1 Caterpillar 1N Engine Sequence Test

The Caterpillar 1N engine sequence test is a 252 hour test that is used to evaluate the performance of the engine oil with respect to piston deposits, ring sticking, piston, ring and liner scuffing, as well as oil consumption. The Caterpillar 1N test uses the Caterpillar 1Y540 single cylinder direct injection diesel test engine with a four-valve arrangement having a 5.4 inch bore and a 6.5 inch stroke, resulting in a displacement of 148.8 cubic inches for this test. Compression ratio is 14.5:1. The engine's aluminium piston has keystone top ring and rectangular second ring. The engine is operated at 2100 rpm, 70 bhp (50 kW), 93.3 °C (200 °F) coolant temperature, 107.2 °C (225 °F) oil temperature, 127 °C (260 °F) inlet air temperature at 71 inches of mercury, 125 grains/lb and 29:1 air fuel ratio for 252 hours. At the end of the test the pistons are rated for top land carbon, top groove carbon and carbon on the top face of the second rectangular ring. The oil consumption experienced in this test is monitored continuously over the entire test time. The pass/fail limits are shown in Table 13.12.

13.18.3.2 GM 6.5L Engine Sequence Test

The GM 6.5L Roller Follower Wear test is a 50 hour test that is used to evaluate the engine oil's ability to protect against soot induced polishing and surface fatigue wear on the engine camshafts. The GM 6.5L Roller Follower Tests uses a General Motors 6.5 litre, indirect injected diesel engine, rated at 160 horsepower at 3400 rpm that is run at 1000 rpm with near maximum load for 50 hours without an oil change. The engine burns 500 ppm sulfur fuel. Make-up oil is added at 25 hours. Oil gallery and coolant out temperatures are controlled to 120 °C (248 °F). New roller followers are installed at the beginning of the test. At the end of the test, the roller follower axles are removed and their wear is measured using a linear profilometer. The pass/fail limits for wear on the roller follower axle have been set at 7.6 microns (0.30 mils) for API CJ-4.

Table 13.12 Pass/fail limits for the Caterpillar 1N test (Gray, 2005)

	1st procedure	2nd procedure	3rd procedure
WDK (Demerits)	286.2	311.7	323
TGF (%)	20	23	25
TLHC (%)	30	4	5
BSOC (g/kW h)	$\leq$0.5	$\leq$0.5	$\leq$0.5

No piston or ring liner distress and no stuck rings are allowed. Based upon the period that each procedure completed, the appropriate parameter adjustment factor is added to the procedure result. For a first procedure run, the adjusted results are compared to the above first procedure limits. For a two- or three- procedure programme, the average of the adjusted procedure results is compared to the appropriate pass limits.

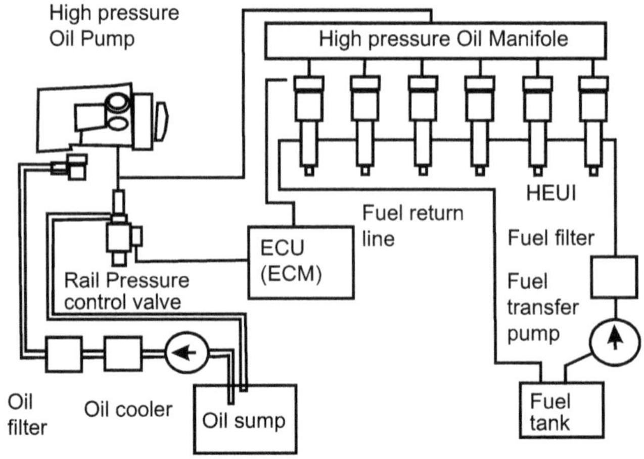

Figure 13.52 Navistar HEUI (Hydraulically Activated Electronically Unit Injector) system

13.18.3.3 Navistar Engine Oil Aeration Test

The Navistar Engine Oil Aeration Test is a 20 hour test that is used to evaluate the effectiveness of the engine oil's ability to minimize air entrainment in large pickups and medium duty trucks. During the development of the previous API CG-4, Navistar reported an oil aeration problem with the new fuel system being developed for its 1994 engines. This system, called the 'hydraulically actuated electronically controlled unit injector' (HEUI), uses oil from the main gallery and pressurizes it up to 20.7 MPa (3 000 psi) in a plunger pump. This oil is used to operate unit injectors that, when used in combination with intensifiers, increase the fuel injection pressure up to 138 MPa (20 000 psi) independent of engine speed. The electronic controls permit varied injection timing and duration to provide the optimum fuel economy and emissions. This system, however, can circulate all the oil in the sump every 8.8 seconds at 3300 rpm, and this can cause oil aeration. Because there is a trend toward extended service intervals, Navistar needed to ensure that rough engine running and misfiring due to aeration did not occur at these longer drains. So the limit for oil aeration was defined. This limit eliminated rough engine running on return to idle after high speed run (Figure 13.52 and Table 13.13) (McGeehan *et al.*, 2006, p. 21).

The test uses a 1994 International Truck 7.3 liter, V-8, four-stroke, turbocharged, compression ignition engine, using the HEUI (hydraulically actuated, electronically controlled unit injector) fuel injection system. The engine is rated at 215 bhp at 3000 rpm. The test is run for 20 hours at rated speed and maximum load conditions with controlled water out, fuel and inlet air temperatures and intake air restriction. At zero, five and 20 test hours, the engine oil is evaluated to determine the amount of entrained air in the oil. To pass this test the engine oil must not, at the end of 20 test hours, contain more than the maximum allowable amount of 8% entrained air.

Table 13.13 Total high oil circulation rates require foaming control – Navistar T44E (HEUI) engine for Ford Trucks (McGeehan *et al.*, 2006, p. 21)

Engine speed (rpm)	Oil capacity (litres) [quart]	Oil flow (litres/min) [gallon/min]	Time for one pass through (seconds)	Circulation of oil in sump (times/min)
3300	13.25 [14]	90 [23.8]	8.8	7

13.18.3.4 Laboratory Bench Tests

In addition to the nine engine tests discussed in above sections, there are an additional six bench tests that are used to qualify the engine oil as meeting the API CJ-4 category. They are:

- ASTM D-892 laboratory bench test for foaming
- High temperature corrosion bench test ASTM D-6594 for corrosion
- Bosch injector test ASTM D-7109 for shear stability
- Noack volatility ASTM D-5800
- High temperature high shear viscosity ASTM D-4683
- Seal compatibility tests.

High Temperature Corrosion Bench Test ASTM D-6594

The High Temperature Corrosion Bench Test was originally developed by Cummins because of concerns Cummins had about the loss of lead in the bearings and bronze pin corrosion that resulted in camshaft failures at extended warranties (Figures 13.53 and 13.54) (McGeehan, 2008).

This test method is intended to simulate the corrosion process of nonferrous metals in diesel engine oils and is used to test diesel engine oils to determine their tendency to corrode various metals, specifically alloys of lead and copper commonly used in cam followers and bearings.

In the test, four metal specimens of copper, lead, tin and phosphor bronze are immersed in a measured amount of engine oil at a test temperature of 135 °C (275 °F). Air is blown into the engine oil sample for a given period. When the test is completed, the copper specimen and the stressed oil are examined to detect corrosion and corrosion by-products, respectively. Reference oil is tested against each engine oil under test to verify test acceptability. At the end of the test the engine oil is analysed for metals and the copper portion of the test coupon is rated for discoloration per the rating system described in the ASTM D-130 Copper Strip Corrosion Test Method. To pass the test the engine oil must exhibit the following limits:

- Copper ppm increase, maximum 20 ppm
- Lead ppm increase, maximum 120 ppm
- Tin ppm increase, maximum 20 ppm
- Copper strip rating 3 maximum.

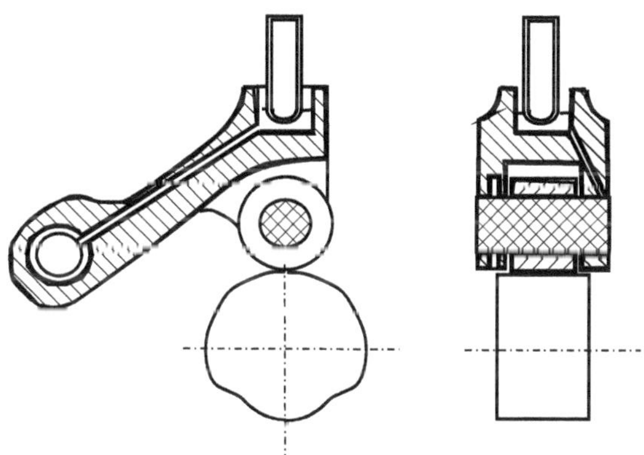

Figure 13.53 Camshaft and roller follower assembly with bronze pin and steel roller

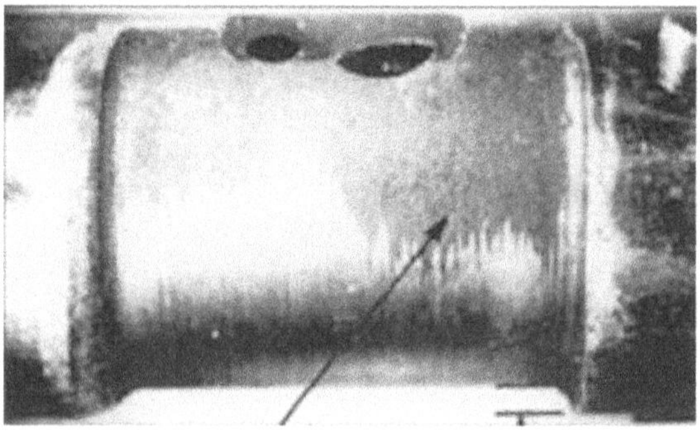

Figure 13.54 Corrosion of roller follower bronze pin from the field

Bosch Injector Test ASTM D-7109

This test method evaluates the percentage viscosity loss for multi-grade heavy duty diesel engine oils due to shearing of the engine oil's viscosity index improvers from the high shear rates and pressures that can be experienced in heavy duty diesel engines. In this test the engine oil's viscosity at 100 °C is measured and it is passed through a diesel injector nozzle at a test temperature of 35 °C (95 °F) for 90 cycles at a shear rate that causes the molecules of the viscosity index improver to degrade. After 90 cycles the engine oil's viscosity at 100 °C is measured again to determine the percentage viscosity loss and to see if it stayed in its viscosity grade. To pass this test the engine oil at the end of 90 cycles must stay in its intended viscosity grade. For example, SAE 15W-40 viscosity grade engine oil must not exhibit a viscosity of lower than 12.5 cSt at 100 °C.

Noack Volatility ASTM D-5800

When an engine oil comes into contact with the high temperature zones in the engine, such as the turbocharger, cylinder walls, valves and the under-crown and ring belt area of the pistons, it can begin to evaporate. Evaporation of the engine oil can contribute to oil consumption in an engine and can lead to a change in the properties of the engine oil and the formation of deposits on critical engine parts. Engine oil's volatility characteristics are directly related to the type of base oil used in the formulation of the engine oil. The volatility characteristics of an engine oil as measured by the ASTM D-5800 Method have been found to correlate with oil consumption in both passenger cars and heavy duty diesel engines.

In this test, a measured quantity of sample is placed in an evaporation crucible or reaction flask that is then heated to 250 °C with a constant flow of air drawn through it for 60 minutes. The loss in mass of the oil is determined. The test method covers three procedures for determining evaporation loss. Procedures A and B use an electrically heated block to heat a steel crucible while Procedure C heats the glass reaction flask with an electric heating element. A vacuum pump is used to maintain a constant flow of air across the surface of the sample. To meet the API CJ-4 requirements an engine oil cannot exhibit greater that a 13% evaporative loss.

High Temperature High Shear Viscosity ASTM D-4683

The High Temperature High Shear Viscosity Test ASTM D-4683 is used to measure an engine oil's ability to provide an effective adequate viscosity in high shear components, such as the journal bearings and between the piston rings and cylinders, under severe operating conditions. The values obtained in this test provide an indication of the temporary shear stability of the viscosity index improver used in the formulation of multi-grade engine oils. If the engine oil is not able to maintain an adequate viscosity when high engine operating temperatures and high shear rates are encountered, wear to critical parts will occur. The test fixture consists of a motor that drives a tapered rotor which is closely fitted inside a matched stator.

The rotor exhibits a reactive torque response when it encounters a viscous resistance from an oil that fills the gap between the rotor and stator. Two oils, a calibration oil and a nonNewtonian reference oil, are used to determine the gap distance between the rotor and stator so that a shear rate of a $1 \times 10^6\,s^{-1}$ is maintained. Additional calibration oils are used to establish the viscosity/torque relationship, which is required for the determination of the apparent viscosity of test oils at 150 °C (302 °F). The apparent viscosity to the nearest 0.01 cP (mPa s) at 150 °C and $1 \times 10^6\,s^{-1}$ shear rate is reported. The API CJ-4 category specifies that the engine oil has a minimum High Temperature High Shear Viscosity of 3.5 cP to ensure that the engine oil provides the proper viscosity to maintain engine durability in high load, severe service applications.

13.18.4 Differences Between CJ-4 and Previous Categories and Benefits of Using CJ-4 Engine Oils

Engine oils that meet the API CJ-4 service classification are designed to function in a more hostile environment than those experienced by previous heavy duty diesel classifications. 2007 low emission compliant engines use increased EGR rates that generate increased soot generation, increased acid generation and increased heat generation. The use of engine oils that meet the API CJ-4 service classification will provide complete protection of both EGR containing and non-EGR containing engines that are used in either on-road or off-road applications in the areas of (Figure 13.55):

- Improved wear protection
- Increased soot handling characteristics
- Improved deposit and oil consumption control
- Improvement of the used oil's ability to pump at low temperatures
- Prevention of viscosity loss from shearing
- Protection from thermal and oxidative breakdown
- Greater oxidation stability at higher engine oil temperatures
- Extra protection by controlling crosshead valve train wear
- Enhanced bearing protection
- Longer seal life
- Increased engine durability.

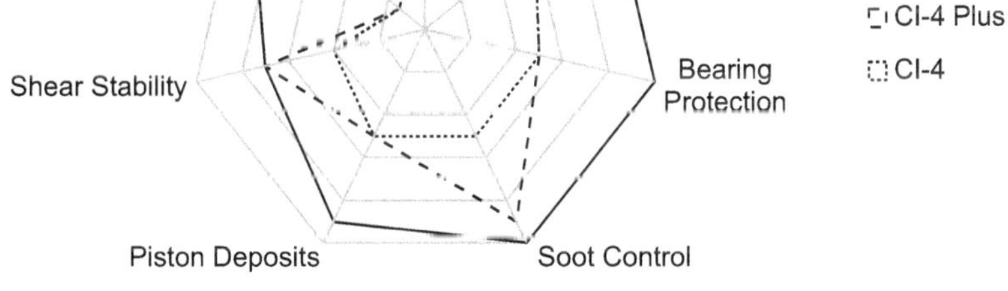

Figure 13.55 OEM specifications

In addition to the API CJ-4 specification, many of the major OEMs have developed their own engine oil specifications to ensure optimal engine oil performance and protection of their engines. These OEM specifications, which are considered to exceed the API CJ-4 specification, are:

- Cummins CES 20081
- Detroit Diesel PowerGuard 93K218
- Mack EO-N Premium Plus-07 (API CJ-4 requirements with more restrictive limits)
- Volvo VDS-4
- Renault RLD-3.

To qualify as meeting the Cummins CES 200081 specification, the engine oil must pass more stringent valve train wear requirements in the Cummins ISB and ISM engine tests, more stringent ring and liner wear requirements in the Mack T-12 engine test and lower MRV limits of 18 000 cP at −25 °C (−13 °F) in the Mack T-11 and T-11A engine tests.

Detroit Diesel PowerGuard 93K218 is similar to the API CJ-4 specification but with the additional requirements to pass the Daimler OM 441LA engine test at the Daimler 228.3 specification levels of no greater than 2% bore polishing. The Daimler OM441LA engine sequence test (CEC L-52-T-97) is a 400 hour test used to evaluate an engine oil's ability to protect against sludge and piston deposits, bore polishing, cylinder liner wear and excessive oil consumption in engines equipped with aluminium pistons. This engine sequence test was included to cover the lubrication needs for the Mercedes Benz engines Detroit Diesel markets in North America. The specification also has lower MRV limits of 18 000 cP at −25 °C (−13 °F) in the Mack T-11 and T-11A engine tests.

To meet the Mack EO-O Premium Plus specification the engine oil must pass more stringent requirements in the Cummins ISB and ISM engine tests, more stringent ring and liner wear requirements in the Mack T-12 engine test and lower MRV limits of 18 000 cP at −25 °C (–13 °F) in the Mack T-11 and T-11A engine tests. The oxidation requirements are also more severe and mirror those required for the API SM service classification for petrol powered passenger car engine oils. The Volvo VD-4 and Renault RLD-4 specifications mirror the performance of the Mack EO-O Premium Plus with the addition of the Volvo D12D Piston Deposit engine sequence test. This is a 400 hour test that uses a Volvo D12D 460 engine. The test is used to evaluate the engine oil's ability to protect against piston deposits, bore polishing and cylinder liner wear. To qualify as meeting Volvo VDS-4 and the Renault RLD-3 specifications the engine oil must not exhibit more than 150 cm^2 (25.32 square inches) of bore polishing wear.

13.19 Future Engine Oil Developments

It is anticipated that in the future that any further developments in heavy duty diesel engine oil technology will not only be driven by further emission reduction by global regulatory bodies but also will be driven by the desire for improved fuel economy, any issues the OEMs or their customers are experiencing in the field, the desire for a globalization of engine oil specifications and the desire of OEMs and their customers to preserve the performance of their older equipment.

With the advent of the emission specifications for increased reduction in NO_x emissions in 2010 for on-road heavy duty diesel engines and the reduction in NO_x and PM emissions for off-road heavy duty diesel engines that began in mid-2010, the use of API CJ-4 quality engine oils is expected to grow in the United States.

Off-road engines from 37–560 kW (50–750 hp) have been regulated for NO_x and PM emissions since 1996 in tiered stages. These emissions reductions have been slowly phased in since 1996 according to horsepower ratings by the following Tiers (Figures 13.56 and 13.57):

- Tier 1 spanned 1996 through 1999
- Tier 2 spanned 2001 through 2004

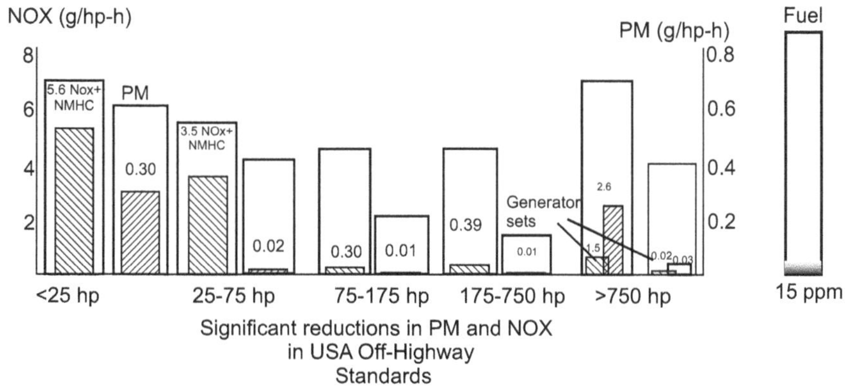

Figure 13.56 Tier 4 off-road emissions 2009–2013

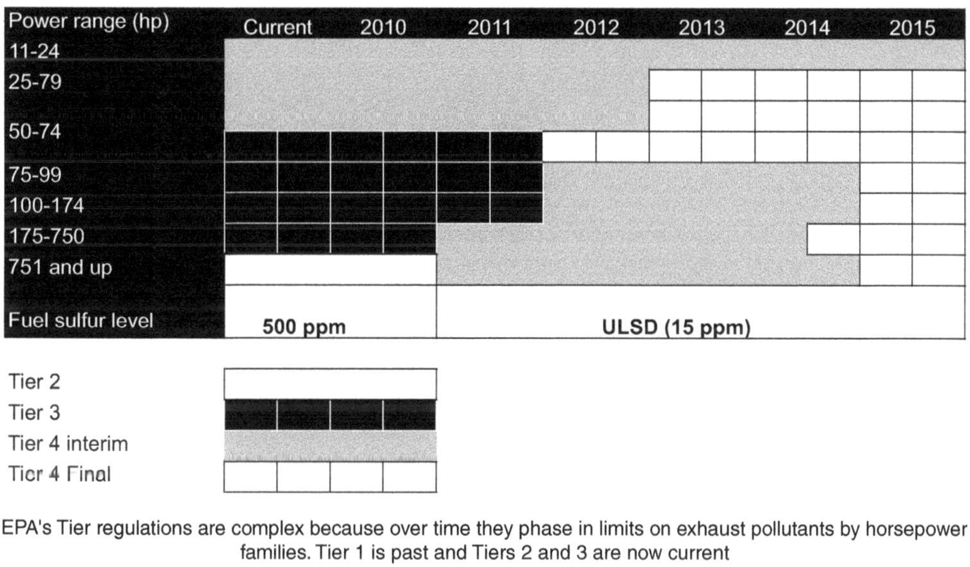

Figure 13.57 Road map of emission standards for off-road engines in the USA

- Tier 3 spanned 2006 through 2008
- Interim Tier 4 spans 2008 through 2012
- Final Tier 4 spans 2012 through 2015.

Current off-road emission regulations cover Tier 3 and Tier 4 Interim engines (Figure 13.58). Tier 3 took effect in 2006 and Tier 4 Interim in January 2011 for engines above 173 hp (129 kW). To meet Tier 3 emissions requirements OEMs have adopted the following emission technologies:

- ACERT (Cat) or Cooled EGR (10% to 15% EGR)
- Variable Geometry Turbocharging

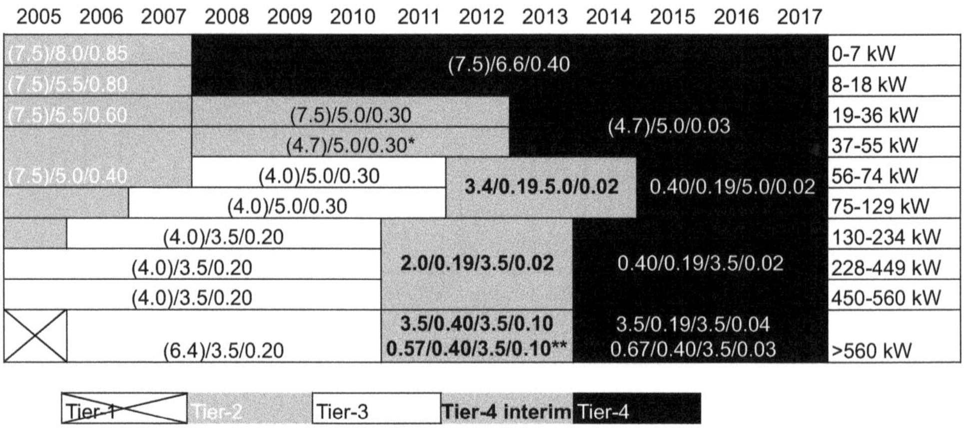

Figure 13.58 US Emissions regulations off-road (Cummins, 2010, p. 11)

- Common rail advanced fuel injection
- Electronically controlled unit injectors
- High injection pressures – up to 35 000 psi.

Tier 4 Interim regulations took effect in North America in January 2011 for engines above 173 hp (129 kW). Equivalent European Union Stage IIIB regulations began at the same time for European Union countries (Figure 13.59), while similar regulations commence in October 2011 for Japan (Figure 13.60). A maximum sulfur content in diesel fuel of 15 ppm will be regulated for these off-road applications in North America and 10 ppm in Europe and Japan. The EPA Tier 4 Interim and European Union Stage IIIB emissions standards and effective dates vary by power category (Cummins, 2010, p. 7).

The regulations require a major reduction of particulate matter (PM) and also require significant oxides of nitrogen (NO_x) reductions. The most stringent reductions apply across the 50–75 hp (37–55 kW) power categories with 90% PM and 45% NO_x reductions. These reductions drive the need for PM exhaust aftertreatment and enhanced engine technology. In the 75–99 hp (56–74 kW) power category, PM aftertreatment will also be required to achieve over 90% PM reduction, although NO_x reductions are less severe than higher power engines. For engines less than 9 hp (7 kW), Tier 4 began in January 2008 and required a PM reduction of 50%. Tier 4 Interim standards began in 2008 for engines greater than 75 hp (56 kW) in the United States and Canada, but are not applicable in Europe or Japan. Emissions levels are less severe than those of engines in the 50–75 hp (37–56 kW) power categories, recognizing the challenge of retaining

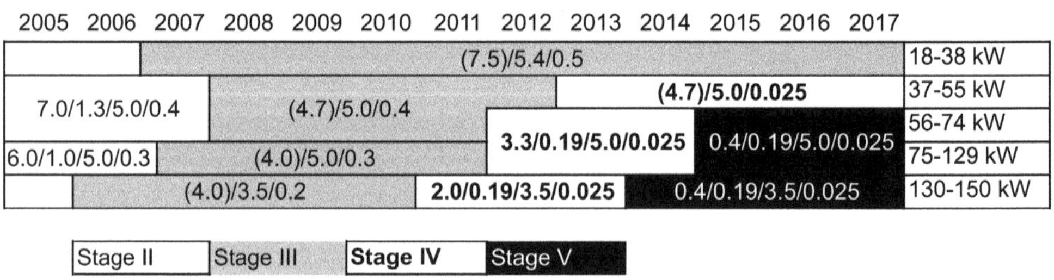

Figure 13.59 EU Emissions regulations off-road (Cummins, 2010, p. 11)

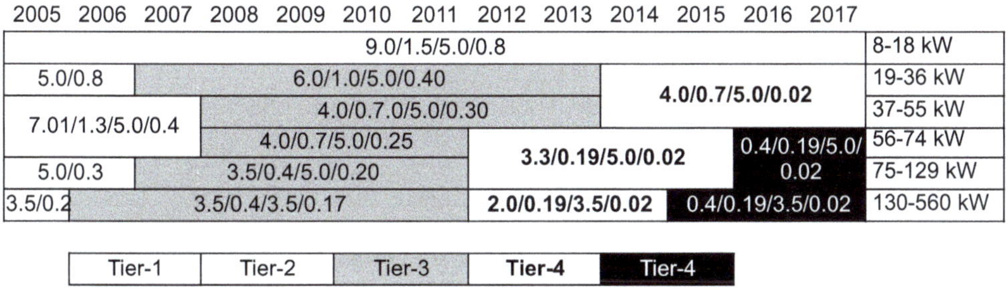

Figure 13.60 Japanese emissions regulations off-road (Cummins, 2010, p. 11)

power density for these larger displacement engines. For large engines used in portable power generation applications, the regulated NO_x emissions levels are set lower than for mobile machines. In January 2011, EPA Tier 4 Final and European Union Stage IV standards reduced NO_x emissions by an additional 45% for engines above 26 hp (19 kW). Similar NO_x reductions will also be required for engines above 75 hp (55 kW) in January 2015. Japanese regulations will closely align with the EPA and European Union emissions standards. At this point, both NO_x and PM emissions will be at near zero levels, comparable with the most stringent on-road applications. With only a three-year gap between the introduction of Tier 4 Interim/Stage IIIB and Tier 4 Final/Stage IV, engine and after-treatment design architecture for 2011 will need to be capable of incremental emissions reduction for the 2014 solution (Cummins, 2010, pp. 7, 8).

To meet these regulations the United States, European Union and Japanese OEM will employ the use of some of the same strategies that are being employed to meet current on-road emission standards. Many will eventually employ the use of DOC, NO_x absorber, DPF, increased EGR rates and SCR. Because of the use of these after-treatment devices and the fact that SCR is an exhaust only device, the use of engine oils meeting API CJ-4 and/or ACEA E-9 will increase in order to protect the durability of these engines.

In the United States as ozone and PM standards are adopted over the next five years, state and regional officials will begin to either regulate or offer incentives that will encourage the retrofitting or the phasing out of existing engines and equipment to reduce emissions. The retrofitting strategies will also increase the use of engine oils meeting API CJ-4 to protect the durability of these engines.

It is expected by the ASTM's Heavy Duty Engine Oil Classification Panel (HDEOCP) and the EMA that the current API CJ-4 Service Classification will last into 2015. Performance changes beginning in 2015 in response to pending fuel economy and Green-House Gas emission regulation for on-road heavy duty diesel engines to be implemented beginning in 2017 in the United States and the availability of 500 ppm sulfur fuel and hardware parts until 2015 for the Navistar Engine Oil Aeration Test, the GM 6.5L Roller-Follower Wear and the Sequence IIIF/IIIG Oxidation Tests may require both the modification of the existing API CJ-4 category and/or the development of a new heavy duty diesel engine oil category. The 6.5L Roller Follower Wear is considered to be redundant with the Cummins ISB and ISM wear tests and its removal from the API CJ-4 Classification is being considered. Caterpillar is working on a replacement for the Navistar Engine Oil Aeration Test and Mack is working on a replacement of Mack T-13 test for the Sequence IIIF/IIIG Oxidation Tests. Additional performance requirements are still being discussed to address issues and requirements regarding TBN depletion and total acidity number (TAN) increase, oxidation stability, turbocharger deposits, fuel economy testing and biodiesel compatibility (ASTM, 2010a).

Fuel economy is another driving factor. New US federal regulations addressing truck fuel economy, which are scheduled to take effect in July of 2012, with the first limits on fuel economy to be established in 2014 and full implementation in 2017 in conjunction with Green-House Gas emission regulations, will mean a new heavy duty diesel engine oil category being implemented. The Green-House Gas Emissions regulations will be based upon grammes of carbon dioxide per ton-mile and fuel consumption will be based upon gallons per 1000 ton-mile. For a well designed and well maintained Class 8 tractor–trailer combination the standard fuel economy is about eight miles per gallon. For this combination hauling 80 000 pounds (40 tons) its fuel economy is 0.0036 gallons per ton-mile.

The ASTM's HDEOCP is investigating the development of a heavy duty fuel economy dynamometer-based test developed by Volvo that uses a D12D Volvo engine run in steady state at 13 speed points of ESC (European Steady state Cycle) and varying speeds of transient cycle. In this test, fuel economy is significantly affected by the engine oil's High Temperature High Shear (HTHS) viscosity. Under operating conditions the engine oil's HTHS viscosity is what occurs in an engine under operating conditions and this is what protects the moving parts of the engine from wear during high temperature high shear operating conditions. Heavy duty diesel OEMs are concerned that the current limit of 3.5 cP minimum be maintained especially in light of the increased use of SAE 5W-40 and SAE 5W-30 viscosity heavy duty diesel engine oils by their customers.

It is well established that engine oils can influence fuel economy by minimizing friction and viscous losses in the engine contact areas and lowering the energy wasted pumping the oil through the engine. It is basically easier to wade through a pool of water than through a pool of honey or molasses. Viscosity of all engine oils varies with temperature. The viscosity of the oil at engine operating temperature is designed to minimize friction and drag, but also protect by separating the moving surfaces. If the oil is too thick in viscosity power loss and drag occur. If the oil is too thin in viscosity high friction and wear result. Low viscosity lubricants are less resistant to flow than conventional lubricants, a property that helps reduce friction and energy losses. Depending upon the application, low viscosity lubricants may also contain additives designed to withstand the extreme pressure that could occur as the lubricant flows between tight-fitting parts. Low viscosity lubricants may be made from synthetic or mineral oil blends with low viscosity and extreme pressure additives. European research demonstrates a three to five percent gain in truck fuel economy using low friction engine lubricants (EPA, 2004). The use of Group II + hydrocracked severely hydrotreated base oils or Group III oils and PAO synthetic base fluids, or combinations of these base oils, improves fuel economy by providing the right viscosity to protect and lubricate the critical parts of the engine while at the same time minimizing the friction by providing lower viscosity, which makes it easy to pump and lubricate and minimizes pumping losses. These base oil combinations along with an increased use of frictional modifiers such as organo-molybdenum compounds will offer substantial fuel economy benefits.

On the European side, ACEA in conjunction with the EMA is working on updates to the present ACEA E-9 classification to address turbocharger deposits, low temperature pumpability issues, engine sludge and piston deposit formation and oil degradation due to the use of biodiesel. This may eventually result in new ACEA Classifications with E-4 going to E-8, E-6 to E-10; E-7 to E-11 and E-9 to E-13 (ASTM, 2010b). This cooperation, in conjunction with the fact that the ACEA E-9 specification uses the API CJ-4 Service Classification as its platform, will eventually lead to a global heavy duty engine oil classification. Globalization has become the norm in many industry sectors and all OEMs sell engines globally that must meet a mire of emission regulations. Further cross-regional mergers or cooperations between OEMs such as Daimler–Detroit Diesel, Scania–Cummins, Volvo–Renault–Mack and Scania–-Hino and the consolidation of lubricant additive manufacturers to a small core of global players has resulted in global cooperation taking place.

13.20 Summary

Changes in diesel engine oil specifications are driven (a) by government through emissions regulations, (b) by OEMs to function with increased soot levels, increased acid levels and increased temperatures at which the engine and oil operate, and (c) by the consumers for long oil life at lower cost. The engine oil is expected to lubricate, prevent wear, reduce friction, protect against rust and corrosion, keep engine parts clean, minimize combustion chamber deposits, cool engine parts, seal combustion pressures, aid in fuel economy and permit easy start up and pumping. The components of engine oil performance are viscosity and protection against wear, deposits and oil deterioration. Engine oil performance standards are developed in the United States systematically starting from Category Request and Evaluation to licensing involving the American Petroleum Institute, engine manufacturers, the American Society of Testing Materials and the Society of Automotive Engineers. The licence contains Service Classifications and Performance Levels, SAE Viscosity Grade and Energy or Resource Conserving or Supplemental Categories. Similar systems exist in Europe through ACEA, which specifies E-sequences of oil. Some OEMs develop their own oil and specify it. The new oils are backward compatible and hence some API Service classifications have become obsolete. Engine oil composes of base oils (mineral or synthetic) and additives.

Most of the base oil stocks are mineral oils, obtained by refining processes. The refining procedures are separation type or chemical conversion type. Distillation processes, propane deasphalting, solvent extraction, dewaxing, hydroprocessing methods, hydrofinishing, hydrocracking, two-stage hydrotreating, isodewaxing and hydrotreating process, raffinate hydroconversion, hydroisomerization are some of the methods. The base oils are grouped one to five with decreasing volatility and increasing high temperature capability along with increasing viscosity index and decreasing sulfur content. Hand in hand, their ability to hold additives in solution reduces and it is taken care of by the chemistry. The additive chemistry consists of improved detergents, dispersants, anti-wear compounds, friction modifiers, rust and corrosion inhibitors, oxidation inhibitors (antioxidants), viscosity index improvers, pour point depressants and foam inhibitors.

Maintaining and changing engine oils are specified by the OEMs as oil change intervals after extensive analysis of used engine oil. For proper engine oil analysis care must be taken while taking the sample. One-off sampling is useless. The quality of engine running can be determined only by oil trends analysis, for which oil must be sampled periodically.

Newer engine design technologies (High Pressure Common Rail fuel system, combustion optimization, advanced turbocharging, exhaust gas recirculation, Advanced Combustion Emissions Reduction Technology (ACERT) and crankcase ventilation) and strategies used to control emissions, namely exhaust after-treatment and diesel particulate filters (catalysed DPFs, fuel additive-catalysed DPFs, continuously regenerating traps, NO_x absorbers, lean NO_x traps, lean NO_x catalysts and selective catalytic reduction using Diesel Exhaust Fluid) have decisively influenced oil development.

The impact of cooled EGR on engine oil is increased soot levels, acid levels and heat levels. Post-injection for regenerating DPFs has increased dilution of engine oils contributing to wear. Engine oils have changed to cope with the demands of low emissions. API CH-4 was upgraded to CH-4 Plus. Now, CJ-4 has been introduced to have reduced impact of sulfated ash on diesel particulate filters with higher soot handling, shear stability and oxidation resistance. Tests were updated and new sequences introduced to check the adaptation of the oils to the new engine and after-treatment technologies.

References

ACEA (2010) ACEA 2010 European Oil Sequences for Service-fill Oils for Gasoline engines, for Light Duty Diesel engines, for Gasoline and Diesel engines. European Automobile Manufacturers Association, Brussels, Belgium.

API (2010) Motor oil guide: which oil is right for you? API, Washington, DC.

API (2011a) Engine Oil Licensing and Certification System, API Bulletin 1509 (16th Edition April 2007; Addendum 1 March 2008). API Publishing Services, Washington, DC, Appendix D – Developing New Diesel Oil Performance Standards for API C Service Classifications, pp. 36–45, January 2011 Version.

API (2011b) Engine Oil Licensing and Certification System, API Bulletin 1509 (16th Edition April 2007; Addendum 1 March 2008). API Publishing Services, Washington, DC, Appendix E – API Base Oil Interchangeability Guidelines For Passenger Car Motor Oils and Diesel Engine Oils, page E-1, Updated January 2011.

ASTM (2010a) Heavy-Duty Engine Oil Classification Panel of ASTM D02.B0.02, Minutes, 29 June 2010, Kansas City, MO.

ASTM (2010b) ASTM Heavy Duty Engine Oil Classification Panel of ASTM D02.B0.02, Minutes, 7 December 2010, Jacksonville, FL.

Cummins (2010) Cummins Mobile Off-Highway Emissions Choosing the Right Technology for Tier 4. Cummins Engine Company, Columbus, IN, pp. 7, 8, 11.

DECSE (Diesel Emission Control–Sulfur Effects Project) (2000) Phase 1 Interim Report No. 4: Diesel Particulate Filters – Final Report, January 2000. National Renewable Energy Laboratory (NREL), US Department of Energy, Golden, CO, p. 3.

Detroit Diesel (2010) BlueTec Emmissions Technology – 2010 Emissions Guide. Detroit Diesel Corporation, Detroit, MI.

EPA (2004) Clean Fleet Strategies Low Viscosity Lubricants. EPA 420-F-04-006, Office of Transportation and Air Quality, Environmental Protection Agency, Washington, DC.

EPA (2000) Regulatory Impact Analysis and Vehicle Standards and Highway Diesel Fuel Control Requirements. EPA 420-R-00-26, Assessment and Standards Division, Office of Transportation and Air Quality, Environmental Protection Agency, Washington, DC. pp. 96–98.

Farania, R. (contributor) (2005) Chevron Global Lubricants GF-4: A New Performance in Passenger Car Motor Oils. Lubrication – A Technical Publication Devoted to the Selection and Use of Lubricants, June 2005, Chevron Products Company, San Ramon, CA, p. 8.

Galligan, J. (2005) 2007 Engines, It's All about DPF. *Light Duty & and Medium Duty Trucking Magazine*, June 2005, 22 and 28.

Gray, C. (2005) Diesel Tests Explained. Abstracts of Procedures Performed by the Fuels and Lubricants Research Division, Southwest Research Institute, San Antonio, TX, p. 5.

ILMA (2010) Forge the Energy of the Future. Presentation on European Engine Oil Specifications, ILMA Annual Meeting, October 2010, Hyatt Grand Champions Resort, Villas and Spa, CA.

Kilcarr, S. (2004) Engine Oils Adding a Plus. *Fleet Owner Magazine*, 1 September 2004 (online).

Lubrizol K2M (2005a) American Petroleum Institute – course notes. The Lubrizol Corporation, Wickliffe, OH.

Lubrizol K2M (2005b) European Engine Oil – course notes. The Lubrizol Corporation, Wickliffe, OH.

Lubrizol K2M (2005c) EGR Engine Course and Attributes of Diesel Engine Oils – course notes. The Lubrizol Corporation, Wickliffe, OH.

(2005) SAPS and Emission Hurdles without End. *Lubes' & n'Greases*, Volume **11**, Issue (5) 10.

McGeehan, J. (2006) Diesel Engine Oils for Ultra Low Emission Diesel Engines. A presentation on Global Heavy Duty Diesel Engine Oil Technology, ChevronTexaco Global Lubricants, San Ramon, CA.

McGeehan, J. (contributor) (2008) API CJ-4: The Most Robust Diesel Engine Oil Category for All Engines. Lubrication – A Technical Publication Devoted to the Selection and Use of Lubricants, March 2008, Chevron Products Company, San Ramon, CA.

McGeehan, J.A., Moritz, J., Shank, G., Kennedy, S., Stehower, D., Urbank, M., Belay, M., Gooder, S., Cassim, A., Runkle, B., DeBaun, H., Harold, S., Chao, K., Herzog, S., Stockwell, R., Passut, C., Fetterman, P., Taber, D., Williams, L., Kleiser, W.M., Zalar, J., Scinto, P., Santos, E., Rutherford, J.A., *et al.* (2006) API CJ-4-Diesel Oil Category for Both Legacy Engines and Low Emission Engines Using Diesel Particulate Filters. Technical Paper 2006-01-3439, SAE, Troy, MI, pp. 4, 7, 21.

Rizvi, S.Q.A. (2003) Additive and Additives Chemistry, in *Fuels & and Lubricants Handbook: Technology, Performance, Properties & and Testing* (ed. G.E. Totten). ASTM Manual Series MNL37WCD, ASTM International, West Conshohocken, PA, Chapter 9, pp. 225, 227, 228, 230.

Whitcare, S., Catalyst Compatible Diesel Engine Oils, DECSE Phase II Presentation at DOE/NREL Workshop 'Exploring Low Emission Diesel Engine Oils', January 31, 2000.

Xinqun, G., Deu, D. and Winsor, R., Non-Road Diesel Engine Emissions and Technology Options for Meeting Them, Product Engineering Center, Deer and Company, 2010, ASAEE Distinguished Lecture #34, Agricultural Equipment Technology Conference, January 2010, Orlando, FL, pages 6, 7, 10, 11, 12 and 16.

Part VI

Fuel Injection Equipment

14

Wear of Fuel Injection Equipment

Fuel injection equipment is at the heart of a diesel engine and has the lowest operating clearances in the engine build, demanding control of fuel quality to protect it from abrasive and corrosive wear. The highly stressed needle seats and the cam drives in the new generation fuel pump depend on the lubricity of fuel for functioning without wear. Worn fuel injection equipment causes loss of power and instability at low loads.

14.1 Introduction

Advanced emission standards require higher injection pressures, which help to bring smoke and consequent particulate matter emissions under control. This enables retarding of the fuel injection timing close to the engine top dead centre (TDC), and hence controls nitric oxides. The fuel injection pressures have progressed from the limits of 60 MPa for the inline block-type pumps and 85 MPa for the distributor-type pumps to 180 MPa for the common rail fuel injection systems. Active engine development is taking place with injection pressures touching 250 MPa. However, higher injection pressures demand very tight clearances (approximately 1 μm) between the metallic parts in the fuel injection pump and the injector, so that the leakage through the clearances is kept low even at these higher fuel pressures. This has pushed up the demand on the fuel filters for higher efficiency to limit small particles. Diesel fuel lubricity and water contamination affect the wear behaviour of fuel injection equipment, as does abrasive dust.

14.2 Wear due to Diesel Fuel Quality

14.2.1 Lubricity of Mineral Diesel Fuel

Lubricity is the oiliness or ability of the fuel to produce a lubricious film between wear parts inside the fuel injection system. Diesel fuel quality is declining throughout the world, a trend that is expected to continue. Due to many factors, 70% of the mineral diesel fuel sold at service stations does not meet the Society of Automotive Engineers (SAE) standard or the European EN 590 standard for diesel fuel lubricity.

14.2.1.1 Measuring the Lubricity of Diesel Fuels

The lubricity of diesel fuels is estimated using the High Frequency Reciprocating Rig (HFRR, similar in principle to the SRV machine, Chapter 12) by measuring the wear scar (in microns) and by measuring the

Critical Component Wear in Heavy Duty Engines, First Edition. P.A. Lakshminarayanan and Nagaraj S. Nayak.

critical load of incipient scuffing (load-carrying capacity) in a High Temperature Oscillating Machine (HiTOM). A good correlation exists between the results obtained using the two methods. To use the fuel safely against scuffing and adhesive wear, the maximum limit for the wear scar in the HFRR method is 400 μm (350 μm is the preferred limit). Apart from the two popular methods, a test method is available in the ASTM standard D6898-03 (2010); it evaluates diesel fuel lubricity in an injection pump rig that uses real components of a Common Rail (CR) or a rotary distributor type fuel injection pump.

14.2.1.2 Factors Affecting the Lubricity of Mineral Diesel Fuels

Crude Oil Sources

As crude oils become scarcer, the ability to select specific crudes carefully to produce high quality diesel is less economically feasible and compromises are made.

Refining Techniques

Two processes at the refineries result in poorer quality of fuels. Refineries now use catalytic cracking to increase the fraction of diesel and petrol that can be refined from crude oil and improve the economy. However, the process is good only for production of high quality petrol and has the side effect of reducing the lubricity of the diesel fuel. For economic reasons this refining process is likely to be the preferred method in the future, resulting in lower quality diesel. The other is that used to reduce the sulfur content in diesel fuel (a) to reduce the particulates formation in engines, (b) to avoid poisoning the diesel oxidation catalyst and the diesel particulate filters (DPF) and (c) prevent clogging of the DPFs. This process reduces the lubricity of the fuel.

Blending to Prepare Arctic Fuel

Diesel fuel has high paraffin (straight chain hydrocarbon, Figure 13.6) content. These can gel and particles of wax are formed which plug the filters at low temperatures. Therefore, for operation in cold weather conditions, 'arctic' fuel is prepared by refiners by blending kerosene with diesel fuel. The blended fuel gels at lower temperatures. Similarly, waxing and fuel filterability problems with the diesel fuel (NATO standard diesel fuel, F54), troubled the engines in army vehicles during cold weather. These problems were solved by 50:50 blending of the diesel fuel with aviation kerosene (aviation turbine fuel either JP5 or JP8) to lower the waxing tendency and decrease the viscosity of the diesel fuel (NATO F65).

This has the undesirable effect of reducing the energy content of the fuel, resulting in a decrease in fuel economy. It also may adversely affect the lubricating properties of the fuel, resulting in potential wear on vital engine components, as in many diesel fuel injection systems the fuel itself is the only lubricant, for example, for the precision engineered injection pump.

Impact of Single Fuel Concept (SFC) of the Military

The requirement for blending of fuels created logistics problems. To simplify fuel operations, JP8 was recommended as the diesel fuel for the military (Le Pera, 2004). One fuel for the army, navy and air force is considerably easier than multiple fuels when managing fuel storage, transport and tailoring distribution for maximum efficiency; it also reduces the possibility of dispensing the wrong fuel. The use of aviation kerosene by the military as the single fuel has enhanced long term storage stability, improved cold weather vehicle operation, reduced engine component wear and reduced fuel system corrosion problems. Although the single fuel concept has merits, it also has shortcomings.

To validate using aviation turbine fuels (kerosene) in diesel engines, many tests on engine dynamometers were conducted in the laboratory as well as field and fleet tests (16 000 km durability) with several commercial vehicles. The tests revealed none of the problems which appeared later in the extraordinary field conditions, namely sustained high load operation, high ambient temperatures, high fuel temperatures and indifferent fuel quality. One of the problems faced with the SFC was heavy white smoke (unburnt diesel vapour in the exhaust) as the Cetane number of kerosene (33) is too low for the

Figure 14.1 A truck at high altitudes (high Himalayas) and low temperatures leaving a large plume of white smoke (unburnt hydrocarbon) when aviation turbine fuel is used without any additives to improve ignition delay. The truck in front is obscured by the huge cloud of white smoke

fuel to auto ignite at high altitudes, where the temperatures are subzero and the ambient pressure is very low (Figure 14.1). A second problem is the severe wear of the fuel injection pump at the engine's operating temperatures.

Improving the Cetane Number of Aviation Turbine Fuel
At high altitudes and subzero temperatures, the Cetane number of aviation turbine fuel is insufficient and the combustion suffers due to an unacceptably long ignition delay in the thermodynamic cycle. To improve the Cetane number, ignition improvers (e. g. dimethyl ether) from respected fuel companies can be added to kerosene (aviation turbine fuel) in amounts up to 0.5%. Some of the mono-acid and ester-based additives used as ignition improvers also enhance lubricity of the fuel.

14.2.1.3 Adhesive Wear Problems with Low Viscosity–Low Lubricity Fuel in Injection Pumps

Of the many types of fuel injection pumps manufactured commercially, such as the single cylinder pump, the inline pump and the distributor pump, the rotary distributor type fuel injection pump is the most sensitive to the lubricating quality of the fuel. This pump is inexpensive and is used in a wide variety of commercial and military equipment powered by diesel engines. In these pumps, the fuel provides the lubrication to the internal moving components. When the lubricity of the fuel becomes insufficient, the pump components wear.

If fuel viscosity is sufficiently high at the operating conditions, the fuel physically separates the sliding components of the fuel injection system, preventing wear. However, with a lower viscosity, the potential for wear increases significantly because the surfaces of the sliding parts can begin to interact (mixed lubrication). The viscosity of kerosene-type fuel could be as low as 1.0 mm^2/s at 40 °C which can be compared to the minimum viscosity specified for low sulfur diesel fuel (ASTM D 975, 1.3 mm^2/s at 40 °C). In addition, the viscosity of the fuel decreases as the fuel temperature increases. The drop in viscosity results in a decrease in the ability of the fuel to lubricate the injection system and increases the dependence on lubricious surface films (due to lubricity) to control component wear (Lacey and Westbrook, 1995; Grigg, 1994). The lubricity of kerosene is of the order of 600–700 μm against the maximum allowable 400 μm for a fuel injection system. Hence, the lower the viscosity the higher is the potential for component wear.

A combination of low viscosity fuel and increased clearances between surfaces due to wear (resulting from insufficient lubricity) can cause increased internal fuel leakage that reduces the amount of fuel delivered to the combustion chamber, leading to gradual loss in power. More internal leakage occurs at low engine speeds, causing hot-starting and hot-idle problems in certain engines in desert operations or tropical climates.

Manufacturers of rotary distribution type fuel injection pumps (National Biodiesel Board, 2009) have mentioned categorically that the rotary fuel injection equipment would survive only when the lubricity of the fuel is assured and that the fuel must satisfy various quality requirements. However, many supply factory-fit or retrofit kits that lessen the potential for wear and hot restart problems when the viscosity of the fuel is low or the ambient temperatures are high or high load conditions are sustained.

When the vehicles are being used more extensively at increased engine power demands (higher load factors) imposed by overload conditions, the rotary distributor type fuel injection pumps have had many problems. Interestingly, the 'Conversion Retrofit Kit' for the use of arctic fuel apparently has done little to solve the problems due to the low lubricity of the fuel (Le Pera, 2005) in desert conditions.

New Thinking on the SFC

In view of the problems faced with rotary fuel injection pumps The United States army is recommending a change to engine types (Erwin, 2004) with fuel injection equipment that can handle aviation turbine fuels at high ambient temperatures.

14.2.1.4 Improving Lubricity

The hydrodesulfurization (refining) process to remove sulfur for reducing engine emissions robs the fuel of its lubricity by 80–200 μm and the load carrying capacity by 1500 N. Sulfur compounds (e.g. benzothiophene) cannot restore the lubricity of a hydrotreated diesel fuel, but can increase the load carrying capacity. Incorporating certain additives in the fuel can generate surface films that provide the necessary wear protection. Nitrogen compounds enhance the lubricity at concentrations as low as 20–100 ppm nitrogen (Matzke *et al.*, 2009). Specific lubricity additives with chemistries based on carboxylic acids, esters and amides can increase the lubricity in a variety of base fuels at a concentration of 200 ppm to meet the lubricity requirement according to EN 590 and can provide further protection to the fuel injection equipment.

It may be noted that 5% biofuel increases the load carrying capacity and the lubricity of kerosene sufficiently.

14.2.1.5 Surface Improvement by Diamond-like Carbon (DLC) Coating

High pressure fuel injection equipment used in common rail system or the wear parts in rotary distribution pumps faces extraordinary contact pressures and the only lubricant is the fuel. The low relative velocity combined with high contact pressures push the operation of the wear surfaces to a mixed lubrication regime.

Modern fuel injection systems consist of precise mechanical parts to hold high pressures in the pumping chambers and the nozzles. Therefore, the surfaces of the parts need solid lubricant coatings to protect them from friction and wear. Diamond-like carbon (DLC) coatings have the crucial advantages of low wear due to their high hardness and a low coefficient of friction, even for the nonlubricated contact of steel against its mating part. The coated surfaces also exhibit resistance to corrosion and slow diffusion (adhesion). DLC coatings offers advanced wear, friction and barrier properties and increase the life and value of the fuel injection equipment (Matzke *et al.*, 2009; Treutler, 2005). Figure 14.2 shows the DLC coated thrust plate in a rotary distributor type fuel injection pump. The application of a DLC coating extends to other areas of engineering, like hip joints, ball joint components, intake valves, wrist and piston pins, retainers, tappets and IR optics (Treutler, 2005) where fretting wear is incipient.

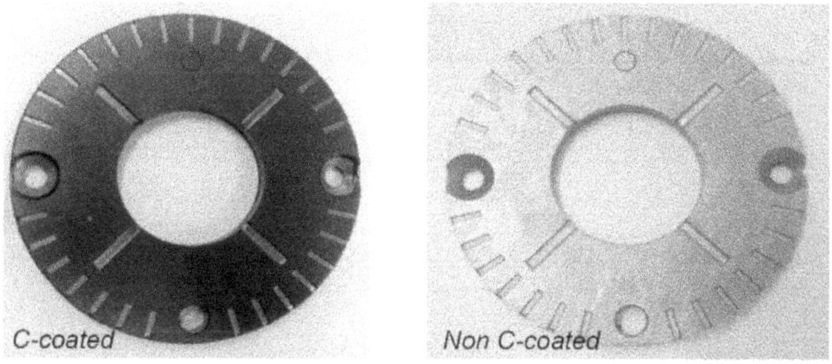

Figure 14.2 DLC coated thrust plate in the pump to control wear

14.2.2 Oxygen Content of Biodiesel

In 1997, the European Committee for Standardization developed a uniform standard, EN 14214:2009, for fatty acid methyl ester (FAME) or biodiesel. The specification enables the engine manufacturers to give warranties to the vehicles run on 7:93 blends of biodiesel–mineral diesel in Europe and for 20:80 blends meeting ASTM D6751-09 in the USA. As mentioned before, biodiesels have high lubricity, and hence the fuel injection equipment survives wear in a mixed lubrication regime.

Biofuels have high Cetane number, hence combustion is smooth. They are oxygenated, hence smoke emission is less. On the other hand, use is accompanied by an increase in NO_x emissions because the oxygen inbuilt in the fuel increases the combustion temperatures locally in the flame. Also, the oxygen in the fuel corrodes fuel injection equipment of all types when the blend exceeds the recommended ratio, especially when the engine is at rest. Therefore, the fuel (in blends) is suitable for taxis or long distance vehicles that are constantly on the move. In the case that the vehicle is to be at rest for a period longer than two days, it is advisable to run the vehicle for about 100 km using neat mineral diesel so that the injection equipment is flushed of biofuel.

14.3 Wear due to Abrasive Dust in Fuel

Depending on the severity of the operating conditions and the logistics of fuel transport and storage (at the stations and on the vehicle) the diesel fuel is loaded with dust and dirt. Based on the concentration of the dust in the fuel (severity of operation), and the type and the expected life of the fuel injection equipment, the filtration efficiency is determined (Figure 14.3). In its lifetime, the fuel injection equipment can tolerate only a certain dosage of dust, similar to the limit for an assembly of piston and liner (Figure 9.16). The critical dosage of dust is three grammes for wearing out the rotary distributor type fuel injection pump or a common rail system which may be expected to cope with 45 000 litres of fuel during its lifetime (Leeming and Hartley, 1989). With higher loading of dust in the fuel, the fuel injection equipment would last for shorter time with the same filtration efficiency. Examples of severe wear at the spill port of a plunger and the ball seat of a common rail injector where the velocity of the fuel is high are shown in Figure 14.4.

14.3.1 Wear of Injector Nozzle due to Heat and Dust

The three-body wear phenomenon (Figure 8.1) takes place between the seat, the abrasives in the fuel entering the nozzle and the needle tip. In addition, micron sized movement by the needle and the seat relative to each other takes place after seating because of elastic distortions of the seat geometries under high spring pressure; and this causes fretting wear (Chapter 3). The seat wear (Figure 14.5) results in an increase in the maximum needle lift and loss of nozzle opening pressure. The flow of fuel

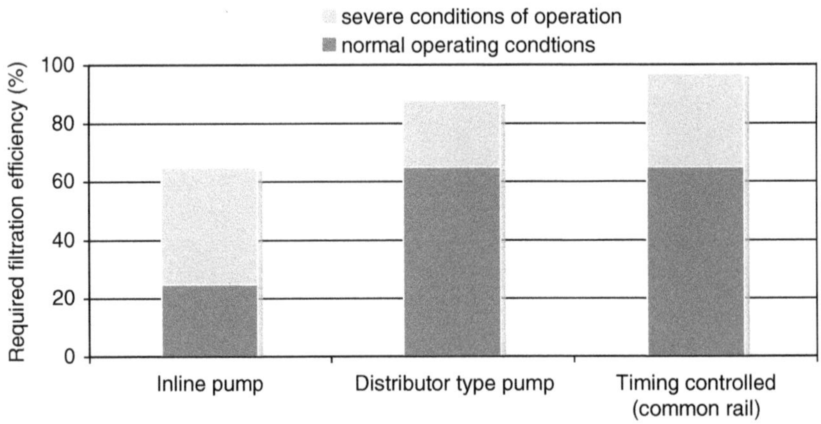

Figure 14.3 Increasing demand on filter efficiency (ISO 19438:2003) with higher injection pressures depending upon the severity of operating conditions

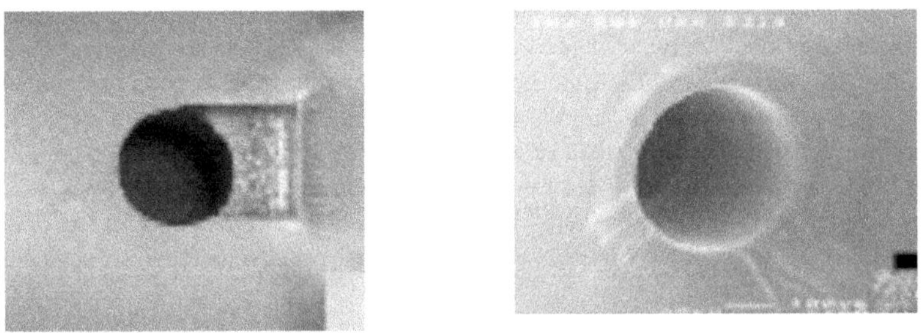

Wear of spill port of distributor plunger Ball seat of common rail injector

Figure 14.4 Examples of severe wear of injection equipment due to dust

Figure 14.5 Needle seat wear

with small amount of contaminant wears out the nozzle hole and distorts the finely machined shape of the injector holes.

Heat flowing from the combustion chamber is taken away by the fuel flowing through the nozzle. However, when the needle is seated there is no flow of fuel and the temperature of the nozzle tip increases; it is cooled again during the next cycle of injection. However, if the heat incident is sufficiently large, the fuel in the nozzle hole, whose diameter is as low as 0.14 mm for modern truck engines, can carbonize and slowly build up on the inner surface of the hole. The allowable minimum size of the hole and the incident heat bear an inverse relationship. Also, techniques to improve spray quality for low emission combustion involve design of converging holes. The difference in the outlet and inlet diameters of these tiny holes, known as the k-factor (usually up to 1.4), intensifies the build up of layers of coke in the hole (Figure 14.6). Increased sac volume in the nozzle tip accelerates clogging as the fuel remaining in the sac distils at high temperature when the needle is closed, allowing carbon to form thin layers on the surface of the holes.

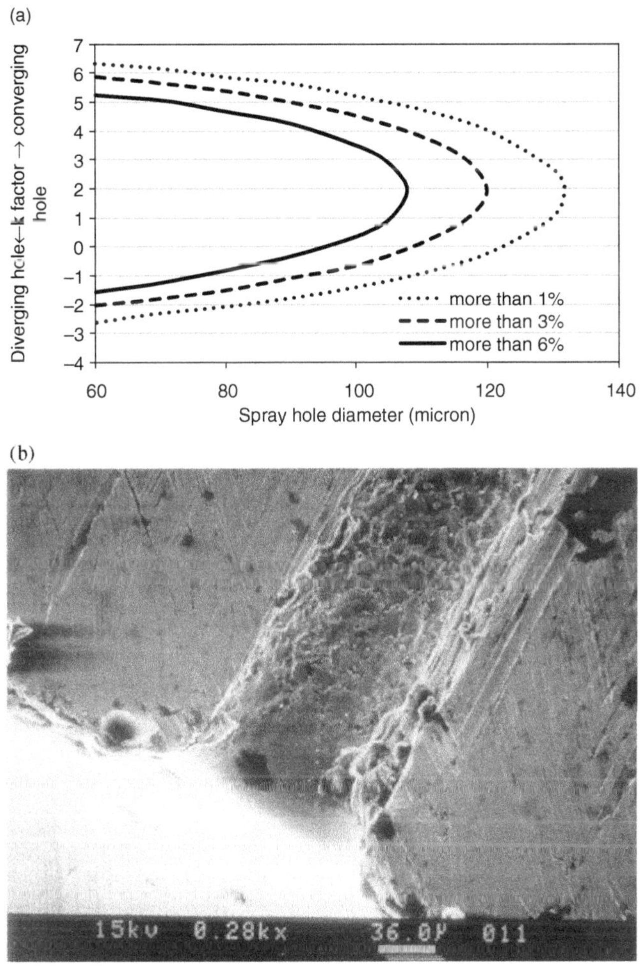

Figure 14.6 (a) Effect of smaller hole size and conical shape of tiny nozzle holes on clogging tendency; clogging of nozzle hole area (in percentage) is contoured. (b) Build up of carbon in the tiny nozzle hole due to high temperature and very small size of the hole

The problem is accentuated in engines rigged with an engine brake, where the inlet and exhaust valves are kept just open to permit the flow of air through them and absorb the kinetic energy of the vehicle as heat in the air. Since the flow of fuel through the nozzle hole is either stopped or kept at a very low level, the build up of carbon in the nozzle hole is rapid. Due to wear and build up of carbon, the shape of the nozzle hole changes from the designed shape and the flow area reduces. Also, the lack of cooling of the nozzle could raise the temperature of the nozzle tip to anneal it permanently when the engine brake is applied for long durations. The resulting loss of hardness accelerates the wear rate of the seat, the needle and the nozzle hole.

The spray pattern produced by a worn nozzle is irregular and prevents optimum combustion. Higher fuel consumption and an increase in emissions of smoke, particulate matter and hydrocarbons are consequences. Worn nozzles can also cause excess smoke during a cold start.

As a wear item, the fuel injector tips (the nozzle) should be regularly replaced. The nozzles in a passenger car could last from 100 000–300 000 km and for a truck engine the life could be 300 000–500 000 km. Replacement will restore the engine performance to its new condition. Worn nozzles would not cause any damage but replacing them will restore lost fuel economy, power, make less smoke and give the vehicle a smoother idle.

14.3.2 Fuel Filters

Dirty fuel fed to the fuel injector pump and the injector nozzles causes high wear rates on closely machined parts within these units. This upsets their efficiency and causes loss of power and an increase in smoke from the exhaust. Because of the machining limits (clearance is approximately 1 μm) the efficiency of the filters must be such that particles of dirt approximately 3 μm and more in size should be removed.

14.3.2.1 Basics of Fuel Filtration

Figure 14.3 shows the required filtration efficiency for different types of fuel injection equipment in normal and severe operating conditions (maximum injection pressures for an inline pump = 70 MPa, for a distributor type pump = 90 MPa, for a common rail system >130 MPa). The higher the injection pressure, the finer is the clearance between wear parts in the fuel injection system. The high pressure common rail systems demand 98% filtration efficiency for the 3–5 μm particles, compared to a high of 65% for an inline fuel injection pump for the same 'severe' operating conditions. It is a general guideline. However, improvement in efficiency may be required based on application, fuel contamination and lifetime. For example, in an extremely dusty area like open mines the concentration of silica dust can be very high. In these areas, the fuel tank breathes the atmospheric air as the tank is emptied during running and, consequently, also takes the dust into the tank along with the air. Also, dust enters the fuel during logistics in these areas. Thus, the fuel is loaded with silica dust and a fraction of it enters the fuel injection equipment depending on the efficiency of filtration. The example given in Table 14.1 shows the calculation

Table 14.1 Selecting the fuel filter for a modern common rail fuel injection system where the timing is controlled electronically

Operating condition	Dust concentration (mg/m^3)	Percentage remaining after filtration × dust concentration	Required minimum fuel filtration efficiency (%)
Tropical roads	20	$2 \times 20 = 40$	98
Tropical cities	5	$8 \times 5 = 40$	92
Open mines	600	$0.067 \times 600 = 40$	99.933

to select the fuel filter for a modern common rail fuel injection system where the timing is controlled electronically.

14.4 Wear due to Water in Fuel

The greatest enemy of diesel fuel injection components is water. Water ingress can take place during either the transport or storage. Water contamination can exist in diesel fuel in three forms:

(i) Emulsified water, where the water is suspended in the fuel, like oil and vinegar in a salad dressing.
(ii) Free water, where the water is separated from the fuel and is usually found on the bottom of fuel/storage tanks.
(iii) Dissolved water, where the water has been chemically dissolved in the fuel, like sugar in coffee. The warmer the fuel, the more water will be dissolved, but as temperatures drop the water will come out of the solution in the form of free water.

Once water enters the fuel system, it will rapidly wear and oxidize steel components, leading to rusting and corrosion of components, failure of governor and metering components, sticky metering components (both pump and nozzle), injection component wear and seizure and formation of colonies of microbes in the fuel living on the hydrocarbons (Figure 14.7).

14.4.1 Corrosive Wear due to Water Ingress

If the water is finely distributed and a conventional water separator using an inertia principle is unable to separate the water better than 20%, serious corrosion may be noticed in the fuel injection pump (Figures 14.8 and 14.9).

14.4.2 Use of Emulsified Water for Reducing Nitric Oxides in Large Engines

Introduction of emulsified water and diesel fuel into the combustion space reduces the maximum temperatures in the combustion process because of its evaporation, and improves atomization of the fuel, thereby reducing the NO_x emission (MAN B&W, 2000, 2001). High speed diesel fuel used in trucks and

Figure 14.7 Microbes colonizing the fuel filter

Figure 14.8 Rusting of parts in the fuel injection equipment

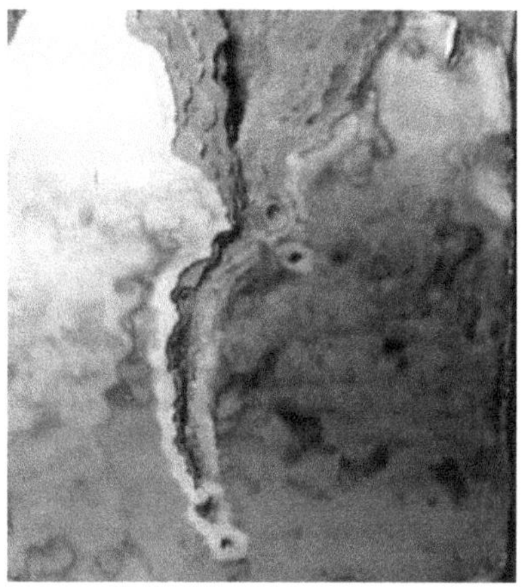

Figure 14.9 Severe corrosion on the surface inside the fuel injection equipment

vehicles cannot hold water in emulsified form without the help of an additive agent and leads to corrosion of fuel injection equipment. On the other hand, heavy fuel oil (HFO) can hold water for a longer period. Before the 1980s the emulsifier was mostly considered for homogenizing HFO to disperse the sludge and water that remained in the fuel after centrifuging (Chapter 15) and filtering. For optimal spray in the combustion chamber, the water droplets in the fuel oil after emulsification should be a maximum of 5 μm in size. This is obtained by using an ultrasonic-type homogenizer, whereas for the earlier mechanical type the size was above this limit. Without fine emulsification corrosion ensues. The addition of water to the HFO by homogenization increases the viscosity. To keep the viscosity of HFO at the engine inlet at 10–15 cSt (limit 20 cSt), the fuel temperature is raised above the usual limit of 150 °C to 170 °C at 50% water and the fuel oil feed pressure raised. The supply (feed) pump pressure can be changed to $\Delta p = 0.9$ MPa instead of the current $\Delta p = 0.4$ MPa.

14.4.3 Microbiological Contamination of Fuel Systems

The cost attributable to microbiological contamination of diesel fuel is a serious threat to marine safety and cannot be easily quantified. The aerobic and anaerobic bacteria of different types produce organic acids such as sulfurous and sulfuric acids. These chemicals are corrosive and cause severe degradation of storage tank coatings, corrosion of metal tank surfaces and injection equipment.

The microbiological contaminants on fuel injection equipment quickly impair performance and dramatically increase wear rates. This brings about poor combustion conditions, loss of engine power and reduced service life of the components. Also, injectors and nozzle orifices will get clogged and wear faster. Restricted fuel flow can result in excessive engine wear, as inconsistent fuel flow will result in combustion variation increasing piston and cylinder wear and an overall increase in maintenance cost (McFarlane, 2011).

The presence of microbiological bacteria causes fuel filter blockage through the growth of bacteria farms in the fuel filter medium. The bacteria colonies passing from the contaminated fuel tank become trapped by the filter element and create an improved environment for bacteria growth within the fuel. The growth accelerates due to its ability to attract more bacteria passing from the tank together with the ability to usze the nutrient and moisture contained in the fuel. This results in an increase in flow resistance and pressure drop across the filter, producing fuel deprivation and rapid loss of engine performance.

The most widespread method of dealing with microbiological contamination is by the addition of a biocide to the contaminated fuel, with special care exercised to protect the health of the treatment operator as the biocide is toxic. It requires special procedures and extreme care in handling.

14.4.4 Water Separators

Free or emulsified water must be removed from the fuel to prevent corrosion and damage to the fuel system. Certain additives that claim to remove water are touted in the market. In fact, they dissolve the water, which eventually passes through the filter and attacks the injection components. However, certain other additives demulsify the water by pulling it out of solution, which helps the separator to remove it effectively.

Checking the effectiveness of the water separator with the right equipment is necessary before approval.

The best way to remove water is with a filter incorporating a water separator. Inertia-type water separators employ centrifugal force generated by a swirling flow to separate water and fuel. These are cheap but are now out-dated, as their efficiency is less than 20%. Modern water separators have a high efficiency of 98%. They have two types of filtering media fitted in series. The first is hydrophobic, which repels water; the water coalesces and flows over the filter media to the bottom of the filter can. The second is hydrophilic, which sponges out the water dissolved in the fuel and collects it at the bottom of the can.

14.5 Summary

Fuel injection equipment is costly and is at the heart of a diesel engine. Increasing injection pressures to satisfying tightening emission norms demands tighter clearances. Water, dust and corrosive fuels wear out the equipment and result in inferior performance of the engine. To protect the equipment, the micron rating and efficiency of filtering are carefully selected.

References

ASTM (2010) D6898 – 03(2010) Standard Test Method for Evaluating Diesel Fuel Lubricity by an Injection Pump Rig, ASTM International, West Conshohocken, PA.

Erwin, S.I. (2004) Army ponders New Diesel Engine for Humvee. *National Defence*, July 2004.

Grigg, H.C. (1994) Reformulated Diesel Fuels and Fuel Injection Equipment. New Fuels and Vehicles for Cleaner Air Conference, 11–12 January 2004, Phoenix, AZ.

Lacey, P. and Westbrook, S. (1995), Diesel Fuel Lubricity. Technical Paper 950248, SAE, Troy, MI.

Le Pera, M.E. (2005) The Reality of the Single-Fuel Concept. *Army Logistician*, **37** (2), 40–43.

Leeming, D.J. and Hartley, R. (1989) *Heavy Vehicle Technology*. Hutchinson, London, p. 93.

MAN B&W (2000) NO_x control in practice and demands made on owners and engine builders. Diesel paper for meeting at the Maritime Museum in Bergen, March 2000, MAN, Copenhagen, Denmark.

MAN B&W (2001) NO_x Emission Reduction with the Humid Motor Concept. 23rd CIMAC Congress, April 2001, Hamburg, Germany.

Matzke, M., Litzow, U., Jess, A. *et al.* (2009) Diesel Lubricity Requirements of Future Fuel Injection Equipment. Technical Paper 2009-01-0848, SAE, Troy, MI.

McFarlane, E. (2011) Problems caused by microbes and treatment strategies, in *Applied Microbiology and Molecular Biology in Oilfield Systems*, Part 3. Springer, 159–167. doi: 10.1007/978-90-481-9252-6_19.

National Biodiesel Board (2009) Fuels for Diesel Engines – Diesel Fuel Injection Equipment Manufacturers Common Position Statement. National Biodiesel Board, Jefferson City, MO.

Treutler, C.P.O. (2005) Industrial use of plasma-deposited coatings for components of automotive fuel injection systems. *Surface and Coatings Technology*, **200** (5–6), 1969–1975.

Part VII

Heavy Fuel Engines

15

Wear with Heavy Fuel Oil Operation

Y.V. Aghav[1], M.N. Kumar[1], M.A. Ravichandran[1] and P.A. Lakshminarayanan[2]
[1]*Kirloskar Oil Engines Limited, Pune, India*
[2]*Ashok Leyland Limited, Hosur, India*

Majority of the goods are transported by sea for trade and commerce. The ships are propelled and powered by large diesel engines running cost efficiently on residual fuel oil. The high viscosity fuel containing a large percentage of carbon, sulfur, chlorine, sodium and catalytic fines can cause low temperature corrosion, high temperature corrosion and abrasive wear. These problems are successfully resolved by purification, heating and special design of critical parts of the engine to assure very long life for the engines. Tighter controls on emissions from on-road vehicles and increasing strength of merchant navy have tilted the balance towards emissions from marine engines. Therefore, the new generation heavy fuel engine will be environment friendly. For the common rail system, finer filtration is required as the operating clearances are less. In addition, loss of lubricity of fuel due to reduction in sulfur demands steps to protect the fuel injection equipment from undue wear.

15.1 Introduction

Fuel oil is classified into six classes, numbered 1 through 6, according to its boiling point, composition and purpose (Table 15.1) (Totten, Westbrook and Shah, 1969). The boiling point (175–600 °C) and carbon chain length (9–70 atoms) of the fuel increases with fuel oil number. Viscosity also increases with the fuel number and the heaviest oil has to be heated to get it to flow. Price usually decreases as the fuel number increases.

No. 1, 2 and 3 fuel oils are variously referred to as distillate fuel oils, diesel fuel oils, light fuel oils, gas oil or just distillate. No. 1 (kerosene) is the fraction that boils off right after petrol. Trucks and diesel cars run on No. 2 (diesel fuel). No. 3 (distillate fuel) is rarely used.

After petrol and the distillate fuel oils are extracted by distillation, the residue (RFO, residual fuel oil) is No. 6 and is also known as heavy fuel oil (HFO). Heavy fuel oil is also called heavy furnace oil. No. 5 fuel oil is a mixture of No. 6 and No. 2 oils in the ratio 80:20. No. 4 is a blend of distillate and residual fuel oils.

Critical Component Wear in Heavy Duty Engines, First Edition. P.A. Lakshminarayanan and Nagaraj S. Nayak.

Table 15.1 Classes of fuel oil

Fuel oil number	Fuel for compression ignition engines	Remark
No. 2 distillate	Distilled from crude oil	High speed diesel fuel
No. 2	Distilled and cracked in general	High speed diesel fuel
No. 3 and No. 4	Blend of distillate, cracked fuel and heavy fuel oil	Rarely used in a compression ignition engine
No. 5 and No. 6	Residual fuel oil (RFO) or Heavy fuel oil (HFO)	No. 5 oil contains 75–80% of No. 6 oil and 25–20% of No. 2 oil No. 6 oil contains very little of No.2 oil All the large 'low' speed engines invariably use HFO

Large diesel generating sets run at 98% load factor for 8000 hours every year. Similarly, large marine engines run at 80% for 6000 hours/year (MAN B&W, 2010a). It is not economical to continuously run them on high speed diesel (HSD) fuel. Therefore, they are run on HFO. Special equipment is used to treat the heavy fuel oil before it is admitted to the engine. Also, care must be taken to maintain the operating parameters, and hence to control the wear induced by the combustion products of the carbon, sulfur and metals present in the fuel.

Table 15.2 gives properties of HFO (ISO8217), its comparison with high speed diesel and the impact of HFO on the engine functioning if the parameters are outside the specifications (Jones, 1983). The effect of the sulfur content of diesel fuel on the wear of diesel engine cylinder liners, the piston assembly and the bearings is considerable. The high sulfur content causes increased low temperature corrosive wear.

Table 15.2 Properties of heavy fuel oil and high speed diesel fuel

Parameter	HFO	HSD	Impact of HFO if the parameters are outside the specifications
Viscosity at 50 °C (cSt)	176–700	3.3	Fuel handling
Density at 15 °C (kg/m^3)	961.7	837.3	Centrifuge operations
Sulfur content (wt-%)	3.33	0.049	Low temperature corrosive wear, piston ring wear, injector deposits
Water content (vol-%)	0.2	<0.1	
Conradson Carbon Residue (wt-%)	10.8	0.0	Gas ways fouling, exhaust turbine deposits, excessive filter clogging, power loss
Ash (wt-%)	0.02	<0.01	Abrasive wear, exhaust valve deposits, piston groove deposits, exhaust turbine deposits
Flash Point (°C)	Above 70	50	
Pour point (°C)	6	<−6	
Vanadium			High temperature corrosion, exhaust turbine deposits, exhaust valve corrosion
Asphaltenes			Scavenge port fouling, compatibility, combustion inefficiency, power loss
Sodium (salt water)			Exhaust turbine deposits, exhaust valve corrosion, excessive centrifuge sludge, injector deposits, piston groove deposits
Alumina/silica (cat fines)			Piston ring wear, piston groove wear, cylinder liner wear, injector pump wear, excessive injector pump wear (abrasive wear)
Incompatibility			Excessive centrifuge sludge, injector deposits, exhaust turbine deposits, exhaust valve deposits

High sulfur is neutralized by specifying a given alkalinity for the engine oil and by maintaining the engine cooling water at a high temperature (Schenk, Hengeveld and Aabo, 2003). High temperature corrosion is caused by catalytic fines.

15.2 Fuel Treatment: Filtration and Homogenization

The standard practice is to have a settling tank, centrifuges and filters (Figure 15.1). The settling tank removes substantial water or solids. Its effectiveness is determined by the viscosity of the fuel bunkered and the temperature (50 °C or 10 °C below the flash point) at which the tank is maintained. Two centrifuges are run in series, one as a purifier and the other as a clarifier. The centrifuges work satisfactorily up to a limiting fuel density of 991 kg/m^3 at 15 °C. The purifier removes water and solids and is situated before the clarifier, which primarily removes solids. A centrifuge is unable to remove the water if it is emulsified. High specific gravity fuels and fuels containing catalytic fines require separation (a) at the highest temperature, 98 °C, to reduce the viscosity as much as possible, and (b) reduction of throughput to increase the efficiency. The concentration of the catalytic fines is reduced from 80 mg/m^3 at the input to 10–15 mg/m^3 after the purifier.

In all residual fuel oil systems, various designs of cold and hot filters are fitted after the centrifuges. They are either cleaned manually or are of the self-cleaning type. Additionally, a homogenizer (Agius *et al.*, 1971) may be used after the filters where the water and other contaminants are intimately mixed just before admission to the engine, as they have a tendency to separate.

The homogenization of the fuel before the separator is not recommended because it leads to a low separation efficiency for both water and particles; both fresh and seawater are emulsified, that is, the water droplets are too small to be removed by centrifugal separation, leaving excessive amounts of seawater (sodium, hydrochloric acid), metals and catalytic fines in the fuel fed to the engine. A concentration of aluminium and silicon (Al and Si) of more than 10 mg/m^3 has an impact on the abrasive wear on injection equipment, piston rings and liners (Aabo, 2003). The chapter considers the different types of contaminants in HFO and their effect on the wear of various critical components of a diesel engine.

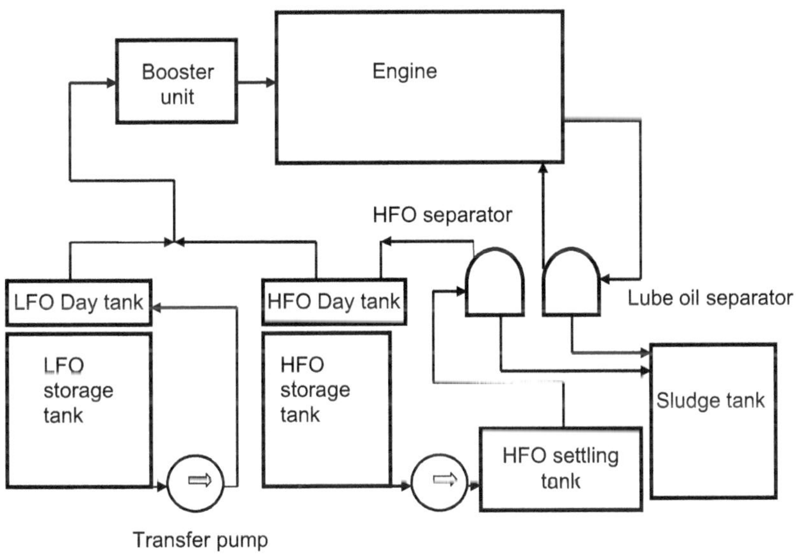

Figure 15.1 Typical fuel oil treatment system

15.3 Water and Chlorine

15.3.1 Fuel Injection Equipment

Water and chlorine in fuel form hydrochloric acid. Delivery valves and plungers in the fuel pump are affected by water.

15.4 Viscosity, Carbon Residue and Dust

The carbon residue of a fuel has the tendency to form carbon deposits at high temperatures in an inert atmosphere. Engine loading, engine tuning and the ignition qualities of the fuel affect such tendencies of a particular fuel. The carbon residue value of a fuel is typically 15–16%. Maintaining 98 °C at the inlet to the purifiers and parallel running of purifiers to reduce fuel oil flow rate to the purifiers help in bringing the viscosity under control.

15.4.1 Fuel Injection Equipment

Above a maximum of 20 wt-% (Conradson carbon residue, Table 15.2) carbon may be problematic and cause increased fouling of injector nozzle holes. High viscosity increases the injection pressures, leading to failure of the injection equipment. Careful control of the nozzle cooling temperature and fuel pre-heating solve this problem. Solidification in the fuel pipes and fittings and vapour lock could also be a problem.

While the viscosity is taken care of by heating the fuel to 150 °C this leads to excessive heating of the nozzle and an increase in the wear rate of guided parts and the needle seat. A special cooling gallery is provided in the nozzle around the needle guide up to the tip to reduce the wear. Also, the needle seat is designed with a 'two cone' seat that is less prone to wear than the 'single cone' seat (Figure 15.2). At the testing rig the single cone nozzle can be heard making strong 'chatter', indicating that it is prone to wear. The double cone seat is geometrically better designed for both sealing and wear.

Leaks from the fuel pump can increase the viscosity of lube oil abnormally, and hence cause problems with lubrication. The oil lubricating the fuel pump is maintained at 0.15 MPa above the fuel feed pressure, so that fuel leakage from the fuel pump to the oil sump is avoided.

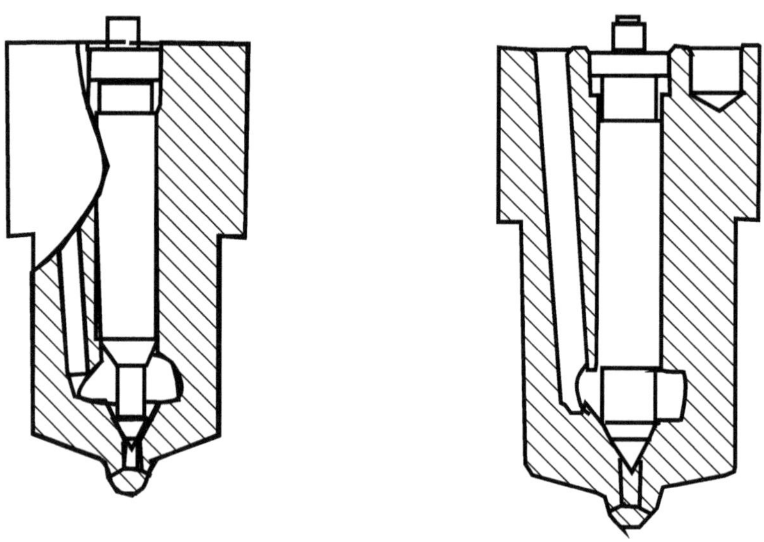

Figure 15.2 Single and double cone needle for the injector nozzle

15.5 Deposit Build Up on Top Land and Anti-polishing Ring for Reducing the Wear of Liner, Rings and Piston

Silica escaping the defective air filter forms a complex compound with oil at high temperatures during combustion and accelerates the formation of thick oil cakes on the top land and the liner top. Also, it creates scratches on the ring faces and, consequently, the ring face peels off. For example, in extreme areas such as desert, a dust concentration of 100 mg/m^3 is not unusual. An air flow of 0.7 m^3/s/cylinder (producing 400 hp per cylinder) means that in 20 hours five kilogrammes of dust is taken in per cylinder. To avoid abrasive wear (Chapter 9) a dry-type air filter (5 μ/99.9%, Chapter 16) must be used.

The sludge formation due to an incompatible mixture in bunker oil (e.g. lubricating oil + HFO + industrial solvents) and excessive carbon (higher viscosity, higher carbon residue) form hard deposits on the top land and reduce the clearance between the liner and the piston land (Figures 15.3a and 15.3c). The carbon build up adds to natural (thermal) growth of the steel crown of the piston to decrease the operating clearance between the liner and the piston top land. The locus of the piston skirt during the up and down stroke is in the shape of an '8' and causes the piston top land to kiss the liner during the middle of the stroke because of the lack of clearance (Figure 15.4). This leads to bore polishing in the middle portion

Figure 15.3 (a) Piston without an anti-polishing ring showing heavy deposits of carbon on the top land and the first ring land. Correspondingly, heavy scoring marks can be seen on the rings. (b) Scoring of liner when run without a polishing ring. (c) Another view showing scoring on rings and heavy deposits on all the lands on a piston running without an anti-polishing ring. (d) Clean top land and rings when an anti-polishing ring is used

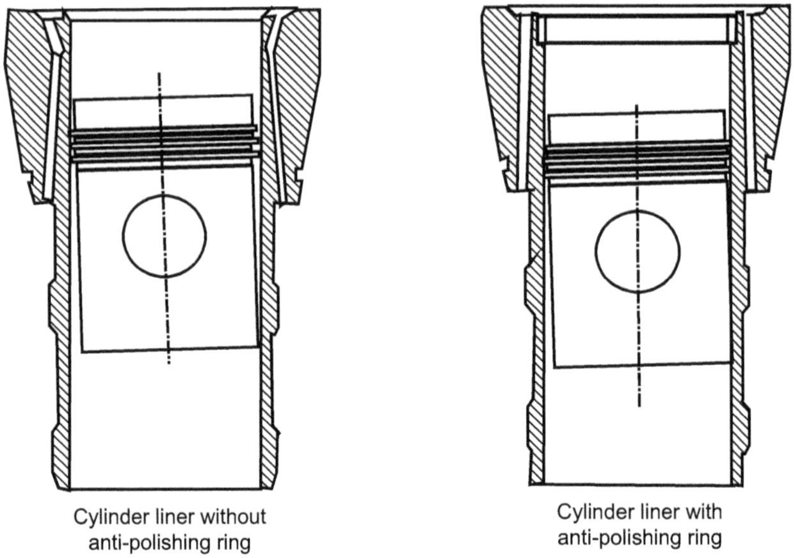

Figure 15.4 Cut-back top land on the piston enables contact with the liner to be avoided in the mid-stroke

of the liner (Figures 15.3a, 15.3b and 15.3c) and scratches on the piston ring. A cut back in piston top land discourages build up of the carbon layer and a bore anti-polishing ring scrapes off whatever cake is built up on the top land of the piston as and when it is formed (Chapter 7; Figure 15.3d). The anti-polishing ring (Figures 15.4 and 15.5) enables no relaxation in the dead volumes during combustion

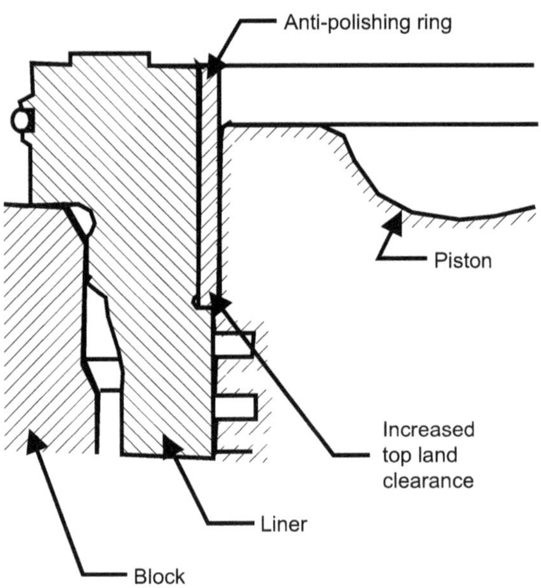

Figure 15.5 Sketch of an anti-polishing ring installed in the liner and the increased piston top land clearance

near the top dead centre, as the clearance between the piston and the liner (antipolishing ring) remains the same as without the ring and cutback on the top land. Hence, the combustion efficiency does not suffer. Later, during expansion, the slightly higher volume between the top land of the piston and the liner does not affect the efficiency. Hard and wear resistant chrome plated rings stabilize oil consumption for a long time, as the geometry of the faces is maintained intact. However, chrome will only work if running-in is correct and air filter efficiency is sufficiently high.

15.6 High Sulfur in Fuel

Sulfur levels in HFO are typically between 1.5 and 4%.

15.6.1 Formation of Sulfuric Acid

By design, it is ensured that the cylinder liner surfaces do not to approach the dew point at the prevailing pressure (Williams *et al.*, 1984; Figure 15.6), as below this water in the intake air and from combustion would condense into liquid during operation. In addition, the outlet temperature of the charge air cooler must be maintained above the dew point (~55 °C) to avoid heavy precipitation of water in the inlet manifold as this would initiate corrosion in the manifold.

In a diesel engine, the sulfur in the fuel first burns to sulfur dioxide, then combines with excess oxygen to form sulfur trioxide. When the sulfur trioxide corrodes the cast iron, the chemical element iron will oxidize, and thus not contribute to abrasive wear. However, in the presence of any water condensed on the cylinder walls, a small portion of the sulfur trioxide is converted to sulfuric acid if the liner temperature is below the dew point inside the combustion chamber.

15.6.2 Mechanism of Corrosive Attack by Sulfuric Acid

The cast iron liner is sensitive to attack by sulfuric acid produced in the combustion of HFO. An investigation showed that acid attack is a major reason for unsatisfactory piston running and sudden high liner wear (Amoser, 2001, 2002). Lower liner wall temperatures lead to dilute acid, which corrodes the

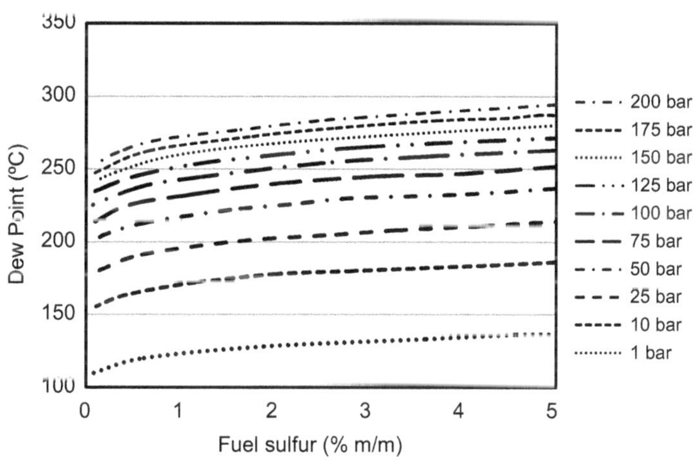

Figure 15.6 Shell acid dew point curves (Schenck, Hengeveld and Aabo, 2003)

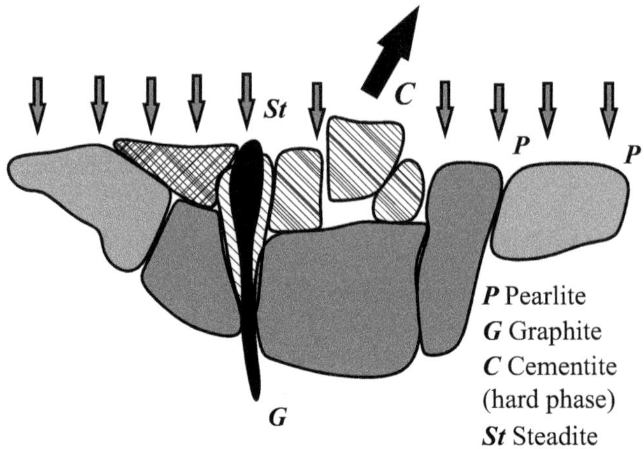

Figure 15.7 Acid attack on the soft and hard portions of the surface microstructure of a cast iron liner (Reproduced from M. Amoser, "Insights into piston-running behaviour," *Marine News*, 23–27, Wärtsilä Corporation, 2001.)

softer matrix in the cast iron (Figure 15.7). As a result, the hard phase particles protrude, break off and act as abrasive particles. At higher wall temperatures, the sulfuric acid is more concentrated. This attacks the cementite (harder phase), leading to high (two-body) wear. At still higher wall temperatures, the sulfuric acid is even more concentrated, but there is less attack of the hard phase, giving good piston running behaviour.

The technique used to control wear is to keep the liner wall temperature above the dew point of water over a sufficiently wide range of engine loads to prevent condensation of water (and thus formation of dilute sulfuric acid) on the liner running surface. To avoid cavitation in the engine (as there are high water flow velocities around the liner) when higher coolant temperature is maintained, the coolant pressure is raised to 0.2 MPa (absolute) at the entry to the water pump. In the longer-stroke two-stroke engines more of the liner length needs insulation for successful control of corrosive wear (Demmerle *et al.*, 2001).

15.6.3 Control of Corrosion by Basicity and Oil Consumption

In a two-stroke engine with a crosshead, the oil for the cylinder is separated from the oil lubricating the rest of the engine. The alkalinity (total base number, TBN) of the oil is maintained 70 mg KOH/g to neutralize the acid and control corrosive wear. In a four-stroke engine where the same lubricant is used both for combustion space and bearings, corrosive wear is caused to all the parts of the engine by sulfur. Figure 15.8 shows the wear rate of an engine liner (Aabo, 2003) against sulfur content in the fuel for two different basicity numbers, namely BN 40 and BN 70. Other TBN oils are being tried in laboratories and may become popular in the future. It should be emphasized that, irrespective of the sulfur content being high or low, the fuels used in low speed engines are usually low quality heavy fuels. Therefore, the cylinder oils must have the right detergency and dispersancy, irrespective of the TBN specified. High sulfur levels will increase the rate of TBN depletion, especially for engine designs that have a low oil consumption rate. Therefore, apart from the TBN (in engines running on HFO) the lubricating oil consumption is maintained at a rate of 1.2–1.5 g/kWh (compared to, for example, the lubricating oil consumption of 0.1 g/kWh in modern truck engines running on costly high speed diesel fuel), so that topping up sweetens the oil. In large two-stroke engines, the oil injection in the cylinder can be further optimized with electronics, so that an oil consumption of 0.7–1.0 g/kWh could still control acids when

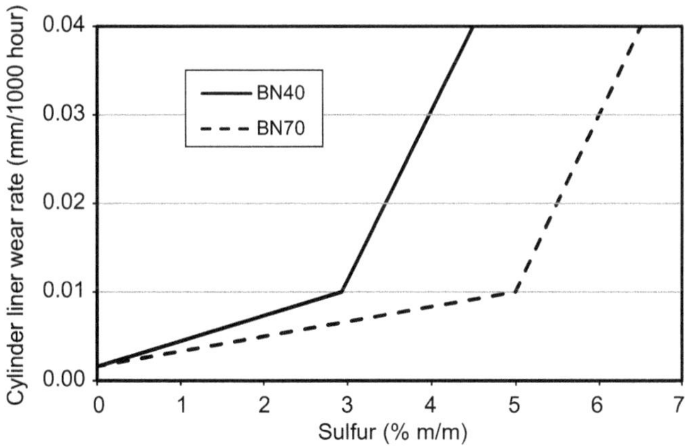

Figure 15.8 Comparison of sulfur content and lubricating oil TBN with respect to cylinder wear with equal cylinder oil feed rates (Aabo, 2003)

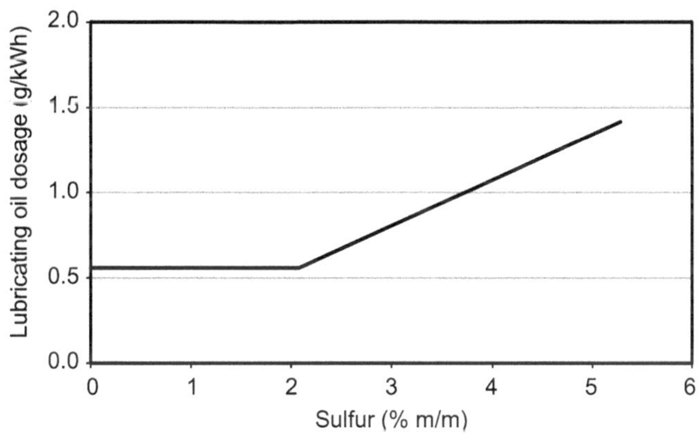

Figure 15.9 Minimum required oil consumption to keep the lubricating oil sweet by topping up (Romih and Zgonik, 2008)

the fuel sulfur is 1.5% (Romih and Zgonik, 2008). Figure 15.9 shows the recommended oil consumption for different sulfur contents in the fuel (Aabo, 2003). When the oil consumption is too little to neutralize the acid formed or to avoid boundary lubrication, the liner wear is high (Figure 15.10). Similarly, wear is high when the oil consumption is very high, as the excess oil is roasted on the piston top land and carbon builds up heavily.

15.6.4 Control of Sulfur Corrosion by Maintaining Cooling Water Temperature High

Corrosion is influenced by the liner wall temperature and thus by the engine design with regard to the cooling of liners. Figures 15.11a, 15.11b and 15.11c show corrosion of the stem, valve guide and exhaust valve seat due to corrosion by sulfuric acid.

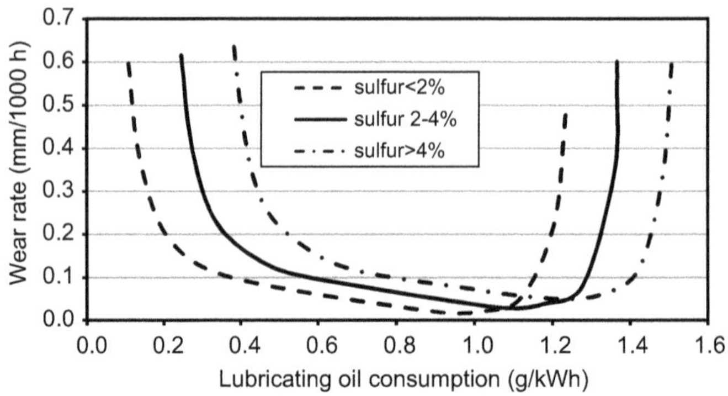

Figure 15.10 Wear rate as a function of lubricating oil consumption: too little causes accumulation of sulfuric acid as sweetening by topping up is less; too much oil flow causes carbonization of oil on the top land and ring grooves (Romih and Zgonik 2008)

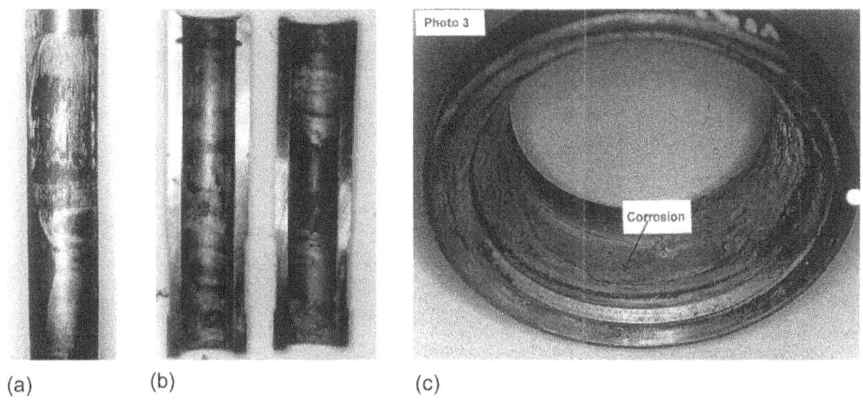

Figure 15.11 Exhaust valve stem, valve guide, valve tip and valve seat corroded by sulfuric acid due to low cooling water temperature

By keeping the liner temperature relatively high level, it is possible to reduce corrosion. However, high liner temperatures also place heavy demands on the thermal stability of the lubricating oil.

Nearly all components of the exhaust port show some low temperature corrosion. Sulfuric acid attacks the valve housing (in two-stroke engines), valve guide, valve stem and valve seat. The corrosion can be controlled by insulating the areas prone to corrosion (Figure 15.22).

15.7 Low Sulfur in Fuel

15.7.1 Lubricity

Sulfur in fuel has a beneficial effect in preventing scuffing of fuel injection components. Very low sulfur fuels are used in environmentally-sensitive areas. Low sulfur worsens lubricity of the fuel and the fuel injection equipment can be damaged in operation (Figure 15.12) with very low sulfur fuel (<0.05%). A lubricity additive must be used to prevent damage to fuel injection components when operating using

Figure 15.12 Low viscosity low sulfur fuel leads to sticky plungers (Reproduced from MAN B&W Diesel A/S, Copenhagen, Denmark, "Operation on low-sulphur fuels, two-stroke engines," © 2009.)

these fuels. Also low viscosity can cause fuel pump and fuel valve wear and, consequently, the risk of sticking.

Considering the situation from a hydrodynamic point of view, if the viscosity is sufficient, and thereby the fuel (oil) film is thick enough, low sulfur fuels can be used.

15.7.2 Lack of Formation of Oil Pockets on the Liner Bore

It has been established that a certain level of controlled corrosion enhances lubrication, when the corrosion (removal of iron) generates small 'pockets' in the cylinder liner running surface which retain oil that can then be used for hydrodynamic lubrication (Aabo, 2003). In addition, corrosion helps remove the roughness peaks.

In some new generation engines, there is no corrosion at all due to a combination of low sulfur in fuel, high TBN oil and high liner temperature. Therefore, these tiny pockets cannot be generated. Also, the roughness peaks (not removed because of lack of corrosion) can fold and close the graphite (that forms oil pockets) in the microstructure. Thus the creation of the necessary oil film on the liner surface is hampered and the liner surface can be mirror-polished (Figure 15.13).

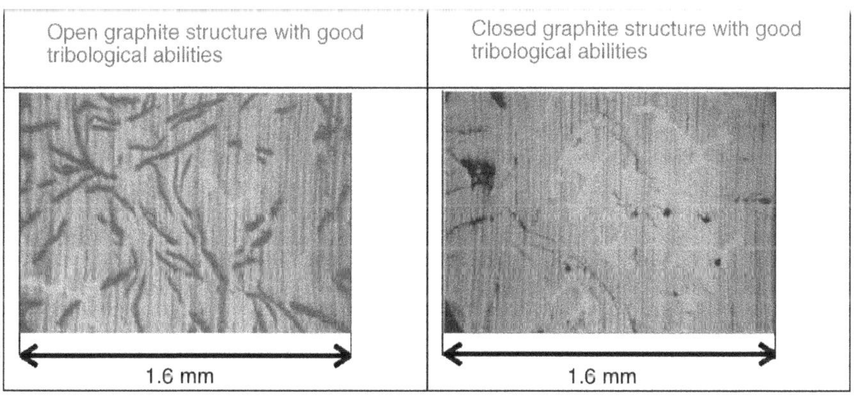

Figure 15.13 Cylinder liner surface with mild corrosion revealing oil-bearing graphite structure; without corrosion they are closed (Reproduced with permission from MAN B&W Diesel A/S, Copenhagen, Denmark, All rights reserved.). Too little corrosion can result in too little wear and in damaging polishing of the liner surface

The oil itself has difficulty in adhering to a mirror-like surface and metal-to-metal contact occurs, often resulting in scuffing, which leads to extremely high liner and piston ring wear. The scuffing is vicious as metal-to-metal contact creates friction and heat, and thereby hardens the liner surface and makes it difficult for the liner and piston rings to work together. The consequence is bore polishing and high oil consumption.

15.7.3 Sudden Severe Wear of Liner and Rings

A low speed diesel engine usually switches from high to low sulfur with no change in piston running behaviour. However, 'sudden severe wear' is more likely if there is an accumulation of so-called risk factors (Table 15.3) (Welsh, 2002).

For example, low sulfur fuel might be used in a temperature region with no problem, but (a) if the ship enters a region with very high water concentration in the scavenge air, (b) if the water separator and air intercooler drainage system are not operating correctly, and (c) if a BN 70 (high TBN for the sulfur content) lubricant is used at a feed rate of greater than 1.3 g/kWh, then these cumulative factors will lead to a high risk of sudden severe wear (Figure 15.14; Welsh, 2002). In other words, if several conditions on the list in Table 15.3 are fulfilled, then some action should be taken to reduce the risk. The lubricant feed rate is a crucial factor. It is often the only parameter that can be easily adjusted to match the fuel sulfur content. Unfortunately, there is (for historical reasons) a lot of resistance to reducing the feed rate. In cases of doubt the feed rate is mostly raised and exactly that can trigger sudden severe wear. A feed rate which is too high will cause instability in the piston ring pack and increase deposits on the piston top land. If too low, there might not be enough oil to form a stable oil film, or more likely piston ring and cylinder liner wear rates will be high due to corrosion (Welsh, 2002). It might be a worthy exercise to keep two tanks of lubricating oil, one with BN 70 and another with BN40, and change from the higher TBN oil to the lower TBN oil if low sulfur fuel is loaded, and vice versa.

Table 15.3 Risk factors for sudden severe wear

Risk factors	Remarks/Impact
Low sulfur fuel	Less acid is available for removing the roughness peaks and forming tiny oil pockets.
High cylinder lubricating oil feed rate	A feed rate which is too high will cause instability in the piston ring pack and increase deposits on the piston top land. If too low, there might not be enough oil to form a stable oil film or, more likely, piston ring and cylinder liner wear rates will be high due to corrosion.
Heavy deposit on the piston crown	The deposits are different, possibly harder or more solid, when less oil TBN additives are neutralized by sulfuric acid. The result is abrasive wear.
No anti-polishing ring	The anti-polishing ring reduces hard coke deposits on the piston top land which destroy the liner's oil film.
High scavenge air humidity	Physical contact of water droplets on the oil film and on the ring surfaces destabilizes the oil film.
Poor scavenge air water removal after intercooler	
Minimal quality of cylinder lubricant	There might not be enough oil to form a stable oil film or, more likely, piston ring and cylinder liner wear rates will be high due to corrosion.
TBN of oil higher than required	A lower TBN results in less calcium deposit in the turbocharger or the exhaust gas boiler and lower feed rate has a similar effect. Calcium forms the main bulk of this deposit and the calcium content of a lubricant is proportional to its TBN.

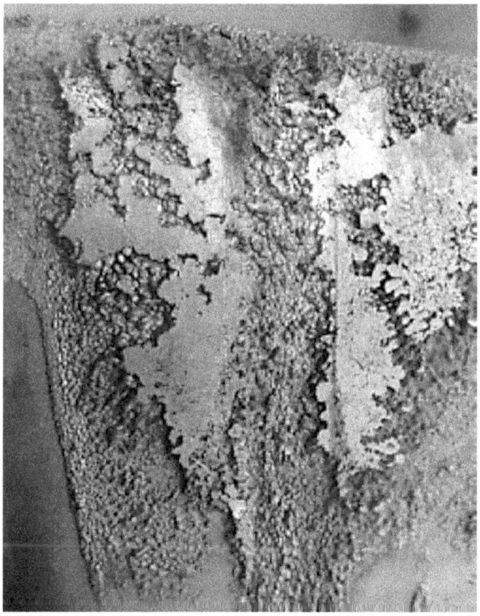

Figure 15.14 Close-up view of piston showing a deposit build up sufficient to contact the liner surface. In this condition there is a high risk of sudden severe wear (Reproduced from Welsh, M. "Considerations for using low-sulphur fuel," Wärtsilä Corporation, 2001.)

15.8 Catalyst Fines

The catalytic cracking process is used in the refinery to maximize the output of diesel and petrol from crude. The fines are particles of spent aluminium and silicon catalysts that are carried away in the residual heavy fuel oil. These fines are in form of complex alumino-silicates and vary in size and hardness depending on the catalyst used. Apart from the catalyst fines, other particles such as iron, magnesium and calcium are also found. These are separated in settling tanks and then in the purifying centrifuges. A reduction in concentration from 80 mg/m^3 to 10 mg/m^3 is normal; to reduce below this concentration it is economically not possible. If the fuel is kept in the settling tank for long, it is possible that the fines will settle at the bottom in large concentration and, if not drained, the settled residue could be stirred during the next cycle of filling and the concentration of catalyst fines could increase suddenly. The purifiers are not able to handle extraordinary concentrations of the catalyst fines, and hence allow higher than permitted concentration of particles in the fuel at the inlet to the engine.

The particles are hard and have sharp edges (Figure 15.15a). The abrasive nature of these fines damages the engine, particularly fuel pumps, injectors, piston rings and liners, if not removed to a concentration less than 15 mg/m^3. Blending used lubricating oil of significant volumes in the fuel or excessive water may also limit the effective removal of fines, which are hydrophilic. Figure 15.15b shows catalyst particles pressed into the surface of the piston ring due to high level of catalyst fines; two-body wear ensues, resulting in heavy wear or scuffing (Figure 15.16) of liners and pistons (Aabo, 2003).

15.9 High Temperature Corrosion

When a fuel is bunkered with a vanadium level greater than that recommended by the engine designer, there is a risk that hot corrosion and fouling may occur. In modern, high load four-stroke engines the exhaust gas temperatures before the turbocharger can easily reach 500–600 °C (e.g. when the air cleaner is

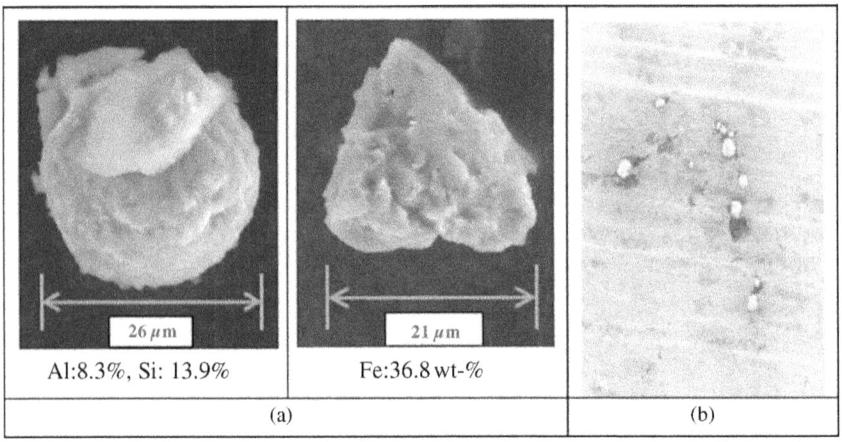

Figure 15.15 (a) Sharp edges on the hard particles from catalyst fines and other sources (b) Catalyst particle pressed into a ring surface (Reproduced with permission from T. Yamada, T. Fuji and H. Ukai, "Development of a new evaluation method for the influences of catalyst fines on abrasive wears of marine diesel engines burning heavy fuel oil," CIMAC Congress 2010. © CIMAC, Frankfurt. All rights reserved.)

fouled, the airflow is restricted and the temperature in the exhaust can rise). In this temperature range, vanadium (V) and sodium (Na) ashes can appear in liquid form by melting, as explained below.

The vanadium and sodium contents in HFO can reach ratios (critical ratio of Na:V of 1:3 corresponding to an eutectic composition; Figure 15.17) up to 600 ppm zanadium and 200 ppm sodium (Pereira and Marschewski, 2001). During combustion in the cylinder, vanadium and sodium oxidize to V_2O_5 and Na_2O. In combination with sulfur, they are able to form different types of sodium-vanadyl-vanadates (ashes). Some of these ashes have low melting points and absorb oxygen while melting:

$$Na_2O.6V_2O_5 \underset{\leftarrow \text{ melt}}{\overset{\rightarrow \text{ crystallise}}{\leftrightarrow}} Na_2O.V_2O_4.5V_2O_5 + \frac{1}{2}O_2$$

In the turbochargers and exhaust valves the ashes solidify (crystallization) and build up very hard deposits. While forming the deposits (i.e. crystallizing) highly reactive atomic oxygen is liberated and it attacks the metal surfaces, commonly called High-Temperature-Corrosion (HTC).

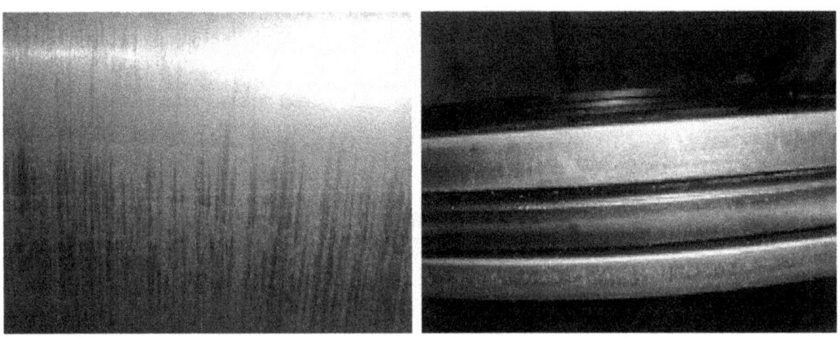

Figure 15.16 Scuffed liner and piston ring surfaces due to catalyst fines (Reproduced with permission from MAN B&W Diesel A/S, Copenhagen, Denmark, All rights reserved.)

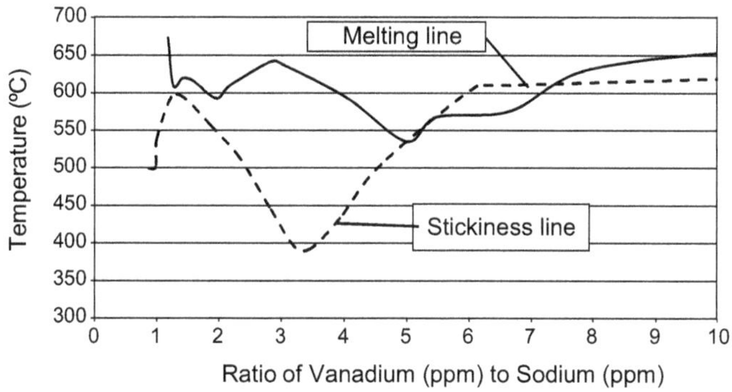

Figure 15.17 Vanadium sodium oxide eutectic (Reproduced with permission from ClassNK (Nippon Kaiji Kyokai), Guidance for measures to cope with degraded marine heavy fuels, 1997.)

15.9.1 Turbocharger

Sulfur in fuel additionally forms sodium sulfate by reacting with sodium in the fuel. The eutectic mixture of sodium sulfate with vanadium oxide has a low melting temperature of 300 °C (Figure 15.18). The mixture of sodium oxide and vanadium oxide and the mixture of sodium sulfate and vanadium oxide melt and deposit, along with the carbon (from HFO of high carbon residue) and dust (from improperly filtered air entering the engine), can together cause deposits to build up quickly when the mixtures are eutectic in composition. The deposits can clog the nozzle in the turbine and reduce the area of gas flow, leading to surge (NKK, 1997) (Figure 15.19a). The mechanical shock due to the surge and the high temperature corrosion can damage the turbine parts (Figure 15.19b).

A known method to suppress the effects of vanadium/sodium ashes and to influence their properties after combustion is the use of additives (Pereira and Marschewski, 2001) and numerous ash-modifying compounds. In addition, sodium and vanadium should not be in the vicinity of the eutectic composition.

15.9.2 Exhaust Valves

Higher exhaust temperature is possible due to air flow lower than the designed level, leaking valves, fouled turbocharger, high inlet air temperature or poor condition of the injector. Exhaust temperature above the

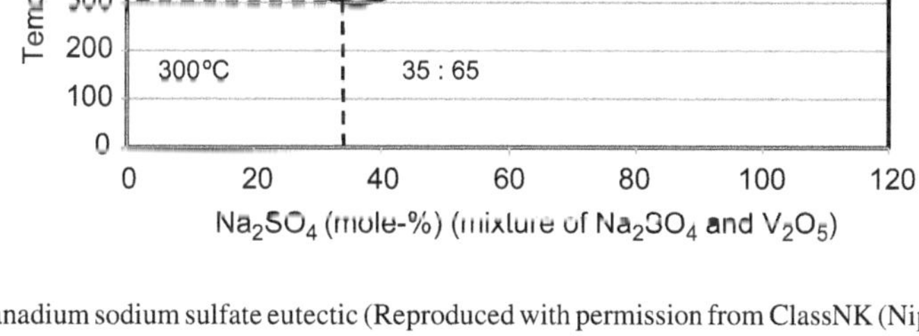

Figure 15.18 Vanadium sodium sulfate eutectic (Reproduced with permission from ClassNK (Nippon Kaiji Kyokai), Guidance for measures to cope with degraded marine heavy fuels, 1997.)

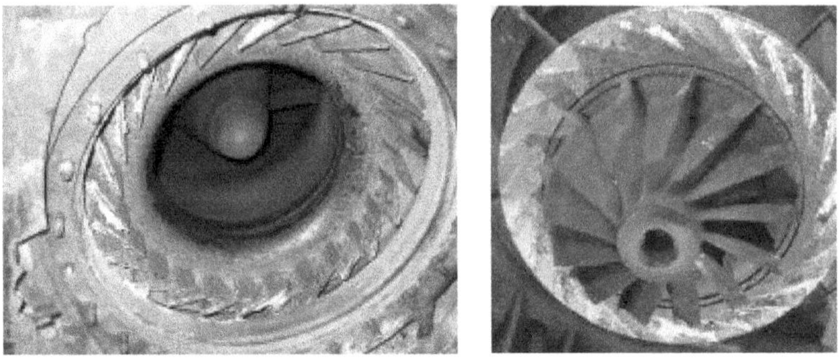

(a) (b)

Figure 15.19 (a) Turbine fouled by deposits of sodium + vanadium eutectic and consequent corrosion during melting leading to (b) damage of turbocharger parts (vane)

eutectic of vanadium and sodium (3:1) mixture, together with sulfur dioxide is risky in terms of valve burning, as the mixture melts into a paste on the valve seat to allow poor closing of the valve. In addition, solid products of combustion, slag particles and hard particles of Al_2O_3 and SiO_2 (catalytic fines and unfiltered dust from intake air) are pressed into the valve seat, forming dents (Nanda, 2003; Fellman, Gross and Ludwig, 2004; Figure 15.20a). Together with high seating pressure, seating velocity and high hardness, this can cause guttering and valve burning. Valve rotators of different designs are used to break

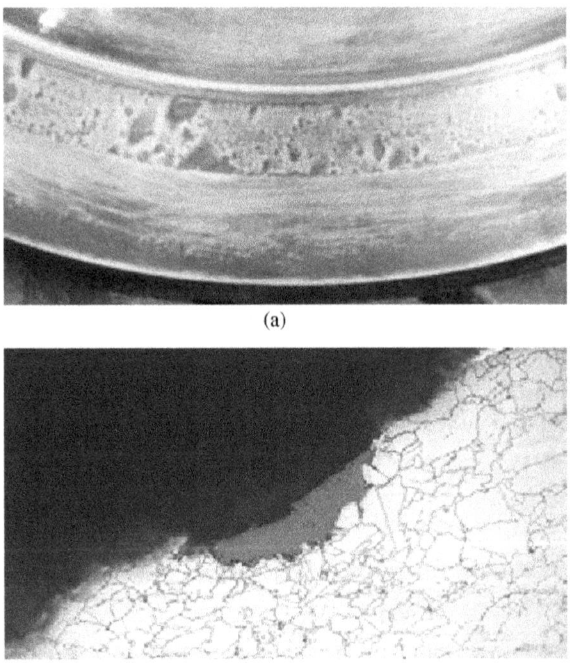

(a)

(b)

Figure 15.20 Wear and tear of inner diameter of the valve head: (a) dent marks; (b) network of cracks on the surface due to low temperature corrosion (Fellman, Gross and Ludwig, 2004) the deposits and cause uniform wear of the valve seats. The amount of deposit build up on the inner diameter depends on the design of the valve guide. When the deposits are in excess, the valve head undergoes corrosion and a network of cracks is seen on the surface under a microscope (Fellman, Gross and Ludwig, 2004; Figure 15.20b). An example of the exhaust valve and seat showing excessive wear and gouging of material leading to breakage of the valve head is shown Figures 15.21a, 15.21b and 15.21c.

At the stem (Figure 15.21a), the chromium plating has completely disappeared due to sulfuric acid corrosion. The loose particles caused secondary abrasion of base steel material. A slow migration of abrasive ash in the small clearance between the valve and the valve guide takes place. Adhesion and

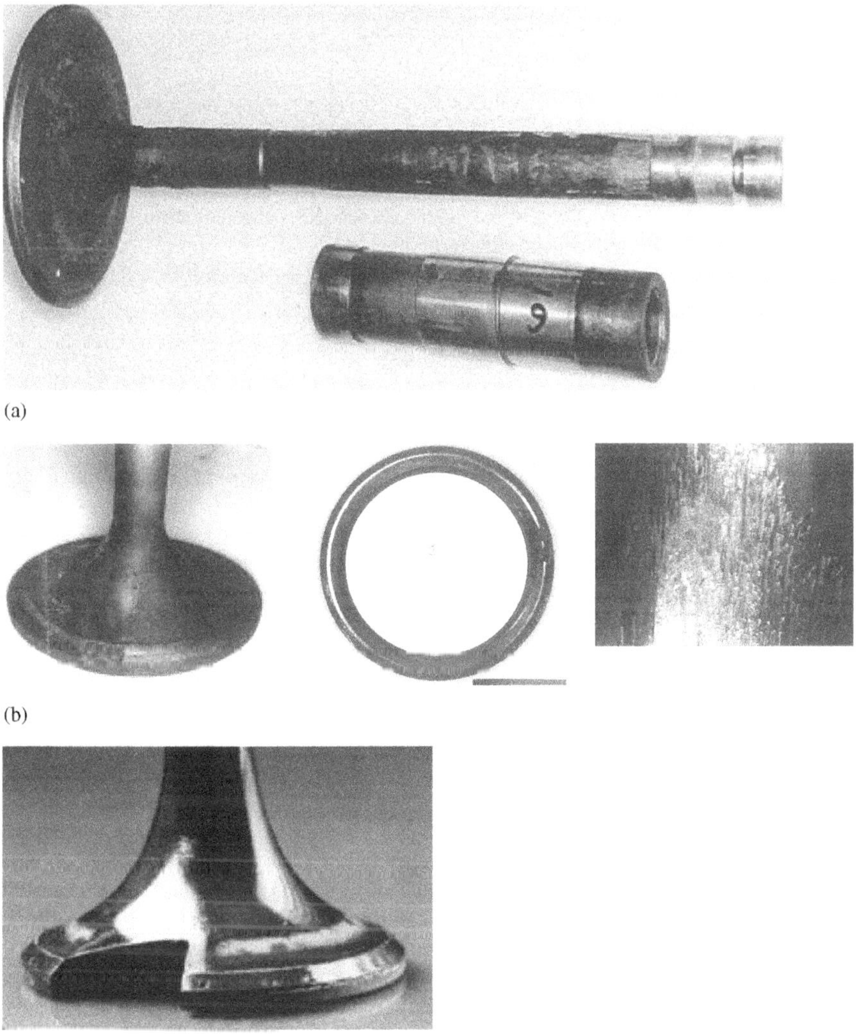

(a) (b) (c)

Figure 15.21 Damage due to wear by high temperature corrosion. (a) Damaged exhaust valve assembly due to eutectic deposits of sodium and vanadium; (b) exhaust valve head and valve seat damaged by low melting eutectic mixture of sodium and vanadium, valve stem shows heavy wear; (c) damage leading to breakage of exhaust valve head

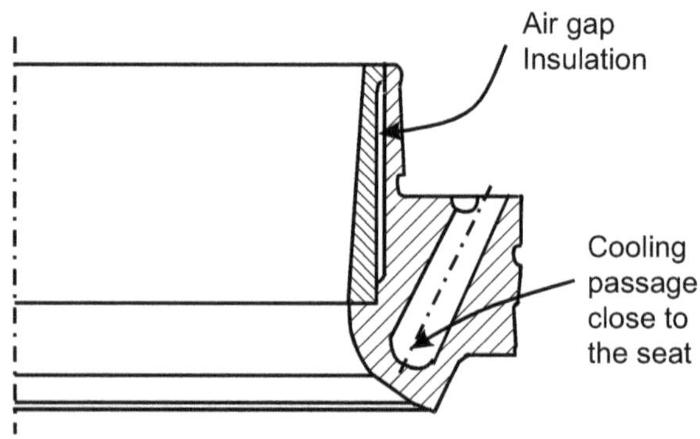

Figure 15.22 Circulation of hot water and insulation of the exhaust port above the seat to avoid low temperature corrosion (Reproduced with permission from ClassNK (Nippon Kaiji Kyokai), Guidance for measures to cope with degraded marine heavy fuels, 1997.)

abrasion cause heavy wear of the valve stem and guides. To solve this problem, the methods are: (a) valve stem seal to control the oil flow, (b) special chrome plating ('pulse' electroplating) of complete stem (with pockets etched by reverse electrolysis to hold oil in traces) and (c) increase in water temperature.

15.9.2.1 High Temperature Corrosion

The valve head undergoes the maximum thermal stress during the warm up period. During this period, the difference between temperatures at the centre of the valve bottom and the seat reaches the maximum (Fellman, Gross and Ludwig, 2004). At steady operating conditions, the temperature distribution is more even. The guttering of the seat areas (Umland and Ritzkopf, 1975; Ritzkopf, 1975) on the valve head due to cracks (induced by high and low temperature corrosion, (Figure 15.2b) and the low cycle thermal fatigue can lead to disastrous failures of the valve head (Figure 15.21c). The same techniques as applied to liner and turbocharger wear could be followed here to solve the problems. The formation of deposits on the valve seat in the cylinder head can be avoided by special cooling galleries, as shown in Figure 15.22 (NKK, 1997).

15.10 Wear Specific to Four-stroke HFO Engines

In a four-stroke engine, as already mentioned, it is not possible to separate the oil in touch with combustion gases from the oil that lubricates the bearings of all the rotating parts. The catalyst fines, hard carbon and dust find their way from the combustion chamber past the piston rings and are washed by the lubricating oil. In the normal course, the connecting rod and main bearings wear out in the zones where mixed lubrication conditions exist during one cycle consisting of two revolutions. Figure 15.23 shows deposits of catalyst fines and ash produced by lubricating oil and hard carbon on the exposed rillen (grooves) on the bearing material in the narrow region of mixed lubrication (Figure 15.23a and 15.23b) after the overlying material has worn out bearing (normally aged). The microscopic view (Figure 15.23c) of the grooves shows the normal wear out of the overlay.

15.10.1 Wear of Bearings

Figure 15.24 shows a typical bearing arrangement at the big end of the connecting rod. The bottom half has an oil groove (Figure 15.25) and the top half is plain without an oil groove. HFO operation demands the coolant temperature, fuel quality and cleanliness of the air are highly controlled. If the hard abrasive

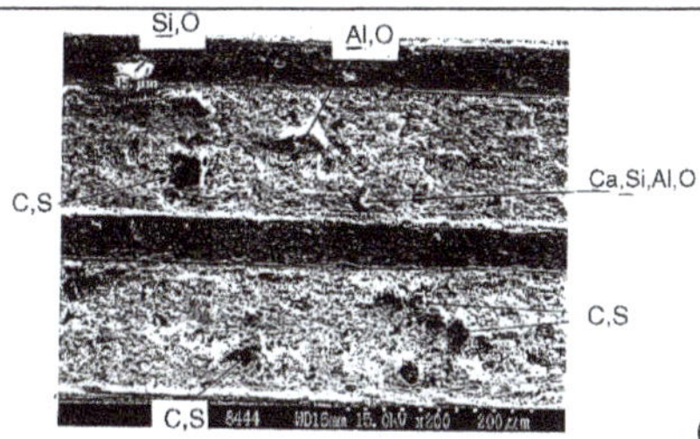

(a)

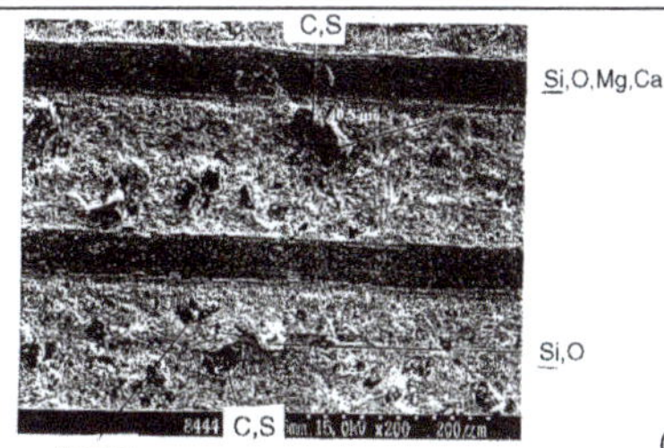

(b)

Normally aged lower shell of the bearing of the connecting rod at the time of replacement. Note the embedded particles of cat fines and carbon residue from the fuel of size 25 -52 μm

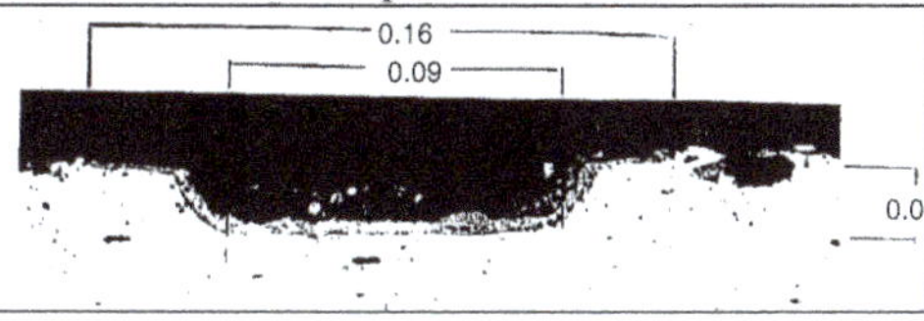

(c) At the end of normal life, in mixed lubrication regime the overlay in the Rillenlager (bearing with grooves for overlay) is completely used up

Figure 15.23 Appearance of bearing rillen (grooves), loss of overlay and embedded particles at the end of normal life of connecting rod bearings

particles entering the lubricating oil are excessive (when control is lacking), they wear the connecting rod big end and main bearings. In the zone of the oil groove the crank pin wears out less than the rest. The zone of less wear has a cam-like appearance (Figures 15.26 and 15.27). However, when the shaft rotates the cam fouls, with the upper bearing destroying the oil film. The wear becomes accelerated for upper half bearings. Replacing the bearing alone is of no help as the cam shape on the shaft still exists and the failure of the shaft could be catastrophic after the bearing is replaced. Re-grinding the crankshaft to remove the cam shape and replacing the bearings is the only way to service such a wear. When the cam formation is severe, the crankshaft has to be replaced. The solution to this problem is to reduce the pore size of the oil filter and improve the efficiency of filtration; a larger filter is recommended to maintain the pressure drop across the filter when the efficiency is increased.

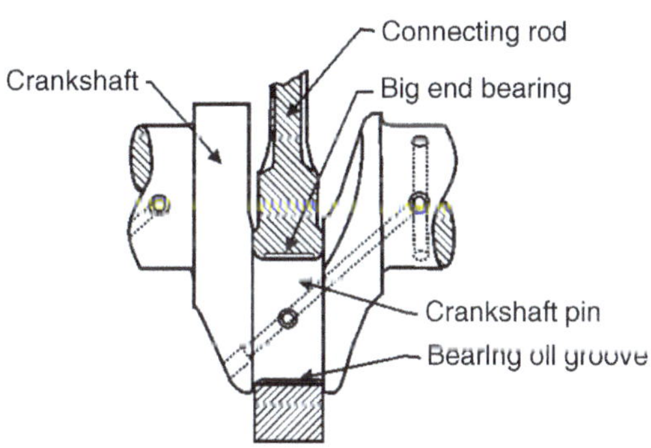

Figure 15.24 Cross-section of connecting rod big end arrangement

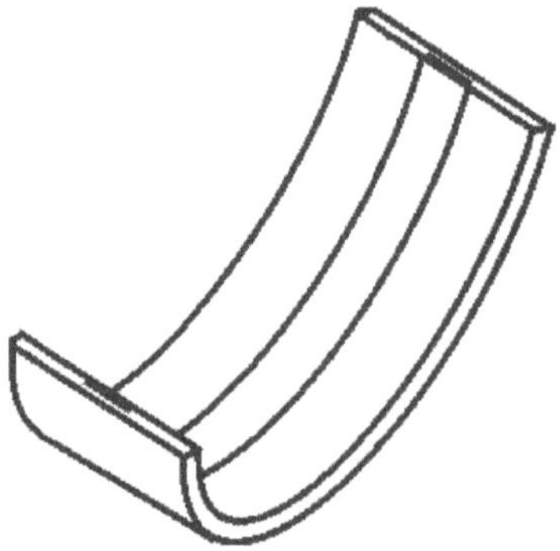

Figure 15.25 Grooved bottom half of the bearing

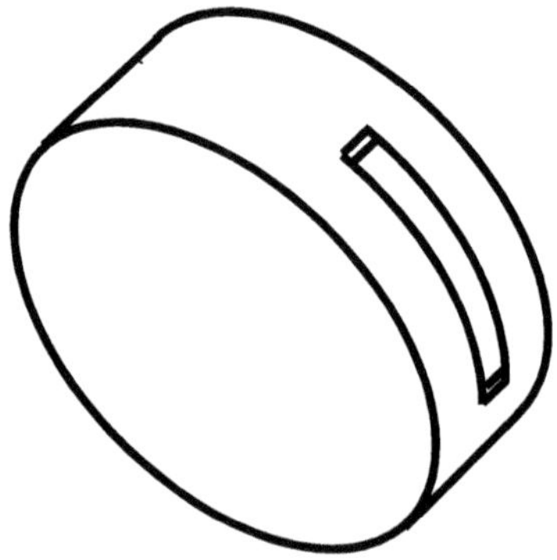

Figure 15.26 Worn out crank pin with a cam shape formed

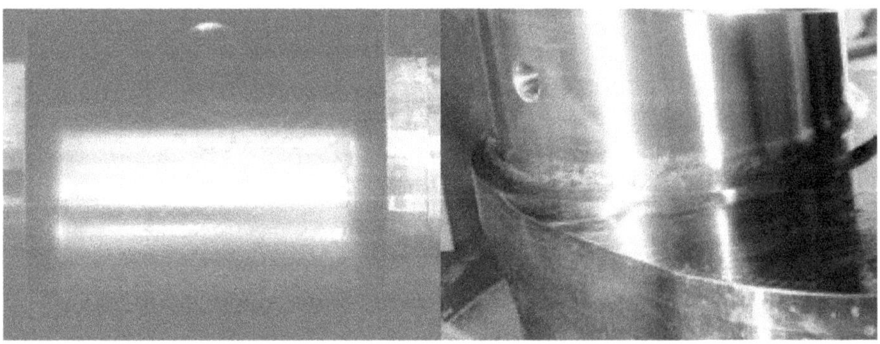

Figure 15.27 Cam formation on the crank pin due to groove only on the bottom half of the bearing when the abrasives concentration is high in the lubricating oil

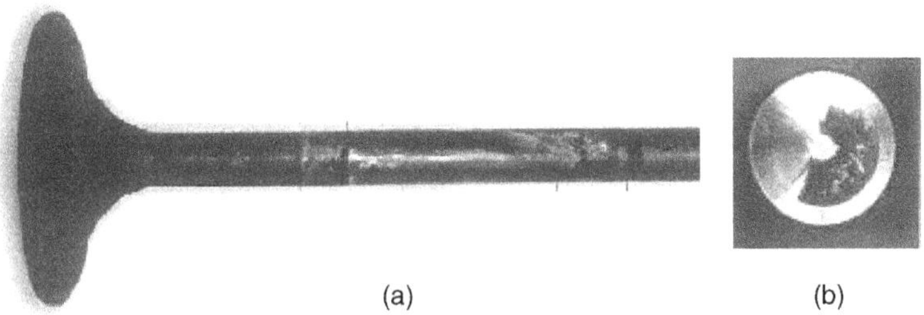

Figure 15.28 (a) Corrosion of valve seat of the inlet valve; (b) valve tip

15.10.2 Inlet Valve

Valve rotators help in maintaining uniform wear of the valve seat on the inlet and exhaust valves and keep them clean. When the rotator of the inlet valve is (low temperature) corroded by sulfuric acid in the oil during a long run of an engine, the seats on the valve and the valve seat suffer wear due to 'micro-beating' or fretting (Chapter 3). In addition, the deposits (sodium and vanadium) are left on the cold seats without cleaning. The exhaust gases flow temporarily to the inlet port past the valves during 'counter' scavenging in certain conditions of operation when the piston is travelling upwards and the inlet valve is open. The seats are (high temperature) oxidized when the deposits melt and release nascent oxygen (Figure 15.28a). The measures taken for increasing resistance of the valve rotators to acid attack improves the situation.

15.10.3 Corrosive Wear of Valve Tips

The lubricating oil circulating in the engine carries sulfuric acid to the valve tip as well and the tip wear takes place where a boundary lubrication condition exists (Figure 15.28b).

15.11 New Engines Compliant to Maritime Emission Standards

Some 90% of world commerce moves on water. Merchant navy ships use residual heavy fuel oil for running. The fuel is inferior in quality to the expensive diesel used in on-road engines. Thus, shipping emissions rates are among the highest for any source category in the world (Marshall, 2005). Figure 15.29 shows the distribution of ozone impact by nitric oxides in ppb (parts per billion) in the atmosphere of the globe in July 2003 (Dalsøren *et al.*, 2010). A similar distribution of sulfur in the atmosphere is observed on the sea routes that are commercially highly active. New emission norms have been evolved by the International Maritime Organization (IMO, 2005) and the United States Environmental Protection Agency (EPA, 1999). In addition, economic incentives (e.g. lower harbouring charges) are given by many maritime countries.

15.11.1 Steps to Satisfy Emission Standards

To satisfy emerging emission norms several steps have been taken by the engine designers (Trozzi and Voccaro, 1998). To reduce nitric oxides, selective catalytic reduction (SCR) using ammonia on board is gaining popularity. On the other hand, exhaust gas recirculation (EGR) could also help reducing nitric oxides. The difficulty posed by the EGR technique is in the after-treatment device, namely the active particulate filter, where the catalyst (platinum) after the engine is easily poisoned by the sulfur in the fuel. In addition, the sulfates formed on the catalysts add to the inventory of particulate matter. What is more, an

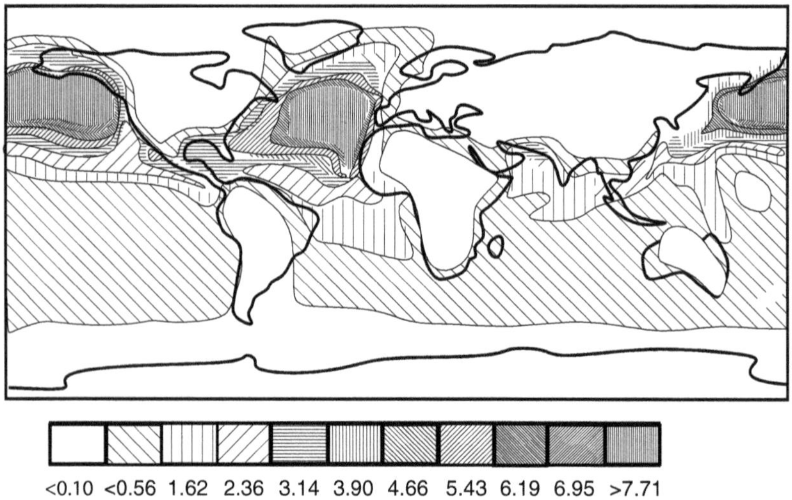

Figure 15.29 Commercially active shipping routes polluted with nitric oxides. Similar contours exist for sulfur (Reprinted with permission from S.B. Dalsøren, M.S. Eide, G. Myhre, Ø. Endresen, I.S.A. Isaksen and J.S. Fuglestvedt, "Impacts of the large increase in international ship traffic 2000–2007 on tropospheric ozone and methane," *Environmental Science and Technology*, **44**, 7, 2482–2489, 2010. © 2010 American Chemical Society.)

EGR engine is a few percentage points less efficient than an SCR engine, and hence is not very popular in large engines where the fuel cost must be controlled. Also, water injection or water homogenized in fuel is used in some engines to achieve lower nitric oxides. This can either create corrosion problems or high loading of fuel injection equipment, which must be suitably addressed.

Fuel sulfur is shown to determine the emission of particulate matter (Figure 15.30) (Sax and Alexis, 2007). The oil scraped up by the top ring to the combustion space also contributes to particulate matter. Therefore, low sulfur fuel and lower lubricating oil consumption (less than 1 g/kW h) is the route to reduce particulate emissions. The problems of low sulfur fuel are solved by reducing the TBN to allow slight corrosion of the roughness peaks on the liner, controlling the viscosity of the fuel above a limit to save the fuel injection equipment and other methods (MAN B&W, 2010b). In large two-stroke engines, oil injection could not be so precisely carried out with conventional mechanical methods. Therefore, the oil

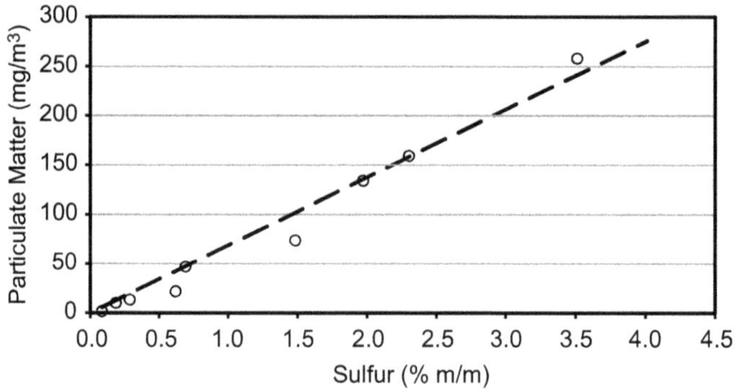

Figure 15.30 Effect of fuel sulfur on particulate matter (Reproduced from T. Sax and A. Alexis, "A critical review of ocean-going vessel particulate matter emission factors," California Environmental Protection Agency Air Resources Board.)

injection is controlled by a computer exactly where and when necessary in the liner so that oil consumption can be reduced below 1 g/kWh. The new generation engines use liners insulated in the middle stroke, chrome ceramic top rings and thick chromium coated oil rings for the pistons and anti-polishing rings at the top of deep honed liners to control wear to less than 0.01 mm/1000 hour and oil consumption to less than 0.7 g/kW h. The wear of the uncoated rings in different cylinders was between 0.07 and 0.15 mm/1000 hour; this dropped to a maximum of 0.02mm/1000mm with coated rings (Aberli, 1999; Schenk, Hengeveld and Aabo, 2003).

Sac-less nozzles are specially developed to reduce hydrocarbon emissions drastically. In addition, it is important to note that the common rail fuel injection system controls the rate of injection pressure and split injection precisely, to reduce nitric oxides and soot emissions (Jorach *et al.*,2002).

15.12 Wear Life of an HFO Engine

Wear of valves, liner and rings in engines running on heavy fuel oil is due to similar causes. Based on the field data 'wear factors' have been estimated. The life of the engine is inversely proportional to the wear factors (Table 15.4). The life in hours can be estimated empirically by a thumb rule: Wear life, hours = 1500/sum of factors in table 15.4.

If an anti-polishing ring is used and the first two points in Table 15.4 are taken care of, then the life of a large engine running on heavy fuel works out as 50 000 hours.

This formula was obtained using the field data. A 50% tolerance is on the estimated life of a wear part.

Table 15.4 Wear factors

Defects	Nominal rate of wear factors and bore polishing rate
Poorly honed liner, smooth finish of liner, low air fuel ratio, poor cooling of piston	1.10
Cold liner, poor running-in and dust during running-in	0.25
Lack of anti-polishing ring	0.07
Use of anti-polishing ring	0.03

15.13 Summary

The heavy fuel engine has a narrower range of operating boundary conditions to play with than a diesel engine because of the high proportion of sulfur, carbon and catalytic fines plus contamination by sea water and incompatible liquids such as industrial solvents as well as oil. Purchase of HFO only from a reputed source with a strict control of quality is recommended as it is not possible for the customer to check the quality; the customer depends on the certificate provided by the vendor (MAN B&W, 2010a).

The design of an engine with (1) an anti-polishing ring for the liner, (2) cut-back top land on the piston, (3) chrome–ceramic top ring in the piston rings, (4) thick plating of oil ring, (5) special higher temperature liner, (6) 'pulse' electroplating of the valves and (7) large fine filters for air and oil helps to increase the wear life of an engine operating on HFO to the same levels as an engine running on costly high speed diesel.

It is important to monitor the thermodynamic parameters and the fuel quality at the inlet to the engine during operation. Firstly, the exhaust temperature should be low to avoid solidification of the eutectic mixture of sodium and vanadium in the exhaust to avoid high temperature corrosion. Secondly, a high air temperature at the inlet to the engine and higher liner temperature help to reduce the low temperature corrosion due to sulfuric acid is recommended.

References

Aabo, K. (2003) (CIMAC Heavy Fuel Oil Working Group) Experience from operation on today's fuels and low-sulphur fuels. 24th International Bunker Conference, 7–9 May 2003, Hilton Hotel, Rotterdam, The Netherlands.

Aeberli, K. (1999) TriboPack for Sulzer engines. *Marine News* (Wärtsilä Corporation), **3-1999**, 4–7.

Agius, P.J., Price R.C., Tiras J.M. *et al.* (1971) Current Trends in Fuel Additives. 8th World Petroleum Congress, 13–18 June 1971, Moscow, USSR.

Amoser, M. (2001) Insights into piston-running behaviour. *Marine News* (Wärtsilä Corporation), **2-2001**, 23–27.

Amoser, M. (2002) Verbesserter Kolbenlauf von langsam laufenden Dieselmotoren, *MTZ*, **63**(4), 288–295.

Dalsøren, B., Eide, M.S., Myhre, G. *et al.* (2010) Impacts of the Large Increase in International Ship Traffic 2000–2007 on Tropospheric Ozone and Methane. *Environmental Science and Technology*, **44**(7), 2482–2489. doi: 10.1021/es902628e.

Demmerle, R., Barrow, S., Terrettaz, F. and Jaquet, D. (2001) New Insights into the piston running behaviour of Sulzer large-bore diesel engines. *CIMAC 2001 Congress*, Hamburg, Germany.

EPA (1999) Final Regulatory Impact Analysis: Control of Emissions from Marine Diesel Engines. EPA420-R-99-026, US Environmental Protection Agency, Washington, DC.

Totten, G.E., Westbrook, S.R.and Shah, R.J. (eds) (1969) *Fuels and lubricants handbook: technology, properties, performance and testing*, Volume 1, ASTM International, West Conshohocken, PA. ISBN 0-8031-2096-6.

Fellman, H., Gross, T. and Ludwig, T. (2004) Typical wear mechanism of 2-stroke exhaust valves. Proceedings of the Marine Propulsion Conference, 28–29 April 2004, Amsterdam, The Netherlands.

Jones, C.H. (ed.) (1983) Marine fuels. A symposium sponsored by ASTM Committee D-2 on Petroleum Products and Lubricants, 7–8 December 1983, Miami, FL, Special Technical Publication 878, ASTM, West Conshohocken, PA.

Jorach, R.W., Ressel, H., Scheibe, W. *et al.* (2002) The upcoming generation of common rail injection systems for large bore engines from L'Orange GmbH, in *Design, application, performance and emissions of modern internal combustion engine systems and components*, Proceedings of the 2002 Fall Technical Conference of the ASME Internal Combustion Engine Division (ed. V.W. Wong). ICE Vol. 39, American Society of Mechanical Engineers, New York.

MAN B&W (2010a) Guidelines for Fuels and Lubes Purchasing Operation on Heavy Residual Fuels, MAN B&W Two-stroke Engines. http://www.mandieselturbo.com/files/news/filesof9981/5510-0041-01ppr.pdf (accessed 7 June 2011).

MAN B&W (2010b), Operation on Low-Sulfur Fuels, MAN B&W Two-Stroke Engines. http://www.mandieselturbo.com/files/news/filesof15012/5510-0075-00ppr_low.pdf (accessed 7 June 2011).

Marshall, D. (2005) An NGO's Perspective on International Shipping. Haagen-Smit Symposium, 19 April 2005, Aptos, CA.

Nanda, S.K., (2003) Exhaust valve failure under residual fuel operation. *Journal of Marine Design and Operations*, No. B2, 23–28.

NKK (1997), Class NK, Guidance for measures to cope with degraded marine heavy fuels. NKK (Nippon Kaiji Kyokai), Chiba, Japan.

Pereira, P. and Marschewski, A. (2001) Overcoming high temperature corrosion. *CIMAC Congress*, 7–10 May 2001, Hamburg, Germany.

Ritzkopf, M. (1975) Untersuchunger von Korrosionerscheinungen an Bauteilen in Dieselmotoren. Dissertation, University of Muenster, Germany.

Romih, D and Zgonik, M. (2008) Sulfur in marine fuel oils. http://www.fpp.edu/~mdavid/TVP/Seminarske%2008-09/ICTS2008CD/papers/Romih,%20Zgonik.pdf (accessed 7 June 2011).

Sax, T. and Alexis, A. (2007) A Critical Review of Ocean-Going Vessel Particulate Matter Emission Factors. California Air Resources Board, Sacramento, CA.

Schenk, C., Hengeveld, J. and Aabo, K. (2003), Wear Processes in Low Speed Diesel Engines, The Role of Temperature and Pressure in Wear Processes in Low Speed Diesel Engines. *Schip en Werf de Zee*, **13** (10), 14–19.

IMO (2005) The IMO Annex VI of MARPOL 73/78, Regulations for the Prevention of Air Pollution from Ships. International Maritime Organization, London, UK.

Trozzi, C. and Vaccaro, R. (1998) Methodologies for estimating future air pollutant emissions from ships. Techne report MEET RF98b, Techne s.r.l., Rome, Italy.

Umland, F. and Ritzkopf, M. (1975) Ventilkorrosion in dieselmotoren, Teil 1. *MTZ-Motortechnische Zeitung*, **36** (H7/8),191–195.

Welsh, M. (2002) Considerations for using low-sulfur fuel. Wärtsilä Corporation, Winterthur, Switzerland.

Williams, R.E., Newbury, P.J., Belcher, P.R. and Hengeveld, J. (1984) Future marine fuels – Prediction and alleviation of potential combustion and lubrication problems. ASME energy sources technology conference and exhibition, 12 February 1984, New Orleans, LA.
Yamada, T., Fuji, T. and Ukai, H. (2010) Development of a new evaluation method for the influences of catalyst fines on abrasive wears of marine diesel engines burning heavy fuel oil. 26th CIMAC World Congress, 14 June 2010, Bergen, Norway.

Part VIII

Filters

16

Air and Oil Filtration and Its Impact on Oil Life and Engine Wear Life

M.V. Ganesh Prasad
Ashok Leyland Limited, Hosur, India

A glimpse of the world of filtration and its relevance to diesel engines is provided. The significance of the air filter and oil filter and their influence on oil life and engine wear life is discussed.

16.1 Introduction

Impurities, solid, semi-solid or liquid, are present in all liquids and gases. Most of these contaminants are of various sizes in the order of micrometres or nanometres and, therefore, are not visible to the naked eye. When the concentration of these impurities is unacceptable, then they have to be removed. A filter is basically a device for separating one substance from the other by interposing a medium through which only the fluid can pass. There are various processes of filtration encountered in the operation of diesel engines which may be classified as:

- Solid–gas filtration
- Solid–liquid filtration
- Liquid–liquid filtration.

A solid–gas filtration process is represented by dust separation from air occurring in an air filter and a solid–liquid filtration process is represented by impurities separation from oil in oil filter and from fuel in fuel filter, while liquid–liquid filtration is represented by a water separation from fuel in water separator.

Critical Component Wear in Heavy Duty Engines, First Edition. P.A. Lakshminarayanan and Nagaraj S. Nayak

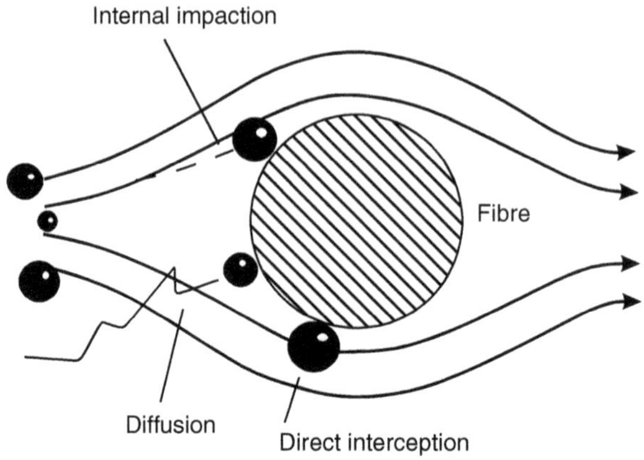

Figure 16.1 Mechanisms of particle filtration (Reprinted from *Filters and Filtration Handbook*, 5th edn., K. Sutherland, Copyright 2008, with permission from Elsevier.)

16.2 Mechanisms of Filtration

A filter works as a porous screen (media), removing and retaining particles too large to pass through openings while allowing the working fluid to pass. Particles are collected from the working fluid by numerous mechanisms and the most important are:

- direct interception
- inertial interception
- diffusion
- adsorption.

Direct interception is a mechanism of filtration in which the particles collide head-on with the filter media and get trapped. Inertial impaction results when the particles suspended in the working fluid fail to follow the tortuous path of air due to their inertia and collide with the filter media. The diffusion mechanism occurs when the particles within the fluid path get in touch with the filter media as a result of Brownian motion. These filtration mechanisms are represented in Figure 16.1. Adsorption filtration is by attraction due to the electrostatic forces between the particles and the filter media.

16.3 Classification of Filtration

Filters are identified by a set of physical characteristics, such as the volume of flow handled, filtration area, thickness, size and material of the media and performance characteristics, such as terminating pressure drop, dust holding capacity (the amount of dust that can be trapped before reaching the terminating pressure drop), burst strength, efficiency and velocity. Dust holding capacity is further quantified (Cummins, 2011) as the ratio of the maximum dust that can be held for a pressure drop to the air flow rate ($g\,m^{-3}$ min). The classification of filtration is done either by the specifications of the filter or by its application. Based on the thickness of filter media it is either surface or depth filtration. Based on the size of the media it is micro, ultra or nano filtration. With reference to the direction of flow it is either a dead-end/full flow filtration or cross flow filtration. In some cases, like oil filters, there is either full flow filtration or bypass flow filtration. By the working principle, it is either magnetic or mechanical filtration. Further explanation on these categories is given below.

16.3.1 Classification by Filter Media

16.3.1.1 Surface Filtration

The simplest type of surface filter is a simple screen working on direct interception. Adsorptive forces, although present, are insignificant. Particles larger than the pore size of the medium are blocked at the upstream surface of the filter. Dirt retention is directly proportional to the surface area, and therefore the higher the surface area the higher is the dust holding capacity. To enhance the surface area the filter paper is pleated. Pleating also increases the mechanical strength of the filter considerably, especially with paper filters. When surface-type filters are exposed to the flow of contaminated fluid, there will be a gradual reduction in pore size owing to the partial clogging of the pores and, as a result, the filter becomes more efficient. When the impurities are deformable, they conform to the shape of the pores thus blocking them to a greater degree as compared to that of hard particles. They can form a slime or gel that can completely clog the filter. In time, a 'cake' or thick bed of particles is formed that itself will act as a filter and cause further improvement in filtration efficiency. Figure 16.2 shows the condition of the filter as it collects impurities from the working fluid. Increase in filtration efficiency is associated with the increase in pressure drop. The life of the filter is given by the time it takes to reach its limiting pressure drop.

16.3.1.2 Depth Filtration

This type of filtration uses a medium with significant thickness. The path followed by the working fluid is complex and longer, thereby increasing the probability of particle retention by direct interception and inertial impaction. Filtration due to the Brownian movement is applicable for those particles with a size of the order one micrometre, causing diffusion which is independent of the direction of fluid flow and retention by adsorption forces. In general, the larger particles tend to get trapped in the surface layers while smaller particles are retained in the deeper layers. Particle size, particle number, fluid viscosity, structure and void volume of media and flow velocity greatly influence the particle retention rates of depth filters. The porous density of the media can vary across the depth to retain particles that are distributed over a wide spectrum of sizes. The only disadvantage with these types of filters is that the pressure drop across the filter is higher than that of surface filters. Resins are used to bind the layers along the depth of the filter element. Cellulose, polymers, micro glass fibres and synthetics are the most commonly used materials. Binder-free media, for example borosilicate, which has a three-dimensional structure, has gained importance in depth filtration, particularly in the filtration of compressed air and gases.

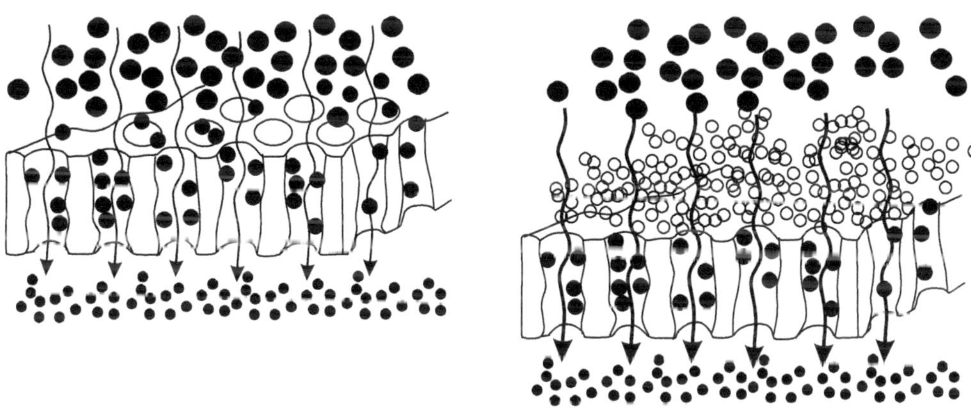

Figure 16.2 Left: Settling of particles on the surface. Right: Bed formation on the surface (Reprinted from *Filters and Filtration Handbook*, 4th edn., T.C. Dickenson, Copyright 1997, with permission from Elsevier.)

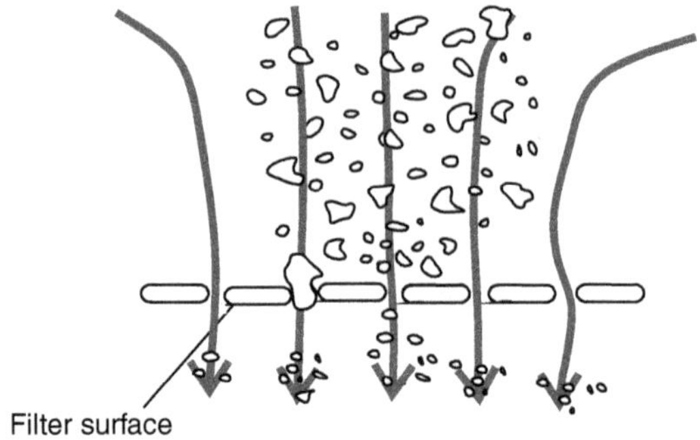

Figure 16.3 Dead-end filtration (Brainerd, 2001)

16.3.2 Classification by Direction of Flow

16.3.2.1 Dead-end Filtration

Dead-end filtration is a process in which the working fluid passes through the medium from one end to the other in a direction normal to the filter surface. The solids are trapped at the inlet side of filter and the working fluid is collected at the other end. This process of filtration is cheap, with high flow rates and needs frequent changes of element. It is represented schematically in Figure 16.3, where particles of different sizes are separated in the filter media. This process is common when the concentration of impurities is lower.

16.3.2.2 Cross Flow Filtration

Cross flow filtration (Figure 16.4) is a process in which the working fluid travels tangentially across the surface of the filter, rather than into the filter. This process is also known as tangential flow filtration. The principal advantage of this is that the filter cake (which can blind the filter) is substantially washed away during the filtration process, increasing the length of time that a filter unit can be operational. It can be a continuous process, unlike batch-wise dead-end filtration. This type of filtration is typically selected for fluids containing a high proportion of small particle size solids, because solid material can quickly block the filter surface with dead-end filtration.

16.3.3 Classification by Filter Size

Micro filtration is a process of separating particles of the order of 0.03–10 μm. This process needs a pressure differential of 0.1–0.5 MPa across the filter element. This process has found its applications in

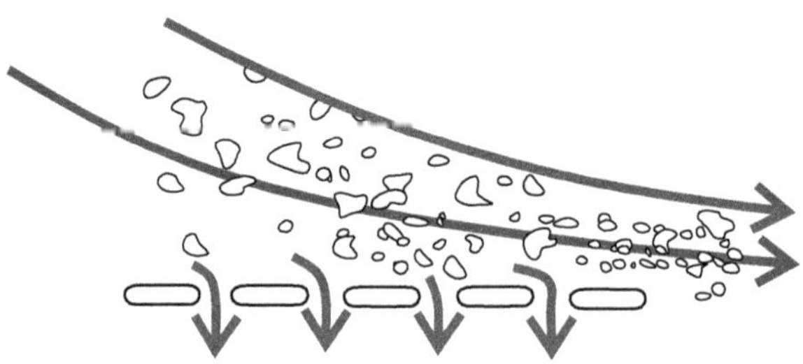

Figure 16.4 Cross flow filtration (Brainerd, 2001)

medicine and electronics industry. Ultra filtration separates particles of the order of 0.005–0.01 μm with a differential pressure of 5–10 bars, while nano filters separate particles of the order of 1 nm with a differential pressure requirement of 3 MPa. Such filtration processes are used in the food, medicine, electronics and paint industries.

16.4 Filter Rating

Filters are rated based on their ability to remove impurities of a specified size (μm). Different types of rating the filter are Absolute rating, Nominal rating, Mean filter rating, β ratio and filter efficiency.

16.4.1 Absolute Rating

The absolute rating or the cut-off point is the diameter (μm) of the largest particle that can penetrate the filter medium. The rating depends on the size and the shape of the pore. It is misleading to say that the rating is the smallest dimension for a square or triangle or slot shaped pores. While test standards define filter rating based on the diameter of spherical glass beads, in reality the impurities are not spherical and their nominal diameter is defined by the largest of the linear dimensions. Sometimes the impurity may be such that its other two linear dimensions are smaller than the nominal diameter, which in turn can be more than the rating, thereby permitting it pass through a much smaller hole. The passage of particles in this way depends on the size and shape of the opening, and the filtration depth. In depth filtration, the probability of trapping such asymmetric particles is high. Rating such filter is further complicated as the pore size is inconsistent and depends on the form of the filter element.

16.4.2 Nominal Rating

Nominal rating is the percentage, by weight, of impurities of a particular material and size that is trapped in a given filter. For example, a filter is rated 95% at 15 μm, if it removes 95% of the impurities by weight which are larger than 15 μm. Nominal rating is often misleading as the results are inconsistent and is thus not widely used.

16.4.3 Mean Filter Rating

The pore size of a filter is not uniform along a particular layer or across layers of the filter element, either because it is both desirable and unavoidable. Mean filter rating is the average pore size of a filter element. This is a better representation of the filter performance than absolute or nominal rating. A bubble point test can be useful to assess the mean filter rating.

16.4.4 β Ratio

Beta (β) ratio is an accurate indication of the performance of the filter. It is defined as the ratio of the number of particles larger than a particular size that are upstream to the number that are downstream of the filter. It is represented as, β_x, and is given by Equation 16.1:

$$\beta_x = \frac{N_{upstream}}{N_{downstream}} \tag{16.1}$$

where β_x is the filter rating for particles larger than x μm, $N_{upstream}$ is the number of particles larger than x μm upstream of filter and $N_{downstram}$ is the number of particles larger than x μm downstream of filter

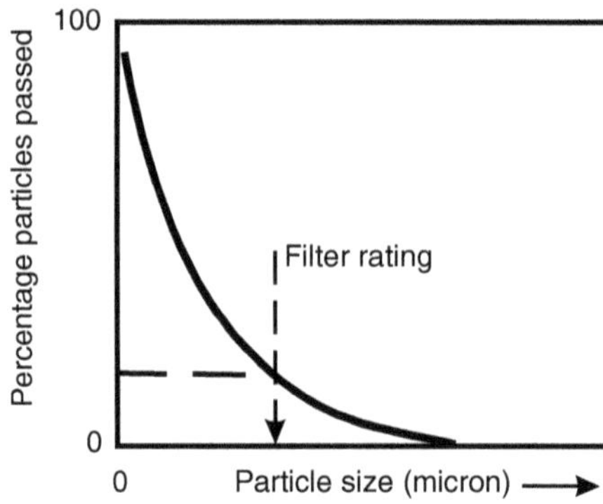

Figure 16.5 Impact of particle size on the number of particles passing the filter (Reprinted from *Filters and Filtration Handbook*, 5th edn., K. Sutherland, Copyright 2008, with permission from Elsevier.)

The higher the β ratio, the better is the particle retention. For a given filter the β ratio increases with size exponentially, that is, if the β ratio is 75 for a 10 μm particle size, then for a particle size of 20 μm the β ratio is much higher than 75.

16.4.5 Efficiency

Filter efficiency, given by Equation 16.2, is a derivative of β ratio and is used in specifying the nominal rating of filter. The higher the β ratio, the higher is the efficiency of the filter:

$$\eta_x = \left(1 - \frac{1}{\beta_x}\right) \cdot 100 \tag{16.2}$$

A filter is considered nominally efficient at a certain micron level if it can remove 50% of particles of that size. In other words, a filter that will consistently remove 50% of particles of size 20 microns or larger is nominally efficient at 20 μm. It is obvious that if 50% of particles pass through at a particular particle size, then the number of particles which are larger that pass through the filter will be less than 50% (Figure 16.5). A filter is considered to achieve absolute filtration efficiency at a certain micron level if it can remove 98.7% of particles of that size.

With time a majority of filters develop a filter bed of impurities which reduce the permeability of particles, thus the performance of a filter can exceed its rating. This is desirable as long as the pressure drop increase is not to such an extent that the flow or mechanical rigidity of the filter is affected beyond acceptable limits. Figure 16.6 shows that with smaller particles the pressure drop is higher or filter clogging is faster, and hence filter life is lower. The above increase in efficiency due to the 'bed effect' is dependent on the operating conditions. In fact, the actual performance is less than the filter rating as the standard tests recommend usage of high concentrations of smaller size impurities and, therefore, the bed is formed quickly, while in actual conditions working fluid is relatively clean. In such cases absolute rating is the true rating for the filter.

16.5 Filter Selection

Selection of a filter is complex as each application has its own challenges. A filter used for one application may be very inefficient for another application. The concentration of impurities, filtration efficiency for the range of contaminant size, pressure drop at required volumetric flow, dust holding capacity, compatibility with the working fluid, filter change interval, operating conditions and cost govern the

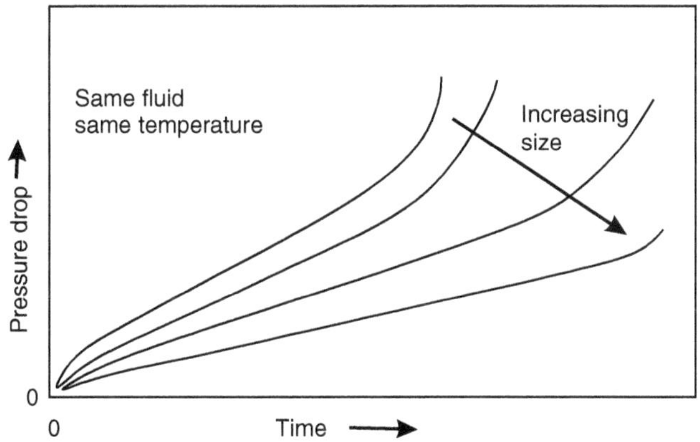

Figure 16.6 Effect of particle size on pressure drop with filter age (Reprinted from *Filters and Filtration Handbook*, 5th edn., K. Sutherland, Copyright 2008, with permission from Elsevier.)

selection of the filter. The dust holding capacity should be sufficient that the filter service interval is not undesirably short. The rating of the filter is so selected that the particles that pass through the filter are of a size that is less than the bearing clearance available for proper functioning of the machine. If the size of the particle is larger than the bearing clearance then the flow may be blocked, causing higher restriction. The filtration media should be compatible with the working fluid and should not be disintegrated or damaged.

For ultra filtration over a wide size range of contaminants it is better to have filters of progressively reducing cut-off ratings instead of a single super fine filter. For applications with a pulsating flow, the fine particles of filter bed are forced to pass through the filter, and hence demand the use of a finer filter than expected.

16.6 Introduction to Different Filters in the Engine

Filtration, wear rate reduction and engine emissions reduction have been the subjects of research for decades. Filters play a significant role in the proper maintenance of an engine and there are different varieties of filters that are used, including air filter, fuel filter, coolant filter, breather element filter, oil filter and centrifugal filter. Contaminants from atmospheric air, combustion and degradation of oil are the most important factors in reducing the life of critical parts of a diesel engine through wear. The performance of filters is tested as per the standards indicated in Appendix 16.A. In the following sections, ambient air quality, the mechanism of dust particle separation, the impact of the selection and maintenance of the air filter on engine wear life are discussed. Subsequently, the requirements of oil filters, performance characteristics of oil, oil life and engine life predictions are examined.

16.6.1 Air Filters

Ambient air contains solid impurities, including dust, organic substances such as bacteria, viruses and pollens, and particulate matter, which come into it due to its movement over land and emissions from industries and automobiles. These impurities, which are of variable size and concentration, either travel along with air or settle/precipitate. Figure 16.7 indicates the range of particle sizes for a few common impurities (Burrows *et al.*, 1997). The distance to which these particles travel depends on their density, shape, size, air velocity and the height at which the particles enter the air. The height that particles reach depends on air turbulence, temperatures, humidity and density. On the other hand, precipitation occurs due to the force of gravity and the particles tend to fall, reaching their terminal settling velocity given by

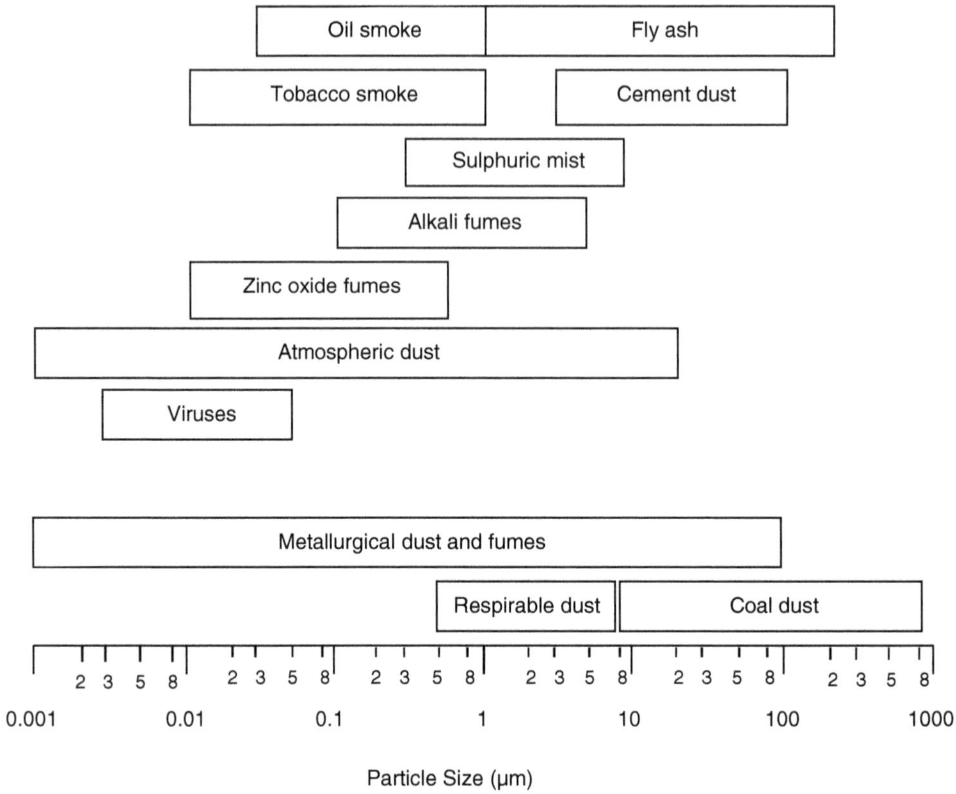

Figure 16.7 Sizes of particles in air (Reproduced with permission from DustWatch CC - Precipitant Dust Monitoring. http://www.dustwatch.com/papers_and_publications.htm. Source: Environmental Engineering in South African Mines, Chapter 12, published in 1989 in association with the Mine Ventilation Society of South Africa.)

Stoke's law (Equation 16.3). Most precipitation occurs when the wind is slow or stops rather than when it is blowing. Stoke's law is applicable for spherical particles that are smaller in size ($<250\ \mu m$) rather than those whose velocity creates turbulence, $Re > 1$. Orientation of the particles can be predicted based on the particle's Reynolds number (Perry and Green, 1984).

$$V_t = \frac{(\rho_s - \rho_a) \quad d^2.g}{18.\eta} \tag{16.3}$$

where d is the geometric diameter of the sphere (m), ρ_s is the density of the sphere (kg/m^3), g is the acceleration due to gravity (m/s^2), ρ_a is the density of air (kg/m^3) and V_t is the terminal particle velocity.

The terminal velocity is a strong function of particle size, particle shape, particle density, ambient air pressure, temperature and moisture content. For a given density, the lower the particle size the lower is the terminal settling velocity. When terminal settling velocities are of the order of the displacement caused by the Brownian motion, they remain permanently suspended, even in still air, due to the bombardment by air molecules. Irregular shaped particles while settling do not take a preferred orientation and may fall edgewise. Consequently, the terminating velocity will be lesser than that of a smooth, regular sphere of equal radius. The higher the density difference between the particle and the ambient air, the higher is the terminal velocity. Therefore ambient pressure, temperature and moisture content influence the terminal velocity.

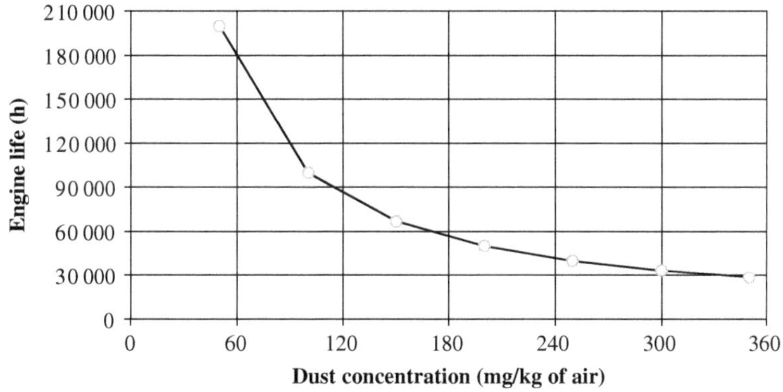

Figure 16.8 Effect of dust concentration on engine life

There are several methods to monitor the concentration of airborne pollutants and dust (Kuhn and Loans, 2003; Rodrigues, 2002; ASTM D1739). The geometric diameters of these particles may vary between 0.001 and 100 μm. Atmospheric dust particles are highly abrasive and have been destroying engines for decades. Studies have shown that engine life is a function of the cleanliness of the air taken in (Sherburn, 1969). Suspended in air there is over 97 mg of dust in 1 m^3 of air. Engines suck in as much as 8 m^3 of air to properly burn 1 kg of fuel and around 800 mg of dust enters the engine per kilogramme of fuel burnt. Thus, the higher the air requirement the higher is the risk, as more impurities enter the engine. In Figure 16.8, the impact of dust concentration in intake air after the air filter on engine life is estimated, and it shows that engine life reduces hyperbolically with the increase in dust concentration. The size of the dust particles influences the wear life. Engine wear is produced by particles in the size range 1–40 μm and the most harmful particles are in the range 1–20 μm (Needleman and Madhavan, 1988; Treuhaft, 1993). Even in small amounts it can significantly increase the wear on piston rings and cylinder walls. By using high efficiency air filters, engine wear can be significantly reduced (Jones and Eleftherakis, 1995). The effect of filtration efficiency on engine life is indicated in Figure 16.9. A new filter in modern engines has an efficiency of 99.8% and, as it ages over a period of about 20 000 km, the filtration efficiency increases to 99.95%. In other words, the wear rate decreases fourfold after 20 000 km.

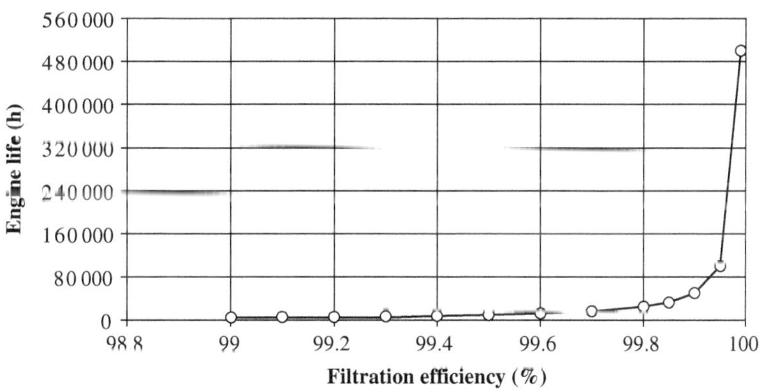

Figure 16.9 Effect of filtration efficiency on engine life

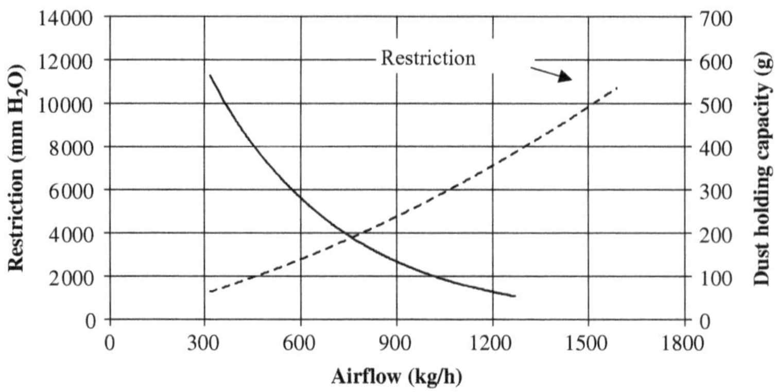

Figure 16.10 Impact of airflow on restriction and dust holding capacity of a filter

Therefore, air filters and the air intake system play a crucial role in controlling the dust concentration and dust particle size of the intake air and control of wear life. An air filter must balance two functions that have a trade-off with each other. Firstly, it has to prevent dirt particles from entering the engine and, secondly, it has to offer least restriction to the air flow. If it filters a lot of dirt it causes a higher restriction, and if it is intended to offer less restriction too much dirt will be allowed into the engine. There is an upper limit given by the engine manufacturers on the air intake restriction offered by the system. With the advent of turbochargers and requirement for high specific power, the air flow demands for diesel engines have increased significantly. Consequently, the filtration quality requirement is higher. The impact of airflow on the restriction and the dust holding capacity are indicated in Figure 16.10. For higher airflow requirements, the restriction is higher, thus the dust holding capacity should be reduced to allow air to pass through. Dust holding capacity of the filter element is selected based on the restriction of the air intake system, pre-cleaning efficiency of the system, engine flow and particle size distribution of the dust in the operating atmosphere.

For handling air a with significant amount of contaminants and with a wide spectrum of particle size, filtration is done in multiple stages where the rating of the filter is progressively increased from the primary stage till the final stage, as indicated in Figure 16.11. For such conditions, filtration in a single stage demands a fine pore size and, therefore, the filter bed is formed very early, thus increasing the flow restriction and demanding an early change in filter element. To reduce the frequency of filter change, dust holding capacity has to be increased, thus increasing filtration area. In multi-stage filtration, the primary filter will remove

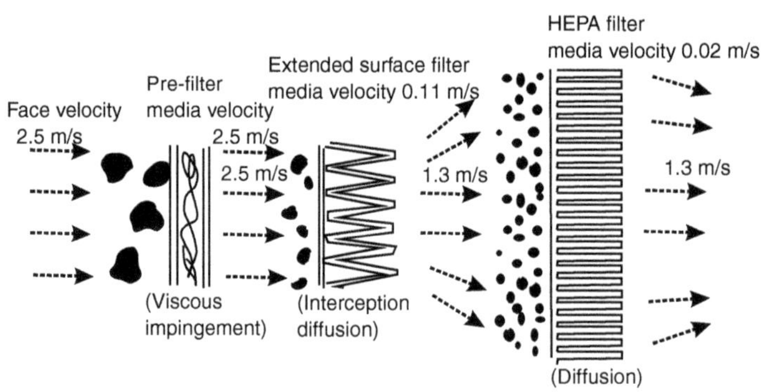

Figure 16.11 Schematic diagram for progressive filtration of dust particles from air (Reprinted from *Filters and Filtration Handbook*, 5th edn., K. Sutherland, Copyright 2008, with permission from Elsevier.)

coarser impurities allowing the finer particles to pass through; these are trapped in the subsequent filters. For a given flow rate, the filtration area is lower for multi-stage than for single-stage filtration.

The air filter may be a paper element type, oil bath type, foam type, cloth type or cyclone type. Paper-type filters are the most widely used owing to their serviceability, cost, efficiency and simplicity of design. Oil bath air cleaners consists of an oil pool at the bottom and an insert made of fibres, mesh and foam fitted above the pool. The insert poses an intricate path to the air with a series of U-turns at the surface of the oil. Larger and heavier dust and dirt particles, due to their inertia, fall into the oil. Lighter and smaller particles get trapped in the insert, which becomes wetted by the oil carried along with the air. Oil bath air cleaners were widely used before the adoption of the paper filter in the early 1960s. However, where dust concentrations are high, such as mines, these are still in use as primary filters after the pre-cleaners, as they can hold dirt without loss of filtration efficiency or airflow. The disadvantages are difficult cleaning and servicing, and they are relatively large to avoid excessive restriction at high airflow rates. With such filters there is a possibility for the oil to travel into the combustion chamber and increase exhaust emissions of unburnt hydrocarbons.

Foam-type filters are made of polyurethane elements which are wetted by oil. They are used predominantly for off-road applications. Cyclone-type separation is a method of removing dust particles by creating a vortex in the flow. Cyclone geometry and flow rate can be used to determine the cut-off point of the cyclone, which is defined as the particle size that can be removed with 50% efficiency. The particle size can be estimated based on the drag, centrifugal and gravitation forces.

The performance of the air cleaner shall be as per ISO 5011/BS 7726/SAE J 726/JIS D1612/GS 565/BIS 11839/SAE J2554. Based on the number of stage the filters are classified as light duty, medium duty or heavy duty. Light or normal duty air cleaners use a paper filter element in a simple housing to provide single-stage filtration. Medium duty air cleaners employ two stages of filtration. The pre-cleaner stage has inertial or centrifugal dust separation and the second stage a filter element. The pre-filter stage increases the life of the filter element by removing a large quantity of contaminants before they reach the main filter. The contaminants removed in the first stage are eliminated through a manual bleed valve or an automatic evacuation system. Some of these air cleaners are available with a 'safety' element (not a second filter) inside the main paper filter element to protect the clean side of the system from contamination when the main element is serviced. Medium duty air cleaners generally have dirt holding capacities of at least 270 g/ m^3 minute of coarse dust. Heavy duty air cleaners employ two filtration stages similar to medium duty cleaners, but are generally designed with more effective first stage separation and larger capacity second stage filter elements. These air cleaners are equipped with a 'safety' element as standard. The dust holding capacity reduces with air flow and heavy duty vehicles have lower dust holding capacity. Heavy duty air cleaners generally have dirt holding capacities of at least 625 g/m^3 minute of coarse dust.

16.6.2 Cleaning Air Filters and Impact on Wear Life

The main reason for cleaning air filter cartridges is to reduce the overall cost of filtration. Servicing disturbs the filter media, seals of the filter and housing, and thereby exposes the clean side of the intake system to contamination. Therefore, it should be serviced only when compelling reasons exist. Even the smallest rupture to the media can make the filter less efficient, causing serious damage to the engine, downtime and rebuilds. Usually one cleaning operation will decrease the efficiency from 99.9 to 95.0%, and hence 50 times more dust is allowed. This will not only reach the dust holding capacity faster and shorten the life of the filter but will also reduce the life of the engine drastically. Therefore, it is strongly recommended by the original equipment manufacturers (OEMs) not to service the filters.

16.7 Oil Filters and Impact on Oil and Engine Life

Early automotive engine designs did not use any oil filtration, as the oil was changed every 1000–3000 km. The development of pressure lubrication brought about a need for some type of filtration to protect the oil pump from damage and excessive wear. In the beginning, only simple wire meshes or screens, which were reusable, were used in the oil pump intake. Sweetland and Greenhalgh (1927) invented the modern oil filter in 1923. The new oil filter was incorporated into the lubricating system after the oil pump and before the main oil gallery. While engine oil as a part of the engine build performs critical functions, it gathers contaminants; these constitute soot, sludge and wear metals. The impurities from the inducted air, particularly silica (quartz), also become absorbed in the oil. When engine oil reaches the critical parts, the contaminants suspended in it initiate wear mechanisms such as soot induced wear (Gautam *et al.*, 1998), two-body wear, three-body wear and corrosive wear and reduce life of the engine. Oil filters play a crucial role in maintaining the cleanliness of the oil and preserve engine performance and life. Due to the entry of solid, liquid and gaseous insolubles formed through oxidation and polymerization reactions, the quantity of ageing products increases. It is impossible to extract many of these ageing products by filtration but their removal by some means is essential to keep an engine in good operating condition. By doing so, the ideal conditions are created for the oil to be in use on a permanent basis. It is established that cleanliness of an engine is a prerequisite for problem-free operation.

16.7.1 Oil Performance and Life

The function of engine oil is to lubricate the sliding parts, clean the deposits, cool the parts, neutralize the acids, transmit forces and seal the piston rings. Thus oil is identified by a set of physical and chemical properties designed to fulfil these functions, such as kinematic viscosity, total base number/total acidity number (TBN/TAN) and concentrations of soot, insolubles, wear metals and silica; these define the condition of the oil. As oil ages, its quality continues to deteriorate and there is an instant when the oil can no longer perform its basic functions and instead adversely affects the engine. Based on the capability and suitability of the oil to perform its basic functions, a set of rejection limits are defined for the properties; beyond these limits the oil will be rejected. From the above, the oil life or the drain interval is defined. The oil life depends on engine design, severity of the operating conditions, environment factors such as dust, ambient temperatures, fuel quality and operating altitude, in addition to the duty cycle.

To reduce nitric oxides and particulate matter emissions, new diesel engines are designed to operate at retarded injection timings and with exhaust gas recirculation. Consequently, engine oils are subjected to higher soot loading and higher temperatures. Higher soot causes increased oil viscosity that could lead to poor oil pumping ability, oil filter plugging, lower engine cleanliness levels and, more importantly, higher engine wear. Higher temperature increases the release of acids from the base oil due to oxidation.

Sulfur in fuel, which is one of the critical parameters used to define fuel quality, leads to the formation of sulfuric acids during combustion. The basicity of engine oil helps to neutralize the acids and prevent corrosion of engine components. Total Base Number (TBN) represents basicity and can be used to understand the oil condition. Conventionally, a drop of around 50% in the TBN of fresh oil indicates that the oil is almost at the end of its useful life. Figure 16.12 shows that reducing the sulfur content in fuel from 1 to 0.05% reduces TBN depletion rate by 40%, thereby increasing oil life and engine life.

The operating conditions of the engine, dust concentration in intake air, ambient temperatures, altitude and the load greatly influence the oil performance. Suspended silica (quartz) in the dusty environment enters the oil circuit through the inducted air. Abrasive quartz causes abnormal wear of components and thereby affects oil life and reduces engine life. Oil temperature depends on the duty cycle and ambient temperature. Higher operating temperatures result in oxidation of oil molecules and affect oil properties. They also result in a less than proper oil film thickness (due to lower viscosities at higher temperatures), leading to boundary lubrication and, hence, metal-to-metal contact. Further, for every 10 °C rise in the oil temperature, the deterioration rate of oil doubles. Therefore, the higher the duty cycle and the higher the ambient temperature, the lower is the oil and engine life. Operating at higher altitude, and therefore lower air availability, results in higher soot levels and temperature causing wear.

Base oil quality has a significant bearing on the performance of lubricant oil. Improvements in refining and after-treatment processes to remove aromatic content and wax produce better quality base oils

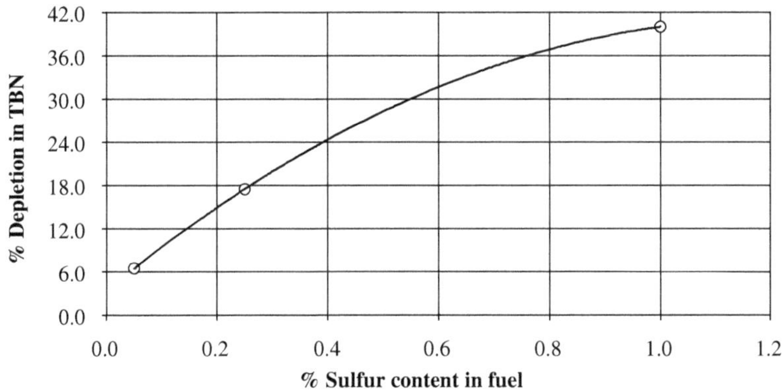

Figure 16.12 Effect of fuel quality on TBN

(Singh, Sahoo and Adhvaryu, 2000). Group II base oils have lower aromatic content and therefore more saturates than Group I base oils.

The sulfur content in oil is lower for Group II oils (Table 16.1). A lower sulfur content in Group II oils implies a lower rate of increase of TAN. A higher percentage of saturates indicates a higher stability of molecules, which implies lower oil consumption. A higher viscosity index indicates that viscosity is stable with varying operating temperatures.

16.7.2 Oil Stress

In the last five decades, several advancements have taken place in engine technology, namely, increasing engine specific power and decreasing oil consumption, thus subjecting oil to higher stress and leading to more rapid oil degradation. Lubricating oil design with adequate performance characteristics to satisfy the above advances is possible through in-depth understanding of tribology. Oil stress can be recognized by its physical and chemical properties, which represent the condition and performance of the oil. There are three kinds of stresses: acid stress due to the formation of acids, thermal stress due to high temperature and asphaltene stress due to fuel dilution (Barnes *et al.*, 2004). Ineffective control of any of the above types of oil stress will lead to shortened oil life.

Acid and thermal stresses are dependent on the amount of fuel burned which carries sulfur and adds heat into the combustion chamber. Asphaltene stress depends on the fuel, oil quality and the engine design in terms of blow-by.

Oil stress is defined as the energy put in per gramme of oil and is represented by oil stress factor (OSF) expressed in kWh/g. The oil stress factor is a function of specific oil consumption, sump volume and oil drain interval. Considering the quantity of sweetening with fresh oil equal to the amount lost by regular oil

Table 16.1 Differences in base oil stocks (Kramer, Lok and Krug, 2001)

Group	Sulfur (%)	Saturates	Viscosity index
I	>0.03	<90	80–119
II	≤0.03	≥90	80–119
III	<0.03	≥90	>120
IV	0	100	120–145

consumption, the OSF may be defined as below (Equation 16.4) (Cannon *et al.*, 1997). As oil ages the oil stress factor increases from zero at time $t = 0$ to '1/R' at $t = \infty$. The OSF can be used to comprehend the condition and define the oil change period if its maximum allowable value is known; this is based on the quality of oil.

$$OSF = \frac{1}{R}(1 - e^{-\frac{Rt}{V}}) \tag{16.4}$$

where OSF = oil stress factor, R = oil consumption (g/kWh), V = specific volume (g/kW) and t = oil drain interval (h).

Therefore, a relationship between OSF and known oil properties needs to be developed. The best property to correlate is TBN, as it represents ability to neutralize acids formed mainly due to sulfur in fuel and oxidation of fuel and oil. The fuel burning rate relates to the amount of energy put into the oil (Equation 16.5). The factor 0.35 converts chemical equivalents of 'S' into chemical equivalents of TBN. A factor 'y' represents the fraction of sulfur that enters the lubrication oil film in the form of oxides. It has to be found experimentally for each type of engine and engine operation, and its value usually lies between 0.06 and 0.1. Rapid TBN depletion is a sign of high oil stress.

$$TBN_t = TBN_{fresh} - 0.35 \times S \times F \times y \times OSF \tag{16.5}$$

where TBN_t = TBN at time t (mg KOH/g), TBN_{fresh} = TBN of fresh oil (mg KOH/g), S = sulfur concentration (wt-%), F = specific fuel consumption (g/kWh) and y = a factor to indicate sulfur entry into the oil film.

With a rejection limit for TBN, the time for oil change can be estimated from Equations 16.4 and 16.5. The rejection limit for TBN is based on engine design in terms of oil throw into the combustion chamber and oil design in terms of detergency or resistance to oxidation. For a given acid loading, when the oil throw is lower the oil film thickness is less and, therefore, oil needs more basicity per unit volume than those engines with higher oil throw. Other than neutralizing acids many times, TBN also represents the residual capacity of properties such as detergency and antioxidation. Therefore, the rejection limit for TBN has to be chosen appropriately. If acid stress is high, it causes high TBN depletion and corrosion can occur. On the other hand, high thermal stress will result in oxidation of oil, deposits, sludge formation, oil thickening, corrosive wear and piston hot corrosion. Piston deposits in the ring groove area influence the sealing of piston rings and deposits under-crown reduce the heat transfer and lead to higher piston temperatures and piston damage.

Asphaltenes, the most polar and heaviest compounds of oil, form complex colloidal structures causing sludge and deposits in fuel tanks and oil lines. Asphaltenes enter the engine oil through blow-by and the fuel injector pumps/pump drive. As fuel injection pressures have increased, so has the risk of fuel contamination of the crankcase oil. Typically, fuel pump plunger seizure and piston seizure or piston burn-through due to piston under-crown deposits are common. It is these compounds, along with other biological sludge, that cause the blockage of filters. Asphaltene solubility is based on the aromaticity and the type (and concentration) of resins in the blend components. When the fuel blend components are mixed, the asphaltenes may precipitate and form sludge. There are asphaltene dispersants and fuel stabilizers to disperse existing sediment and sludge, and prevent new formations of sludge. They also improve the storage stability of the fuel with a good filtration and priming set-up.

There is another model available to predict the global oil stress factor; it articulates oil stress as dependent on the design of both the vehicle and the engine. It is a function of the specific power (kW/litre), engine fuel economy, engine and vehicle operating speeds and oil volume for the drain interval (including the volume of sweetening by top-up required as a result of oil consumption) (Equation 16.6) (Thorn,

Warnecke and Frend, 1995). Higher specific power and longer drain intervals lead to higher oil stress. Higher fuel economy and higher oil capacity will result in lower stress. The higher the oil stress index, the faster is the ageing of the oil, which leads either to reduced oil drain intervals or the need for higher performance oils. The value of the stress index that is allowable is maintained such that sludge, physical and chemical properties, wear metals and silica contaminants are within their corresponding rejection limits over the desired interval. However, oil stress index is not the exact measure of oil drain interval (ODI), as it does not take into account the quality of the additive package, EGR loading, wear metals and dust loading, and load factor. Therefore, ODI has to be finalized by taking a comprehensive view of oil performance in a given type of engine for an application.

$$O_Z = \frac{1}{V_{max}} \times \frac{N_{V_{max}}}{N_{P_{e\,max}}} \times \frac{P_e}{V_h} \times \frac{ODI}{O_{vol}} \tag{16.6}$$

where O_z = the oil stress index, V_{max} = vehicle top speed (km/h), NV_{max} = engine rpm at top speed (in top gear), $N_{Pe\,max}$ = engine rpm at maximum engine power, P_e = maximum engine power (kW), V_h = engine displacement (litres), ODI = oil drain interval (km) and O_{vol} = oil volume including top-ups (litres).

16.7.3 Application of the Concept of Oil Stress

Real world field tests to determine the oil drain interval are particularly time consuming and very expensive to run. Therefore, validation tests are to be conducted on the dynamometer. The above oil stress concepts can also be used to accelerate the oil validation tests on the rig. For example, an engine with 100 g/kW oil volume run for 500 h will have the same oil stress level as the engine with 200 g/kW for 1000 hours, oil consumption remaining same for both (0.6 g/kWh).

16.7.4 Advances in Oil Filter Technology

Several advancements have taken place in oil additive technology to address challenges in meeting the performance requirements, namely, dispersants and anti-wear additives for handling soot, antioxidants for the oxidation, over-based detergents for the acids and special additives to minimize internal wear. Engines which adopt EGR to reduce nitric oxide emissions face additional problems due to increased levels of soot, acids and temperature. Recent advances in oil filter technology assist in extending the oil life by a linear release of additives, rather than a high dose early. The release mechanism was developed based on two main principles. Firstly, the stagnation pressure is determined by the density and velocity of the oil through the filter, which is a function of the engine speed and temperature of the oil. Secondly, the metering tube, which is based on Darcy's formula, to factor the stagnation pressure and additive viscosity, using the tube's dimensions to control the rate at which the additives are released. Taking advantage of the flow dynamics inside the lubrication oil filter, stagnation pressure is created to drive the additive out, then the release rate of the additive is controlled using a metering tube. Metering tube diameter can be varied to suit different engine applications (Evans, 2008).

16.8 Engine Wear

With the latest generation low sulfur fuels, higher quality of base oils, additive package and base detergents, the acid stress on oil is under control and, therefore, the wear due to manifold corrosion is reduced. With the addition of soot dispersants, the ill effects of soot accumulation are prevented. The oil temperatures are kept under control by the addition of oil cooler and oxidation is prevented by the addition

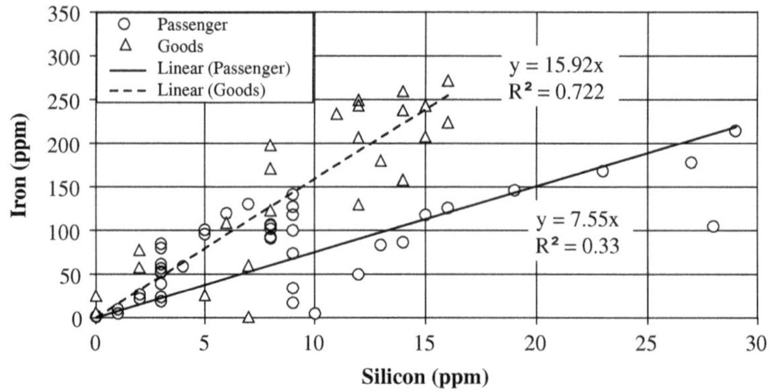

Figure 16.13 Iron accumulation with silicon for different applications

of antioxidants. Therefore, the air filter and the air duct play a crucial role in protecting the engine by preventing abrasive wear due to silica.

Abrasive wear is the most common wear mechanism (Gautam *et al.*, 1998) in a diesel engine. Abrasive wear is either due to projections or abrasive particles stuck on any one of the mating surfaces (two-body wear, e.g. grinding) or due to the presence of foreign (abrasive) particles trapped between the mating surfaces (three-body wear). Three-body wear is one order of magnitude less than the two-body wear (Zum Gahr, 1998; Tewari, 2003; Chapter 8). Two-body and three-body wear are induced mostly by silica particles, and the wear caused by silica is higher than that caused by other contaminants, such as soot present in oil, owing to the higher hardness of silica (Archard and Hirst, 1956). Engine wear can be represented by the concentration of wear metals in engine oil. Typically, due to the presence of iron in almost all the components, engine wear can be represented by the concentration of iron in oil in parts per million by weight/volume. Engine oil drain interval is governed more by the concentration of iron than other physical and chemical properties. The effect of silica contamination on iron concentration for engines under passenger and goods application operating in real world is shown in Figure 16.13. The wear rate for engines in goods applications is higher than those in passenger applications. With the above hypothesis that silica (silicon) is mainly responsible for engine wear, it is expected that the rates of accumulation of wear metals are independent of whether it is passenger bus or truck application. This means silicon contamination of oil affects engine components proportionately. However, it can be observed that for the same concentration of silicon (ppm), the concentration of iron (ppm) is different for passenger bus and truck application. This can be explained as the effect of the duty cycle. The wear life of the engine in addition to the concentration of impurities depends on load factor (less than unity), which is the ratio of the average power required to cruise the vehicle at a speed to the rated power of the engine (Equation 16.4). For the same concentration of impurities, the higher the load factor the lower is the wear life (Chapter 2).

$$Load\,factor = \frac{Average\ power}{Rated\ power} \tag{16.7}$$

The concentration of silicon multiplied by the load factor is plotted against the rate of accumulation of iron (Figure 16.14). The graph shows the wear rate is truly a function of the product of silicon and load factor, and is independent of the type of operation. As already mentioned, oil life is governed by the set of physical and chemical properties identified with rejection limits. Wear metals and then kinematic viscosity reach the rejection limits faster, and therefore the wear rate, oil stability to soot accumulation and oxidation govern the oil life.

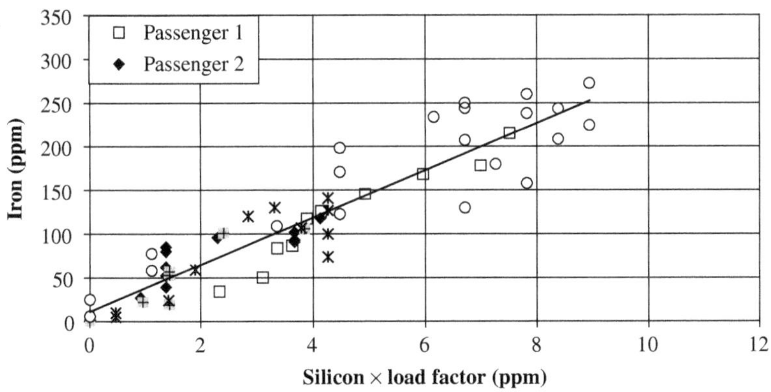

Figure 16.14 Iron accumulation with silicon × load factor for different applications

16.8.1 Method to Predict Wear of Critical Engine Components

Archard's law (Archard and Hirst, 1956) based on the theory of asperity contact is an elegant model to describe sliding wear. The law states that wear volume is directly proportional to normal load and sliding length, and inversely proportional to the hardness of the material (Equation 16.5).

$$Q = \frac{KWL}{H} \tag{16.8}$$

where Q is the total volume of wear debris produced, W is the total normal load, H is the hardness of the softest contacting surfaces, K is a dimensionless constant depending on operating condition and L is the sliding length.

As a case study, estimation of wear volume, and thereby wear metal concentration, was attempted based on data from vehicles running in the field. Wear metals like iron, aluminium, chromium, copper and lead originate from the liner, piston, piston pin, piston rings connecting rod bush, connecting rod and main bearings. For the mileage covered by a vehicle, the sliding length, L, was calculated for reciprocating and rotating parts from the transmission ratio, tyre radius and driving cycle. Normal load, W, is estimated from the peak gas force. The calculations are limited to the power stroke alone, as significant wear happens mainly during this stroke. Taking into account the important changes in piston velocity and gas force during a stroke and, as a result, the change in the friction coefficient, the liner is divided into three sections along its length (Figure 16.15). The wear volumes for rings and liner were estimated separately for each section. The dimensionless constant, K, referred to as Archard's constant, depends on operating conditions and the type of contact. With the load, sliding length, hardness values and experimental wear metal concentration from oil analysis, K was estimated (Table 16.2). It is assumed that K is same for the rings and liner for a segment and wear rate will be uniform for the entire life of the components. Results showed that the values of K for section 2 are much lower when compared to those of sections 1 and 3. This is attributed to the high friction coefficients in the boundary lubrication regime in sections 1 and 3. Based on the wear volume and the material composition, the wear metal accumulation was estimated against the experimental values of wear metals (Figure 16.16).

16.9 Full Flow Oil Filters

Based on the oil flow through the filter, oil filters are categorized as primary and secondary filters. Primary oil filters are also called full flow filters, as 100% of the engine oil passes through it in normal operation.

Figure 16.15 Segmentation of liner

The filters must work without introducing a lot of restriction, or else oil will not flow into the engine during cold start-ups owing to higher viscosities at lower temperatures. If blockage occurs in the filter, bypass valves open which allow unfiltered oil to the engine. In this situation, lubrication with unfiltered oil is better than none at all. This is one reason the (particle size) cut-off rating of a full flow filter is lower. These filters are not very effective for particles smaller than 20 μm.

Table 16.2 Archard's constant for wear components

Component	Archard's Constant (K)
Piston	1.50
Piston pin	6.00
Connecting rod bush	1.50
Connecting rod bearings	0.04
Main Bearings	0.04
Section 1	
Ring 1	100
Ring 2	100
Ring 3	100
Liner	100
Section 2	
Ring 1	1
Ring 2	1
Ring 3	1
Liner	1
Section 3	
Ring 1	40
Ring 2	40
Ring 3	40
Liner	40

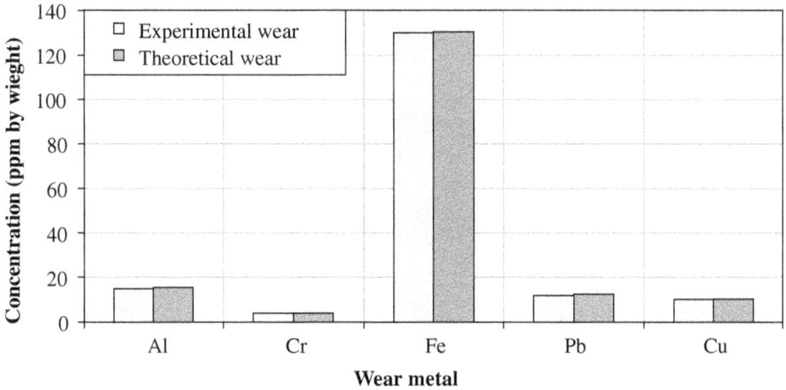

Figure 16.16 Estimated and experimental wear metal concentrations

16.9.1 Bypass Filters

With the latest demands for reduced down time and extended oil drain intervals several advances have happened in oil and oil filter technology. The filter media materials have changed from steel wool, wire meshes and metal screens to bulk cotton or various woven fabrics such as linen to cellulose papers to synthetic papers. Cellulose-type filter media is the most widely used. Special synthetic glass micro-glass fibres are about ten times smaller than conventional cellulose filter fibres (Figure 16.17) and have 99% efficiency at 10 µm, 98% efficiency at 7 µm and 95% efficiency at 5 µm size.

In addition to the primary filters (full flow/partial flow), secondary filters (better known as bypass filters) with ultra-fine filtration capacity take a small portion of the normal oil flow, usually less than 10%, and subject it to additional cleaning by removing solids including soot and wear metals (Andrews *et al.*, 1999). They act separately from the primary filter and have nothing to do with its bypass valves. A schematic diagram of the oil filters in the oil circuit is shown in Figure 16.18. They reduce the concentration of wear metals in oil which are formed due to wear and reduce the increase in kinematic viscosity due to the accumulation of soot in oil (McGeehan *et al.*, 1993; Iwakata *et al.*, 1993; Bardasz *et al.*, 1997a, 1997b).

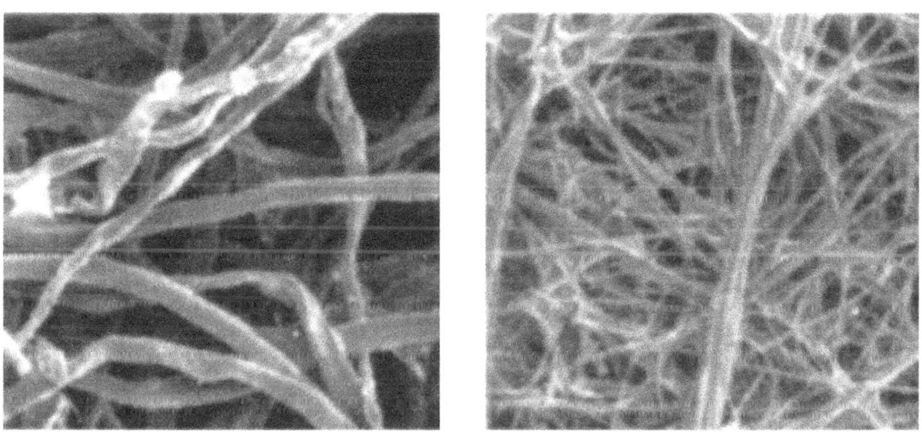

Figure 16.17 Left: Cellulose filter media. Right: Micro-glass filter media

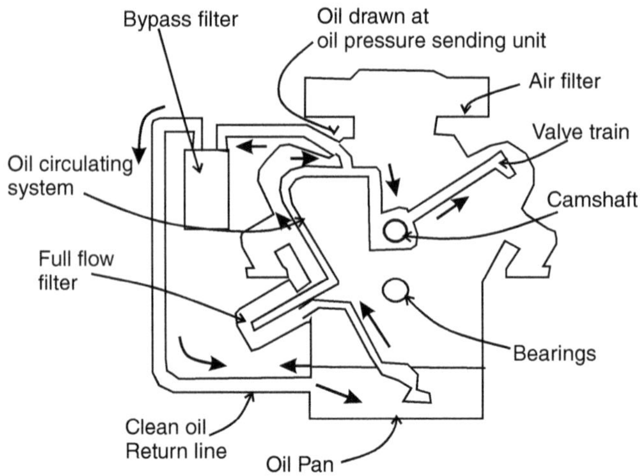

Figure 16.18 Schematic diagram of oil circuit with full flow and bypass filters

Soot accumulation is at the rate of 4% of engine particulate emissions (Andrews, Abdelhalim and Williams, 1993; Sun, Kittelson and Blackshear, 1991). Therefore, the use of bypass filters can enhance the oil drain interval and filter change interval, since oil life is governed by concentration of wear metals and increase in kinematic viscosity.

The bypass filters have high filtration efficiency and a high particulate storage capacity (Loftis and Lanius, 1997). They reduce the load on the main filter by removing the fine particulate matter. Of the contaminants collected 80% are organic, including sludge, resin and soot (Stehouwer, 1996). Thus the bypass filters reduce the soot content by almost 50% (Andrews *et al.*, 1999). Zinc dialkyl-dithiophosphate (ZDDP) is a part of the additive package to prevent oxidation and wear. It decomposes during engine operation and forms products such as sulfides and phosphates, which behave as solid lubricants at high loads and prevent metal-to-metal contact (Corso and Adam, 1985). Oil recycling using bypass filters results in the reduction of ZDDP depletion by 30% (Andrews *et al.*, 1999). Therefore, when introducing a bypass filter into the engine design the loss of ZDDP must be weighed against the benefit of removing wear metals and soot.

Three times the reduction in the contaminants and two to three times enhancement in the engine wear life could be observed by using a three micron bypass filter (this rating is not absolute retention value, because of the characteristic of depth-type filter) in conjunction with the main filter, which was rated at 40 micron (Butler, Stewart and Teasley, 1971). Abrasive engine wear can be substantially reduced with an increase in filter efficiency. Compared to a 40 micron filter, engine wear was reduced by 50% with 30 micron filtration. Likewise, wear was reduced by 70% with 15 micron filtration (Staley, 1988). It is usual to have a 15–20 micron full flow filter with a tighter bypass filter. It was also reported by Staley that, for the same oil quality, the wear of the main bearing was proportional to the contaminants and that improving the filtering improved the wear rate of main and connecting rod bearings by better than 95% and the piston ring wear rate by 90%. In other words, if the contaminants are completely eliminated from the oil, the wear life of engine parts is enhanced substantially, corresponding to natural aging which is many times higher.

16.9.2 Centrifugal Filters

These bypass filters can use a filter element or fast moving rotor that is either self or externally powered. The self-powered centrifugal bypass filters are in common use where the pressurized oil passes into a drum rotor, free to spin on a bearing and seal. The rotor has two jet nozzles at the inner housing to rotate the drum at high speeds. Centrifugal force introduces higher acceleration (g), which separates the contaminants from the oil because of the density difference (Equation 16.3, Stoke's law). The waste forms a hard cake on the rotor and can be cleaned periodically. If the rotor is not maintained, particles accumulate to such a thickness as to stop the drum from rotating and the filtration ceases.

16.10 Summary

Mechanisms of filtration are physical. Classification of filtration by filter media, direction of flow and filter size have been discussed. The air filter must be properly selected for a given engine flow and the dust concentration. Cleaning the air filter decreases the efficiency and dust holding capacity.

The main oil filters are full flow type and are 20 micron sized. A bypass filter with tighter specification allows a fraction of the oil to be cleaned of wear metals and may enable enhancement of oil drain life. However, care must be taken to study if the important additives are also removed by the bypass filter.

Wear of critical parts of the engine is directly correlated with the silica content and the load factor.

Appendix 16.A Filter Tests and Test Standards

	Filter Test	Test Standard
Air Filter	Minimum dust holding capacity	ISO 5011
	Minimum efficiency, full life	ISO 5011
	Minimum efficiency, initial	ISO 5011
	Flow rate at standard conditions	ISO 5011
Oil Filter	Dust holding capacity	ISO 4548-12
	Efficiency	ISO 4548-12
	Terminal differential pressure	ISO 4548-12
	Element collapse/burst	ISO 4548-3
	Hydrostatic burst pressure	ISO 4548-6
	Hydrostatic pulse durability	ISO 4548-5
	Vibration fatigue test	ISO 4548-7/JIS 1601D

References

Archard, J.F. and Hirst, W. (1956) The Wear of Metals under Unlubricated Conditions. *Proceedings of the Royal Society London, A*, **236**, 397–410.

Andrews, G.E., Abdelhalim, S. and Williams, P.T. (1993) The Influence of Lubricating Oil Age on Emissions from an IDI Diesel. Technical Paper 931003, SAE, Troy, MI.

Andrews, G.E., Li, H., Xu, J. *et al.* (2000) Oil quality in diesel engines with on line oil cleaning using a heated lubricating oil recycler. Technical Paper 1999-01-1139, SAE, Troy, MI.

ASTM, D1739 – 98 (2010) *Standard Test Method for Collection and Measurement of Dustfall (Settleable Particulate Matter)*. ASTM, West Conshohocken, PA.

Bardasz, E.A., Carrick, V.A., George, H.F. *et al.* (1997a) Understanding Soot Mediated Oil Thickening Through Designed Experimentation – Part 4: Mack T-8 Test. Technical Paper 971693, SAE, Troy, MI.

Bardasz, E.A., Carrick, V.A., George, H.F. *et al.* (1997b) Understanding Soot Mediated Oil Thickening Through Designed Experimentation – Part 5: Knowledge Enhancement in the General Motors 6.5L. Technical Paper 972952, SAE, Troy, MI.

Barnes, J., Hengeveld, J., Foster, S. *et al.* (2004) Oil stress investigations in Shell's medium speed laboratory engine. 24th CIMAC Congress, 7–10 June 2004, Kyoto, Japan.

Brainerd, E. L. (2001) Caught in the cross flow. *Nature*, **412**, 387–388.

Burrows, J *et al.* (1997) *Environmental Engineering in South African Mines*. Mine Ventilation Society of South Africa, Johannesburg, South Africa, Chapter 12.

Butler, J.L., Stewart, J.P. and Teasley, R.E. (1971) Lube Oil Filtration Effect on Diesel Engine Wear. Technical Paper 710813, SAE, Troy, MI.

Cannon, M.J., Hengeveld, J., Scheele, M.J and Foster, L.J.S (1997) Interactions between engine design, oil consumption, and lubricant performance. Fifth CEC International Symposium, May 1997, Gothenburg, Sweden, paper CEC97-EL09.

Corso, S., and Adam, R. (1985) Incipient scuffing detection by ferrography in diesel valve train systems. Technical Paper 852124, SAE, Troy, MI.

Cummins (2011) Air filtration. Cummins Filtration. http://www.fleetguard.com/pdfs/product_lit/americas_brochures/LT32599_main_cover.pdf (accessed in March 2011).

Dickenson, T.C. (1997) *Filters and Filtration Handbook*, 4th edn. Elsevier, Oxford, UK.

Evans, H. (2008) Diesel oil filter extends lubricant life. www.sae.org/mags/sohe/5498 (accessed 20 June 2011).

Gautam, M., Durbha, M., Chitoor, K. *et al.* (1998) Contribution of soot contaminated oils to wear. Technical Paper 981406, SAE, Troy, MI.

Iwakata, K., Onodera, Y., Mihara, K. and Ohkawa, S. (1993) Nitro-Oxidation of Lubricating Oil in Heavy-Duty Diesel Engine. Technical Paper 932839, SAE, Troy, MI.

Jones, G. and Eleftherakis, J. (1995) Correlating Engine Wear with Filter Multi-pass Testing. Technical Paper 952555, SAE, Troy, MI.

Kramer, D.C., Lok, B.K. and Krug, R.R. (2001) The evolution of base oil technology, in *Turbine Lubrication in the 21st Century* (eds W.R. Herguthand T.M. Warne). ASTM STP #1407, American Society for Testing and Materials, West Conshohocken, PA, 25–38.

Kuhn, G. and Loans, C. (2003) Practical determination and location of windblown dust sources and how to establish a good monitoring program. *Chemical Technology*, November 2003, 27–29.

Loftis, T.S. and Lanius, M.B. (1997) A new method for combination full-flow and bypass filtration: venturi combo. Technical Paper 972957, SAE, Troy, MI.

McGeehan, J.A., Shmah, E., Couch, M.C. and Parker, R.A. (1993) Selecting Diesel Crankcase Oils to Use with Low-Sulfur Fuel. Technical Paper 932845, SAE, Troy, MI.

Needleman, W. M. and Madhavan, P. V. (1998) Review of Lubricant Contamination and Diesel Engine Wear. Technical Paper 881827, SAE, Troy, MI.

Perry, R.H. and Green, D. (1984) *Perry's Chemical Engineers' Handbook*, 6th edn. McGraw-Hill, New York, pp. 5–64.

Projahn, I.U. (2002) Requirements for diesel fuel injection equipment and fuel quality under consideration of global market aspects. Proceedings of the Fifth International Filtration Conference, October 2002, Stuttgart, Germany.

Rodrigues, M. (2002) New dust-monitoring technology developed. *Engineering News* (Creamer Media), November 2002.

Sherburn, P. E. (1969) Air Cleaner Design – Present and Future. Technical Paper 690007, SAE, Troy, MI.

Singh, I.D., Sahoo, S.K., and Adhvaryu, A. (2000) Recent trends in the total characterization of new generation base fluids. ISFL-2000, 10-12 March 2000, New Delhi, India.

Staley, D.R. (1988) Correlating Lube Filtration Efficiencies with Engine Wear. Technical Paper 881825, SAE, Troy, MI.

Stehouwer, D.M. (1996) Effects of extended service intervals on filters in diesel engines. Proceedings of the International Filtration Conference – The Unknown Commodity, Southwest Research Institute.

Sun, R., Kittelson, D. and Blackshear, P. (1991) Size Distribution of Diesel Soot in the Lubricating Oil. Technical Paper 912344, SAE, Troy, MI.

Sutherland, K., Filters and Filtration Handbook, 5th edition, 2008.

Sweetland, E. and Greenhalgh, G.H. (1927) Renewable filter unit. United States Patent (filed 27 November 1923) 1 646 378, 18 October 1927.

Tewari, U.S. (2003) Two-body and Three-body abrasive wear behaviour of polyaryletherketone composites. *Polymer Testing*, **22**, 403–418.

Thorn, R., Warnecke, W. and Frend, M. (1995) Extended Oil Drain Intervals: Conservation of Resources or Reduction of Engine Life. Technical Paper 951035, SAE, Troy, MI.

Treuhaft, M. B. (1993) The Use of Radioactive Tracer Technology to Measure Engine Ring Wear in Response to Dust Ingestion. Technical Paper 930019, SAE, Troy, MI.

Zum Gahr, K.H. (1998) Wear by hard particles. *Tribology International*, **31**, 587–596.

Index

Critical Component Wear in Heavy Duty Engines, First Edition. P.A. Lakshminarayanan and Nagaraj S. Nayak.

www.ingramcontent.com/pod-product-compliance
Ingram Content Group UK Ltd.
Pitfield, Milton Keynes, MK11 3LW, UK
UKHW050943290926
15470UKWH00005B/1598

9 780470 828823